I0038710

Nanobiotechnology Handbook

Nanobiotechnology Handbook

Editor: Tiffany Gardner

www.callistoreference.com

Callisto Reference,
118-35 Queens Blvd., Suite 400,
Forest Hills, NY 11375, USA

Visit us on the World Wide Web at:
www.callistoreference.com

ISBN: 978-1-64116-089-6 (Hardback)

Cataloging-in-Publication Data

Nanobiotechnology handbook / edited by Tiffany Gardner.
p. cm.
Includes bibliographical references and index.
ISBN 978-1-64116-089-6
1. Nanobiotechnology. 2. Nanotechnology. 3. Biotechnology. I. Gardner, Tiffany.
TP248.25.N35 N36 2019
660.6--dc23

Table of Contents

Preface

Every book is initially just a concept; it takes months of research and hard work to give it the final shape in which the readers receive it. In its early stages, this book also went through rigorous reviewing. The notable contributions made by experts from across the globe were first molded into patterned chapters and then arranged in a sensibly sequential manner to bring out the best results.

Nanobiotechnology is an evolving interdisciplinary study of nanotechnology and biology. It is concerned with the solution of complex biological and medical problems with the application of nanomaterials and nanoparticles. Synthesizing new protein structures, bimolecular tissue and membranes, application of nanotechnology for efficient chemical delivery in agriculture are some of the areas of investigation in nanobiotechnology. Current research in nanobiotechnology investigates the biosynthesis of nanomaterials, development of green nanobiotechnology. It also studies the use of advanced nanomaterial implements like nanorobots as alternative treatments for certain diseases like cancer in nanomedicine. This book contains some path-breaking studies in the field of nanobiotechnology and its applications. Different approaches, evaluations, methodologies and advanced studies on nanobiotechnology have been included in this book. It will help new researchers by foregrounding their knowledge in this field.

It has been my immense pleasure to be a part of this project and to contribute my years of learning in such a meaningful form. I would like to take this opportunity to thank all the people who have been associated with the completion of this book at any step.

Editor

1

Beyond the passive interactions at the nano-bio interface: evidence of Cu metalloprotein-driven oxidative dissolution of silver nanoparticles

Daniel N. Freitas[1], Andrew J. Martinolich[1,2], Zoe N. Amaris[1] and Korin E. Wheeler[1*]

Abstract

Background: In a biological system, an engineered nanomaterial (ENM) surface is altered by adsorbed proteins that modify ENM fate and toxicity. Thus far, protein corona characterizations have focused on protein adsorption, interaction strength, and downstream impacts on cell interactions. Given previous reports of Ag ENM disruption of Cu trafficking, this study focuses on Ag ENM interactions with a model Cu metalloprotein, Cu(II) azurin. The study provides evidence of otherwise overlooked ENM-protein chemical reactivity within the corona: redox activity.

Results: Citrate-coated Ag ENMs of various sizes (10–40 nm) reacted with Cu(II) azurin resulted in an order of magnitude more dissolved ionic silver (Ag(I)(*aq*)) than samples of Ag ENMs only, ENMs mixed Cu(II) ions, or control proteins such as cytochrome c and horse radish peroxidase. This dramatic increase in ENM oxidative dissolution was observed even when Cu(II) azurin was combined with a diverse mixture of *Escherchia coli* proteins to mimic the complexity of the cellular conona. SDS PAGE results confirm that the multiprotein ENM corona includes azurin. A Cu(I)(*aq*) colorimetric indicator confirms Cu(II) azurin reduction upon interaction with Ag ENMs, but not with the addition of ionic silver, Ag(I)(*aq*).

Conclusions: Cu(II) azurin and 10–40 nm Ag ENMs react to catalyze Ag ENM oxidative dissolution and reduction of the model Cu metalloprotein. Results push the current evaluation of protein-ENM characterization beyond passive binding interactions and enable the proposal of a mechanism for reactivity between a model Cu metalloprotein and Ag ENMs.

Keywords: Nanomaterials, Nanoparticles, Silver nanoparticles, Protein corona, Redox chemistry, Cu metalloprotein, Azurin

Findings

Background

The design of metal and metal oxide engineered nanomaterials (ENMs) are complicated by chemical and physical changes in a physiological system. Protein adsorption, for example, forms a "corona" and leaves ENM surfaces with little resemblance to the original material [1–3] to permanently alter ENM reactivity [2, 4–8]. Current ENM protein corona studies focus on characterization of protein adsorptions [9–11] and mediation of downstream biological uptake and toxicity [5, 6, 12, 13]. Proteins and metal ENMs are, however, likely to undergo chemical reactions and alter biological and environmental reactivity of ENMs. Despite the evidence that ENMs are biochemically reactive [14–16] and some even demonstrate enzyme-like activity [16], characterization of ENMs chemically reacting with proteins (beyond protein binding and unfolding) has been mostly overlooked.

With this in mind, Cu metalloproteins were chosen as a case study characterization of protein-ENM biochemical

*Correspondence: kwheeler@scu.edu
[1] Department of Chemistry and Biochemistry, Santa Clara University, Santa Clara, CA 95053, USA
Full list of author information is available at the end of the article

reactivity because recent studies by Armstrong et al. [17] and others [18] have demonstrated that Ag ENMs can disrupt copper trafficking. Importantly, free silver ions (Ag(I)(*aq*)), by contrast, were found to have no impact on Cu trafficking. Results suggest that Cu-metalloenzymes react uniquely with Ag ENMs within the organisms studied and establish the need to specifically evaluate the biochemical interactions between Cu metalloproteins and Ag ENMs.

Cu(II) azurin, a model Cu metalloprotein, is extensively characterized, redox active, and structurally simple with one metal center [19]. Importantly, previous studies of Cu(II) azurin–Ag ENM interactions align with findings by Armstrong et al. [17]; when directly interacting with the Ag ENM surface Cu(II) azurin forms biologically inactive, but fully folded apo- and Ag(I) azurin [20]. Ag(I)(*aq*), however, cannot displace the tightly bound Cu(II) within azurin, a result consistent with previous work [21]. The reactivity of Cu(II) azurin with Ag ENMs, but not with Ag(I)(*aq*), justifies further study in the elucidation of biochemical reactivity between Cu metalloproteins and Ag ENMs.

Here, we evaluate the hypothesis that Cu(II) azurin–Ag ENM are a redox pair. Evidence is presented for Cu(II) azurin binding to Ag ENMs and increasing oxidative dissolution to form Ag(I)(*aq*), even in the presence of a complex mixture of other bacterial proteins. Evidence is also presented for reduction of Cu(II) azurin. Taken together, these experiments explain the unique behavior of Ag ENMs with Cu(II) azurin and, by extension, provide biochemical reactivity as a foundation for Ag ENM modification of Cu homeostasis. More broadly, these data introduce the importance of considering the biochemical reactivity of ENMs beyond passive binding interactions.

Methods

Sample preparation

Citrate-coated ENMs were purchased from Nanocomposix Inc (San Diego, CA). Cu(II) azurin was overexpressed and purified as previously described [20, 22]. Other chemicals were purchased from Fisher Scientific, unless otherwise noted. Eppendorf Centrifuge 5424 and Molecular Devices SpectraMax M2 were used for centrifugation and UV–Vis spectra, respectively.

Unless otherwise stated, all samples included 50 μM Cu(II) azurin, horseradish peroxidase (HRP, Sigma), or equine cytochrome c (cyt c, US Biologicals) reacted with Ag ENMs in nanopure water (18 mΩ). ENM concentrations ensured equal surface area and protein binding sites [23] at 3.73, 0.955, 0.416 and 0.233 nM for 10, 20, 30, and 40 nm ENMs, respectively. Soluble protein extract (SPE) was taken from *E. coli Migula Castellani* and *Chalmers* (ATCC) and used at 0.7 and 0.07 mg/ml, which is equivalent to 50 and 5 μM azurin. Procedures for SPE extraction and purification are provided in supplemental materials.

ICP-MS quantification of Ag ENM dissolution

Samples were centrifuged (30 min, 21 K RCF) to remove ENMs from solution after 6 h incubation. 85 % of the supernatant was removed and re-centrifuged. Again, 85 % of the supernatant was removed for preparation and analysis of Ag(I) concentration using an Agilent 7500CE ICP-MS (Agilent Technologies, Palo Alto, CA, USA) by the Interdisciplinary Center for Plasma Mass Spectrometry (University of California at Davis, CA, USA). The samples were introduced using a MicroMist Nebulizer (Glass Expansion, Pocasset, MA, USA) into a temperature-controlled spray chamber. Instrument standards diluted from Certiprep 2A (SPEX CertiPrep, Metuchen, NJ, USA) encompassed the range 0, 0.5, 1, 10, 50, 100, 200, 500, 1000 parts per billion (ppb) in 3 % trace element grade HNO_3 (Fisher Scientific, Fair Lawn, NJ, USA) in 18.2-MΩ water. A separate 100 ppb Certiprep 2A standard was analyzed as every tenth sample as a quality control. Sc, Y and Bi Certiprep standards (SPEX CertiPrep) were diluted to 100 ppb in 3 % HNO_3 and introduced by peripump as an internal standard.

BCA analysis for Cu(I) detection

The reactions for Cu(I) detection were executed as described above. After 6 h, 100 μL of each sample was combined with 1 mM bicinchoninic acid (BCA, MP Biomedicals, LLC) and analyzed immediately via UV–Vis spectrophotometry using a Shimadzu UV-1800 for Cu(I) detection at 562 nm ($\varepsilon_{562} = 14{,}150\ M^{-1}\ cm^{-1}$). To enable peak analysis, ENMs were centrifuged out of solution (30 min, 15 K RPM) and the supernatant was analyzed for Cu(I) using BCA. To ensure BCA was not reacting with Cu(II) from Cu(II) azurin, Cu(II)(*aq*), or Ag(I)(*aq*), an array of controls were also run as described in supplemental materials.

Results and discussion

To assess the role of proteins in catalysis of oxidative dissolution, Ag ENM oxidative dissolution was measured across 10-40 nm ENMs with the addition of Cu(II) azurin and two other well-characterized, redox active proteins: heme-containing proteins cytochrome c (cyt c) and horse radish peroxidase (HRP). Cu(II) azurin was also mixed with SPE in varying concentrations to evaluate catalysis of Ag ENM oxidative dissolution within a complex protein mixture like that in the cell. Independent evidence of redox activity was measured through quantification of Cu(II) azurin reduction and Cu(I) release. With multiple sources of evidence for Ag ENM- Cu(II) azurin redox activity, a mechanism of Ag ENM-Cu(II) azurin reactivity is proposed.

Cu(II) azurin catalyzes Ag ENM dissolution

Formation of Ag(I)(*aq*) by oxidative dissolution of 10–40 nm Ag ENMs was quantified by ICP-MS (Fig. 1). Oxidative dissolution was small at 2 μM Ag(I)(*aq*) or less for Ag ENMs alone, or with $CuSO_4$, cyt c, and HRP. Addition of Cu(II) azurin, however, increased Ag(I)(*aq*) concentrations by roughly an order of magnitude. Protein driven oxidative dissolution has been previously reported for Ag ENMs, as well as other metal and metal oxide ENMs.

When azurin was mixed with varying concentrations of SPE, Cu(II) azurin still catalyzed Ag ENM dissolution. When the contribution of SPE is removed, Ag(I)(*aq*) concentrations are similar to those found in samples of ENM and azurin alone (Fig. 1, see Additional file 1: Figure S2 for raw concentrations). Even at lower, 5 μM azurin concentrations, where azurin does not dominate the protein corona (SDS PAGE results, Additional file 1: Figure S3) oxidative dissolution is two to four times that measured in control samples. SDS PAGE gels indicate that Cu(II) azurin is within the hard corona even when mixed with SPE (Additional file 1: Figure S3).

Notably, although Ag ENM reduction potentials are size dependent, there is not a clear size dependence in Ag ENM oxidative dissolution within this sample set; the sole exception is that highest dissolved Ag(I)(*aq*) concentrations were consistently observed in the smallest, 10 nm ENM samples.

Ag ENMs reduce Cu(II) azurin

Reduction of Cu(II) azurin was assessed with bicinchoninic acid (BCA) as a colorimetric indicator of Cu(I)(*aq*) (Fig. 2). The Cu(I)-BCA absorption peak appears as a shoulder on the Cu(II)-thiol ligand to metal charge transfer (LMCT) band of Cu(II) azurin. As expected, samples with the strongest Cu(I)-BCA shoulder have smaller LMCT bands from Cu(II) azurin, indicative of reduction or loss of Cu(II) from azurin. Control studies demonstrate that the BCA was reactive only with Cu(I)(*aq*), but not with azurin or Ag(I)(*aq*) alone (Additional file 1: Figure S3). In addition, 10-nm citrate-coated Au ENMs reacted with Cu(II) azurin show no evidence of Cu(I) and ESI–MS analysis of the resulting azurin did not reveal any apo- or Au- azurin (data not shown). Consistent with previous reports [21] that the very strong Cu(II)-thiol bond must be reduced before copper release, these results confirm both ENM and redox activity are necessary for Cu(II) azurin-ENM reactivity.

Electronic absorbance spectra were deconvoluted (Additional file 1: Figure S3.b) and spectral contributions were used to calculate the respective

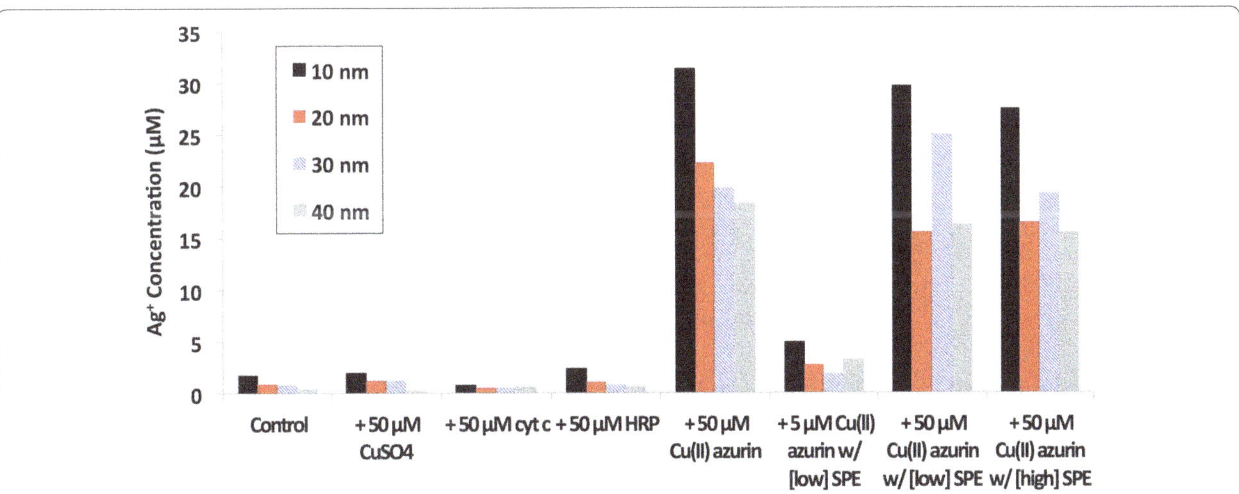

Fig. 1 ICP-MS measurements of Ag(I)(*aq*) concentrations from Ag ENM oxidative dissolution. Oxidative dissolution of Ag ENMs was compared across four sizes: 10 (*black*), 20 (*red*), 30 (*blue*), and 40 nm (*grey*). Dissolution of Ag ENMs alone (labeled control) is compared to dissolution in the presence of 50 μM copper sulfate, cyt c, HRP, and Cu(II) azurin. In addition, both 5 and 50 μM Cu(II) azurin with 0.07 mg/ml SPE were reacted with Ag ENMs (labeled [low] SPE), as well as a higher concentration sample with 50 μM Cu(II) azurin with 0.7 mg/ml SPE (labeled [high] SPE). Data for samples with SPE present in solution are shown with the contribution of SPE subtracted from the total Ag(I)(*aq*) concentration. Raw data for samples with SPE are given in Additional file 1: Figure S1

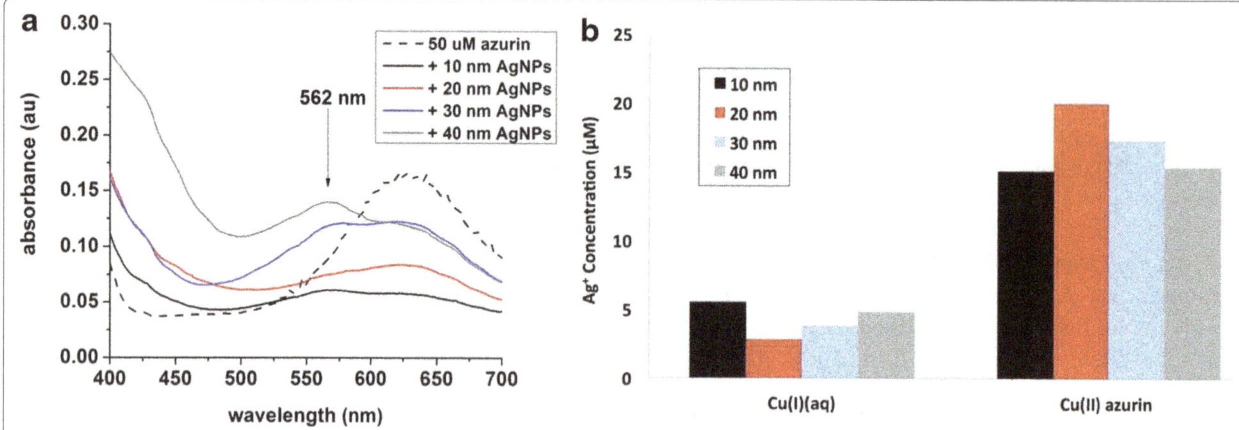

Fig. 2 Colorimetric BCA assay of Cu(I)(*aq*) concentrations. **a** Spectra of BCA added to 50 μM Cu(II) azurin (*black dashes*) and added to a reacted mixture of 50 μM Cu(II) azurin with 10 (*black*), 20 (*red*), 30 (*blue*), and 40 nm Ag ENMs (*grey*). The BCA-Cu(I) complex (λmax = 562 nm) presents as a shoulder on the Cu(II)-thiol LMCT band from Cu(II) azurin (λmax = 630 nm). **b** Deconvolution of UV–Vis spectra enables comparison of Cu(I)(*aq*) and Cu(II) azurin concentrations after reaction with Ag ENM of various sizes. Control spectra for BCA assays and sample deconvoluted spectra are given in Additional file 1: Figure S2

concentrations of Cu(I)(*aq*) and Cu(II) azurin (Fig. 2b). Monovalent copper concentrations were uniformly low at 5 μM or less. Consistent with oxidative dissolution results to form Ag(I)(*aq*), no size-dependence was observed in formation of Cu(I)(*aq*), but 10 nm Ag ENMs did react to form the most Cu(I). Concentrations of Cu(II) azurin decreased in accordance with Cu(I)(*aq*) formation; 10 nm Ag ENMs resulted in the lowest amount of Cu(II) in azurin.

Notably, measured total concentrations of copper from Cu(I)(*aq*) and Cu(II) azurin were consistent across samples, at 21.1 ± 1.2 μM. This concentration is 25–30 % lower in than the starting concentration of copper in azurin and has been confirmed by ICP-MS. The remaining copper may be bound to Ag ENMs and Ag ENM-azurin complexes pelleted out during sample processing. These results suggest a small percentage of released copper or Cu azurin may be adsorbing to the ENM surface; a conclusion consistent with evidence of strong azurin–Ag ENM complex formation previously reported [20] and SDS PAGE results indicating azurin is in the hard corona. The calculated Cu(I)(*aq*) concentrations from deconvoluted electronic spectra do not follow an Ag ENM size trend. This may be due to the loss of copper during sample processing, but it may also be explained by variations in Ag ENM curvature with size that alter protein-Ag ENM interactions.

Conclusions

Evidence of a protein-Ag ENM redox reaction is twofold. First, Cu(II) azurin increases Ag ENM oxidative dissolution by an order of magnitude over other model redox proteins, even in the presence of a complex mixture of proteins within the corona. Second, Ag ENMs, not Ag(I)(*aq*) or Au ENMs, reduce and displace Cu(II) from azurin. We propose a mechanism of Cu(II) azurin–Ag ENM interactions wherein complex formation results in oxidation of surface Ag on the ENM and reduction of Cu(II) in azurin (Fig. 3). After redox, the Cu(I) is displaced from azurin to either form apo-azurin, or to bind the Ag(I) and form Ag(I) azurin. Either way, the fate of both reactants is altered by increasing Ag ENM dissolution and disrupting the physiological function of the Cu metalloprotein. This mechanism provides a biochemical explanation for Cu(II) azurin reactivity with Ag ENMs, but not with Ag(I)(*aq*) or when Ag ENM-protein complex formation is prevented [20].

This work provides an example of an adventitious protein-ENM redox reaction that alters both metal ENM and protein reactivity. Although many consider dissolution the main mechanism of silver ENM toxicity [14, 24, 25], few have considered the role of the protein corona in redox activity of ENMs and potential role in metal homeostasis [17]. More broadly, this work emphasizes

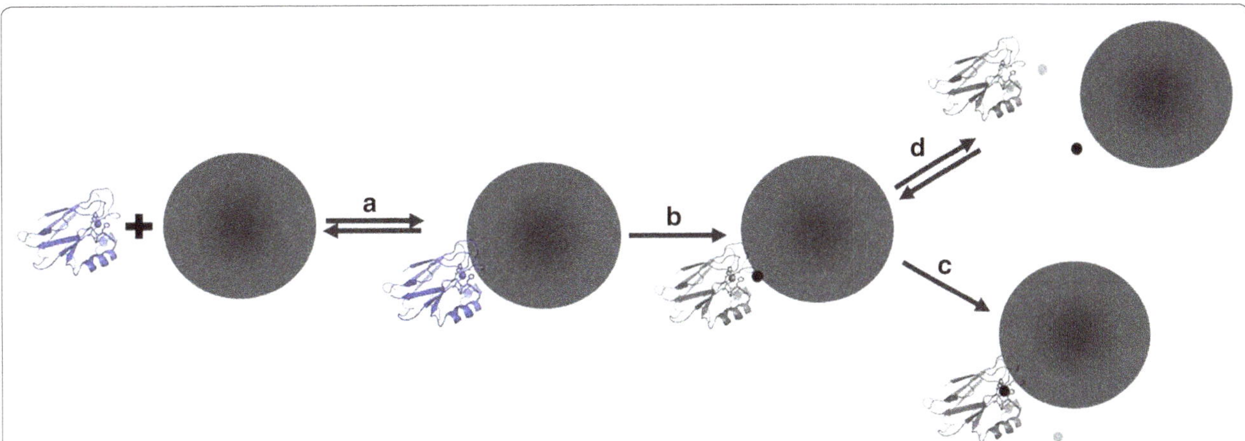

Fig. 3 Schematic for proposed Cu(II) azurin–Ag ENM reaction mechanism. **a** Cu(II) azurin (blue protein with Cu(II) shown as *blue sphere* in active site) and Ag ENM (*large grey sphere*) bind to form a complex. **b** Redox reaction oxidizes Ag ENM surface to form Ag(I) (*black sphere*) and reduces Cu(II) azurin to Cu(I) azurin (*light grey* protein with Cu(I) azurin shown with a *light grey* sphere in active site). Weakly bound Cu(I) is displaced from the active site via one of two mechanisms, either via **c** direct displacement to form Ag(I) azurin and dissolved Cu(I)(*aq*) or via **d** two dissolution equilibria to form apo-azurin with dissolved Cu(I)(*aq*) and Ag ENMs with dissolved Ag(I)(*aq*)

the need to further study adventitious protein redox reactivity with ENMs, especially considering the enzyme-like reactivity of some ENMs, and broadly demonstrates the chemical reactivity of the protein corona at the ENM surface.

Abbreviations
ENMs: engineered nanomaterials; BCA: bicinchoninic acid; ICP-MS: inductively coupled plasma mass spectrometry; SPE: soluble protein extract from *E. coli*; SDS PAGE: sodium dodecyl sulfate polyacrylamide gel electrophoresis; LMCT: ligand metal charge transfer; cyt c: cytochrome c; HRP: horse radish peroxidase.

Authors' contributions
DNF designed and performed all ICP-MS sample preparation and analysis and made all figures. AJM carried out the BCA analysis of ENM-azurin reactions, and participated in the design of the study; ZA carried out experimental controls; DNF and AJM worked with KEW to design experiments and in manuscript preparation; KEW conceived of, designed, and coordinated the study, and drafted the manuscript. All authors gave final approval for publication. All authors read and approved the final manuscript.

Author details
[1] Department of Chemistry and Biochemistry, Santa Clara University, Santa Clara, CA 95053, USA. [2] Present Address: Department of Chemistry, Colorado State University, Fort Collins, CO 80523-1872, USA.

Acknowledgements
We would like to thank Paul Abbyad for providing assistance in fitting the electronic spectra; we also thank Kyle Lancaster and Harry Gray for fruitful discussions on azurin purification and reactivity. We thank Joel Commisso and Austin M. Cole (UC Davis Interdisciplinary Center for Plasma Mass Spectrometry) for technical assistance and ICP-MS data.

Competing interests
The authors declares that they have no competing interests.

Funding
This research was supported by an award from Research Corporation for Science Advancement, with additional financial assistance from Santa Clara University.

References
1. Monopoli MP, Aberg C, Salvati A, Dawson K. A. Nat Nanotechnol. 2012;7(12):779.
2. Docter D, Westmeier D, Markiewicz M, Stolte S, Knauer SK, Stauber RH. Chem Soc Rev. 2015;44(17):6094.
3. Hellstrand E, Lynch I, Andersson A, Drakenberg T, Dahlback B, Dawson KA, et al. Complete high-density lipoprotiens in nanoparticle corona. FEBS J. 2009;276:3372.
4. Arvizo RR, Giri K, Moyano D, Miranda OR, Madden B, McCormick DJ, et al. Identifying new therapeutic targets via modulation of protien corona formation by engineered nanoparticles. PLoS One. 2012;7(3):e33650.
5. Walkey CD, Olsen JB, Song F, Liu R, Guo H, Olsen DWH, et al. Protein corona fingerprinting predicts the cellular interaction of gold and silver nanoparticles. ACS Nano. 2014;8(6):5515.
6. Walkey CD, Chan WCW. Understanding and controlling the interaction of nanomaterials with proteins in a physiological environment. Chem Soc Rev. 2012;41(7):2780.
7. Setyawati MI, Tay CY, Docter D, Stauber RH, Leong DT. Understanding and exploiting nanoparticles' intimacy with the blood vessel and blood. Chem Soc Rev. 2015;44(22):8174.
8. Walczyk D, Bombelli FB, Monopoli MP, Lynch I, Dawson KA. What the cell "sees" in bionanoscience. J Am Chem Soc. 2010;132:5761.
9. Tenzer S, Docter D, Kuharev J, Musyanovych A, Fetz V, Hecht R, et al. Rapid formation of plasma protein corona critically affects nanoparticle pathophysiology. Nat Nanotechnol. 2013;8(10):772.

10. Eigenheer R, Castellanos ER, Nakamoto MY, Gerner KT, Lampe AM, Wheeler KE. Environ Sci Nano. 2014; 1(3):238.
11. Zhou JD, Mortimer G, Martin D, Minchin RF. Differential plasma protein binding to metal oxide nanoparticles. Nanotechnology. 2009;20(45):455101.
12. Fleischer CC, Kumar U, Payne CK. Biomater. Sci. 2013;1(9):975.
13. Durán N, Silveira CP, Durán M, Martinez DST. Silver nanoparticle protein corona and toxicity: a mini-review. J Nanobiotechnology. 2015;13:55.
14. Behra R, Sigg L, Clift MJD, Herzog F, Minghetti M, Johnston B, et al. Bioavailability of silver nanoparticles and ions: from a chemical and biochemical perspective. Soc Interface. 2013;10(87):20130396.
15. Yokel RA, Hussain S, Garantziotis S, Demokritou P, Castranova V, Cassee FR. The Yin: An adverse health perspective of nanoceria: uptake, distribution, accumulation, and mechanisms of its toxicity. Environ Sci Nano. 2014;1(5):406.
16. Hirst SM, Karakoti AS, Tyler RD, Sriranganathan N, Seal S, Reilly CM. Anti-inflammatory properties of cerium oxide nanoparticles. Small. 2009;5(24):2848.
17. Armstrong N, Ramamoorthy M, Lyon D, Jones K, Duttaroy A. Mechanism of silver nanoparticles action on insect pigmentation reveals intervention of copper homeostasis. PLoS One. 2013;8(1):e53186.
18. Wang B, Feng W, Zhao Y, Chai Z. Metallomics insights for in vivo studies of metal based nanomaterials. Metallomics. 2013;5(7):793.
19. Gray HB, Malmström BG, Williams RJP. Copper coordination in blue proteins. J Biol Inorg Chem. 2000;5(5):551.
20. Martinolich AJ, Park G, Nakamoto MY, Gate RE, Wheeler KE. Structural and functional effects of Cu metalloprotein-driven silver nanoparticle dissolution. Environ Sci Technol. 2012;46(11):6355.
21. Tordi MG, Naro F, Giordano R, Silvestrini MC. Silver binding to Pseudomonas aeruginosa azurin. Biometals. 1990;3(2):73.
22. Piccioli M, Luchinat C, Mizoguchi TJ, Ramirez BE, Gray HB, Richards JH. Inorg Chem. 2002;34(3):737.
23. Mattoussi H, Mauro JM, Goldman ER, Anderson GP, Sundar VC, Mikulec FV, et al. J Am Chem Soc. 2000;122(49):12142.
24. Liu J, Sonshine DA, Shervani S, Hurt RH. Controlled release of biologically active silver from nanosilver surfaces. ACS Nano. 2010;4(11):6903.
25. Yang X, Gondikas AP, Marinakos SM, Auffan M, Liu J, Hsu-Kim H, et al. Mechanism of silver nanoparticle toxicity is dependent on dissolved silver and surface coating in Caenorhabditis elegans. Environ Sci Technol. 2012;46(2):1119.

2

Interaction of silver nanoparticles with algae and fish cells: a side by side comparison

Yang Yue[1,2,4], Xiaomei Li[1,2], Laura Sigg[1,3,5], Marc J-F Suter[1,3], Smitha Pillai[1,3], Renata Behra[1,3*] and Kristin Schirmer[1,2,3*]

Abstract

Background: Silver nanoparticles (AgNP) are widely applied and can, upon use, be released into the aquatic environment. This raises concerns about potential impacts of AgNP on aquatic organisms. We here present a side by side comparison of the interaction of AgNP with two contrasting cell types: algal cells, using the algae *Euglena gracilis* as model, and fish cells, a cell line originating from rainbow trout (*Oncorhynchus mykiss*) gill (RTgill-W1). The comparison is based on the AgNP behavior in exposure media, toxicity, uptake and interaction with proteins.

Results: (1) The composition of exposure media affected AgNP behavior and toxicity to algae and fish cells. (2) The toxicity of AgNP to algae was mediated by dissolved silver while nanoparticle specific effects in addition to dissolved silver contributed to the toxicity of AgNP to fish cells. (3) AgNP did not enter into algal cells; they only adsorbed onto the cell surface. In contrast, AgNP were taken up by fish cells via endocytic pathways. (4) AgNP can bind to both extracellular and intracellular proteins and inhibit enzyme activity.

Conclusion: Our results showed that fish cells take up AgNP in contrast to algal cells, where AgNP sorbed onto the cell surface, which indicates that the cell wall of algae is a barrier to particle uptake. This particle behaviour results in different responses to AgNP exposure in algae and fish cells. Yet, proteins from both cell types can be affected by AgNP exposure: for algae, extracellular proteins secreted from cells for, e.g., nutrient acquisition. For fish cells, intracellular and/or membrane-bound proteins, such as the Na^+/K^+-ATPase, are susceptible to AgNP binding and functional impairment.

Keywords: AgNP, *Euglena gracilis*, RTgill-W1 cell line, Nanoparticle uptake, Nanoparticle toxicity, Nanoparticle-protein interactions

Background

Owing to their unique antimicrobial properties, silver nanoparticles (AgNP) are among the most widely used engineered nanoparticles in a variety of consumer products and medical applications, such as textiles and paints. With washing, rain and through other routes, these nanoparticles can be released into the environment, especially into the aquatic environment [1]. This raises concern about potential adverse effects in aquatic organisms. On this background, the toxicity of AgNP to aquatic organisms has been tested on a variety of organisms, ranging from bacteria, to plants, fungi, algae, invertebrates and fish [2–4]. However, with few exceptions [5, 6], most studies did not clearly attribute toxicity to either direct effects of AgNP or to indirect effects of dissolved silver, which includes all the silver species in oxidized state Ag(I) in aqueous solution, such as Ag^+, $AgCl_n$ (aq) and AgOH (aq), stemming from AgNP.

Among aquatic organisms, algae and fish are two important models. As autotrophic organisms, algae are primary producers, i.e. they fix CO_2 to produce oxygen in the presence of light. They are at the base of the food

*Correspondence: renata.behra@eawag.ch; kristin.schirmer@eawag.ch
[1] Department of Environmental Toxicology, Eawag, Swiss Federal Institute of Aquatic Science and Technology, 8600 Dübendorf, Switzerland
Full list of author information is available at the end of the article

chain, serving as food to, e.g. water flea but also fish. Microalgae are single cell organisms surrounded by an inner plasma membrane and an outer semi-permeable cell wall of various compositions. The pores in such cell walls have a size estimated to be 5–20 nm. It helps the algae to maintain integrity and constitutes a primary site for interaction with the surrounding environment [7]. Algae connect with their environment by releasing, e.g. digestive enzymes, for nutrient acquisition. Whether algae have sophisticated mechanisms of particle uptake, such as via endocytosis (see below), is still a matter of debate. Accordingly, internalization of nanoparticles in algae was suggested in only a few studies [8, 9]. There was no evidence of nanoparticle uptake into algae in many other studies using electron microscope imaging and/or analysis of internalized metal in cells [10–14]. These findings emphasize the role of the algal surface as a potential barrier against nanoparticle entry into the cells, with the limitation likely being the pore size in the cell wall.

In contrast to microalgae, fish are heterotrophic, multiple organ- and tissue-based organisms. Fish are at a higher trophic level than algae but depend on the oxygen that algae and other autotrophic organisms produce. Depending on the species, fish can be consumers of algae or of other heterotrophs. With respect to environmental exposure to chemicals or nanoparticles, the fish gill is an important interface due to its large surface. The gill affords gas exchange between the external water environment and internal environment of the organism. In this exchange process, other substances, like metal nanoparticles and organic compounds, can interact with fish gill cells and eventually pass into the blood stream. Therefore, the fish gill can be considered a target of fish-nanoparticle interactions. Accordingly, AgNP were found to be most highly concentrated within gill and liver tissue of rainbow trout (*Oncorhynchus mykiss*) after a 10-day exposure [15]. In contrast to algae cells, fish gill cells, like all animal cells, are cell wall-free. Several kinds of endocytic pathways were proposed for nanoparticle incorporation into animal cells: clathrin-mediated endocytosis, caveolae-mediated endocytosis, macropinocytosis and phagocytosis [14, 16]. Once the vesicles carrying nanoparticles are internalized and detach from the plasma membrane, the vesicles are sorted and transported to different endocytic compartments. By these processes, nanoparticles are delivered to other subcellular compartments in endocytic pathways, from early endosome and multi-vesicular bodies to late endosomes and lysosomes [17].

Independent of the mechanism of particle uptake, nanoparticles tend to bind molecules from the surrounding environment owing to their big surface-to-mass ratio. During nanoparticle interaction with cells, proteins are an important class of biomolecules that are prone to binding to nanoparticles, leading to a protein corona [18, 19]. With regard to extracellular proteins, such as the digestive proteins excreted by algae and bacteria, a so-called "eco-corona" can form [20, 21]. Intracellular proteins, on the other hand, can bind to particles upon uptake into cells. With the binding to nanoparticles, the properties and functions of proteins can change compared to unbound proteins. Thus, it is also important to understand to what extent nanoparticle-protein complexes impact on the properties of the proteins. Studies on the nanoparticle-protein interactions initially focused on single proteins. For example, Wigginton [22] found that AgNP inhibited tryptophanase (TNase) activity in the interaction with *E. coli* proteins and a dose-dependent inhibition of enzyme activity was observed for the incubation of citrate-coated AgNP with firefly luciferase [23]. In contrast to single protein-nanoparticle interactions, only few studies have thus far focused on identifying proteins that bind out of a complex mixture, especially in an intact intracellular environment [24, 25]. Such studies not only help identify the proteins most susceptible to particle binding but can also guide future research on single protein-particle interactions.

In order to shed light on the detailed mechanisms of interaction between AgNP and cells of algae and fish, we explored different aspects of AgNP-cell interactions, spanning AgNP behavior in exposure media, toxicity to cells, uptake and interaction with proteins. We aimed to critically compare the interaction of AgNP with contrasting cell types belonging to autotrophic vs. heterotrophic organisms in order to support a rational assessment of risks based on our previous studies [26–29]. A species of algae, *Euglena gracilis*, and a fish gill cell line, RTgill-W1 [30], originating from rainbow trout (*Oncorhynchus mykiss*), were selected to represent an autotrophic and a heterotrophic aquatic cellular system. The *Euglena gracilis* has no rigid cell wall but a flexible glycoprotein-containing pellicle, which aligns on the surface in longitudinal articulated stripes [31]. It was selected on purpose because nanoparticle uptake was thought to more likely occur in such an algae compared to one with a rigid cell wall. The RTgill-W1 cell line can survive in a simplified exposure medium, which provides the possibility to expose cells in medium that more closely mimics the aqueous environment a fish gill would face [32, 33]. Both algae and fish gill cell exposures were performed in minimal media supporting cell survival but not proliferation, in order to provide better controllable exposure and effect assessment for mechanistic studies. Here we focus on the comparative aspects of the outcome of our research. Unless noted otherwise, we will refer to *E.*

gracilis as "algal cells" and to the RTgill-W1 fish gill cell line as "fish cells".

Results and discussion

The composition of exposure media significantly influences AgNP behavior

The size, zeta potential and dissolution of AgNP were tested over time in exposure media for algae and fish cells (Table 1). To avoid silver complexation, only 10 mM 3-morpholinopropanesulfonic acid (MOPS, pH 7.5) was used as exposure medium in algae experiments [26]. In the stock solution, the initial Z-average size and zeta potential of AgNP were 19.4 nm and −30 mV, respectively. AgNP were stable in this medium with an average size of 38–73 nm and a zeta potential of −23 to −28 mV up to 4 h of incubation [26]. For the fish cells, three kinds of exposure media were selected: L-15/ex, a regular, high ionic strength and high chloride cell culture medium based on Leibovitz' 15 (L-15) [32, 34]; L-15/ex w/o Cl, a medium without chloride to avoid the formation of AgCl and study the role of chloride in silver ion and AgNP toxicity; and d-L-15/ex, a low ionic strength medium that more closely mimics freshwater [27]. The AgNP moderately agglomerated (average size: 200–500 nm; Zeta potential: −15 mV) in L-15/ex medium. In L-15/ex w/o Cl medium, AgNP strongly agglomerated with an average size of 1000–1750 nm and a zeta potential of −10 mV. In d-L-15/ex medium, AgNP dispersed very well (average size: 40–100 nm; Zeta potential: −20 mV). Even though the size of AgNP increased up to 1750 nm, we found that large size AgNP were due to agglomeration [27], which is a reversible process and AgNP can easily be dispersed again [35]. The UV–Vis absorbance of AgNP in exposure media confirmed the different behavior of AgNP in the different media [26, 27]. Transmission electron microscopy (TEM) images of fish cells showed that single or slightly agglomerated AgNP were located in endosomes and lysosomes in fish cells, which indicates that fish cells took up AgNP in nanoscale [28].

The dissolution of AgNP, expressed as percentage of free to total silver, was comparable in MOPS and L-15/ex (~1.8%); dissolution was somewhat lower in L-15/ex w/o Cl and d-L-15/ex medium (~0.5%). Depending on the applied concentrations, this amounts to dissolved silver in the range of 1 nM to 2 μM (assuming 1–2% dissolution in 0.1–100 μM AgNP suspension). Upon contact with algae or fish cells, the uptake of dissolved silver shifts the AgNP/silver ion equilibrium and more silver ions are released. Furthermore, previous work reported that AgNP accumulated in mammalian cell endosomes and lysosomes displayed higher dissolution in these acidic environments than in a neutral environment [17, 36]. Therefore, we expect significant dissolution of AgNP in this process and used $AgNO_3$ as a dissolved silver control throughout.

The diverse behavior of AgNP in the different exposure media demonstrates the importance of accounting for nanoparticle characteristics in the respective exposure environments. The composition of the exposure media showed a strong influence, especially in terms of particle agglomeration but also in terms of dissolution. In high ionic strength medium, high concentrations of ions can break the electrical double layers surrounding the AgNP and thereby decrease the surface charge, which leads to AgNP agglomeration. In the presence of chloride, AgNP were more stable (compare L-15/ex medium to L-15/ex w/o Cl), which means chloride ions can stabilize AgNP, likely by binding to AgNP surfaces and contributing to a negative surface charge. In terms of AgNP dissolution, a higher percentage was found in L-15/ex with high chloride: chloride shifts the equilibrium of AgNP dissolution by complexing the dissolved silver.

AgNP adsorb to the algal cell surface but can be taken up by fish cells

To quantitatively relate $AgNP/AgNO_3$ exposure to the toxicity seen in algal and fish cells, cell-associated silver was quantified by inductively coupled plasma mass spectrometry (ICP-MS). Upon exposure to

Table 1 AgNP behavior in exposure media for algae and fish cells

	Algae exposure medium [26]	Fish cell exposure media [27]		
		L-15/ex	L-15/ex w/o Cl	d-L-15/ex
Medium ionic strength (mM)	3.44	173.0	177.1	72.0
Size of AgNP (nm)	38–73	200–500	1000–1750	40–100
Zeta potential of AgNP (mV)	−23 to −28	−15	−10	−20
Dissolution of AgNP (% of total Ag)[a]	1.7%	1.89%	0.67%	0.40%

[a] The level of dissolution of AgNP represents the mean of dissolution data obtained using two different methods to separate dissolved silver from particles: ultrafiltration and ultracentrifugation. Values given are the mean of the average data obtained for each method, carried out three independent times

similar concentrations of AgNP or $AgNO_3$, the cell-associated silver in algae cells was comparable with the cell-associated silver which was reported for the alga *Chlamydomonas reinhardtii* [11]. Similarly, the cell-associated silver in RTgill-W1 cells was also comparable with the silver content in other vertebrate cell types, such as mouse erythroleukemia cells [37] and HepG2 cells [38].

At comparable external $AgNO_3$ exposure concentrations (0.1–0.5 μM), the silver content associated with algal cells was 2.4–4.2 times higher than in the fish cells (Fig. 1). This was probably due to the different compositions of the exposure media and the resulting different dissolved silver species. In the algal exposure medium, MOPS, almost all dissolved silver was present as free silver ions (Ag^+) as predicted by Visual MINTEQ (V3.1, KTH, Sweden). Free silver ions are taken up via copper transporters in algae, as suggested in *C. reinhardtii*, *Pseudokirchneriella subcapitata* and *Chlorella pyrenoidosa* [39–41]. On the contrary, in fish cell exposure medium, only around 60% of dissolved silver was in the form of Ag^+. The other 40% reacted with chloride and formed neutral or negatively charged complexes ($AgCl_n^{(n-1)-}$) [27]. Earlier research showed that Ag^+ has a higher bioavailability than $AgCl_n^{(n-1)-}$ complexes in rainbow trout and Atlantic salmon [42], since Ag^+ enters into gill cells via copper transporters and sodium channels, while $AgCl_0$(aq) may be taken up by simple diffusion [43].

In the case of AgNP exposure, the algal cells again had 2.5–4 times more cell-associated silver than the fish cells at 2.5–5 μM of external AgNP concentration (Fig. 1). We attribute this difference to a higher overall exposure of the algal cells. There might be various factors influencing the level of cell-associated silver, e.g. kinetics of internalization into fish cells, sorption differences, ongoing dissolution at the interface between AgNP and cell surface, and abundance of metal transporters. Indeed, algae cells were exposed in suspension, allowing AgNP and $AgNO_3$ to interact from all sides with the cell surface (643 μm^2/cell). In contrast, the fish cells were exposed as a cell monolayer sitting on a cell culture surface, which means only one side of the fish cells (half of the cell surface: 286 μm^2/cell) was in immediate contact with AgNP or $AgNO_3$.

AgNP and silver ions elicit toxicity to algae and fish cells

The photosynthetic yield was assessed to study the time-dependent toxicity of AgNP and $AgNO_3$ in algae. The photosynthetic yield is an important parameter for evaluating the viability of algal cells as autotrophic organisms. In the fish cells, the overall metabolic activity was used as an endpoint upon AgNP and $AgNO_3$ exposure. Effective concentrations causing a 50% decline (EC50s) in photosynthetic yield and metabolic activity were calculated from dose–response curves derived with algal and fish cells. The EC50s ranged from 1.5 to 1.9 μM (0.16–0.21 mg/L) AgNP in algal cells and from 12.7 to 70.3 μM (1.37–7.59 mg/L) AgNP in fish cells (Fig. 2). In $AgNO_3$ exposures, EC50s were 0.09 μM (0.01 mg/L) in algae and 0.8–9.7 μM (0.09–1.05 mg/L) in the fish cells

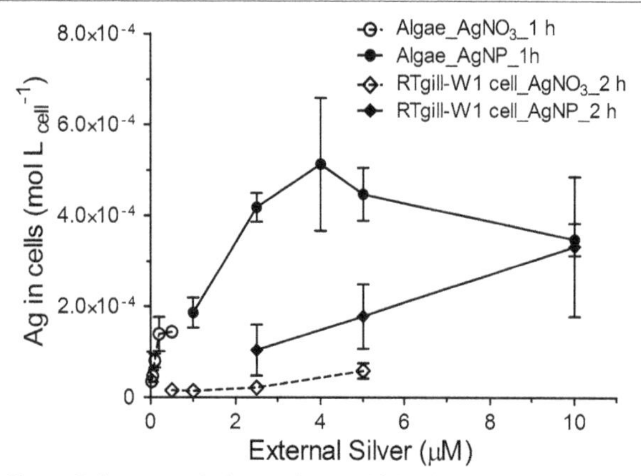

Fig. 1 Cell-associated silver in algae and fish cells. Cell-associated silver levels (mol/L_{cell}) were quantified by ICP-MS after exposure to AgNP and $AgNO_3$ for 1 h (algae) and 2 h (fish cells). The exposure of the algal cells was in MOPS; that of the fish cells in d-L-15/ex medium. The concentrations of silver (AgNP, $AgNO_3$) were selected based on the concentration response curves obtained for algae [26] and fish cells [28]. Cells were washed with cysteine solution to remove any loosely bound silver prior to extraction and analysis. Data presented as mean ± SD; n = 3

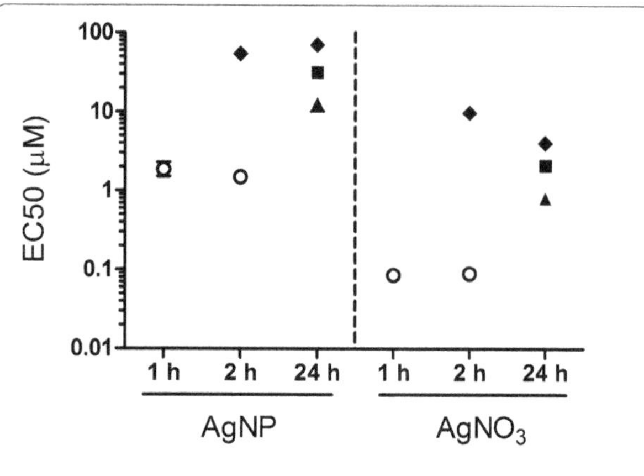

Fig. 2 EC50 values of AgNP and $AgNO_3$ in algae and fish cell exposures as a function of total silver. Times of exposure were selected based on the response of the respective cell type, with algal cells responding quickly with no further change in EC50 after 1 h whereas EC50 further decreased for fish cells over a 24 h period. Data presented as mean ± SD; n = 3. The *error bars* are smaller than the symbols due to the exponential scale in Y-axis

(Fig. 2). In the algae cell model, the EC50 values of AgNP determined in our study were comparable with EC50 values reported for other algal species [3, 44]. In the fish cell model, the EC50 values were similar to the EC50 values measured in other fish cell types [45, 46]. According to the categorization of toxic or non-toxic concentrations to aquatic organisms (<0.1 mg/L = extremely toxic; 0.1–1 mg/L = very toxic; 1–10 mg/L = toxic; 10–100 mg/L = harmful; <100 mg/L = non-toxic [47, 48]), we conclude that AgNP and $AgNO_3$ are toxic to both algae and fish cells.

When comparing these EC50s, both similarities and differences were found for algal and fish cells. In terms of similarities, AgNP induced significantly lower toxicity than $AgNO_3$ in both cell types if EC50s are expressed as a function of total silver present (Fig. 2). In terms of differences, algae were 10 to 100 times more sensitive than fish cells to both AgNP and $AgNO_3$ exposure. On the other hand, if EC50s are expressed as a function of dissolved silver in the respective exposure media, it was found that the toxicity of AgNP to algae is mediated solely by dissolved silver [26] while in fish cells, AgNP were found to induce toxicity by a nanoparticle specific effect as well as dissolved silver [27].

The media composition also had a strong effect on the toxicity of AgNP to fish cells (Fig. 2) [27]. Among the three fish cell exposure media, AgNP yielded highest toxicity in L-15/ex w/o Cl and lowest toxicity in d-L-15/ex. This difference correlates with the degree of AgNP agglomeration in the media: the strongly agglomerated AgNP (size: 1000–1750 nm) in L-15/ex w/o Cl induced a 2-times higher toxicity than moderately agglomerated AgNP (size: 200–500 nm) in L-15/ex and 5-times higher toxicity than weakly agglomerated AgNP (size: 40–100 nm) in d-L-15/ex medium. Likely, the different degrees of agglomeration resulted in differing degrees of deposition of AgNP onto the fish cells, with the strongest agglomeration leading to highest cell exposure. This trend indicated that agglomeration and deposition could increase the interaction of AgNP with RTgill-W1 cell and induce higher toxicity. In the present exposure model, fish cells formed a cell layer on the bottom of the wells and AgNP suspensions were added on top. Previous modeling work showed that large size nanoparticles are transported faster than small nanoparticles due to deposition [49, 50]. Because of particle deposition on the cell monolayer, AgNP agglomeration may increase the interaction of AgNP with cells and thereby AgNP toxicity. Among the fish cell exposure media, d-L-15/ex medium maintains AgNP stability and better reflects the freshwater environment to which gill cells of freshwater fish would be exposed. Therefore, d-L-15/ex was selected to study the interaction of AgNP with fish cells in more detail.

The effects of AgNP and $AgNO_3$ to algae and fish cells were recalculated as a function of cell-associated silver (Fig. 3; Additional file 1: Table S1). The EC10s (concentrations leading to 10% effect compared to unexposed control) were used for comparison because effects were quantifiable based on experimental data under all conditions for this level of effect (see horizontal dashed line in Fig. 3). The EC10 of AgNP and $AgNO_3$ in algae are 1.40×10^{-4} and 3.55×10^{-4} mol/L_{cell}, respectively. The EC10 of AgNP and $AgNO_3$ in fish cells are 1.80×10^{-5}

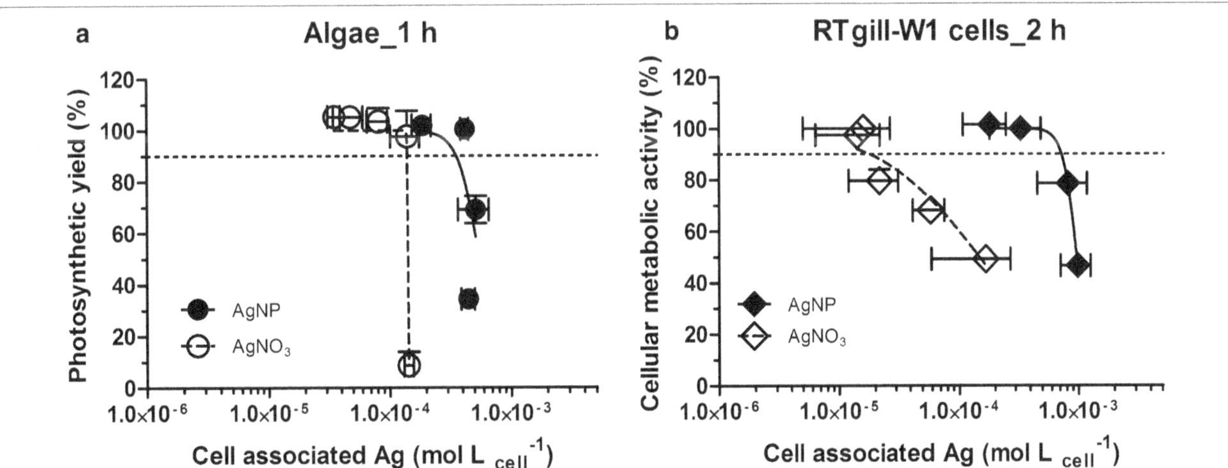

Fig. 3 Recalculation of the toxicity of AgNP and $AgNO_3$ to algae in MOPS **a** and fish cells in d-L15/ex medium **b** as a function of cell-associated silver. The *dashed horizontal lines* show the EC10 level. Each data point presents the mean of three independent experiments with the *horizontal lines* indicating the variation in cell-associated silver and the *vertical lines* the variation in effect (mean ± SD, n = 3). All data are expressed as % of the respective unexposed control. Data presented as mean ± SD; n = 3

and 7.22×10^{-4} mol/L_{cell}, respectively. In both cell models, the $AgNO_3$ concentration response curve is left of the AgNP concentration response curve, indicating stronger effects of $AgNO_3$. This can be interpreted as AgNP inducing toxicity via different mechanisms compared to $AgNO_3$. However, this difference between $AgNO_3$ and AgNP concentration response curves is much greater in the fish cells, demonstrating that fish cells respond strongly to a particle-specific impact, whereas in algae, dissolved silver is the dominant cause of toxicity.

The fact that no particle-specific effect was seen in algae suggested that AgNP were not incorporated into algae but that they may adhere to the algal surface from which Ag^+ may dissolve and as such be taken up in the cells. To follow up on this hypothesis, the localization of cell-associated silver in algae was investigated by time-of-flight secondary ion mass spectrometry (TOF–SIMS), a qualitative and quantitative surface analysis. Indeed, the silver intensity from algae exposed to AgNP and $AgNO_3$ indicated a strong sorption of AgNP onto the algal surface [26]. In contrast, fish gill cells take up AgNP in an energy—dependent process: as demonstrated by TEM coupled with energy-dispersive X-ray analysis that localized the NPs in the endocytic compartments of the cells [28]. This latter finding corresponds to the work of others who confirmed metal nanoparticles uptake by vertebrate cells via endocytic pathways [51–53].

AgNP can bind cellular proteins and inhibit enzyme activity

Considering the contrasting finding that algal cells do not take up AgNP while fish cells do, and the importance of the AgNP-protein binding in nano-bio interaction, we postulate that extracellular proteins are more important in terms of AgNP exposure for algae while intracellular proteins should be considered for fish cells. One important extracellular protein is alkaline phosphatase, an enzyme responsible for phosphorus acquisition. This enzyme is highly abundant in aquatic environments, and is produced by a wide range of organisms including bacteria, fungi, zooplankton and algae [54–56]. Previously, alkaline phosphatase activity in periphyton was shown to be unaffected [57] or stimulated [58] by AgNP. However, considering that algae within periphyton are embedded in a matrix of biomolecules forming a biofilm [59], other factors might influence direct interaction of AgNP with the enzyme, e.g. periphyton community continuously synthesizes and secrets enzymes, it could not be determined whether the absence of inhibitory effects was due to a lack of interaction with AgNP or whether any impact was masked by de-novo synthesis of the enzyme.

Indeed, studying the interaction of AgNP or $AgNO_3$ with isolated alkaline phosphatase showed that AgNP have a particle—specific, inhibitory effect on alkaline phosphatase activity and that this inhibition depends on the sequences of addition of enzyme substrate or AgNP [29]. Other studies have reported on an inhibitory effect on extracellular enzymes by nanoparticles. For example, in the same study on periphyton cited above [57], inhibition of β-glucosidase and L-leucine aminopeptidase was attributed to dissolved silver and particle specific effects. Inhibitory effects of the same proteins were also seen in heterotrophic biofilm exposed to titanium dioxide nanoparticles [60]. Thus, understanding the mechanisms of interaction of extracellular proteins and nanoparticles in general is an important future direction.

In order to study the interaction of AgNP with intracellular proteins in fish cells, the AgNP-protein corona was recovered from intact endocytic compartments by a newly established method with subcellular fractionation. Proteins acquired from the AgNP-protein corona were identified by mass spectrometry and analyzed with Gene Ontology. A total of 383 proteins were identified in this way and broadly classified as belonging to cell membrane functions, uptake and vesicle-mediated transport and stress-response pathways [28]. These proteins regulate substance transport across the plasma membrane or play key roles in cell metabolic processes, such as Na^+/K^+-ATPase, Ca^{2+}-ATPase, adaptor-related protein complex 1 (AP-1B1), caveolin 1, flotillin 1/2, EH-domain containing protein 1/2/4 and Rab Family Small GTPases (RAB5A, RAB7A, RAB18) [28]. Based on the identified proteins, the processing of AgNP in fish cells was reconstructed: AgNP were taken up by fish cells via endocytic processes and stored in endosomal/lysosomal compartments [28]. Binding to AgNP could impair the function of these proteins and subsequently disrupt the normal cell activity, which would relate to the decline of cell viability in RTgill-W1 cells exposed to AgNP. Some of these proteins were also identified in the corona of magnetic nanoparticles exposed to human lung epithelial cells (A549) and HeLa cells [24, 25]. This indicated that vertebrate cells take up metal nanoparticles via common pathways regardless of elemental composition or coating of particles, exposure conditions and cell types.

Among the proteins identified from fish cells, Na^+/K^+-ATPase was selected to study the effect of AgNP on corona proteins. Experiments on the isolated, single protein showed that the inhibition of enzyme activity is attributable primarily to a particle-specific rather than a dissolved silver ion effect (Fig. 4). Schultz reported that citrate coated AgNP, i.e. the same type of particle used in this work, led to a particle-specific inhibition of Na^+/K^+ ATPase activity in juvenile rainbow trout gill in vivo [61]. Thus, our in vitro study was confirmative of the findings

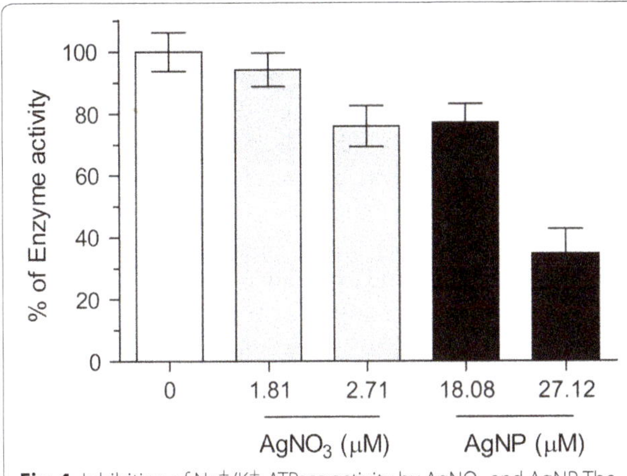

Fig. 4 Inhibition of Na^+/K^+-ATPase activity by $AgNO_3$ and AgNP. The concentration of Na^+/K^+-ATPase was 0.5 U/mL (19.5 µg/mL) in all experiments; the concentration of $AgNO_3$ was selected based on the concentration of dissolved silver in AgNP suspension. Both a silver ion as well as a particle-specific effect was found with the latter being more dominant. Figure was reproduced from Yue et al. [28] with permission from the Royal Society of Chemistry. Data presented as mean ± SD; $n = 3$

in vivo and signifies a strategy to further investigate AgNP and other nanoparticles for their interaction with corona proteins, based on the protein list established as described above.

Conclusion

The results of the side by side comparison of AgNP-cell interactions for algal and fish cells are summarized in Table 2. The composition of exposure media influenced AgNP behavior and toxicity, highlighting once more the importance of characterizing nanoparticle speciation in the risk assessment of nanomaterials. Because of the barrier surrounding the cell membrane, AgNP cannot be taken up by algal cells but adsorb onto the cell surface instead, and toxicity is thus induced by dissolved silver. On the other hand, in fish cells, AgNP are taken up via endocytic pathways, causing toxicity by both dissolved silver and a nanoparticle specific effect. Thus, the cell type and structure are important features to be considered in nanotoxicity research. AgNP can bind extracellular and intracellular proteins and inhibit enzyme activities via nanoparticle specific effects. Current work provides a first concrete attempt to study the interaction of AgNP with extracellular and intracellular proteins from aquatic organisms. For future assessment, this kind of knowledge not only aids in mechanism-based aquatic risk assessment but also helps designing safer nanoparticles.

Methods

AgNP preparation and characterization

The citrate coated AgNP were purchased from NanoSys GmbH (Wolfhalden, Switzerland) as aqueous suspension

Table 2 Summary of the interaction of AgNP with algae and fish cells

	Algae	Fish cell
NP behavior	Slight agglomeration in MOPS medium	Slight agglomeration in d-L-15/ex medium and strong agglomeration in L-15/ex and L-15/ex w/o Cl media
Toxicity	Dissolved silver[a]	Dissolved silver[a], nanoparticle-specific effect
Cellular uptake	No, adsorbed on algal surface	Yes, uptake via endocytic pathways
AgNP-protein interaction	Adsorption of extracellular enzyme on AgNP, inhibition of enzymatic activity	Cell membrane proteins and endocytic proteins binding to NP, inhibition of enzymatic activity
Cell structure and AgNP association		

[a] Dissolved silver indicates all the silver species in oxidized state Ag(I) in aqueous solution, such as Ag^+, $AgCl_n$ (aq) and AgOH (aq), stemming from AgNP

with a concentration of 1 g/L (9.27 mM referring to the total silver, pH 6.46). The Z-average size and Zeta potential of AgNP in the stock solution were 19.4 nm and −30 mV, respectively. The stock AgNP solution was kept in the dark. A stock solution (10 mM) of $AgNO_3$ (Sigma-Aldrich, Switzerland) was prepared by dissolving $AgNO_3$ in nanopure water (16–18 MΩ/cm; Barnstead Nanopure Skan AG, Switzerland). The experimental solutions of AgNP and $AgNO_3$ were freshly prepared by adding AgNP and $AgNO_3$ stock solution into the respective exposure media and vortexing for 10 s. Unless specifically indicated, all chemicals were purchased from Sigma-Aldrich. As the work focused on the impacts of AgNP on algae and fish cells, we chose the respective AgNP and $AgNO_3$ concentrations to observe significant effects in the toxicity experiments. In the uptake experiments, the exposure concentrations were selected to meet the ICP-MS detection limit.

The AgNP were characterized in nanopure water and under experimental conditions in each exposure media. The Z-average size and zeta potential of the AgNP were measured by dynamic light scattering (DLS) and electrophoretic mobility using a Zetasizer (Nano ZS, Malvern Instruments, UK) [26, 27]. To measure the dissolution of AgNP in exposure medium, dissolved silver was separated by two methods: centrifugal ultrafiltration with a nominal molecular weight cut-off of 3 kDa (Amicon ultra-4 centrifugal filter units, Millipore, Germany) centrifuged at 3000×*g* for 0.5 h (Megafuge 1.0R, Heraeus Instruments, Germany) and by ultra-centrifugation (CENTRIKON T-2000, KONTRON Instruments, Switzerland) at 145,000×*g* for 3 h. The silver concentration was measured by Inductively Coupled Plasma Mass Spectrometry (ICP-MS, Element 2, Thermo Finnigan, Germany). The dissolution of AgNP was calculated by dividing the measured dissolved silver concentration to the related nominal total silver concentration. The average of the AgNP dissolution (Table 1) obtained by ultra-filtration and ultra-centrifugation was used for recalculation of the concentration–response curves as a function of dissolved silver (Fig. 3).

Algal culture and exposure of cells

The alga *E. gracilis* strain Z (Culture Collection of Algae, Göttingen, Germany) was cultured in the Talaquil medium (pH 7.5) supplemented with vitamins B1 and B12 [62] at 20 °C under light–dark cycles of 12 h each on a shaker (Infors, Switzerland) with 90 rpm. The cell number and volume were measured by a particle counter (Beckman Coulter Z2, USA).

Before exposure to $AgNO_3$ and AgNP, exponentially grown algae were centrifuged at 2000×*g* for 10 min and then re-suspended in MOPS. The final cell density for toxicity assessment was 1.5×10^4 cell/mL. After exposure to $AgNO_3$ (0–400 nM) and AgNP (0–40 μM) for 1 and 2 h, to study toxic effects, the photosynthetic yield was measured by fluorometry using a PHYTO-PAM (Heinz Walz GmbH, Germany) [26]. The values were presented as percentage of controls, and were plotted as a function of measured total silver and cell-associated silver [26].

RTgill-W1 cell culture and exposure of cells

RTgill-W1 cells were routinely cultivated in L-15 medium (Invitrogen, Switzerland), supplemented with 5% fetal bovine serum (FBS, Gold, PAA Laboratories GmbH, Austria) and 1% penicillin/streptomycin (Sigma-Aldrich, Switzerland) in 75 cm^2 flasks. The L-15 medium containing these supplements is termed "complete L-15". Cells are routinely cultured in the dark in normal atmosphere at 19 °C.

For exposure to $AgNO_3$ (0–5 μM) and AgNP (0–100 μM), cells were seeded in 24-well microtiter plates or 25 cm^2 flasks, and cultured in complete L-15 medium. After being fully confluent, cell monolayers were washed with either L-15/ex, L-15/ex w/o Cl or d-L-15/ex. Then, 1 mL/well of AgNP or $AgNO_3$ suspension in the respective media was added to culture wells. Exposure proceeded for 2–24 h at 19 °C. AlamarBlue (AB, Invitrogen, Switzerland) was used to measure the cellular metabolic activity to assess the toxicity of AgNP to fish gill cells [27, 32]. Before incubation with AlamarBlue, the AgNP suspension was removed and exposed cells were carefully washed with PBS to removed loosely adsorbed AgNP. Control experiments showed no interference of the silver with the AlamarBlue assay.

Uptake of AgNP by algae and fish cells

The algae were exposed to AgNP (0–10 μM) and $AgNO_3$ (0–500 nM) at a cell density of 1×10^5 cell/mL in order to meet the detection limit of the ICP-MS. After 1 h of exposure, the algae were washed to remove loosely bound AgNP with fresh medium or adsorbed silver ions by cysteine using the following protocol: algae exposed to AgNP or $AgNO_3$ were first centrifuged (2000×*g*, 10 min) and resuspended in MOPS. After 2 wash cycles, the algae were re-suspended in cysteine–MOPS and gently stirred for 5 min. After washing, the algae were filtered (SM 16510, Sartorius) and digested for metal analysis [26].

To quantify the fish cell-associated silver, RTgill-W1 cells were cultured in 25 cm^2 flasks until confluency and then exposed to AgNP (2.5–10 μM) or $AgNO_3$ (0.5–5 μM) in d-L-15/ex medium. After exposure, the medium with AgNP or $AgNO_3$ was removed and cells were washed twice with cysteine for 5 min. Cells were then trypsinized. Detached cells were re-suspended in complete L-15 medium. Cell suspensions were centrifuged at $1000 \times g$ for 3 min to pellet the cells. Cell pellets were re-suspended in 550 μL PBS and the cell density determined by an electronic cell counter (CASY1 TCC, Schärfe System, Germany). A volume of 500 μL cell supernatant was digested for metal analysis [28].

Samples from algae and fish cell exposures were digested with 4.5 mL of 65% HNO_3 in a high-performance microwave digestion unit (MLS-1200 MEGA, Switzerland) at a maximum temperature of 195 °C for 20 min. The digests were diluted 50-times and measured by ICP-MS. The detection limit for ICP-MS quantitation of silver was 10 ng/L (1.0×10^{-5} mol/L_{cell} in the current work). The reliability of the measurements was determined using specific water references (M105A, IFA-Tull, Austria). As the volume of algal cells is larger than that of the fish cells, the measured cell associated silver was related to cell volume and expressed as mol/L_{cell} in order to be able to directly compare the cell-associated silver in algae and fish cells. The associated silver was also related to the cell number and expressed as mol per cell to be able to compare with other reports.

The localization of cell-associated silver in algae was checked by time-of-flight secondary ion mass spectrometry analysis (TOF–SIMS, ToF.SIMS 5 instrument, ION-TOF GmbH). The primary ion was 25 keV Bi^+ to ensure high sensitivity to silver and the sputtering time was 5.2 s leading to ablation of a few nanometers of the surface layer of the cell [26]. The AgNP uptake in fish cells was investigated by transmission electron microscopy (TEM, FEI Morgagni 268, 100 kV) and energy dispersive X-ray (EDX) spectroscopy analyses in a scanning transmission electron microscope (STEM, Hitachi HD-2700) [28].

Interaction of AgNP with proteins

Alkaline phosphatase (Sigma-Aldrich) was select as a representative extracellular algal enzyme to study the interaction with AgNP. The effect of AgNP to alkaline phosphatase was assayed in MOPS by determining enzyme activity, using fluorescently linked 4-methylumbelliferyl phosphate disodium salt as substrate [29].

In fish cell exposures, to identify the proteins binding to AgNP in cells, AgNP-protein corona complexes were recovered from intact subcellular compartments isolated by subcellular fractionation and proteins lysed from the AgNP to be detected by mass spectrometry. The identified proteins were analyzed by DAVID (http://david.abcc.ncifcrf.gov/), a protein ontology analysis tool. Among the identified proteins, Na^+/K^+-ATPase (Sigma-Aldrich, No. A7510) was selected to study the interaction of AgNP with intracellular proteins. The effect of AgNP on Na^+/K^+-ATPase activity was measured in a buffer containing 20 mM Tris–HCl, 0.60 mM EDTA, 5 mM $MgCl_2$, 3 mM KCl and 133 mM NaCl (pH 7.8) and substrate, ATP (Sigma-Aldrich, No. A9062) [28].

Data analysis

All data were analyzed by GraphPad Prism (version 5.02 for Windows, USA). Fluorescent units obtained in the cell assays were converted to percent viability of control cells. Concentrations leading to 10% and 50% effect (EC10s, EC50s) were determined by nonlinear regression sigmoidal dose–response curve fitting using the Hill slope equation, and were presented as the mean of three independent experiments, with a 95% confidence interval. Data presented as mean ± standard deviation, n = 3.

Authors' contributions

YY and XL carried out the study and analyzed data. YY and KS wrote the manuscript. LS, MJFS, SP, RB, and KS participated in the design of the study and revised the manuscript. All authors read and approved the final manuscript.

Author details

[1] Department of Environmental Toxicology, Eawag, Swiss Federal Institute of Aquatic Science and Technology, 8600 Dübendorf, Switzerland. [2] School of Architecture, Civil and Environmental Engineering, École Polytechnique Fédérale de Lausanne, 1015 Lausanne, Switzerland. [3] Department of Environmental Systems Science (D-USYS), ETH-Zürich, 8092 Zürich, Switzerland. [4] Present Address: Department of Basic Sciences and Aquatic Medicine, Norwegian University of Life Sciences (NMBU), Oslo 0454, Norway. [5] Present Address: Wattstrasse 13a, 8307 Effretikon, Switzerland.

Acknowledgements

We thank D. Kistler (Eawag, Swiss Federal Institute of Aquatic Science and Technology, Dübendorf, Switzerland) for ICP-MS measurements.

Competing interests

The authors declare that they have no competing interests.

Funding

This research was part of "Interaction of metal nanoparticles with aquatic organisms (MeNanoqa)" project (Project Number: 406440-131240), which was supported by the Swiss National Science Foundation in the framework of the Swiss National Research Program 64 (NRP 64) "Opportunities and Risks of Nanomaterials".

References

1. Kaegi R, Sinnet B, Zuleeg S, Hagendorfer H, Mueller E, Vonbank R, Boller M, Burkhardt M. Release of silver nanoparticles from outdoor facades. Environ Pollut. 2010;158:2900–5.
2. Collin B, Auffan M, Johnson AC, Kaur I, Keller AA, Lazareva A, Lead JR, Ma X, Merrifield RC, Svendsen C, et al. Environmental release, fate and ecotoxicological effects of manufactured ceria nanomaterials. Environ Sci Nano. 2014;1:533–48.

3. Fabrega J, Luoma SN, Tyler CR, Galloway TS, Lead JR. Silver nanoparticles behaviour and effects in the aquatic environment. Environ Int. 2011;37:517–31.
4. Schultz AG, Boyle D, Chamot D, Ong KJ, Wilkinson KJ, McGeer JC, Sunahara G, Goss GG. Aquatic toxicity of manufactured nanomaterials: challenges and recommendations for future toxicity testing. Environ Chem. 2014;11:207–26.
5. Yang X, Gondikas AP, Marinakos SM, Auffan M, Liu J, Hsu-Kim H, Meyer JN. Mechanism of silver nanoparticle toxicity is dependent on dissolved silver and surface coating in *Caenorhabditis elegans*. Environ Sci Technol. 2011;46:1119–27.
6. Park E-J, Yi J, Kim Y, Choi K, Park K. Silver nanoparticles induce cytotoxicity by a Trojan-horse type mechanism. Toxicol In Vitro. 2010;24:872–8.
7. Fleischer A, O'Neill MA, Ehwald R. The pore size of non-graminaceous plant cell walls is rapidly decreased by borate ester cross-linking of the pectic polysaccharide rhamnogalacturonan II. Plant Physiol. 1999;121:829–38.
8. Taylor NS, Merrifield R, Williams TD, Chipman JK, Lead JR, Viant MR. Molecular toxicity of cerium oxide nanoparticles to the freshwater alga *Chlamydomonas reinhardtii* is associated with supra-environmental exposure concentrations. Nanotoxicology. 2016;10:32–41.
9. Wang Y, Miao AJ, Luo J, Wei ZB, Zhu JJ, Yang LY. Bioaccumulation of CdTe quantum dots in a freshwater alga *Ochromonas danica*: a kinetics study. Environ Sci Technol. 2013;47:10601–10.
10. Leclerc S, Wilkinson KJ. Bioaccumulation of nanosilver by *Chlamydomonas reinhardtii*—nanoparticle or the free ion? Environ Sci Technol. 2014;48:358–64.
11. Piccapietra F, Allué CG, Sigg L, Behra R. Intracellular silver accumulation in *Chlamydomonas reinhardtii* upon exposure to carbonate coated silver nanoparticles and silver nitrate. Environ Sci Technol. 2012;46:7390–7.
12. Röhder LA, Brandt T, Sigg L, Behra R. Influence of agglomeration of cerium oxide nanoparticles and speciation of cerium(III) on short term effects to the green algae *Chlamydomonas reinhardtii*. Aquat Toxicol. 2014;152:121–30.
13. Hoecke KV, Quik JTK, Mankiewicz-Boczek J, Schamphelaere KACD, Elsaesser A, Meeren PVd, Barnes C, McKerr G, Howard CV, Meent DVD, et al. Fate and effects of CeO2 nanoparticles in aquatic ecotoxicity tests. Environ Sci Technol. 2009;43:4537–46.
14. Behra R, Sigg L, Clift MJD, Herzog F, Minghetti M, Johnston B, Petri-Fink A, Rothen-Rutishauser B. Bioavailability of silver nanoparticles and ions: from a chemical and biochemical perspective. J R Soc Interface. 2013;10:20130396.
15. Scown TM, Goodhead RM, Johnston BD, Moger J, Baalousha M, Lead JR, van Aerle R, Iguchi T, Tyler CR. Assessment of cultured fish hepatocytes for studying cellular uptake and (eco)toxicity of nanoparticles. Environ Chem. 2010;7:36–49.
16. Stern S, Adiseshaiah P, Crist R. Autophagy and lysosomal dysfunction as emerging mechanisms of nanomaterial toxicity. Part Fibre Toxicol. 2012;9:20.
17. Canton I, Battaglia G. Endocytosis at the nanoscale. Chem Soc Rev. 2012;41:2718–39.
18. Nel AE, Madler L, Velegol D, Xia T, Hoek EMV, Somasundaran P, Klaessig F, Castranova V, Thompson M. Understanding biophysicochemical interactions at the nano-bio interface. Nat Mater. 2009;8:543–57.
19. Monopoli MP, Walczyk D, Campbell A, Elia G, Lynch I, Baldelli Bombelli F, Dawson KA. Physical–chemical aspects of protein corona: relevance to in vitro and in vivo biological impacts of nanoparticles. J Am Chem Soc. 2011;133:2525–34.
20. Lynch I, Dawson KA, Lead JR, Valsami-Jones E. Chapter 4-macromolecular coronas and their importance in nanotoxicology and nanoecotoxicology. Front Nanosci. 2014;7:127–56.
21. Schirmer K. Chapter 6-mechanisms of nanotoxicity. Front Nanosci. 2014;7:195–221.
22. Wigginton NS, Wigginton NS. Binding of silver nanoparticles to bacterial proteins depends on surface modifications and inhibits enzymatic activity. Environ Sci Technol. 2010;44:2163–8.
23. Aleksandr K, Feng D, Pengyu C, Monika M, Anne K, Pu Chun K. Interaction of firefly luciferase and silver nanoparticles and its impact on enzyme activity. Nanotechnology. 2013;24:345101.
24. Bertoli F, Davies G-L, Monopoli MP, Moloney M, Gun'ko YK, Salvati A, Dawson KA. Magnetic nanoparticles to recover cellular organelles and study the time resolved nanoparticle-cell interactome throughout uptake. Small. 2014;10:3307–15.
25. Hofmann D, Tenzer S, Bannwarth MB, Messerschmidt C, Glaser S F, Schild H, Landfester K, Mailänder V. Mass spectrometry and imaging analysis of nanoparticle-containing vesicles provide a mechanistic insight into cellular trafficking. ACS Nano. 2014;8:10077–88.
26. Li X, Schirmer K, Bernard L, Sigg L, Pillai S, Behra R. Silver nanoparticle toxicity and association with the alga *Euglena gracilis*. Environ Sci Nano. 2015;2:594–602.
27. Yue Y, Behra R, Sigg L, Fernández Freire P, Pillai S, Schirmer K. Toxicity of silver nanoparticles to a fish gill cell line: role of medium composition. Nanotoxicology. 2015;9:54–63.
28. Yue Y, Behra R, Sigg L, Suter MJF, Pillai S, Schirmer K. Silver nanoparticle-protein interactions in intact rainbow trout gill cells. Environ Sci Nano. 2016;3:1174–85.
29. Li X. Interactions of silver and polystyrene nanoparticles with algae. EPFL Ph.D. Thesis. 2015.
30. Bols NC, Barlian A, Chirinotrejo M, Caldwell SJ, Goegan P, Lee LEJ. Development of a cell-line from primary cultures of rainbow-trout, *Oncorhyrzchus mykiss* (Walbaum), gills. J Fish Dis. 1994;17:601–11.
31. Nakano Y, Urade Y, Urade R, Kitaoka S. Isolation, purification, and characterization of the pellicle of *Euglena gracilisz*. J Bioch. 1987;102:1053–63.
32. Schirmer K, Chan AGJ, Greenberg BM, Dixon DG, Bols NC. Methodology for demonstrating and measuring the photocytotoxicity of fluoranthene to fish cells in culture. Toxicol In Vitro. 1997;11:107–19.
33. Dayeh VR, Schirmer K, Bols NC. Applying whole-water samples directly to fish cell cultures in order to evaluate the toxicity of industrial effluent. Water Res. 2002;36:3727–38.
34. Leibovitz A. The growth and maintenance of tissue–cell cultures in free gas exchange with the atmosphere. Am J Epidemiol. 1963;78:173–80.
35. Sigg L, Behra R, Groh K, Isaacson C, Odzak N, Piccapietra F, Rohder L, Schug H, Yue Y, Schirmer K. Chemical aspects of nanoparticle ecotoxicology. Chimia. 2014;68:806–11.
36. Setyawati MI, Yuan X, Xie J, Leong DT. The influence of lysosomal stability of silver nanomaterials on their toxicity to human cells. Biomaterials. 2014;35:6707–15.
37. Wang Z, Liu S, Ma J, Qu G, Wang X, Yu S, He J, Liu J, Xia T, Jiang G-B. Silver nanoparticles induced RNA polymerase-silver binding and RNA transcription inhibition in erythroid progenitor cells. ACS Nano. 2013;7:4171–86.
38. Liu W, Wu Y, Wang C, Li HC, Wang T, Liao CY, Cui L, Zhou QF, Yan B, Jiang GB. Impact of silver nanoparticles on human cells: effect of particle size. Nanotoxicology. 2010;4:319–30.
39. Fortin C, Campbell PGC. Silver uptake by the green alga *Chlamydomonas reinhardtii* in relation to chemical speciation: influence of chloride. Environ Toxicol Chem. 2000;19:2769–78.
40. Lee D-Y, Fortin C, Campbell PGC. Influence of chloride on silver uptake by two green algae, *Pseudokirchneriella subcapitata* and *Chlorella pyrenoidosa*. Environ Toxicol Chem. 2004;23:1012–8.
41. Pillai S, Behra R, Nestler H, Suter MJF, Sigg L, Schirmer K. Linking toxicity and adaptive responses across the transcriptome, proteome, and phenotype of Chlamydomonas reinhardtii exposed to silver. Proc Natl Acad Sci USA. 2014;111:3490–5.
42. Bury NR, Hogstrand C. Influence of Chloride and Metals on Silver Bioavailability to Atlantic Salmon (Salmo salar) and Rainbow Trout (*Oncorhynchus mykiss*) Yolk-Sac Fry. Environ Sci Technol. 2002;36:2884–8.
43. Wood CM. 1-Silver. In: Chris M, Wood APF, Brauner CJ, editors. Homeostasis and toxicology of non-essential metals. Cambridge: Academic Press; 2011. p. 1–65.
44. Bondarenko O, Juganson K, Ivask A, Kasemets K, Mortimer M, Kahru A. Toxicity of Ag, CuO and ZnO nanoparticles to selected environmentally relevant test organisms and mammalian cells in vitro: a critical review. Arch Toxicol. 2013;87:1181–200.
45. Connolly M, Fernandez-Cruz ML, Quesada-Garcia A, Alte L, Segner H, Navas MJ. Comparative cytotoxicity study of silver nanoparticles (AgNPs) in a Variety of rainbow trout cell lines (RTL-W1, RTH-149, RTG-2) and primary hepatocytes. Int J Environ Res Public Health. 2015;12:5386–405.
46. Minghetti M, Schirmer K. Effect of media composition on bioavailability and toxicity of silver and silver nanoparticles in fish intestinal cells (RTgutGC). Nanotoxicology. 2016;10:1526–34.
47. Sanderson H, Johnson DJ, Wilson CJ, Brain RA, Solomon KR. Probabilistic hazard assessment of environmentally occurring pharmaceuticals

toxicity to fish, daphnids and algae by ECOSAR screening. Toxicol Lett. 2003;144:383–95.
48. Jemec A, Kahru A, Potthoff A, Drobne D, Heinlaan M, Böhme S, Geppert M, Novak S, Schirmer K, Rekulapally R, et al. An interlaboratory comparison of nanosilver characterisation and hazard identification: harmonising techniques for high quality data. Environ Int. 2016;87:20–32.
49. Hinderliter P, Minard K, Orr G, Chrisler W, Thrall B, Pounds J, Teeguarden J. ISDD: a computational model of particle sedimentation, diffusion and target cell dosimetry for in vitro toxicity studies. Part Fibre Toxicol. 2010;7:36.
50. Teeguarden JG, Hinderliter PM, Orr G, Thrall BD, Pounds JG. Particokinetics in vitro: dosimetry considerations for in vitro nanoparticle toxicity assessments. Toxicol Sci. 2007;95:300–12.
51. Busch W, Bastian S, Trahorsch U, Iwe M, Kühnel D, Meißner T, Springer A, Gelinsky M, Richter V, Ikonomidou C, et al. Internalisation of engineered nanoparticles into mammalian cells in vitro: influence of cell type and particle properties. J Nanopart Res. 2011;13:293–310.
52. García-Alonso J, Khan FR, Misra SK, Turmaine M, Smith BD, Rainbow PS, Luoma SN, Valsami-Jones E. Cellular internalization of silver nanoparticles in gut epithelia of the estuarine polychaete nereis diversicolor. Environ Sci Technol. 2011;45:4630–6.
53. Greulich C, Diendorf J, Simon T, Eggeler G, Epple M, Köller M. Uptake and intracellular distribution of silver nanoparticles in human mesenchymal stem cells. Acta Biomater. 2011;7:347–54.
54. Jansson M, Olsson H, Pettersson K. Phosphatases; origin, characteristics and function in lakes. In Phosphorus in freshwater ecosystems. Berlin: Springer; 1988. p. 157–75.
55. Rier ST, Kuehn KA, Francoeur SN. Algal regulation of extracellular enzyme activity in stream microbial communities associated with inert substrata and detritus. J North Am Benthol Soc. 2007;26:439–49.
56. Rose C, Axler RP. Uses of alkaline phosphatase activity in evaluating phytoplankton community phosphorus deficiency. Hydrobiologia. 1997;361:145–56.
57. Gil-Allué C, Schirmer K, Tlili A, Gessner MO, Behra R. Silver nanoparticle effects on stream periphyton during short-term exposures. Environ Sci Technol. 2015;49:1165–72.
58. Tlili A, Cornut J, Behra R, Gil-Allué C, Gessner MO. Harmful effects of silver nanoparticles on a complex detrital model system. Nanotoxicology. 2016;10:728–35.
59. Stewart TJ, Traber J, Kroll A, Behra R, Sigg L. Characterization of extracellular polymeric substances (EPS) from periphyton using liquid chromatography-organic carbon detection—organic nitrogen detection (LC-OCD-OND). Environ Sci Pollut Res. 2013;20:3214–23.
60. Schug H, Isaacson CW, Sigg L, Ammann AA, Schirmer K. Effect of TiO_2 nanoparticles and uv radiation on extracellular enzyme activity of intact heterotrophic biofilms. Environ Sci Technol. 2014;48:11620–8.
61. Schultz AG, Ong KJ, MacCormack T, Ma G, Veinot JGC, Goss Goss GG. Silver nanoparticles inhibit sodium uptake in juvenile rainbow trout (*Oncorhynchus mykiss*). Environ Sci Technol. 2012;46:10295–301.
62. Le Faucheur S, Behra R, Sigg L. Phytochelatin induction, cadmium accumulation, and algal sensitivity to free cadmium ion in Scenedesmus vacuolatus. Environ Toxicol Chem. 2005;24:1731–7.

3

Quantification and visualization of cellular uptake of TiO_2 and Ag nanoparticles: comparison of different ICP-MS techniques

I-Lun Hsiao[1,2], Frank S. Bierkandt[3], Philipp Reichardt[1], Andreas Luch[1], Yuh-Jeen Huang[2], Norbert Jakubowski[3], Jutta Tentschert[1†] and Andrea Haase[1*†]

Abstract

Background: Safety assessment of nanoparticles (NPs) requires techniques that are suitable to quantify tissue and cellular uptake of NPs. The most commonly applied techniques for this purpose are based on inductively coupled plasma mass spectrometry (ICP-MS). Here we apply and compare three different ICP-MS methods to investigate the cellular uptake of TiO_2 (diameter 7 or 20 nm, respectively) and Ag (diameter 50 or 75 nm, respectively) NPs into differentiated mouse neuroblastoma cells (Neuro-2a cells). Cells were incubated with different amounts of the NPs. Thereafter they were either directly analyzed by laser ablation ICP-MS (LA-ICP-MS) or were lysed and lysates were analyzed by ICP-MS and by single particle ICP-MS (SP-ICP-MS).

Results: All techniques confirmed that smaller particles were taken up to a higher extent when values were converted in an NP number-based dose metric. In contrast to ICP-MS and LA-ICP-MS, this measure is already directly provided through SP-ICP-MS. Analysis of NP size distribution in cell lysates by SP-ICP-MS indicates the formation of NP agglomerates inside cells. LA-ICP-MS imaging shows that some of the 75 nm Ag NPs seemed to be adsorbed onto the cell membranes and were not penetrating into the cells, while most of the 50 nm Ag NPs were internalized. LA-ICP-MS confirms high cell-to-cell variability for NP uptake.

Conclusions: Based on our data we propose to combine different ICP-MS techniques in order to reliably determine the average NP mass and number concentrations, NP sizes and size distribution patterns as well as cell-to-cell variations in NP uptake and intracellular localization.

Keywords: Nanoparticles, Single particle ICP-MS, Laser ablation ICP-MS, Cellular internalization, Neurons

Background

Due to their enhanced optical, electronical and antimicrobial properties, nanoparticles (NPs) are used in a variety of different consumer products like cosmetics, textiles or packaging and also in nanomedicine or electronics [1]. Titanium dioxide nanoparticles (TiO_2 NPs) and silver nanoparticles (Ag NPs) belong to the group of highly commercialized NPs [2–5]. TiO_2 NPs are frequently used as UV filters in cosmetics due to their absorption and scattering properties. However, TiO_2 E171 is also used as a food additive, with a fraction of E171 being nanoscaled. Some types of TiO_2 materials, in dependence on their crystal structure and chemical surface coating, have photocatalytic properties and are therefore used in sensors and self-cleaning surfaces [6, 7]. Ag NPs are often used because of their antimicrobial properties, e.g. in textiles but also for several medical applications such as wound dressings [8, 9]. Thus, human exposure to these kinds of NPs is increasing. Numerous in vivo and in vitro studies have evaluated the toxicity and toxic mechanisms of these NPs, which is summarized in several review articles [10, 11].

*Correspondence: andrea.haase@bfr.bund.de
†Jutta Tentschert and Andrea Haase contributed equally to this work
[1] Department of Chemical and Product Safety, German Federal Institute for Risk Assessment (BfR), Max-Dohrn-Strasse 8-10, 10589 Berlin, Germany
Full list of author information is available at the end of the article

However, for safety assessment it is also important to analyze and quantify cellular uptake. Electron microscopy-based methods such as scanning electron microscopy (SEM) or transmission electron microscopy (TEM) can be used to localize NPs in intracellular ultrastructures with high spatial resolution. In combination with energy-dispersive X-ray spectroscopy (EDX) they also allow to identify the materials [12–14]. However, these techniques require extensive, time-consuming sample preparation and typically provide qualitative information [15]. Quantification might be also possible but requires time-consuming 3-D reconstruction and automated image analysis.

In cell biology, flow cytometry and/or fluorescent microscopy are often used for quantification of cellular uptake. In principle, both techniques are also applicable for fluorescent-labelled NPs [16, 17]. These techniques are easy to use and also allow for time resolved analysis. Possible drawbacks are dye leakage or possible surface modification due to dye-labelling, both of which may strongly influence the results and hence need to be considered beforehand [18]. An absolute quantification is not possible via these approaches.

Inductively coupled plasma mass spectrometry (ICP-MS) is a common technique for absolute quantification of the cellular uptake of metal or metal oxide NPs. It offers high sensitivity and selectivity for elemental analysis [19, 20]. Routinely it requires that cells are lysed before analysis. The absolute uptake per cell may be calculated after the measurement taking into account the cell number of the sample. However, this procedure reveals an average number only. In addition this approach will yield the total amount of the particular metal only. It will not allow to differentiate between single particles and agglomerated or ionic species. Also, it will not give information about the sizes of NPs. Despite these drawbacks ICP-MS is an established approach to quantify NP uptake and has been used in many studies. For instance it has been repeatedly found that the cellular uptake depends on the sizes of NPs [21, 22]. Mostly these studies present only calculated number concentrations and average amounts. Due to these limitations, little is known about real numbers of NPs per cell or about cell-to-cell heterogeneity. In the meantime, several modifications of ICP-MS became available that allow for single cell or single particle analysis [e.g. single cell ICP-MS (SC-ICP-MS) and single particle ICP-MS (SP-ICP-MS)] [14, 23, 24]. In SP-ICP-MS, highly diluted samples are measured. Since each signal in SP-ICP-MS corresponds to a single particle, the frequency of ICP-MS signals can be used to estimate the NP number concentration. On the other side, the intensity of signals is related to the amount of the chemical element and thus to the sizes of the respective NPs. When assuming spherical particle shapes, NP sizes and size distributions can be derived using SP-ICP-MS. For instance, the uptake of gold NPs in human umbilical vein-derived endothelial cells (HUVECs) was analyzed by SP-ICP-MS and a concentration-dependent increase in uptake was detected without agglomeration [14]. ICP-MS can be also coupled to laser ablation (LA-ICP-MS), which then allows the analysis of solid materials. This method has been applied to quantify gold, silver and aluminum oxide NPs at the single cell level [25–27]. The high spatial resolution of the LA-ICP-MS images also allows for subcellular localization of NPs or for detection of NP aggregation in cells [28].

The work flow of our study is depicted in Fig. 1. The aim of the present study was to apply and to compare conventional ICP-MS, SP-ICP-MS, and LA-ICP-MS in the quantification of the cellular uptake of two types of TiO_2 NPs (diameter of 7 and 20 nm, as determined by X-ray, respectively) and two types of Ag NPs (diameter of 50 and 70 nm, as determined by TEM, respectively). While the average NP mass concentration was determined by ICP-MS, average NP number concentration and size distribution was available from SP-ICP-MS. Cell-to-cell heterogeneity of NP uptake and intracellular distribution was analyzed by LA-ICP-MS. A differentiated Neuro-2a mouse neuroblastoma cell has been chosen because several studies indicated that NPs might be taken up into the central nervous system thereby exerting neurotoxic effects [29–32].

Results and discussion

Characterization of NPs

We used two sizes of TiO_2 NPs with diameters of 7 nm and 20 nm (as determined by X-ray), respectively. NP sizes were also analyzed by TEM (Fig. 2a, b). Their average hydrodynamic diameters in pure water was determined by dynamic light scattering (DLS). Size distribution in complete cell culture medium (CCM) containing serum was assessed by DLS (Fig. 2e) and nanoparticle

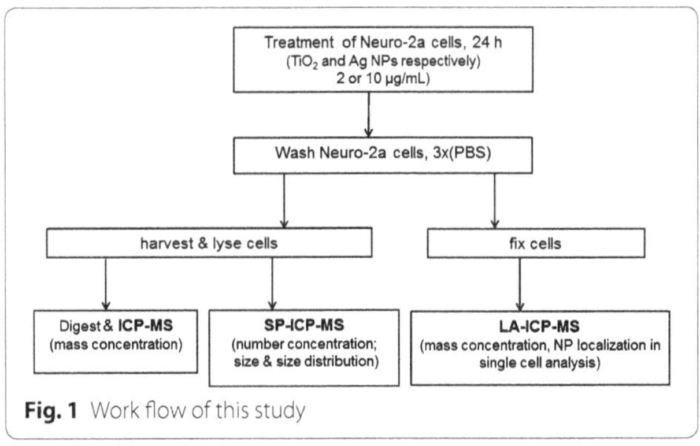

Fig. 1 Work flow of this study

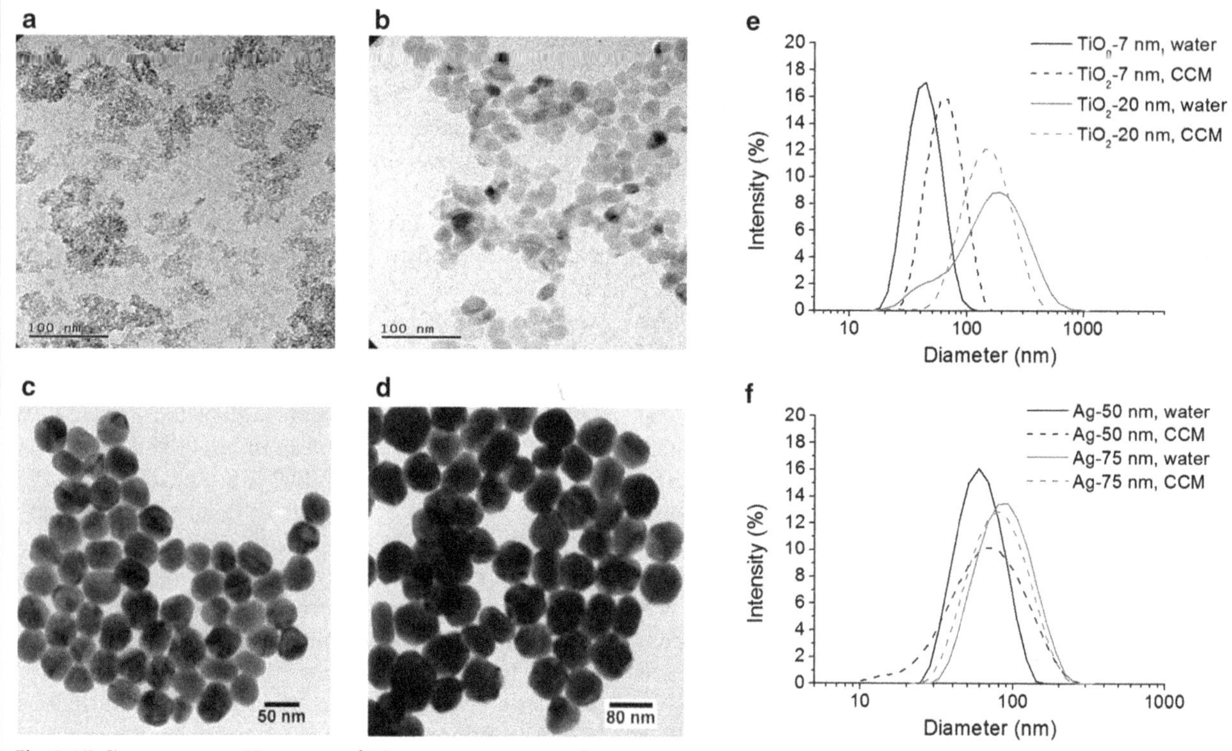

Fig. 2 NP Characterization. TEM images of TiO_2 7 nm (**a**); TiO_2 20 nm (**b**); Ag 50 nm (**c**); Ag 75 nm (**d**). Images **c** and **d** were taken from NanoComposix with kind permission. Size distribution of TiO_2 (**e**) and Ag NPs (**f**) in water and complete cell culture medium (CCM) as measured by DLS

tracking analysis (NTA) (Additional file 1: Figure S1). Data indicate some agglomeration for both types of TiO_2 NPs but dispersion quality was still reasonably good. The alkali modification process for TiO_2 NPs increases the density of surface hydroxyl groups and enhances their dispersion [33]. Furthermore, we used two sizes of Ag NPs with TEM diameters of 50 and 75 nm, respectively (Fig. 2c, d). Sizes determined by TEM was reported by the manufacturer. In addition, DLS was used to assess the average hydrodynamic diameter in pure water. Size distribution in CCM was investigated by DLS (Fig. 2f) and NTA (Additional file 1: Figure S1). Ag NPs were very well dispersed with only little sign of agglomeration visible using NTA. Material characterization is summarized in Table 1.

Assessment of NP-mediated cytotoxicity

Prior to the cellular uptake experiments, we carefully analyzed the cytotoxicity of the TiO_2 and Ag NPs after 24 h of incubation (Additional file 1: Figure S2). The results show that both sizes of TiO_2 NPs display no toxicity up to 25 μg/mL. Conversely, both sizes of Ag NPs showed a minimal toxicity at 25 μg/mL, with cell viabilities of 87 % for 50 nm and 90 % for 75 nm Ag NPs. According to these results, we used clearly non-cytotoxic doses in all subsequent experiments (i.e. 2 and 10 μg/mL of both Ag NPs and TiO_2 NPs).

Cellular uptake of NPs depends on particle size and concentration

We first performed calibrations using Ti and Ag ionic standards and In ions as internal standard. These resulted in linear correlations with squared correlation coefficients of (R^2) >0.995 for both, Ti and Ag. The limit of detection (LOD) (as 3× STD+ background, n = 3) was determined at 0.05 pg Ag/cell and 0.31 pg TiO_2/cell, respectively.

Using these calibrations, we quantified the cellular uptake by measuring the total amounts of Ti and Ag after digestion of Neuro-2a cells upon treatment with two concentrations, i.e. 2 and 10 μg/mL (Fig. 3a, b).

Higher mass amounts of TiO_2 and Ag were taken up after incubation with larger particles compared to the smaller ones. Furthermore, uptake was increased at higher concentrations. For the TiO_2 NPs this resulted in a 1.9 and 2.6-fold higher cellular Ti amount for the 7 and 20 nm NPs at 10 μg/mL when compared to 2 μg/mL, respectively (Fig. 3a). In the case of Ag NPs only the 75 nm (TEM) showed a clear increase in uptake after treatment with the higher incubation

Table 1 Charaterization of the NPs used in this study

	X-ray (nm)	TEM (nm)	Dispersion in water	Dispersion in CCM		Dissolution in water (%)
			DLS (nm)	DLS (nm)	NTA (nm)	
TiO_2 (7 nm), Anatase	7[a]	8 ± 2	48 ± 0.5	60 ± 0.5	43/78/143/283	NA
TiO_2 (20 nm), Anatase	20[a]	35 ± 6	130 ± 4	128 ± 1.5	88/228	NA
Ag Cit (50 nm)		47 ± 5[b]	55 ± 0.6	57 ± 0.3	83 (Additional small peaks at 133 and 193)	1 ± 0.1 (10 μg/mL) 3.8 ± 0.9 (2 μg/mL)
Ag Cit (75 nm)		74 ± 8[b]	79 ± 0.5	71 ± 0.2	113 (Additional small peak at 188)	1 ± 0.03 (10 μg/mL) 3.1 ± 0.7 (2 μg/mL)

DLS data were presented as means including standard deviation. NTA data were reported as mode values. Dissolution was determined at 2 and 10 μg/mL. Dissolution of TiO_2 NPs was not analyzed (NA)

[a] X-ray sizes of TiO_2 NPs were reported by Ishihara Sanyo Kaisha, Ltd., Japan

[b] TEM values of Ag NPs were reported by NanoComposix, Prague, Czech Republic

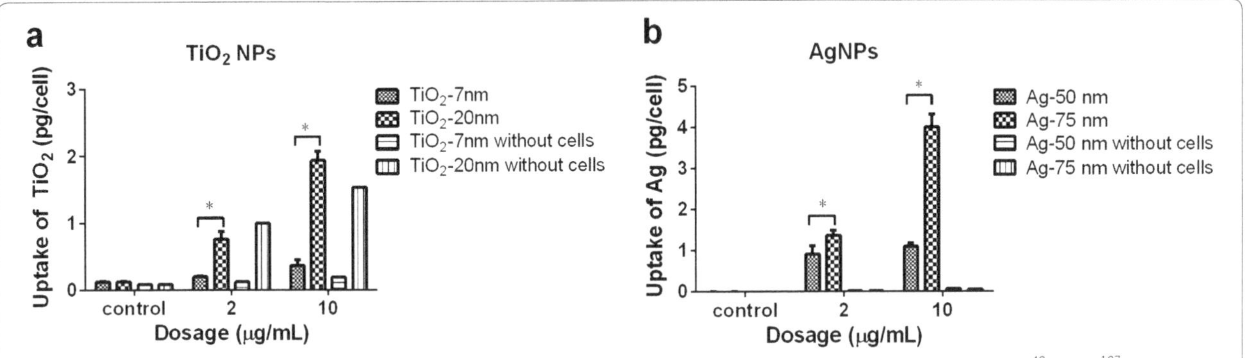

Fig. 3 Cellular uptake of NPs in Neuro-2a cells as analyzed by ICP-MS. Depicted are data for TiO_2 NPs (**a**) and Ag NPs (**b**). The ^{49}Ti and ^{107}Ag isotopes were selected for Ti and Ag detection, respectively. The internal standard was monitored using ^{115}In. Analysis was conducted using a quadrupole ICP mass spectrometer, operated in standard mode. TiO_2 was also detected in samples without cells, which indicates that NPs physically attach to cell culture plates and cannot be completely removed by subsequent washing with PBS. (*) indicates a significant difference between two treatment groups as determined by student's t test ($p < 0.05$)

concentration (2.9-fold) (Fig. 3b). To rule out the possibility that the detected signals were due to NPs that physically attached to the surface of the cell cultures plates, we performed control treatments without any cells, which were analyzed accordingly (Fig. 3a, b). After identical sample preparation in the controls almost no Ag could be detected (Fig. 3b). However, some Ti was detected, which is likely to represent the fraction of NPs that unspecifically attached to the cell culture plates and which could not be completely removed by washing (Fig. 3a). In the case of 10 μg/mL of 7 nm TiO_2 NPs (X-ray), the numbers were 109 ng total Ti in the presence of cells (equals 0.36 pg/cell) versus 46 ng in the absence of cells (equals 0.19 pg/cell when assuming the same cell number as in the treatment group). Similarly, for 10 μg/mL of 20 nm TiO_2 NPs (X-ray) 559 ng of Ti was detected when cells were present (equals 1.93 pg/cell), whereas 443 ng were detected in controls (equals 1.53 pg/cell).

Furthermore, we used SP-ICP-MS to quantify the NP numbers in Neuro-2a cells directly, which was, according to our knowledge, investigated here for the first time. First, we performed a calibration experiment by using NP suspensions. The LOD_{size} and $LOD_{numberNP}$ of Ag NPs was 22 nm and 2.26 × 10^5 particles, respectively. The LOD_{size} and $LOD_{numberNP}$ of TiO_2 NPs was 69 nm and 1.92 × 10^5 particles, respectively. Therefore, the 7 nm TiO_2 NPs with a secondary size of 48 nm (DLS) could not be detected. However, the 20 nm TiO_2 NPs (X-ray) as well as both types of Ag NPs could be quantified via SP-ICP-MS. The 20 nm TiO_2 NPs (X-ray) showed a concentration-dependent uptake into Neuro-2a cells as indicated by the SP-ICP-MS measurements (Fig. 4a). In addition to the concentration dependency, for Ag NPs we were able to confirm a size dependency as well, as the number of NPs per cell increased with decreasing particle size (Fig. 4b). Therefore, the smaller Ag particles can penetrate into cells more efficiently.

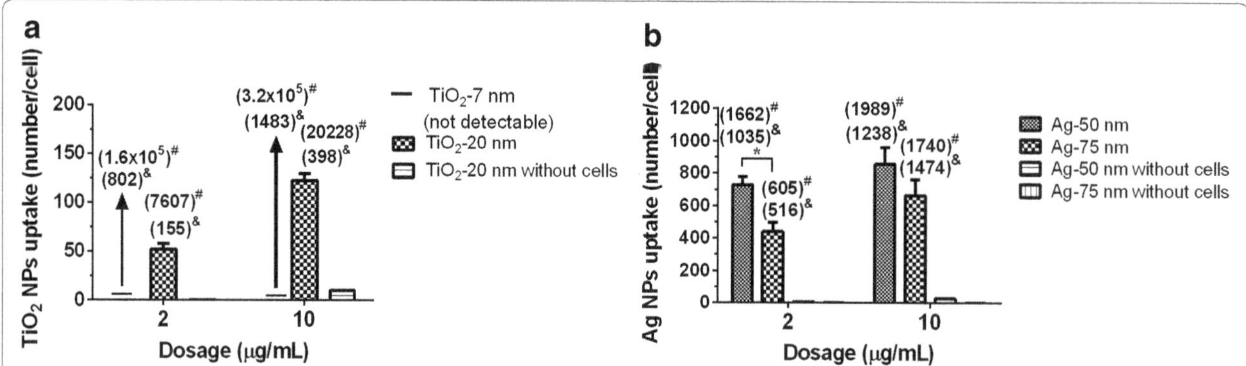

Fig. 4 Number-based cellular uptake of NPs in Neuro-2a cells as analyzed by SP-ICP-MS. Depicted are data for TiO_2 NPs (**a**) and Ag NPs (**b**). The ^{48}Ti and ^{107}Ag isotopes were selected for Ti and Ag detection, respectively. Analysis was conducted using a quadrupole ICP mass spectrometer, operated in spike mode (1 min per run, dwell time of 3 ms per reading). The small TiO_2 NPs cannot be detected by SP-ICP-MS because their size was below the limit of detection size (LOD_{size}) of TiO_2 (69 nm). #Calculated number of NPs per cell using ICP-MS (mass) data. &Calculated number of NPs per cell using ICP-MS (mass) data based on DLS size. (*) indicates a significant difference between two treatment groups as determined by student's t-test ($p < 0.05$)

Assuming ideal spheres and no dissolution for both types of NPs, the NP numbers per cell can also be calculated using the results from ICP-MS. These results confirmed that the treatment with the smaller TiO_2 and Ag NPs resulted in higher particle numbers per cell (Fig. 4a, b). However, assuming the TEM size as single particle diameter, the particle numbers calculated based on ICP-MS data were considerably higher when compared to the particle numbers directly measured by SP-ICP-MS using the same lysates. Between calculated and measured NP numbers the differences were 100-fold for TiO_2 and two-fold for Ag NPs (Fig. 4a, b). By contrast, assuming DLS size as single particle diameter, the differences between calculated and measured NP numbers decreased to about three- to fourfold for TiO_2 and 1.1- to 1.4-fold for Ag NPs. This result indicated that NP sizes determined by DLS were more suitable for the transformation of NP amounts into NP numbers, when compared to NP sizes determined via TEM or X-ray. Indeed, the differences between estimated and measured NP numbers may be explained, at least partially, by NP agglomeration in cells, which was found higher in the case of TiO_2 NPs. This aspect will be discussed in the following section in more detail.

LA-ICP-MS also confirms the size-dependent cellular uptake of NPs

For NP quantification by LA-ICP-MS we first analyzed the calibration curve, which was based on adding NPs to a cell lysate of untreated cells (Additional file 1: Figure S3a). A laser spot size of 250 μm (line distance of 240 μm) and a scan speed of 200 μm/s were considered to be optimal LA parameters. For silver, the calibration range comprised five Ag-containing spots and one control spot without Ag. The latter was pipetted at around X = 2.5 mm but could not be distinguished from the background of the glass slide. A background correction for each drop was performed. Since the upper and lower area showed a different background signal, both replicates were corrected separately. A good correlation ($R^2 = 0.995$) between the amount of silver and the integrated signal intensity was observed and a limit of detection of 78.5 pg was calculated (Additional file 1: Figure S3b).

LA-ICP-MS in general is able to provide high spatial localization information (1 μm pixel size in *x*-direction and 6 μm in *y*-direction) as reported before by Drescher and co-workers analyzing Au and Ag NP aggregates in the mouse fibroblast 3T3 cell line [25]. A laser spot size of 4 μm (line distance of 6 μm) and a scan speed of 5 μm/s was used. As Neuro-2a cells were smaller compared to 3T3 cells this may result in a much lower number of pixels to represent the NPs in a single cell but does not affect the integrated sensitivity under the same conditions. Therefore, for Neuro-2a cells a reasonable sub-cellular localization requires further optimization of the ablation parameters. However, by decreasing the line distance to 5 μm (1 μm pixel size in *x*-direction and 5 μm in *y*-direction in spatial resolution) we could show that no NPs were localized in the cell neurites. Neurites propagate the signals between neural cells and are usually not responsible for uptake. However, we may not entirely exclude that NP concentrations in neurites were below the detection limits of LA-ICP-MS.

The contour plots for the ^{107}Ag signal intensities of the dried cells nicely corresponded with bright field images of Neuro-2a cells before ablation (compare Fig. 5a, b). We found that cells after incubation with 75 nm Ag NPs

(TEM) often revealed more ^{107}Ag spots with a higher intensity compared to cells treated with 50 nm Ag NPs (TEM). Furthermore, 75 nm Ag NPs (TEM) could frequently be detected outside the cells while 50 nm Ag NPs (TEM) were mainly detected within the cell areas only.

Signals from 99 and 109 cells from several ablated areas were integrated after treatment with 50 and 75 nm Ag NPs (both TEM sizes), respectively. Using the calibration curve, the calculated amounts of silver per cell after a 2 μg/mL incubation ranged from 0.006 to 0.528 ng for 50 nm Ag NPs (0.081 ± 0.067 ng in average), and from 0.001 to 1.332 ng for 75 nm Ag NPs (0.205 ± 0.188 ng in average), respectively (Fig. 5c). This again confirmed a higher NP mass per cell for the bigger 75 nm particles and at the same time demonstrates a high cell-to-cell variability, which seems increasing at higher treatment concentrations.

Assuming ideal spheres, DLS size as particle diameter, consisting purely of silver for both nanoparticles types, the total silver amounts per cell were converted into NP numbers (Fig. 5d). Again, a higher NP number per cell was found after incubation of the cells with the smaller 50 nm Ag NPs (TEM) (see Table 2).

Using the same strategy as for the Ag NPs, we could not achieve an according calibration for the TiO_2 NPs though. The ^{48}Ti signal spots for the pipetted calibration spots could not be detected and integrated clearly. Instead the spot areas showed a depleted signal intensity compared to a relatively high and irregular background (data not shown). Higher spots within these areas might indicate some flow mechanism for the TiO_2 NPs similar to the coffee stain effect [34, 35]. As our calibration was not successful, we can only provide ^{48}Ti signal intensities in this study but no quantification. For further studies one may consider to use aqueous Ti standards instead of TiO_2 NPs.

Figure 6a–d show exemplarily the overlaps of the contour plots for the ^{48}Ti signals with the corresponding bright field images of the dried Neuro-2a cells. In agreement with prior results, more ^{48}Ti signal spots in each cell were observed after higher concentrated incubations (10 μg/mL) compared to lower concentrations (2 μg/mL).

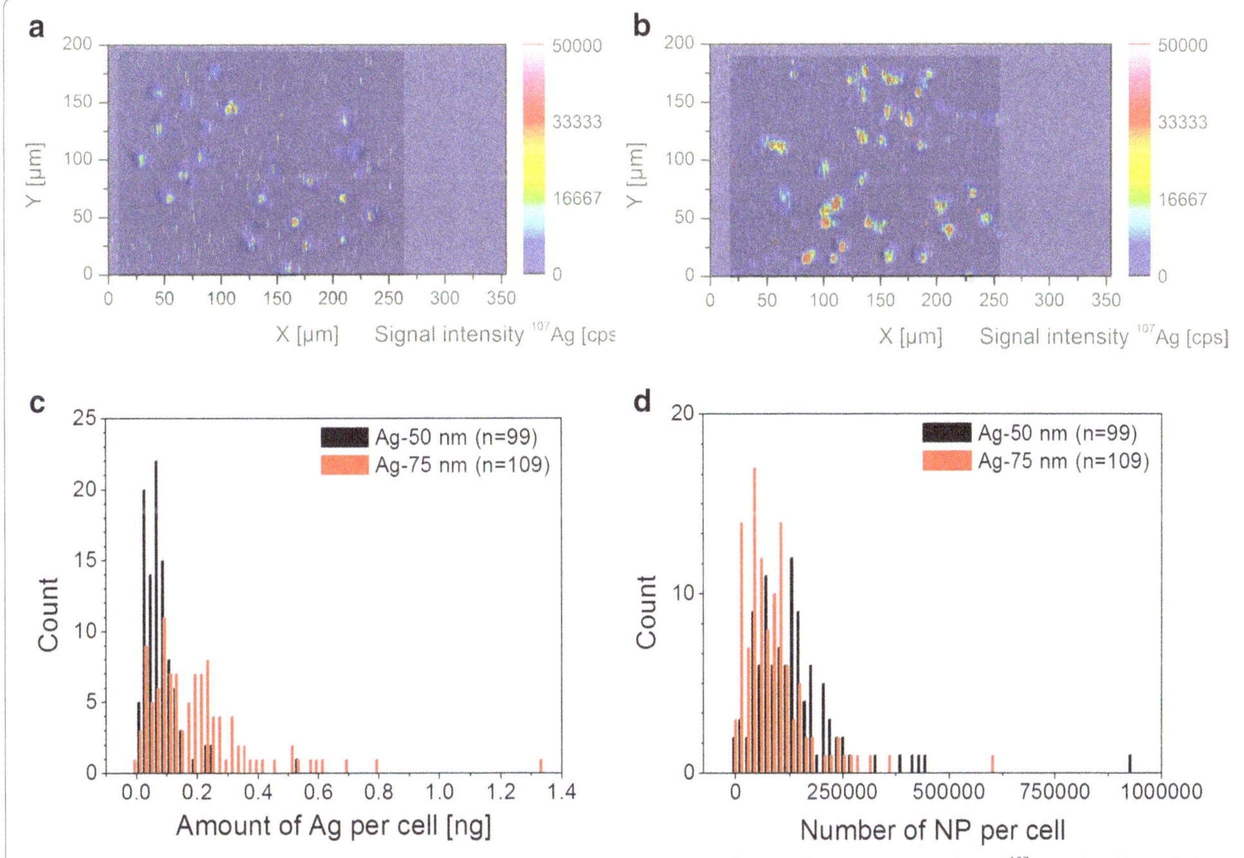

Fig. 5 Cellular uptake of Ag NPs at single cell level as analyzed by LA-ICP-MS. Depicted are overlapping contour plots of ^{107}Ag with cell morphology of Neuro-2a cells after incubation with 2 μg/mL 50 nm (**a**) and 75 nm (**b**) Ag NPs. Histogram for the amount of Ag determined in single cells upon incubation with Ag NPs (bin width: 0.02 ng) (**c**). Histogram of the numbers of Ag NPs per cell (bin width: 15,000 NPs) (**d**). Histograms (**c**) and (**d**) were calculated from three ablated areas on the slide

Table 2 Average and median values of NP uptake in single Neuro-2a cells

Sample/unit	Median	Average ± SD
50 nm Ag (pg per cell)	72	81 ± 67
50 nm Ag (NPs per cell)	126,085	142,394 ± 117,729
75 nm Ag (pg per cell)	164	205 ± 188
75 nm Ag (NPs per cell)	73,684	92,152 ± 84,502
7 nm TiO_2 (integrated signal intensity cps per cell)	107,242	136,234 ± 103,648
7 nm TiO_2 (relative NPs per cell)	4.8×10^{11}	$6.0 \times 10^{11} \pm 4.6 \times 10^{11}$
20 nm TiO_2 (integrated signal intensity cps per cell)	281,341	310,732 ± 203,387
20 nm TiO_2 (relative NPs per cell)	6.3×10^{10}	$6.9 \times 10^{10} \pm 4.6 \times 10^{10}$

The cellular uptake was analyzed by LA-ICP-MS. Exposure concentration: 2 μg/mL

Similarly, more ^{48}Ti signal spots could be detected for larger TiO_2 NPs (20 nm, X-ray). Part of the titanium signal detected using conventional ICP-MS may stem from particles that unspecifically attached to the surface of the cell culture plate, which is causing an overestimation of cellular uptake. Using LA-ICP-MS allows to selectively integrate only signals within the cells thereby correcting any background and eliminating this error.

Integration of the dried cells showed that after incubation with a higher NP concentration or larger NPs, a higher signal intensity of ^{48}Ti can in average be found per cell which equals a higher amount (Fig. 6e). The 10 μg/mL incubation resulted for both NP sizes in an approximately 4.5-fold increased signal intensity. The larger 20 nm TiO_2 NPs led to twice the signal intensity, independent of the used NP concentration (Fig. 6e).

Despite the fact that calibration failed for the TiO_2 NPs, we could use our data to obtain the relative numbers of particles per cell. Utilizing the DLS size ratio between both TiO_2 NP types, the integrated signals of ^{48}Ti per cell were converted to relative NP numbers—only missing the calibration factor (cps/ng)—and could be compared. Figure 6f shows the distribution of the relative numbers

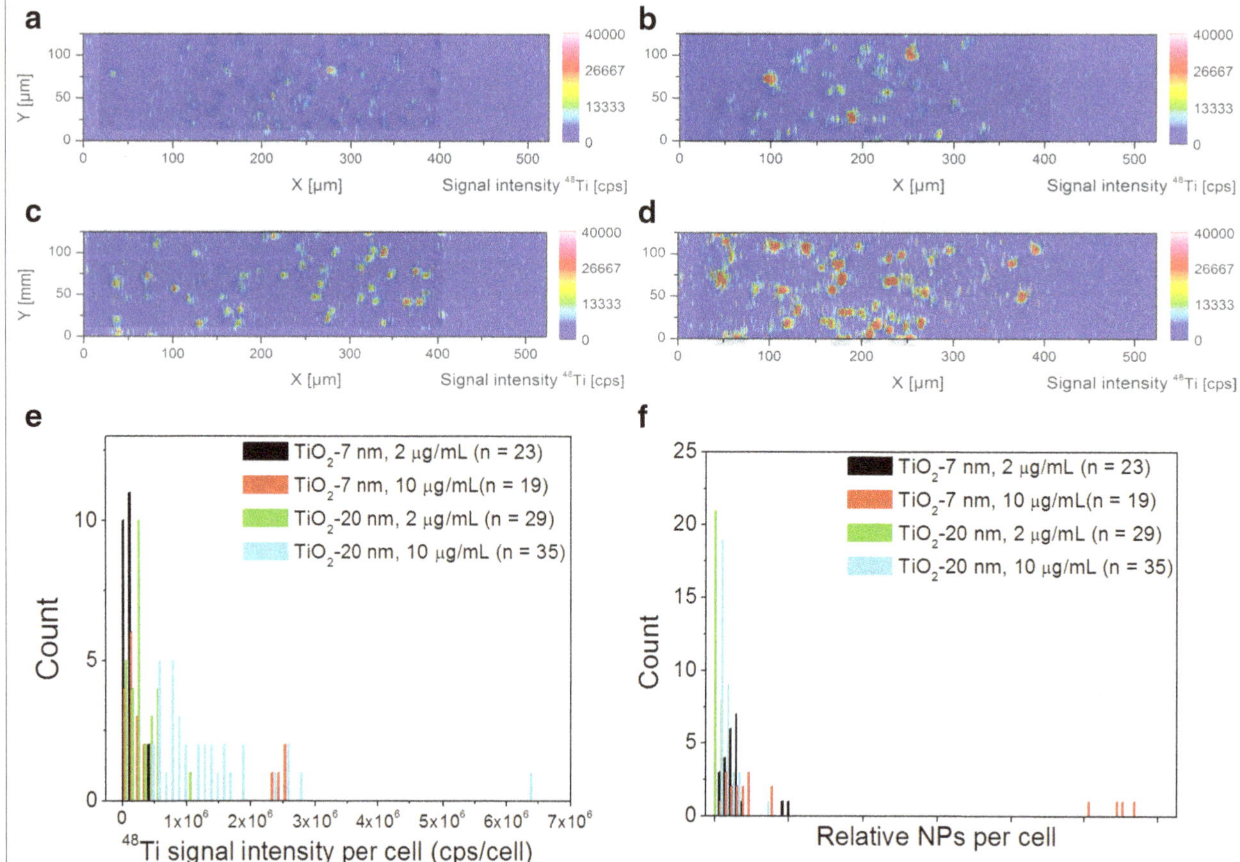

Fig. 6 Cellular uptake of TiO_2 NPs at single cell level as analyzed by LA-ICP-MS. Depicted are overlapping contour plots of ^{48}Ti with cell morphology of Neuro-2a cells after incubation with TiO_2 NPs: 7 nm, 2 μg/mL (**a**); 7 nm, 10 μg/mL (**b**); 20 nm, 2 μg/mL (**c**); and 20 nm, 10 μg/mL (**d**). Histogram for the cps of TiO_2 determined in single cells upon incubation with TiO_2 NPs (bin width: 0.02 ng) (**e**). Histogram of the relative numbers of TiO_2 NPs per cell (bin width: 2000 NPs) (**f**)

of particles per cell for each exposure condition. On average, the relative number of the smaller TiO_2 NPs per cell was found to be four times the number of the larger TiO_2 NPs (Table 2; Fig. 7). This again confirms that smaller particles are more easily incorporated into cells. At the same time a big variance in the numbers of NPs per cell was seen especially after incubation with the higher concentration (10 µg/mL). This indicates a cell-to-cell diversity, which may be even increased at higher treatment concentrations.

Using LA-ICP-MS we could detect even the smaller TiO_2 NPs with a diameter of 7 nm (X-ray) above background levels, which was not the case in the corresponding SP-ICP-MS experiments.

NP agglomerates in cells can be detected using SP-ICP-MS

In addition to particle numbers, SP-ICP-MS can also be applied to estimate NP size distributions. A cell-free 20 nm TiO_2 NP suspension in ultrapure water revealed an average diameter of 115 nm and a size range of 76–260 nm (Additional file 1: Figure S4a, b). This size value was close to the size value determined by DLS (130 nm). However, in cell lysates the average NP size value increased to 134 nm with size ranging from 76 to 300 nm, especially for the 10 µg/mL incubation (Additional file 1: Figure S5a, b). Cell-free 50 and 75 nm Ag NP suspensions (both TEM sizes) in pure water showed 46 and 72 nm in average diameter, respectively (Additional file 1: Figure S4c–f). Again, in cell lysates, average sizes of both Ag NPs also increased to 55 and 84 nm at 10 µg/mL because more particles could be detected in the range of 150–200 nm (Additional file 1: Figure S5c-f). This indicates that NP agglomerates are being formed in Neuro-2a cells after particle uptake. Such agglomerates of NPs, ranging from submicron to micron size, can frequently be found in cells by different electron microscope techniques like TEM [36, 37]. These data correspond well with lower numbers of NPs measured by SP-ICP-MS as theoretically calculated based on conventional ICP-MS data. One reason for this discrepancy is likely the agglomeration of both NP types. Agglomeration is more pronounced for TiO_2, which nicely conforms with the higher difference between calculated and measured NP numbers for TiO_2 NPs. In addition, Ag NPs could dissolve into silver ions. We observed a higher background noise for cell samples treated with Ag NPs compared to control cells in pure water (data not shown), which may be due to Ag NP dissolution. Finally, for TiO_2 the LOD_{size} hinders the detection of small TiO_2 particles by SP-ICP-MS and may thereby also affect the results.

Studies applying conventional ICP-MS have already found a size-dependent uptake of TiO_2 and Ag NPs in mammalian cells [38, 39]. However, in these studies frequently smaller NPs were reported to show a higher mass concentration in cells, which is in contrast to the results in our study. Zhu et al. analyzed uptake of silica NP by atomic absorption spectrometry (AAS) in HeLa cells, and found a higher uptake efficiency for smaller SiO_2 NPs in both mass and estimated number concentrations [40]. Furthermore, some studies reported no size-dependency of the uptake, as shown for citrate-coated Ag NPs and human lung cells analyzed by AAS [13]. These studies rely on indirect, average results. Our study clearly showed a size-dependent uptake for both Ag and TiO_2 NPs in Neuro-2a cells. However, we find a higher mass concentration for the larger particles but a higher particle number for the smaller NPs, consistently across different techniques. The different results between all these studies may also be caused by different uptake mechanisms of NPs in different cell types. Furthermore, differences may arise from different surface modifications. Therefore a general conclusion might not be possible at the moment until the underlying mechanisms of cellular uptake are fully understood. The main features of each technique applied in our study is summarized in Table 3.

Conclusions

In the present study, ICP-MS, SP-ICP-MS, and LA-ICP-MS were used to quantify the uptake of TiO_2 NPs or Ag NPs in neuroblastoma cells. While cellular uptake was clearly NP size-dependent, we consistently found a higher mass concentration in the case of the larger NPs but a higher particle number for the smaller NPs. ICP-MS provides average mass quantification data, which can be converted into NP numbers. The results may, however, be confounded by NP agglomeration inside cells, resulting in an overestimation of NP numbers. Furthermore, NPs may unspecifically attach to the surface of the cell culture dishes and withstand the regular washing steps prior to analysis. This again would lead to an overestimation of the NP amounts in conventional ICP-MS. In particular this was observed for the cellular uptake of TiO_2 NPs in our study. SP-ICP-MS directly measures NP numbers. Furthermore, using SP-ICP-MS it is possible to estimate size distributions of spherical particles directly in cell lysates. This allows to analyze possible NP agglomeration inside cells. However, smaller TiO_2 particles cannot be measured due to the high LOD_{size} value. By LA-ICP-MS it becomes possible to directly quantify NP mass concentrations at single cell level as done here for Ag NPs. On the other side, we observed difficulties in the calibration of TiO_2 NPs, so that we only could derive

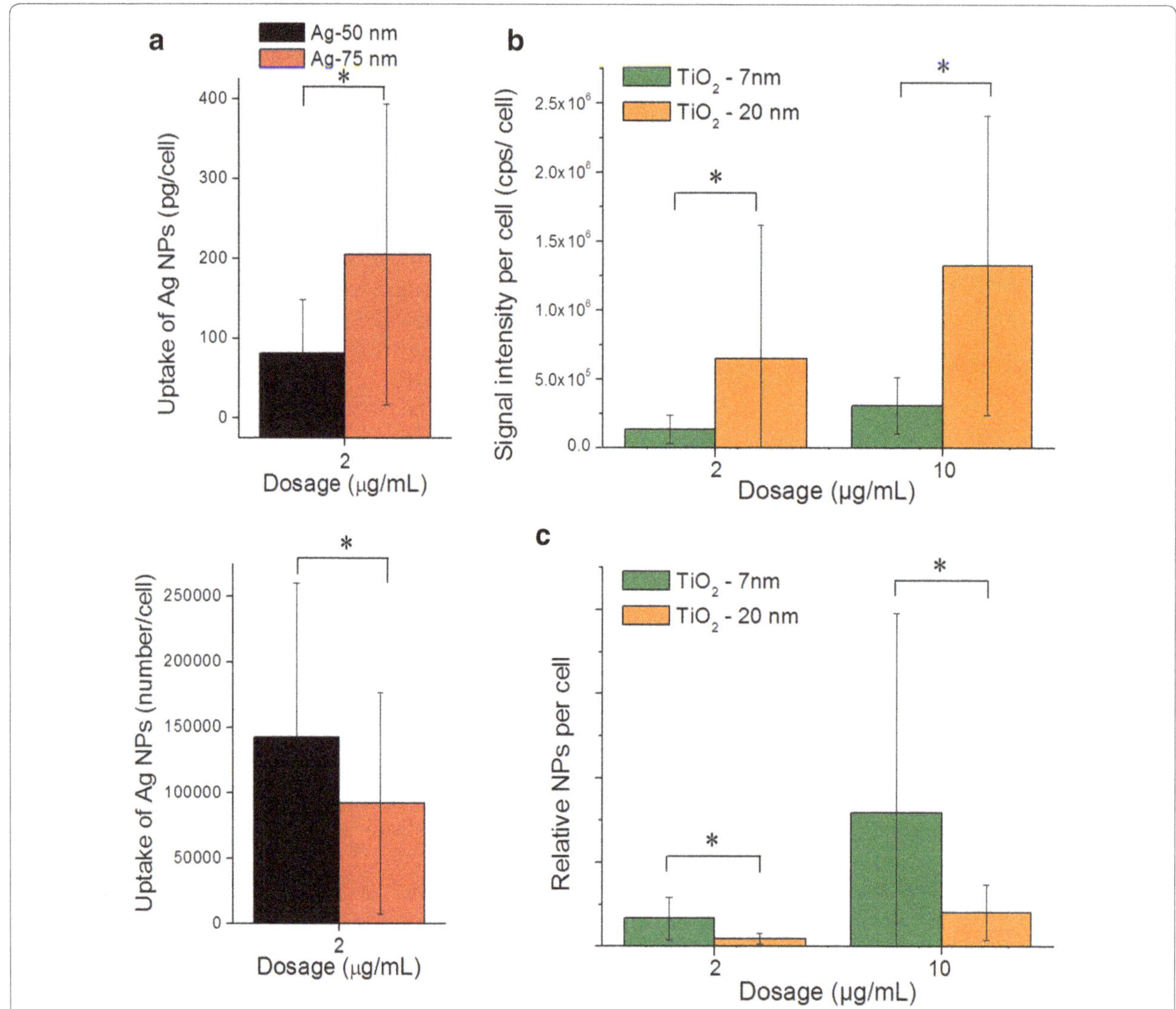

Fig. 7 Quantification of cellular uptake of Ag and TiO_2 NPs in Neuro-2a cells as analyzed by LA-ICP-MS. Average cell uptake based on particle mass (pg/cell) and particle number (Ag NPs/cell) upon exposure to 2 μg/mL Ag NPs (**a**). Average signal intensity of ^{48}Ti per cell upon exposure of cells to 2 or 10 μg/mL TiO_2 NPs (**b**), and corresponding average relative TiO_2 NP numbers per cell (**c**). (*) indicates a significant difference between two treatment groups as determined by student's t-test s ($p < 0.05$)

Table 3 Specific features of the different ICP-MS techniques applied in the present study

Technique	Parameter	Size	Localization	Limitations	Limit of detection (LOD)
ICP-MS	Mass	No	No	Error from NPs attached to culture plates	Ag: 0.05 pg/cell TiO_2: 0.31 pg/cell
SP-ICP-MS	Number	Yes	No	Spherical NPs only, high LOD_{size} for TiO_2	Ag: 2.26 × 10^5 particles (22 nm)[a], TiO_2: 1.92 × 10^5 particles (69 nm)[a]
LA-ICP-MS	Mass in single cell (relative measure)	No	Yes	Localization for small cells difficult	Ag: 78.5 pg per pippetted spot of 0.5 µl TiO_2: not available in this study

[a] Diameter of NPs was calculated from LOD

relative signal intensities. Notwithstanding applying LA-ICP-MS it is possible to obtain information on the sub-cellular localization of particles. As each technique offers specific advantages we propose to combine them in order to successfully elucidate different aspects of NP's cellular uptake.

Methods

Materials

Ultrapure water (18.2 MΩ/cm) was generated by a Milli-Q Plus system from Merck Millipore (Darmstadt, Germany). The Ag elemental standard as well as nitric acid (p. A., 65 %) were purchased from Merck (Darmstadt, Germany). The Ti and In elemental standard were purchased from Sigma-Aldrich (TraceCERT®, Steinheim, Germany). Hydrogen peroxide (H_2O_2) solution (p.A., 30 %), sodium hydroxide (NaOH), Triton X-100 solution, forskolin, 3-isobutyl-1-methylxanthine (IBMX), and paraformaldehyde were purchased from Sigma-Aldrich (Steinheim, Germany). Hydrochloric acid (p.A., 37 %) was bought from Applichem (Darmstadt, Germany). Hydrofluoric acid (p.A., 5 %) was bought from Carl Roth (Karlsruhe, Germany). SuperFrost®Plus Slides were used as glass slides for cell culture LA-ICP-MS experiments (Thermo Fisher Scientific GmbH, Dreieich, Germany).

Nanoparticles

Commercially available 50 and 75 nm Ag nanospheres, carrying a citrate modification, were utilized (NanoComposix, Prague, Czech Republic). Alkali-modified TiO_2 NPs were synthesized as described in a previous study [33]. Briefly, 1 g TiO_2 commercial nanopowder (ST-01, 7 nm or ST-21, 20 nm, 100 % anatase, Ishihara Corporation, Japan) was dispersed in a 100 mM H_2O_2 and 8 M NaOH mixed aqueous solution. Then, the suspension was heated to 50 °C and kept for 24 h. After centrifugation at 8000 rpm for 20 min, the particles were washed twice with de-ionized water and re-dispersed in a 1.6 M HNO_3 solution. The suspension was stirred for 3 h and then centrifuged at 13,500 rpm for 20 min. After three times wash with de-ionized water, the TiO_2 NPs were dried at 60 °C and re-dispersed in water for subsequent experiments. Final concentrations of the suspension were determined by inductively coupled plasma optical emission spectrometry (ICP-OES) (Agilent 725, Santa Clara, USA). Stock solutions of Ag NPs and TiO_2 NPs were first ultrasonicated for 5 min and then diluted to 100 µg/mL in the indicated medium. All NP suspensions were freshly prepared before each experiment.

Characterization of nanoparticles

The particle size and morphology of the alkali-modified TiO_2 NPs in suspension were analyzed using a JEM2100 transmission electron microscope at an acceleration voltage of 200 kV (TEM) (JEOL, Japan). The hydrodynamic size of the Ag NPs and TiO_2 NPs in stock solution and cell culture medium (CCM) were monitored using a Zetasizer Nano ZS apparatus (Malvern Instruments GmbH, Herrenberg, Germany). Particle solutions were diluted for the DLS measurement to a final concentration of 100 µg/mL for TiO_2 NPs and 10 µg/mL for Ag NPs. For nanoparticle tracking analysis (NTA) we used a NanoSight LM20 system (Malvern, Germany) equipped with a green laser (532 nm) at RT (22 °C). Sizes were calculated from videos (60 s, 30 frames per s) by NTA3.1 software. Fetal calf serum (FCS) contained numerous agglomerates, which disturbed analysis. Therefore, NP dispersions were prepared in CCM with final concentrations of 10 µg/ml for nanosilver and 100 µg/ml for TiO2 and then diluted 1:40 into protein-free MEM such that the serum-agglomerates were not detectable any more. Proteincorona of NPs were considered as rather stable, not likely to disintegrate during the time period of NTA measurements. Nanosilver dissolution in water was determined using ICP-MS as described below. Two different concentrations were tested, i.e. 2 and 10 µg/ml. Suspensions were incubated for 24 h at 37 °C and were then centrifuged at a table top centrifuge at 16,000×*g* for 30 min to remove NPs. Supernatants were filtered through Amicons filters (cut off 30 kDa) and then processed as described below for ICP-MS analysis.

Cell culture

Mouse neuroblastoma (Neuro-2a) cells (Cell Lines Service GmbH, Eppelheim, Germany) were cultured in MEM medium (Gibco, Darmstadt, Germany) supplemented with 10 % fetal calf serum (FCS) (Pan Biotech, Aidenbach, Germany), 2 mM L-glutamine, 0.1 mM non-essential amino acids, and 1.0 mM sodium pyruvate (Gibco, Darmstadt, Germany). Cells were cultivated at 37 °C, 5 % CO_2 and 95 % relative humidity. Twenty four hours after seeding, cells were differentiated using 30 µM forskolin and 200 µM 3-isobutyl-1-methylxanthine (IBMX) (both obtained from Sigma-Aldrich, Steinheim, Germany) in MEM/1 % FCS medium for 2 days into neuronal-like cells.

Cytotoxicity

WST-1 cell viability assay was used to evaluate the toxicity of TiO_2 NPs and Ag NPs according to manufacturer's instructions (Roche Diagnostics, Mannheim, Germany). Neurite-bearing cells (1.8×10^4 cells/cm^2) were treated with 5, 10 and 25 µg/mL TiO_2 NPs or Ag NPs, respectively, in 96-well plates for 24 h. Interfering NPs were removed in a table top centrifuge by centrifugation with maximum speed prior to spectrophotometric read-out (TECAN, Crailsheim, Germany) at 450 nm.

Cell incubation and sample preparation

For analysis by ICP-MS and SP-ICP-MS, cells were seeded and differentiated in 12-well plates (1.8×10^4 cells/cm^2). They were exposed to 2 or 10 µg/mL NPs in MEM/5 %

FCS medium for 24 h. It should be noted, that in vitro test concentrations in the range from 1 to 10 µg/cm^2 correlate very well to test concentrations usually used in in vivo inhalation studies and in particular they correlate well to the overload dose, i.e. the dose where toxic effects become detectable. Therefore, in vitro test concentrations in the range from 1 to 10 µg/cm^2 are useful for comparing the data later on to results obtained in in vivo experiments.

Before analysis cells were washed three times with DPBS (Dulbecco's Phosphate Buffered Saline) before being trypsinized and harvested by centrifugation (250×*g*, 5 min). In each case, 10 µL were used to count cells. Cells were lysed in 1 mL 0.1 % Triton X-100 (Sigma, Taufkirchen, Germany). Half of each lysate was analyzed by SP-ICP-MS. The other half was digested for ICP-MS. The cell lysates and digested solutions were stored at 4 °C before analysis. Control samples without any cells were prepared using the same protocol.

For LA-ICP-MS, the cells (1 × 10^4 cells/cm^2) were grown on polymer coated glass slides and incubated with NPs (2 µg/mL for Ag NPs or 2 and 10 µg/mL for TiO_2 NPs) for 24 h. Cells were washed three times with DPBS and fixed with 4 % paraformaldehyde in DPBS. Therafter they were dehydrated in a series of ethanol (15, 30, 50, 70, 90 and 99.5 %). The slides were stored at 4 °C before analysis. Control cells without NPs were prepared similarly.

ICP-MS analysis of digested cells

For Ag NPs, 10 mL 65 % HNO_3 was used for digesting cell lysates. For TiO_2 NPs, 2 mL 65 % HNO_3 and 4 mL 5 % HF was used. Digestion was performed in open vials at 60 °C using a heating block for 30 h. Total amounts of titanium and silver in the cells were determined by ICP-MS using a quadrupole ICP mass spectrometer (XSeries 2, Thermo Fisher Scientific GmbH, Dreieich, Germany). The ICP-MS parameters were optimized and are given in Additional file 1: Table S1. For the Ti and Ag detection, the ^{49}Ti and ^{107}Ag isotopes were selected. As an internal standard ^{115}In was used. Digested samples were filled up with ultrapure water to 50 mL with 10 ppb of ^{115}In as internal standard before analysis. Calibrations were performed using ionic Ti and Ag standards in 6.5 % HNO_3 solution ranging from 0 to 20 µg/L for both elements. All experiments were conducted in three independent replicates for each treatment and control group. For the calculation of the number of NPs per cell, a perfect sphere was assumed for the particles and either TEM or DLS particle diameters were used.

SP-ICP-MS analysis of cell lysates

SP-ICP-MS was used to analyze number and size of NPs in the cell lysates (XSeries 2, Thermo Fisher Scientific GmbH, Dreieich, Germany). The SP-ICP-MS parameters were optimized and are given in Additional file 1: Table S1. Before analysis, the lysates were homogenized by vortexing and then diluted in ultrapure water to 50–100 ng/L Ag or Ti based on the ICP-MS measurements of digested cells, which were performed beforehand. For quality control of the nebulizer transport efficiency a freshly prepared 60 nm AuNP standard (NIST-RM 8013, Gaithersburg, MD, USA) was used at a concentration of 100 ng/L in ultrapure water, which yielded an average size of 61 nm (Additional file 1: Figure S6). The nebulizer transport efficiency ranged from 0.039 to 0.045.

For all measurements the instrument was operated in spike mode separately detecting ^{197}Au, ^{107}Ag and ^{48}Ti isotopes one at a time. A constant sample flow rate of 0.34 mL/min was utilized. The duration time for each run was set at 1 min with a dwell time of 3 ms per reading. To determine the detector sensitivity in SP-ICP-MS for both elements, one point of each calibration series established by ICP-MS for digested cells (10 µg/L for Ag and Ti ion standard) was measured in spike mode. The limits of detection regarding the NP size (LOD_{size}) of the TiO_2 NPs and Ag NPs were calculated as described as follows with a 3σ threshold to identify the particle signal from the background noise [41]:

$$\mathrm{LOD_{size}} = \left(\frac{6 \times 3\sigma}{R \times f \times \rho \times \pi}\right)^{1/3}$$

Where σ is the standard deviation of the signal background (here: control group cell lysates); R is the sensitivity of the detector for the element of the analyte (cps/µg); *f* is the mass fraction of analyzed metal element in the NPs; ρ is the density of the NPs.

NP number limits of detection ($LOD_{numberNP}$) were calculated by:

$$\mathrm{LOD_{numberNP}} = 3 \times \frac{1}{\delta_{neb}\beta_{sam}t_i}$$

Where δ_{neb} is the nebulizer transport efficiency; β_{sam} is the sample flow rate; and t_i is the total acquisition time.

LA-ICP-MS of single cells

LA-ICP-MS was performed using an NWR 213 laser system (Electro Scientific Industries, Huntingdon, UK) coupled to an Element XR sector field ICP-MS (Thermo Fisher Scientific GmbH, Dreieich, Germany). The system was warmed up before analysis and tuned by ablating line scans with 200 µm spot size, 10 µm/s scan rate, 20 Hz repetition rate and 100 % laser energy from a microscope glass slide while optimizing the parameters for high signal intensities.

Glass slides were fixed in the ablation cell which mechanically moves the samples in xyz-direction under the fixed laser. At first, ablation parameters for dried cells were optimized to ensure complete ablation of the cells and a total coverage of the analyzed area which resulted in a scan speed of 5 μm/s, a spot size of 4 μm, a repetition rate of 10 Hz, a laser fluence of about 2.5 mj/cm^2 and a lane distance of 5 μm, respectively. Details on the used equipment and the optimized parameters are given in Additional file 1: Table S2. To analyze sufficient numbers of cells, three different areas for each treatment and control group were evaluated. The aerosol was transported by a He carrier gas flow from the ablation cell to the ICP-MS where an additional Ar gas flow was introduced prior to the atomization and ionization in the plasma.

For quantification, two dilution series made from TiO_2 NPs and Ag NPs in an untreated cell lysate matrix were prepared. The cell concentration was adjusted to theoretically 2 cells per μl (Additional file 1: Table S3). These solutions were then digested by adding nitric acid and hydrochloric acid (1:1) in an ultrasonic bath over night at 60 °C. 0.5 μl of each point of these dilution series were pipetted in duplicate onto a glass slide, dried and analyzed by LA-ICP-MS. For the quantification of this calibration series, ablation under the aforementioned parameters would be too time-consuming. Therefore, laser ablation for the calibration series was done with a higher scan speed of 200 μm/s, a bigger spot size of 250 μm and corresponding lane distance of 240 μm, and a higher laser fluence of about 2.5 mJ/cm^2 (Additional file 1: Table S4).

In parallel, adjacent lines were ablated and the resulting time-dependent ICP-MS scans merged to one data set using lane distance, scan speed and time to calculate the dimensions. Integration of the signal intensity in the pipetted drops and in the cells was done using ImageJ software. Only clearly separated cells were used for integration and analysis. For the calculation of the number of NPs per cell a perfect sphere was assumed for the particles and the DLS size was used. Background correction for the cells and pipetted calibration was done for each slide using areas not containing any cells, cell compartments or droplets.

Monitored isotopes

^{107}Ag and/or ^{109}Ag were detected for the Ag NPs. Despite its low abundance of 5.5 %, ^{49}Ti was selected to conduct ICP-MS because it shows a low background noise and polyatomic interferences such as $^{33}S^{16}O$ or $^{31}P^{18}O$ can be resolved by the collision cell technique (CCT) of quadrupole ICP-MS [20]. Higher abundant and therefore more sensitive ^{48}Ti (73.8 %) was selected to conduct SP-ICP-MS [42] and LA-ICP-MS. For SP-ICP-MS, even though polyatomic interferences such as $^{32}S^{16}O$ and $^{36}Ar^{12}C$ or the isobaric interference with ^{48}Ca are possible, these interferences only contribute to the continuous background while the TiO_2 particles have distinct discontinuous signals and can thus be distinguished and isolated from the background. Our LOD_{size} for ^{48}Ti particles was 57 nm (equivalent to 69 nm TiO_2), which was derived from background noise of cell lysates. Our LOD_{size} was lower than for ^{49}Ti (75 nm) and ^{47}Ti (93 nm) particles, which was derived from water background by using the same instrument [41]. For LA-ICP-MS, using a sector field ICP-MS allows resolving the polyatomic interferences for ^{48}Ti but still does not suffice to distinguish it from ^{48}Ca.

Statistics

Two-tailed Student's unpaired t–test was used for two-group comparison. $p < 0.05$ was considered statistically significant.

Abbreviations

AAS: atomic absorption spectrometry; CCM: cell culture medium; CCT: collision cell technique; DLS: dynamic light scattering; DPBS: Dulbecco's phosphate buffered saline; EDX: energy-dispersive X-ray spectroscopy; HUVECs: human umbilical vein endothelial cells; IBMX: 3-isobutyl-1-methylxanthine; ICP-MS: inductively coupled plasma mass spectrometry; ICP-OES: inductively coupled plasma optical emission spectrometry; LA-ICP-MS: laser ablation ICP-MS; LOD: limit of detection; NPs: nanoparticles; SC-ICP-MS: single cell ICP-MS; SP-ICP-MS: single particle ICP-MS; SEM: scanning electron microscopy; TEM: transmission electron microscopy.

Authors' contributions

The study was planned by AH and JT, with input from all co-authors. ILH prepared all samples and performed cytotoxicity studies. ICP-MS and SP-ICPMS analysis was done by ILH with support of PR. LA-ICP-MS analysis was performed by FSB with support from ILH. Data analysis was done by ILH and FSB with input of NJ, JT and AH. The manuscript was written by ILH, FSB, AH with input of all co-authors. All authors read and approved the final manuscript.

Author details

[1] Department of Chemical and Product Safety, German Federal Institute for Risk Assessment (BfR), Max-Dohrn-Strasse 8-10, 10589 Berlin, Germany. [2] Department of Biomedical Engineering and Environmental Sciences, National Tsing Hua University, Hsinchu, Taiwan. [3] Division of Inorganic Trace Analysis, German Federal Institute for Materials Research and Testing (BAM), Berlin, Germany.

Acknowledgements

The authors acknowledge funding from their institutes. In addition, I-Lun Hsiao thanks DAAD and the Ministry of Science and Technology Taiwan (MOST) for financial support. The authors thank Julian Tharmann for support with sample preparation and Tony Bewersdorff for help with nanoparticle characterization using NTA.

Competing interests

The authors declare that they have no competing interests.

References

1. Kessler R. Engineered nanoparticles in consumer products: understanding a new ingredient. Environ Health Perspect. 2011;119:a120–5.
2. Josset S, Keller N, Lett MC, Ledoux MJ, Keller V. Numeration methods for targeting photoactive materials in the UV-A photocatalytic removal of microorganisms. Chem Soc Rev. 2008;37:744–55.
3. Li WR, Xie XB, Shi QS, Zeng HY, Ou-Yang YS, Chen YB. Antibacterial activity and mechanism of silver nanoparticles on *Escherichia coli*. Appl Microbiol Biotechnol. 2010;85:1115–22.
4. Kowal K, Cronin P, Dworniczek E, Zeglinski J, Tiernan P, Wawrzynska M, Podbielska H, Tofail SAM. Biocidal effect and durability of nano-TiO_2 coated textiles to combat hospital acquired infections. Rsc Adv. 2014;4:19945–52.
5. Veronovski N, Lesnik M, Lubej A, Verhovsek D. Surface treated titanium dioxide nanoparticles as inorganic UV filters in sunscreen products. Acta Chim Slov. 2014;61:595–600.
6. Chen XX, Cheng B, Yang YX, Cao AN, Liu JH, Du LJ, Liu YF, Zhao YL, Wang HF. Characterization and preliminary toxicity assay of nano-titanium dioxide additive in sugar-coated chewing gum. Small. 2013;9:1765–74.
7. Dubey A, Zai J, Qian X, Qiao Q. Metal oxide nanocrystals and their properties for application in solar cells. In: Bhushan B, Luo D, Schricker SR, Sigmund W, Zauscher S, editors. Handbook of nanomaterials properties. Berlin: Springer; 2014. p. 671–707.
8. Wilkinson LJ, White RJ, Chipman JK. Silver and nanoparticles of silver in wound dressings: a review of efficacy and safety. J Wound Care. 2011;20:543–9.
9. Larguinho M, Baptista PV. Gold and silver nanoparticles for clinical diagnostics—from genomics to proteomics. J Proteom. 2012;75:2811–23.
10. Reidy B, Haase A, Luch A, Dawson KA, Lynch I. Mechanisms of silver nanoparticle release, transformation and toxicity: a critical review of current knowledge and recommendations for future studies and applications. Materials. 2013;6:2295–350.
11. Shi HB, Magaye R, Castranova V, Zhao JS. Titanium dioxide nanoparticles: a review of current toxicological data. Part Fibre Toxicol. 2013;10:15.
12. Brown AP, Brydson RMD, Hondow NS. Measuring in vitro cellular uptake of nanoparticles by transmission electron microscopy. In: Nellist P, editor. Journal of physics conference series: electron microscopy and analysis group conference 2013, vol. 522. 2014. p. 250–4.
13. Gliga AR, Skoglund S, Wallinder IO, Fadeel B, Karlsson HL. Size-dependent cytotoxicity of silver nanoparticles in human lung cells: the role of cellular uptake, agglomeration and Ag release. Part Fibre Toxicol. 2014;11:11.
14. Klingberg H, Oddershede LB, Loeschner K, Larsen EH, Loft S, Moller P. Uptake of gold nanoparticles in primary human endothelial cells. Toxicol Res. 2015;4:655–66.
15. Schrand AM, Schlager JJ, Dai L, Hussain SM. Preparation of cells for assessing ultrastructural localization of nanoparticles with transmission electron microscopy. Nat Protoc. 2010;5:744–57.
16. Thurn KT, Arora H, Paunesku T, Wu A, Brown EM, Doty C, Kremer J, Woloschak G. Endocytosis of titanium dioxide nanoparticles in prostate cancer PC-3M cells. Nanomedicine. 2011;7:123–30.
17. Braun GB, Friman T, Pang HB, Pallaoro A, de Mendoza TH, Willmore AMA, Kotamraju VR, Mann AP, She ZG, Sugahara KN, et al. Etchable plasmonic nanoparticle probes to image and quantify cellular internalization. Nat Mat. 2014;13:904–11.
18. Salvati A, Aberg C, dos Santos T, Varela J, Pinto P, Lynch I, Dawson KA. Experimental and theoretical comparison of intracellular import of polymeric nanoparticles and small molecules: toward models of uptake kinetics. Nanomedicine. 2011;7:818–26.
19. Rashkow JT, Patel SC, Tappero R, Sitharaman B. Quantification of single-cell nanoparticle concentrations and the distribution of these concentrations in cell population. J R Soc Interface. 2014;11:5.
20. Krystek P, Kettler K, van der Wagt B, de Jong WH. Exploring influences on the cellular uptake of medium-sized silver nanoparticles into THP-1 cells. Microchem J. 2015;120:45–50.
21. Lu F, Wu SH, Hung Y, Mou CY. Size effect on cell uptake in well-suspended, uniform mesoporous silica nanoparticles. Small. 2009;5:1408–13.
22. Huang K, Ma H, Liu J, Huo S, Kumar A, Wei T, Zhang X, Jin S, Gan Y, Wang PC, et al. Size-dependent localization and penetration of ultrasmall gold nanoparticles in cancer cells, multicellular spheroids, and tumors in vivo. ACS Nano. 2012;6:4483–93.
23. Zheng LN, Wang M, Wang B, Chen HQ, Ouyang H, Zhao YL, Chai ZF, Feng WY. Determination of quantum dots in single cells by inductively coupled plasma mass spectrometry. Talanta. 2013;116:782–7.
24. Laborda F, Bolea E, Jimenez-Lamana J. Single particle inductively coupled plasma mass spectrometry: a powerful tool for nanoanalysis. Anal Chem. 2014;86:2270–8.
25. Drescher D, Giesen C, Traub H, Panne U, Kneipp J, Jakubowski N. Quantitative imaging of gold and silver nanoparticles in single eukaryotic cells by laser ablation ICP-MS. Anal Chem. 2012;84:9684–8.
26. Drescher D, Zeise I, Traub H, Guttmann P, Seifert S, Büchner T, Jakubowski N, Schneider G, Kneipp J. In situ characterization of SiO_2 nanoparticle biointeractions using BrightSilica. Adv Func Mat. 2014;24:3765–75.
27. Bohme S, Stark HJ, Meissner T, Springer A, Reemtsma T, Kuhnel D, Busch W. Quantification of Al_2O_3 nanoparticles in human cell lines applying inductively coupled plasma mass spectrometry (neb-ICP-MS, LA-ICP-MS) and flow cytometry-based methods. J Nanopart Res. 2014;16:2592.
28. Buchner T, Drescher D, Traub H, Schrade P, Bachmann S, Jakubowski N, Kneipp J. Relating surface-enhanced Raman scattering signals of cells to gold nanoparticle aggregation as determined by LA-ICP-MS micromapping. Anal Bioanal Chem. 2014;406:7003–14.
29. Liu SC, Xu LJ, Zhang T, Ren GG, Yang Z. Oxidative stress and apoptosis induced by nanosized titanium dioxide in PC12 cells. Toxicology. 2010;267:172–7.
30. Haase A, Rott S, Mantion A, Graf P, Plendl J, Thunemann AF, Meier WP, Taubert A, Luch A, Reiser G. Effects of silver nanoparticles on primary mixed neural cell cultures: uptake, oxidative stress and acute calcium responses. Toxicol Sci. 2012;126:457–68.
31. Lee JH, Kim YS, Song KS, Ryu HR, Sung JH, Park JD, Park HM, Song NW, Shin BS, Marshak D, et al. Biopersistence of silver nanoparticles in tissues from Sprague-Dawley rats. Part Fibre Toxicol. 2013;10:36.
32. Huang CL, Hsiao IL, Lin HC, Wang CF, Huang YJ, Chuang CY. Silver nanoparticles affect on gene expression of inflammatory and neurodegenerative responses in mouse brain neural cells. Environ Res. 2015;136:253–63.
33. Wu CY, Tu KJ, Wu CH. High hydroxyl group density on the surface of TiO_2 pretreated with alkaline hydrogen peroxide. In: Abstracts of papers of the American chemical society 2013; 246.
34. Deegan RD, Bakajin O, Dupont TF, Huber G, Nagel SR, Witten TA. Capillary flow as the cause of ring stains from dried liquid drops. Nature. 1997;389:827–9.
35. Bhardwaj R, Fang XH, Somasundaran P, Attinger D. Self-assembly of colloidal particles from evaporating droplets: role of DLVO interactions and proposition of a phase diagram. Langmuir. 2010;26:7833–42.
36. Ahlinder L, Ekstrand-Hammarstrom B, Geladi P, Osterlund L. Large uptake of titania and iron oxide nanoparticles in the nucleus of lung epithelial cells as measured by Raman imaging and multivariate classification. Biophys J. 2013;105:310–9.
37. Sengstock C, Diendorf J, Epple M, Schildhauer TA, Koller M. Effect of silver nanoparticles on human mesenchymal stem cell differentiation. Beilstein J Nanotech. 2014;5:2058–69.
38. Liu W, Wu Y, Wang C, Li HC, Wang T, Liao CY, Cui L, Zhou QF, Yan B, Jiang GB. Impact of silver nanoparticles on human cells: effect of particle size. Nanotoxicology. 2010;4:319–30.
39. Allouni ZE, Hol PJ, Cauqui MA, Gjerdet NR, Cimpan MR. Role of physicochemical characteristics in the uptake of TiO_2 nanoparticles by fibroblasts. Toxicol In Vitro. 2012;26:469–79.
40. Zhu J, Liao L, Zhu L, Zhang P, Guo K, Kong J, Ji C, Liu B. Size-dependent cellular uptake efficiency, mechanism, and cytotoxicity of silica nanoparticles toward HeLa cells. Talanta. 2013;107:408–15.
41. Lee S, Bi X, Reed RB, Ranville JF, Herckes P, Westerhoff P. Nanoparticle size detection limits by single particle ICP-MS for 40 elements. Environ Sci Technol. 2014;48:10291–300.
42. Peters RJ, van Bemmel G, Herrera-Rivera Z, Helsper HP, Marvin HJ, Weigel S, Tromp PC, Oomen AG, Rietveld AG, Bouwmeester H. Characterization of titanium dioxide nanoparticles in food products: analytical methods to define nanoparticles. J Agric Food Chem. 2014;62:6285–93.

4

Magnetocontrollability of Fe_7C_3@C superparamagnetic nanoparticles in living cells

Irina B. Alieva[1†], Igor Kireev[1,2†], Anastasia S. Garanina[2], Natalia Alyabyeva[3], Antoine Ruyter[3], Olga S. Strelkova[1], Oxana A. Zhironkina[1], Varvara D. Cherepaninets[1], Alexander G. Majouga[4,5], Valery A. Davydov[6], Valery N. Khabashesku[7], Viatcheslav Agafonov[3,5*] and Rustem E. Uzbekov[8,9*]

Abstract

Background: A new type of superparamagnetic nanoparticles with chemical formula Fe7C3@C (MNPs) showed higher value of magnetization compared to traditionally used iron oxide-based nanoparticles as was shown in our previous studies. The in vitro biocompatibility tests demonstrated that the MNPs display high efficiency of cellular uptake and do not affect cyto-physiological parameters of cultured cells. These MNPs display effective magnetocontrollability in homogeneous liquids but their behavior in cytoplasm of living cells under the effect of magnetic field was not carefully analyzed yet.

Results: In this work we investigated the magnetocontrollability of MNPs interacting with living cells in permanent magnetic field. It has been shown that cells were capable of capturing MNPs by upper part of the cell membrane, and from the surface of the cultivation substrate during motion process. Immunofluorescence studies using intracellular endosomal membrane marker showed that MNP agglomerates can be either located in endosomes or lying free in the cytoplasm. When attached cells were exposed to a magnetic field up to 0.15 T, the MNPs acquired magnetic moment and the displacement of incorporated MNP agglomerates in the direction of the magnet was observed. Weakly attached or non-attached cells, such as cells in mitosis or after cytoskeleton damaging treatments moved towards the magnet. During long time cultivation of cells with MNPs in a magnetic field gradual clearing of cells from MNPs was observed. It was the result of removing MNPs from the surface of the cell agglomerates discarded in the process of exocytosis.

Conclusions: Our data allow us to conclude for the first time that the magnetic properties of the MNPs are sufficient for successful manipulation with MNP agglomerates both at the intracellular level, and within the whole cell. The structure of the outer shells of the MNPs allows firmly associate different types of biological molecules with them. This creates prospects for the use of such complexes for targeted delivery and selective removal of selected biological molecules from living cells.

Keywords: Superparamagnetic nanoparticles, Living cells, Magnetocontrollability, Endocytosis, Cytoskeleton, Cell adhesion

Background

The studies of interaction mechanisms between various types of MNPs and living cells, as well as internalization routes and intracellular motility of individual magnetic particles or their agglomerates are very important for development of various biotechnological applications of magnetic nanomaterials [1–4]. One of the applications of MNPs in cell biology is magnetofection—magnetic field-driven delivery of cargo-loaded MNPs through the cellular membrane. The magnetofection is already a widespread approach to an accelerated transport of nucleic acids associated with MNPs into cells by a magnetic field [3, 5, 6]. However, the precise details of MNPs

*Correspondence: viatcheslav.agafonov@univ-tours.fr; rustem.uzbekov@univ-tours.fr
†Irina B. Alieva and Igor Kireev contributed equally in this work
[3] GREMAN, UMR CNRS 7347, Université François Rabelais, 37200 Tours, France
[8] Laboratoire Biologie Cellulaire et Microscopie Electronique, Faculté de Médecine, Université François Rabelais, 37032 Tours, France
Full list of author information is available at the end of the article

behavior in living cells and the possibility of their subsequent removal from the cells under the effect of magnetic field still remains an open question.

Generally, the behavior of MNPs in homogeneous liquids under the effect of external magnetic fields is currently well characterized [7, 8]. Since superparamagnetic MNPs are mono-domain magnets, the magnetic dipole–dipole interaction between them in the constant homogeneous magnetic field should induce their alignment along the magnetic field lines of the permanent magnet. The magnetic attraction force in anisotropic magnetic field causes MNPs acceleration in the direction of increasing magnetic field strength. This force, which is typically a few pN for MNPs [8], is proportional to the magnetic field gradient, MNP volume and its magnetic moment. The alignment and movement speed of MNPs also depend on the viscosity of the medium and hydrodynamic radius of the nanoparticle [7, 8].

In contrast to liquid, the behavior of MNPs in living cells is also affected by anisotropic viscosity and resilience of the cytoplasmic structures, such as cytoskeleton and different membranous compartments. The main process of MNPs internalization into the cells is known to be an endocytosis [9, 10]. The intracellular MNPs may exist in various states like free individual particles or their agglomerates of 100–200 nm in diameter, or enclosed in membrane vesicles—so-called «magnetic endosomes» [11]. Free cytoplasmic MNPs may originate from the «magnetic endosomes» through the process termed « endosomal escape » [12, 13]. Besides the anisotropic viscosity of cytoplasm and the resilience of the cytoskeleton filaments, the movements of magnetic endosomes can be affected by the activity of cytoskeleton-associated protein motors acting in two opposite directions—kinesin-like motors translocating vesicles from the center of the cell to the periphery (centrifugally) and dynein-like ones acting towards the center (centripetally), as well as actin-associated myosins [14].

It was shown that the relatively low value of magnetization of traditionally used SPIONs creates difficulties for the control of their magnetic behavior in a number of applications. A new type of superparamagnetic nanoparticles with chemical formula Fe_7C_3@C was recently obtained by high pressure and high temperature process and studied by physico-chemical and biological methods [15, 16] (Table 1). The in vitro biocompatibility tests demonstrated that Fe_7C_3@C MNPs display high efficiency of cellular uptake and do not affect cyto-physiological parameters of cultured pig kidney epithelia (PK) cells [16].

In present work we performed a study of Fe_7C_3@C MNPs behavior in living cells cultured in vitro in the presence or absence of a constant magnetic field to evaluate the impact of cytoskeleton architecture and cell-substrate interactions on their magneto-controllability at cellular and subcellular levels.

Table 1 Magnetic properties of various types of MNPs

NP	Size (nm)	Ms (emu/g)	Source
(Fe_7C_3@C)	25	54	Our data [16]
Fe_3O_4 (+PEG or DOX)	7–10	1.12	[25]
γ-Fe_2O_3 @C	15	28	[26]
γ-Fe_2O_3 @Si	50–200	15–35	[27]
γ-Fe_2O_3 (pure)	15	35	[28]

Methods

Cell culture and experimental treatments

Human fibrosarcoma cells HT1080 (kindly provided by Russian Collection of cell lines, St. Petersburg) were cultured in DMEM culture media (Sigma, USA) supplemented with 10 % fetal calf serum (HyClone, USA) and antibiotic–antimycotic (100 units/ml penicillin G, 100 mg/ml streptomycin sulfate and 0.25 mg/ml amphotericin B) (Sigma, USA). For microscopic experiments cells were plated onto cover slips at a concentration of 10,000 cells/cm^2 and grown for 48 h to reach 50 % confluency before the addition of Fe_7C_3@C MNPs to a final concentration of 20 μg/ml. Kinetics of cell interaction with Fe_7C_3@C was studied by TEM on serial ultrathin sections, and also by optical microscopy using time lapse video recording of living cells.

Live cell experiments

For live imaging, human fibrosarcoma cells were plated on glass-bottomed Petri dishes (LabTek, USA) at a density of 10^5 cells/ml and incubated with Fe_7C_3@C MNPs for 24 h. Imaging was performed in an environmental chamber kept at 37 °C under 5 % CO_2. The chamber was mounted on a Ti-E inverted microscope (Nikon, Japan) equipped with EMCCD-camera iXon (Andor) operating under control of NIS-Elements 4.0 software. Illumination conditions (ND filters, lamp voltage, and exposure time) were set to minimize photo toxicity. To generate a magnetic field we used in our experiments a gold-plated NdFeB permanent magnet with the size of 5 × 5 × 5 mm and B_z = 0.15 T. The magnet was placed directly inside the dish, therefore cells located up to 2 mm from the magnet edge have been recorded. Images were taken every 10 min for 72 h for long recording or every 1 min during short recording. Image sequences were analyzed and time-lapse movies of cells loaded with MNPs were assembled using ImageJ software.

Immunofluorescent staining

For immunofluorescent staining, cells were fixed with 4 % formaldehyde (Sigma) in physiological phosphate buffered saline (PBS), pH 6.8, for 10 min and then rinsed with three changes of PBS (for 10 min each). The fixed cells were permeabilized with 0.1 % Triton X_100 (Sigma) in PBS for 15 min with subsequent washing out with PBS (three times for 10 min). For elimination of background fluorescence, prior to labeling with antibodies, the cells were treated with 0.2 % $NaBH_4$ (Sigma) in PBS (three times for 10 min) and rinsed with PBS (three times for 10 min). The cells were then incubated with primary anti-Rab5 (C8B1) rabbit monoclonal antibodies (Cell Signaling, US, dilution 1:100) (30 min, 37 °C) and secondary antibodies conjugated with Texas Red fluorescent dye (Molecular Probes, dilution 1:1000) (30 min, 37 °C). Cells were mounted in Mowiol and observed in Eclipse Ti-E fluorescent microscope (Nikon, Japan) with CFI Plan Apo VC 60X/NA 1.4 lens, equipped with EMCCD-camera iXon (Andor) under the control of NIS-Elements 4.0 software.

Transmission electron microscopy

For transmission electron microscopy (TEM) experiments, cells were washed three times with fresh prewarmed media to remove free particles, fixed in 2.5 % glutaraldehyde in 0.1 M phosphate buffer (pH 7.4) for 2 h with subsequent post-fixation in 1 % OsO_4 and embedding in Epon (Sigma, USA). Serial ultrathin sections (70 nm) were cut with Leica Ultracut-E ultramicrotome and observed with JEM 1011 (JEOL, Japan) equipped with a Gatan digital camera driven by Digital Micrograph software (Gatan, Pleasanton, USA) at 100 kV.

Correlative Magnetic force microscopy and Transmission electron microscopy (CMFM-TEM)

For CMFM-TEM we used a Solver microscope (NT-MDT) under air conditions to investigate magnetic properties of Fe_7C_3@C MNPs. Semi-thin (500 nm) Epon sections of MNPs-loaded cells fixed after 24 h incubation in magnetic field were attached to 5 × 5 mm glass slide (1 mm thick) and measured by the two-pass MFM method using a silicon cantilever with Co-Cr coating (resonant frequency is 62.7 kHz, spring constant (k) is 3 N/m, created magnetic moment is about 10^{-13} emu) [17]. During first pass, the morphology of the cell is determined. Then, the cantilever was lifted from the surface and kept at a constant distance of $\Delta z = 100$ nm, then by following the surface morphology profile the response of vertical component of magnetic gradient was recorded (i.e. phase shift ($\Delta\phi$) of the cantilever oscillation). During second pass, the cantilever oscillated with 50 %, reduced amplitude the vibration system quality factor (Q) was 10, Δz-magnetic field was around 200 Oe. All 512 × 512 pixels (40 nm/pixel) images were recorded with a scan rate of 0.5 Hz.

After MFM imaging semi-thin sections were reembedded in Epon, ultrathin sections (70 nm) of the same cells were made and imaged with JEM 1011 (JEOL, Japan) equipped with a Gatan digital camera driven by Digital Micrograph software (Gatan, Pleasanton, USA) at 100 kV. Image alignment and scaling was performed with Photoshop CS3 (Adobe, USA).

Results

Endocytosis of MNPs by human fibrosarcoma cell

In the present work we performed live-cell imaging to study the kinetics of interactions between MNPs and transformed cells isolated from biopsies of human fibrosarcoma (line HT1080). The cells grown on solid substrate acquire fibroblast-like shape and demonstrate rather high motility; cell movement speed was measured as about 0.2 μm/min (Fig. 1; Additional file 1: Movie 1). This cell line also demonstrates high proliferative activity typical for transformed cells. During cell division, the cells round up and lose contacts with the substrate, performing all the phases of mitosis (including ana- and telophase) in this rounded state.

After administration of MNPs suspension to the culture media, the cells actively internalize the agglomerates of MNPs formed in solution and on the cell surface by endocytosis, similar to what we described earlier for non-transformed cells [16]. Internalized MNPs move from the cell membrane into the cytoplasm and form one or several agglomerates of various sizes.

Live-cell imaging demonstrated that the cells can actively collect MNPs agglomerates laying on the substrate (Fig. 1; Additional file 1: Movie 1) as well as on the surface of neighboring cells (Additional file 2: Movie 2) during their movement.

The mitotic activity of transformed MNPs-treated fibrosarcoma HT1080 cell line remained the same as in control untreated cells. Abnormal mitotic figures, colchicine-like mitotic cells and cells with chromosome segregation anomalies as well as with cytokinesis defects, were not observed in these experiments. All observations described here allowed us to conclude that MNPs have no cytotoxicity effect on cultured HT1080 cells, similarly to our experiments with MNPs-loaded non-transformed PK cells [16].

Immunofluorescence analysis of MNPs and endosome co-localization in the cells

In our previous work we suggested that at least part of MNPs is localized inside the endosomes [16, 18]. To confirm these observations we studied colocalization

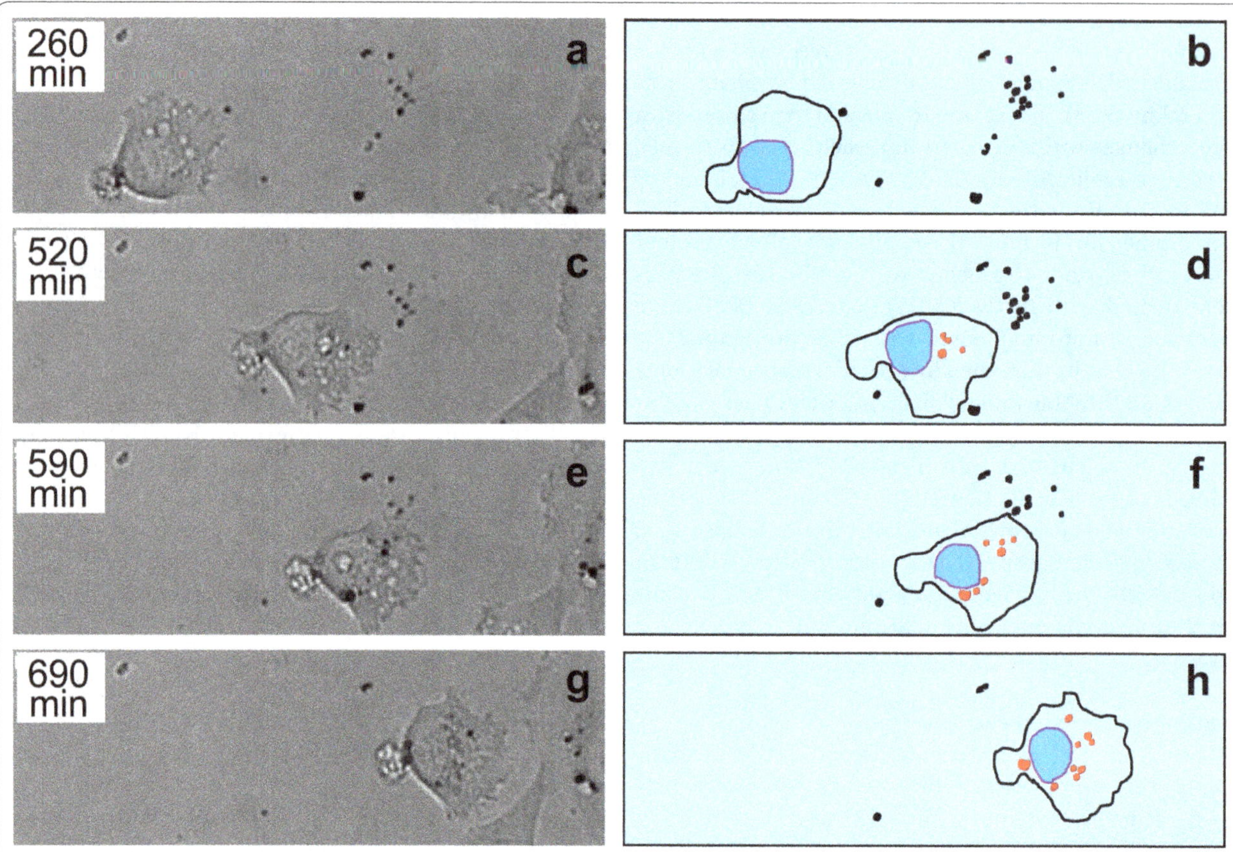

Fig. 1 Live imaging of moving HT-1080 cell that actively uptakes MNPs from the glass surface. Time scale from the beginning of the recording is indicated in the *upper-left corner* of each image (*left column*). *Left column* (**a**, **c**, **e**, **g**) represents successive photos of the cell, *right column* (**b**, **d**, **f**, **h**) represents a sketch of the movie with free MNPs shown in *black* and internalized MNPs in *red* (see also Additional file 1: Movie 1)

of cytoplasmic agglomerates of MNPs with endosomes immunostained for endosomal marker Rab5 (Fig. 2). Immunofluorescence analysis showed us that the regions of cytoplasm where endosomes are preferentially localized match rather well the area of MNPs agglomerates distribution with some small agglomerates of MNPs located inside the endosomes. However, the majority of endosomes are free of detectable MNP agglomerates and many of the latter, especially big ones, did not colocalize with endosomes either. This observation may suggest that the "endosome escape" occurs rather early, after MNPs internalization, before formation of secondary lysosomes. Otherwise, one would observe high cell mortality due to the membrane destruction and cytoplasmic release of activated lysosomal enzymes.

Effects of magnetic field on intracellular MNPs positioning and movements

The main motivation of using superparamagnetic nanoparticles in current study was the possibility to manipulate their localization and movement by external magnetic field. Relatively small size of the magnet used allowed its positioning inside a glass-bottomed Petri dish utilized for live imaging, so the cells can be placed in close vicinity to the magnet where the intensity of magnetic field is sufficiently high. Direct measurement of the magnetic fields showed typical exponential attenuation from 0.15 T near the surface to 0.01 T at the distance of 25 mm. All experimental cells we observed were located within 1 mm from the magnet surface, thus the magnetic field intensity at this distance ranged from 0.15 to 0.1 T.

As has been already demonstrated earlier [16], internalized MNPs move from the cell surface into the cytoplasm where they form one or several agglomerates or stay as individual particles. Upon applying an external magnetic field these agglomerates are capable of moving in the direction of the source of magnetic field, i.e. permanent magnet, along the magnetic field lines (Fig. 3; Additional file 3: Movie 3).

This movement is rather slow relative to cell motility so it was impossible to measure momentary speeds. However, after prolonged observations of cells located close to the magnet surface, gradual accumulation of MNPs in the part of the cells facing the magnet has become

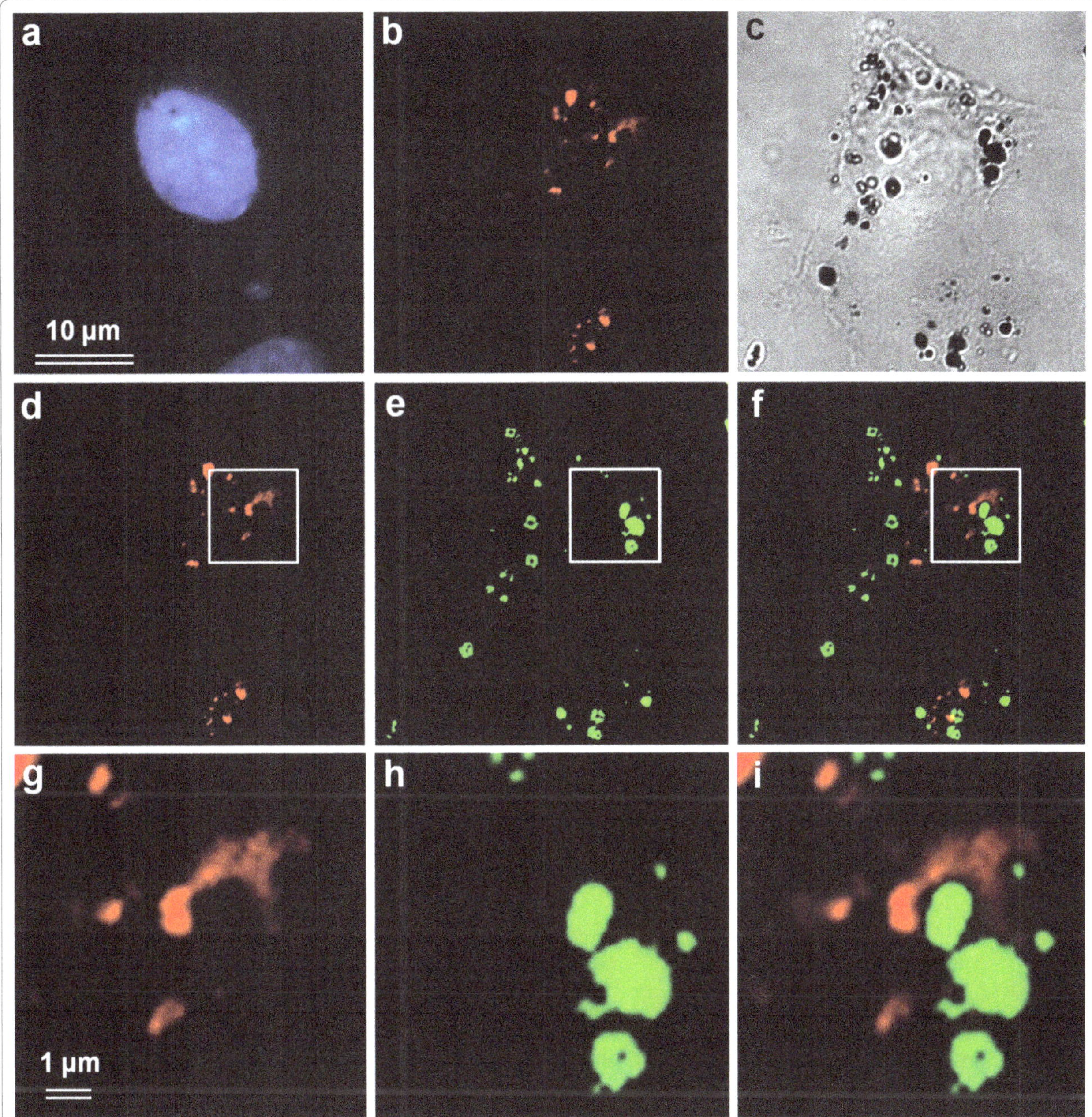

Fig. 2 Immunofluorescence analysis of MNPs and endosomes co-localization in the cells. **a** DAPI nuclear labeling, **b**, **d**, **g** endosome visualization with antibodies against Rab5 (*red*); **c** phase contrast, **e**, **h** pseudo color presentation of MNPs localization (*green*), **f**, **i** overlay of endosomes and MNPs. **g**–**i** Show enlarged areas indicated on **d**–**f** with a frame. *Bar* 10 μm (**a**–**f**), 1 μm (**g**–**i**)

obvious (Fig. 3). The degree of MNP agglomerate alignment critically depends on cell movement activity: the more actively the cell changes its position and shape, the less MNPs alignment towards the magnet occurs.

Along with a slow drift of internalized MNPs towards the magnet, numerous agglomerates of MNPs tend to orient themselves in magnetic field so they form highly extended agglomerates of smaller MNPs agglomerates with their long axis becoming parallel to the magnetic field lines. This orientation does not typically affect cell motility but is usually preserved upon changes in cellular shape, cytoskeleton rearrangement and changes of the direction of cell migration (Additional file 4: Movie 4).

Live imaging has also showed that, while a control cells move chaotically with respect to the orientation of external magnetic field, the MNPs-loaded cells display

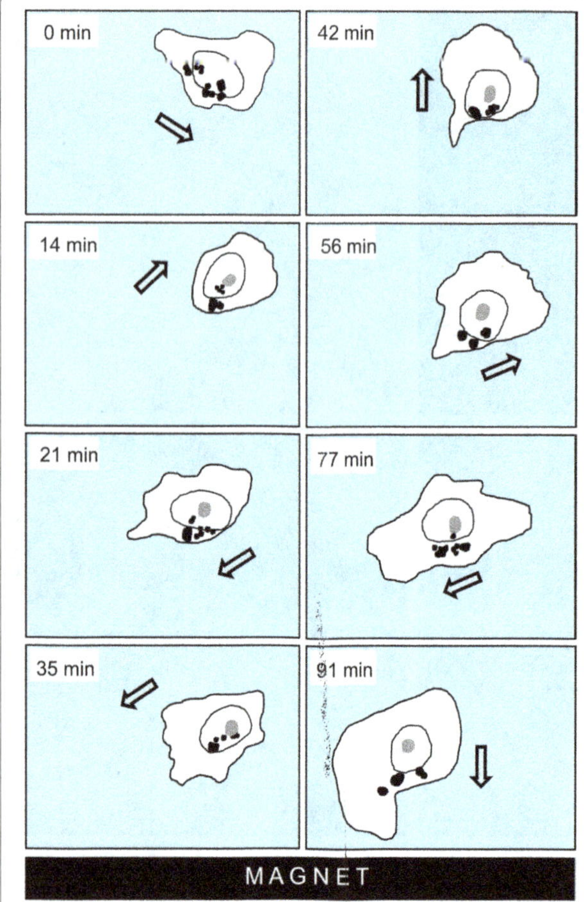

Fig. 3 As a cell changes the direction of its movement (the movement direction indicated by an *arrow* on this drawing for each time point), MNPs (indicated by *black dots*) in the cell can keep their position on the side facing the magnet (its position is down in this case). See also Additional file 3: Movie 3

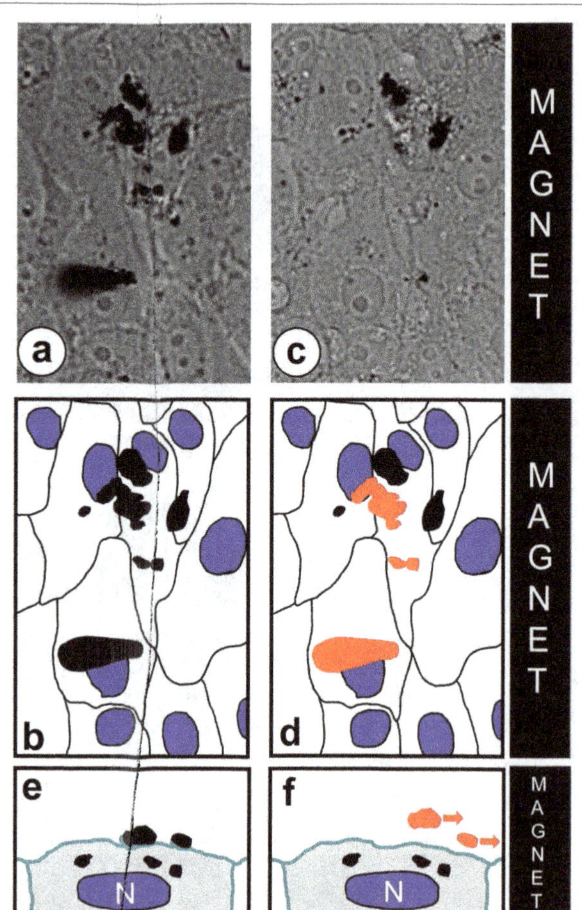

Fig. 4 The immediate effect of magnet introduction on MNPs agglomerates on the surface of the cells. **a–c** cells after 24 h of incubation with MNPs, **d–f** cells 5 min after magnet introduction, **a**, **d** phase contrast images, **b**, **e** schematic drawing of the cells shown in *figure* "**a**, **d**", **c**, **f** schemes illustrating the effect of magnet on MNPs agglomerates on the cell surface

different behavior. The cells with high concentration of cytoplasmic MNPs tend to move in the direction of the magnet (Additional file 5: Movie 5). Apparently, the magnetic force in the vicinity of the magnet is high enough to counteract the forces generated by cytoskeleton in moving cells and deviate their trajectory towards the magnet.

Exocytosis of MNPs out of cell cytoplasm under the action of magnetic field

As shown previously, the MNPs, which adhered to the cell membrane were internalized within 12 h after MNPs addition [16]. Live cell imaging provided several examples of MNPs agglomerates located in cytoplasm, which move towards cell periphery under the effect of magnetic field and eventually pass through a plasma membrane and leave the cell. This process is apparently exerted through exocytosis pathway. Once outside the cell, the agglomerates are rapidly translocated towards the magnet (Fig. 4).

Long-term observations of interactions between HT1080 cells and MNPs have also showed many examples of MNPs «recycling». After losing most of internalized MNPs under the effect of magnetic field the cell can absorb new MNPs, collecting them again either from the substrate or from the surface of neighboring cells (Additional file 2: Movie 2). This «recycling» apparently reflects intermittent or constant endocytotic and exocytotic activities, which involve MNPs available until all of them are removed from culture media by magnetic field.

Ultrastructure and subcellular distribution of incorporated MNPs

To better understand the behavior of MNPs inside the cell and estimate the effect of cytoplasmic environment on the structure of MNPs themselves, the cells loaded with MNPs were subjected to TEM analysis. We applied correlative light-electron microscopy approach, selecting those cells which contained large MNPs agglomerates oriented in the direction of magnet. It was found that after internalization, MNPs can be associated with endosomes (Fig. 5d), but more often they break free from endosomes and lay intact in the cytoplasm as single MNPs or groups of various sizes (Fig. 5h, i, arrows). These observations confirmed the immunofluorescent data on incomplete colocalization between MNPs and endosomal marker Rab5 (Fig. 2). MNPs preserve their typical structure of electron-dense metal core surrounded by less dense carbon shell, showing no indications of defects in their shells (Fig. 5f, j). Thus, cytoplasmic environment seems to be permissive to Fe_7C_3@C MNPs structure and shell integrity, which explains their extremely low toxicity.

In situ measurements of magnetic properties of MNPs by CMFM-TEM

In order to be able to move in a magnetic field, the superparamagnetic Fe_7C_3@C MNPs should acquire magnetic polarization. To estimate a degree of this process for MNPs located inside the cells, we performed correlative MFM-TEM microscopy. The magnetic phase contrast image shows the gradient of magnetic fields and allows us to conclude that Fe_7C_3@C MNPs agglomerates clearly exhibit magnetic properties (Fig. 6a–c). Superimposed images do not show any significant correlation between the morphology and the magnetic response (Fig. 6c) indicating that each agglomerate contributes individually to the resulting magnetic moment. The force gradient was found to be $k\Delta\phi/Q \approx 2.01$ N/m and the resulting magnetic moment was estimated to be about $8.4\cdot10^{-14}$ emu (evaluation based on [19, 20]). The preliminary application of an external magnetic field thus led to the orientation and alignment of Fe_7C_3@C MNPs agglomerates.

Are the components of the cell cytoskeleton—microtubules and microfilaments—involved in the response of MNPs to magnetic field influence?

A closer look at MNPs in magnetic field-oriented agglomerates showed their close apposition to microtubules (Fig. 5h, i). We can speculate that, provided our earlier observations of saltatory movement of MNPs [16] and relatively minor effects of magnetic field on their intracellular movement compared to free MNPs in solution (this study), both membrane-encircled and non-endosomal MNPs can interact with microtubules through some cross-linkers or motor proteins. Our speculations led us to a hypothesis that restricted motility of MNPs in the cytoplasm, caused by their binding to cytoskeletal structures rather than sole viscosity of the cytoplasm, can be facilitated upon selective disruption of microtubules or actin filaments. To test this hypothesis, we performed live cell imaging of MNPs-loaded cell placed in magnetic field in the presence of microtubule-depolymerizing drug nocodazole or cytochalasin D, which disrupts actin cytoskeleton. Quite unexpectedly, we did not observe a dramatic increase in MNPs distribution under the effect of magnetic field after disassembly of either type of the cytoskeleton. This suggests that both microtubules and actin filaments are involved in specific or non-specific interaction with MNPs in the cytoplasm. However, simultaneous treatment with both nocodazole and cytochalasin D caused a cells round-up and loss of contact with the substrate. Under these conditions, application of the magnetic field led to rapid and massive displacement of entire cells (either individual cells or cellular agglomerates) towards the magnet (Additional file 6: Movie 6). Speed of this replacement was critically depend on degree of detachments of the cells from support and vary from 2 to 13 μm/min.

Similar effect has been observed after treatment with Ca^{2+} chelator EDTA which stimulates detachment of cells from the substrate (Fig. 7). It must be noted that cells lacking internalized MNPs rounded up the same way as other cells but did not move towards the magnet (Fig. 7c).

Discussion

The main idea of using MNPs as a tool for intracellular manipulations requires the MNPs to comply certain requirements, including cell permeability, low toxicity and magnetocontrollability. This latter property, apart from achieving principle goals of positioning and/or moving MNPs inside the cell, is particularly useful for removing MNPs at the end of their action, thus further improving their biocompatibility on the organismal level.

Since unmodified carbide MNPs are efficiently internalized by non-transformed cells and are non-toxic for them [16], we anticipated that the same properties are characteristic for their interaction with transformed cells as well. Indeed, the kinetics of MNPs internalization by human fibrosarcoma cells in vitro and cell viability tests gave results practically identical to previously described experiments on PK cells.

As has been reported in our previous work, the main mechanism of MNPs internalization is endocytosis. However, detailed ultrustructural analysis of intracellular localization of MNP agglomerates demonstrated that the majority of MNPs are either only partially encircled by the membrane or lay free in cytoplasm. These results were confirmed by the immunostaining for late endosome marker Rab5. The absence of colocalization apparently suggests that the vesicles containing MNPs are disrupted shortly after internalization. Observations

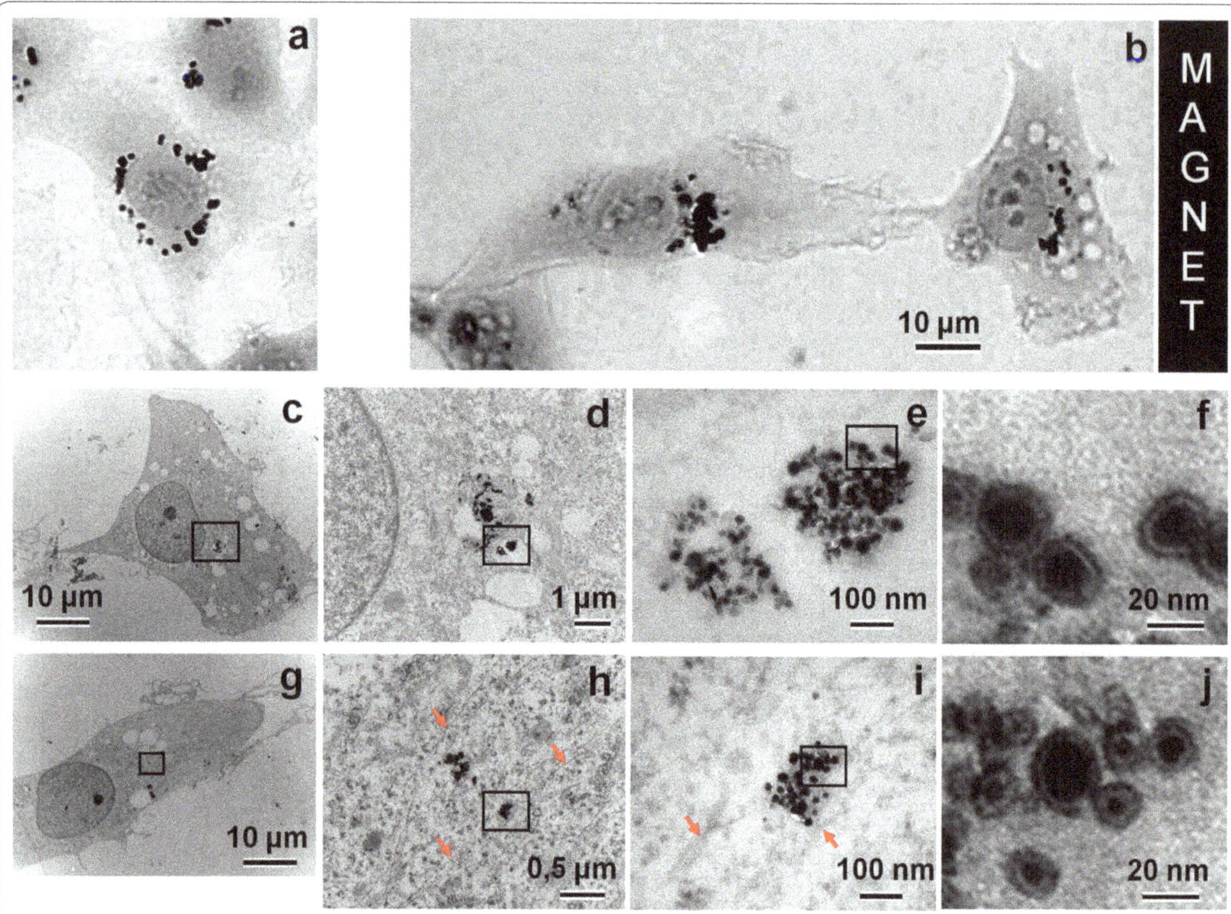

Fig. 5 The effect of prolonged exposure (24 h) of magnet on MNPs agglomerates in the cells. **a** Phase contrast images of cells cultivated without magnet, **b** phase contrast images of cells cultivated with magnet placed at *right side* (distance near 2 mm), **c–j** correlative electron microscopy pictures of two cells shown on "**b**" at different magnifications. MNPs agglomerates preferably lie freely in the cytoplasm of the cells, a part of agglomerates is in contact with microtubules (*red arrows*). Photos "**f**, **j**" show a primary ultrastructure of MNPs

of endosome escape with similar kinetics have been reported for nanodiamonds [21]. Since the nanodiamonds and Fe_7C_3@C MNPs possess identical carbon surface structure, this behavior is not surprising. However, in contrast to the hypothesis explaining fast endosome escape of nanodiamonds by the mechanical damage of membranes due to prickly shape of nanodiamonds [21], we believe that fast MNPs release into the cytoplasm depends on the chemical structure of the particles but not their shape. The mechanism of carbon nanoparticles release ([21] and this study) apparently differs from phagocytosis pathway typical for internalization of biological substances (bacteria, cell debris, etc.) [12, 22]. Early MNPs release from immature endosomes does not provoke spillage of lysosome enzymes in the cytoplasm, which explains the absence of detectable cytotoxicity of Fe_7C_3@C MNPs.

Measurements of Fe_7C_3@C MNPs magnetic properties demonstrated their superparamagnetic capacity [16]. On the other hand, low intrinsic toxicity of carbon surface of Fe_7C_3@C MNPs does not require additional protective coating being necessary for SPION nanoparticles [23]. Therefore, the magnetic properties of Fe_7C_3@C MNPs are far superior compared to biocompatible SPIONs, thus opening a new window for intracellular magnetocontrollability. Here we tested the effect of constant magnetic field on distribution and movement of the MNPs in living cells. Relatively low intracellular mobility of the MNPs compared to their in vitro behavior in magnetic field can obviously be explained by high viscosity of cytoplasm; however, cytoplasmic environment cannot be approximated as merely homogenous concentrated solution of biopolymers. The behavior of MNPs depends to a great extent on their subcellular localization, overall cytoskeleton organization and a mode of MNPs interaction with main cytoskeletal systems (microtubules, actin and intermediate filaments). Importantly, this slow movement of cytoplasmic MNPs compared to MNP agglomerates in

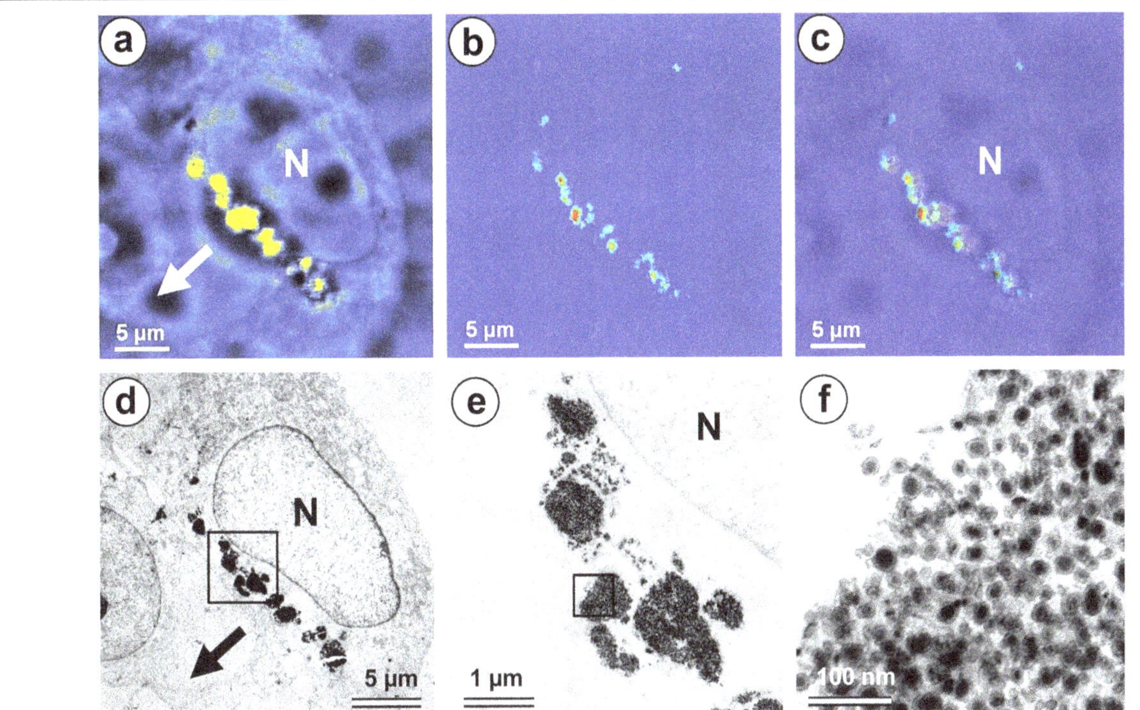

Fig. 6 CMFM-TEM microscopy images of the cell with Fe_7C_3@C MNPs taken after the 24 h exposure to magnetic field. **a** Morphology image of the cell in AFM; **b** phase shift map; **c** overlay image of superimposed topographic and phase shift map; **d** the same field of view in TEM; **e** group of Fe_7C_3@C MNPs agglomerates; **f** fine ultrastructure of agglomerate region with highest magnetic moment. *N* the cell nucleus. An *arrow* indicates the direction towards the magnet. *Scale bar* **a–d** 5 µm, **e** 1 µm, **f** 100 nm

solution, and variations in the mode of MNPs motility inside the cell (Brownian motion and salutatory movements observed in the same cell) suggest involvement of cytoskeleton. Indeed, an association of cytoplasmic MNP agglomerates has been observed in our study at the ultrastructural level at least with microtubule cytoskeleton system. These observations suggest the involvement of microtubule-associated protein motors (dyneins and kinesins) in directional movement of Fe_7C_3@C MNPs. However, whether these movements involve vesicle associated MNPs or direct interaction of MNPs with microtubules requires further investigation. We also cannot exclude that other cytoskeletal systems, including intermediate filaments which are more rigid and hard to experimentally disassemble in living cells, may affect passive or active microrheology of nanoparticles in living cells.

Nevertheless, even moderate magnetic fields (less than 0.15T) applied to the internalized agglomerates of Fe_7C_3@C MNPs, induce their magnetic polarization, as seen by correlative MFM-TEM, and are capable of alignment of the MNPs along magnetic field lines and concentration at magnet-proximal side of the cell. Similar alignment of natural "magnetosomes", made by magnetotactic bacteria has been described earlier [24]. Although the preliminary application of an external magnetic field leads to localization of nanoparticles in the cell and does not participate in magnetic properties of individual superparamagnetic nanoparticles imaged by MFM, formation of large agglomerates of Fe_7C_3@C MNPs causes residual magnetization. This effect depends on the size of the agglomerate, which explains more visible phase shift for larger agglomerates when analyzed by CMFM-TEM (see Fig. 6).

The specific order of MNPs distribution imposed by external magnetic field can be often counteracted by active cellular motility. This balance of forces was critically dependent of the amount of internalized MNPs. These effects hold true primarily for the cells attached to a substrate, and the situation dramatically reverses when cells loose adhesion, due to either cytoskeleton depolymerization or modification of Ca^{2+}-dependent adhesive properties of cellular membrane, as shown in our experiments. In these conditions, the magnetic force was sufficient to fast and immediate translocation of entire cell containing MNP agglomerates towards the magnet.

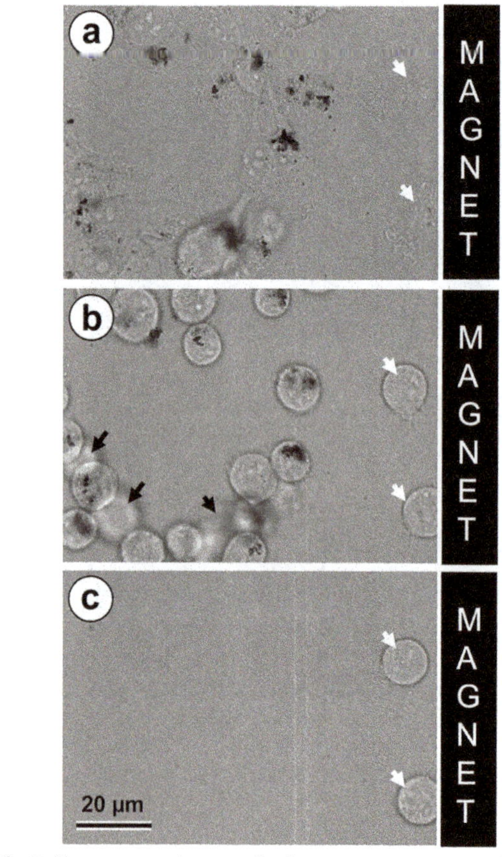

Fig. 7 Phase contrast images of cells movement in magnetic field after EDTA treatment. MNP-containing cells detached and moved to the magnet. Cells without MNP (marked with *white arrows*) detached but kept their positions. Cells shown in the fig. **b** (marked with *black arrows*) were moving to the magnet from left part of the sample. **a** Cells before EDTA treatment; **b** 30 min of EDTA treatment; **c** 100 min of EDTA treatment. *Scale bar* 20 µm

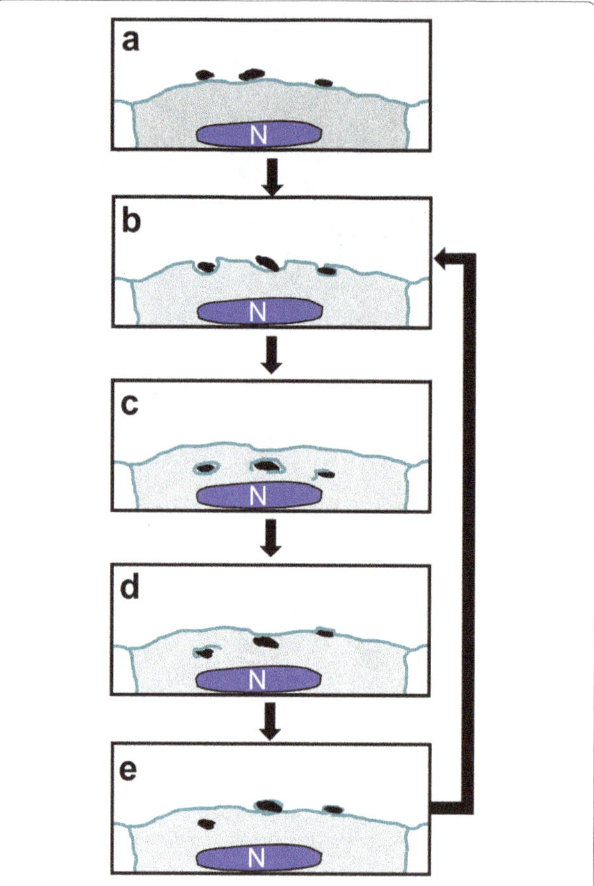

Fig. 8 Scheme of endocytosis-exocytosis cycle. **a** MNPs interaction with cell membrane; **b** cell membrane invagination; **c** internalization on MNPs agglomerates in membrane vesicles; **d** lost of membrane and beginning of exocytosis; **e** exit of MNPs agglomerates to the cell surface. After last phase MNPs agglomerates can be re-absorbed by this or a neighboring cell and the cycle repeats (*arrow*)

Similar effect was observed for MNP agglomerates positioned on the cell surface immediately upon introduction of magnetic field (Fig. 4).

Long-term live imaging of cells with MNPs allowed us to discover that a part of internalized MNPs with time reappears at the cell surface where they can be re-captured by the same or neighboring cell (Additional file 2: Movie 2). This MNPs "recycling" becomes more apparent when the external magnetic field was applied. In this case, MNPs agglomerates appearing at the cell surface are readily dragged towards the magnet (Fig. 4), similar to the whole MNP-loaded cells that lost contact with the substrate. Ultimately, after one-week incubation in magnetic field, the majority of cells had lost cytoplasmic MNPs (data not shown). The scheme of MNPs turn-over through endocytosis-exocytosis cycle is presented on Fig. 8.

Conclusions

We demonstrated superparamagnetic properties and magnetocontrollability of Fe_7C_3@C MNPs in living cells. In combination with low cytotoxicity and high potential for chemical modification of carbon shell shown in our previous studies, these properties make the MNPs a promising candidate as a platform for targeted drug delivery. Enhanced magnetic properties of Fe_7C_3@C MNPs make it possible to control and concentrate MNPs efficiently at least at the centimeter scale, thus opening the opportunities to manipulate MNPs not only at cellular but also organismal level. Additional advantages of Fe_7C_3@C MNPs seem to be their early endosome escape into the cytoplasm that does not require additional efforts to facilitate drug release into cytoplasm and assures protection of carried substances from lysosomal degradation. The MNPs "recycling" would potentially allow magnetic field-assisted tissue clearance after drug release that would decrease side effects of therapeutic application of MNPs.

Additional files

Additional file 1: Movie 1. Live-cell imaging of HT1080 cells collecting MNPs agglomerates from glass surface. No magnetic field applied.

Additional file 2: Movie 2. Live-cell imaging of HT1080 cells collecting MNPs agglomerates from the surface of neighboring cell. No magnetic field applied.

Additional file 3: Movie 3. In moving HT1080 cell MNPs agglomerates keep their position in the magnet-proximal side of the cell independently of the direction of cell motility. Magnet is positioned below.

Additional file 4: Movie 4. In moving HT1080 cell MNPs agglomerates keep their orientation along the magnetic field lines independently of the direction of cell motility. Magnet is positioned below.

Additional file 5: Movie 5. The magnetic field favors the directionality of movement for cells heavily loaded with MNPs. Magnet is positioned below.

Additional file 6: Movie 6. HT1080 cells dragged by the magnetic field force towards the magnet as a consequence of disassembly of microtubule and actin cytoskeletons. Magnet edge visible in the upper part of the frame.

Authors' contributions

Study design: VA, VD, RU Data collection: IA, IK, AG, NA, OS, OZ, VC, VD, RU. Data analysis and interpretation: IA, IK, AG, NA, OS, OZ, VC, AR, AM, VK, VA, RU. Drafting manuscript: IA, IK, AR, AM, VK, VA, RU. All the authors critically reviewed the manuscript. All authors read and approved the final manuscript.

Author details

[1] A.N. Belozersky Institute of Physico-Chemical Biology, Moscow State University, Moscow, Russia 119992. [2] Biology Faculty, Moscow State University, Moscow, Russia 119992. [3] GREMAN, UMR CNRS 7347, Université François Rabelais, 37200 Tours, France. [4] Chemistry Faculty, Moscow State University, Moscow, Russia 119992. [5] MISiS, Leninskiy prospekt 2, Moscow, Russia 119049. [6] Institute of High Pressure Physics RAS, Troitsk, Moscow region, Russia 142190. [7] Center for Technology Innovation, Baker Hughes Inc., Houston, TX 77040, USA. [8] Laboratoire Biologie Cellulaire et Microscopie Electronique, Faculté de Médecine, Université François Rabelais, 37032 Tours, France. [9] Faculty of Bioengineering and Bioinformatics, Moscow State University, Moscow, Russia 119992.

Acknowledgements

We thank Prof. Elena Kornilova for providing the Rab5 (C8B1) rabbit monoclonal antibodies. Our data were obtained with the assistance of the IBiSA Electron Microscopy Facility of François Rabelais University and CHRU de Tours and Nikon Center of excellence at A.N. Belozersky Institute of Physico-Chemical Biology.

Competing interests

The authors declare no competing financial interests. All the authors declare no competing interests.

Funding

I. Kireev and I. Alieva were supported by University of Tours as invited researches. Russian Foundation for Basic Research (15-03-04490 to V.D., 15-04-08550 to I.A., 15-54-77078 to I.K.), Lomonosov Moscow State University Development Program (PNR 5.13 to I.K.). One of the authors (V.A.) gratefully acknowledge the financial support of the Ministry of Education and Science of the Russian Federation in the framework of Increase Competitiveness Program of NUST «MISiS» (№ K3-2016-010) implemented by a governmental decree dated 16th of March 2013, N 211.

References

1. Robert D, Aubertin K, Bacri J-C, Wilhelm C. Magnetic nanomanipulations inside living cells compared with passive tracking of nanoprobes to get consensus for intracellular mechanics. Phys Rev E. 2012;85:011905–9.
2. Scherer F, Anton M, Schillinger U, Henke J, Bergemann C, Kruger A, Gansbacher B, Plank C. Magnetofection: enhancing and targeting gene delivery by magnetic force in vitro and in vivo. Gene Ther. 2002;9:102–9.
3. Lau AW, Hoffman BD, Davies A, Crocker JC, Lubensky TC. Microrheology, stress fluctuations, and active behavior of living cells. Phys Rev Lett. 2003;91:198101–4.
4. Ito A, Takizawa Y, Honda H, Hata KI, Kagami H, Ueda M, Kobayashi T. Tissue engineering using magnetite nanoparticles and magnetic force: heterotypic layers of cocultured hepatocytes and endothelial cells. Tissue Eng. 2004;10:833–40.
5. Plank C, Schillinger U, Scherer F, Rosenecker J. The magnetofection method: using magnetic force to enhance gene delivery. Biol Chem. 2003;384:737–47.
6. Kolosnjaj-Tabi J, Wilhelm C, Clément O, Gazeau F. Cell labeling with magnetic nanoparticles: opportunity for magnetic cell imaging and cell manipulation. J Nanobiotech. 2013;11:S7.
7. Hofmann-Amtenbrink M, von Rechenberg B, Hofmann H. Superparamagnetic nanoparticles for biomedical applications. In: Tan MC, editor. Nanostructured materials for biomedical applications. Research Signpost; 2009. p. 119–43.
8. Nacev A, Beni C, Bruno O, Shapiro B. The behaviors of ferromagnetic nano-particles in and around blood vessels under applied magnetic fields. J Magn Magn Mater. 2011;323:651–68.
9. Vercauteren D, Rejman J, Martens TF, Demeester J, De Smedt SC, Braeckmans K. On the cellular processing of non-viral nanomedicines for nucleic acid delivery: mechanisms and methods. J Control Release. 2012;161:566–81.
10. Zaki NM, Tirelli N. Gateways for the intracellular access of nanocarriers: a review of receptor-mediated endocytosis mechanisms and of strategies in receptor targeting. Expert Opin Drug Deliv. 2010;7:895–913.
11. Rivière C, Wilhelm C, Cousin F, Dupuis V, Gazeau F, Perzynski R. Internal structure of magnetic endosomes. Eur Phys J E. 2007;22:1–10.
12. Martens TF, Remaut K, Demeester J, De Smedt SC, Braeckmans K. Intracellular delivery of nanomaterials: how to catch endosomal escape in the act. NanoToday. 2014;9:344–64.
13. Varkouhi AK, Scholte M, Storm G, Haisma HJ. Endosomal escape pathways for delivery of biologicals. J Controlled Release. 2011;151:220–8.
14. Malik R, Gross SP. Molecular motors: strategies to get along. Curr Biol. 2004;4:R971–82.
15. Davydov V, Rakhmanina A, Allouchi H, Autret C, Limelette P, Agafonov V. Carbon-encapsulated iron carbide nanoparticles in the thermal conversions of ferrocene at high pressures. Fuller Nanotub Carbon Nanostruct. 2012;20:451–4.
16. Davydov V, Rakhmanina A, Kireev I, Alieva I, Zhironkina O, Strelkova O, Dianova V, Samani TD, Mireles K, Yahia LH, Uzbekov R, Agafonov V, Khabashesku V. Solid state synthesis of carbon-encapsulated iron carbide nanoparticles and their interaction with living cells. J Mater Chem B. 2014;2:4250–61.
17. Kaidatzis A, Garcia-Martin JM. Torsional resonance mode magnetic force microscopy: enabling higher lateral resolution magnetic imaging without topography-related effects. Nanotechnology. 2013;24:165704.
18. Alieva I, Kireev I, Rakhmanina A, Garanina A, Strelkova O, Zhironkina O, Cherepaninets V, Davydov V, Khabashesku V, Agafonov V, Uzbekov R. Magnet-induced behavior of iron carbide (Fe7C3@C) nanoparticles in the cytoplasm of living cells. Nanosyst Phys Chem Math. 2016;7:158–60.
19. Manalis S, Babcock K, Massie J, Elings V, Dugas M. Submicron studies of recording media using thin-film magnetic scanning probes. Appl Phys Lett. 1995;66:2585–7.
20. Grutter P, Mamin HJ, Rugar D. Magnetic force microscopy (MFM). In: Wiesendanger R, Guntherodt HJ, editors. Scanning tunneling microscopy II. Berlin: Springer; 1992. p. 151–207.

21. Chu Z, Miu K, Lung P, Zhang S, Zhao S, Chang H-C, Lin G, Li Q. Rapid endosomal escape of prickly nanodiamonds: implications for gene delivery. Sci Rep. 2015;5:11661.
22. Kinchen JM, Ravichandran KS. Phagosome maturation: going through the acid test. Nat Rev Mol Cell Biol. 2008;9:781–95.
23. Bychkova AV, Sorokina ON, Rosenfeld MA, Kovarskii AL. Multifunctional biocompatible coatings on magnetic nanoparticles. Russ Chem Rev. 2012;81:1026–50.
24. Schüler D. Genetics and cell biology of magnetosome formation in magnetotactic bacteria. FEMS Microbiol Rev. 2008;32:654–72.
25. Shkilnyy A, et al. Synthesis and evaluation of novel biocompatible superparamagnetic iron oxide nanoparticles as magnetic anticancer drug carrier and fluorescence active label. J Phys Chem C. 2010;114:5850–8.
26. Wu W, Xiao XH, Zhang SF, Peng TC, Zhou J, Ren F, Jiang CZ. Synthesis and magnetic properties of maghemite (γ-Fe_2O_3) short-nanotubes. Nanoscale Res Lett. 2010;5(9):1474–9.
27. Serna CJ, Morales MP. Maghemite (γ-Fe_2O_3): a versatile magnetic colloidal material. In: Matijevic E, Borkovec M, editors. Surface and colloid science, chap 2, vol. 17. New York: Kluwer Academic/PlenumPublishers; 2011. p. 27–81.
28. Drbohlavova J, et al. Preparation and properties of various magnetic nanoparticles. Sensors. 2009;9:2352–62.

5

Intracellular trafficking and cellular uptake mechanism of PHBV nanoparticles for targeted delivery in epithelial cell lines

Juan P. Peñaloza[1,2], Valeria Márquez-Miranda[3], Mauricio Cabaña-Brunod[1,2], Rodrigo Reyes-Ramírez[1,2], Felipe M. Llancalahuen[1,2], Cristian Vilos[1,3], Fernanda Maldonado-Biermann[4], Luis A. Velásquez[1], Juan A. Fuentes[5], Fernando D. González-Nilo[3], Maité Rodríguez-Díaz[6] and Carolina Otero[1*]

Abstract

Background: Nanotechnology is a science that involves imaging, measurement, modeling and a manipulation of matter at the nanometric scale. One application of this technology is drug delivery systems based on nanoparticles obtained from natural or synthetic sources. An example of these systems is synthetized from poly(3-hydroxybutyrate-co-3-hydroxyvalerate), which is a biodegradable, biocompatible and a low production cost polymer. The aim of this work was to investigate the uptake mechanism of PHBV nanoparticles in two different epithelial cell lines (HeLa and SKOV-3).

Results: As a first step, we characterized size, shape and surface charge of nanoparticles using dynamic light scattering and transmission electron microscopy. Intracellular incorporation was evaluated through flow cytometry and fluorescence microscopy using intracellular markers. We concluded that cellular uptake mechanism is carried out in a time, concentration and energy dependent way. Our results showed that nanoparticle uptake displays a cell-specific pattern, since we have observed different colocalization in two different cell lines. In HeLa (Cervical cancer cells) this process may occur via classical endocytosis pathway and some internalization via caveolin-dependent was also observed, whereas in SKOV-3 (Ovarian cancer cells) these patterns were not observed. Rearrangement of actin filaments showed differential nanoparticle internalization patterns for HeLa and SKOV-3. Additionally, final fate of nanoparticles was also determined, showing that in both cell lines, nanoparticles ended up in lysosomes but at different times, where they are finally degraded, thereby releasing their contents.

Conclusions: Our results, provide novel insight about PHBV nanoparticles internalization suggesting that for develop a proper drug delivery system is critical understand the uptake mechanism.

Background

Nanotechnology is the science of engineering materials and systems on a molecular scale. Its application to medicine, "nanomedicine", has enabled the development of nano-sized drug-delivery vehicles. These nanocarriers are generally <200 nm in size and have the ability to carry and deliver therapeutics to discrete sites into the cells [1].

*Correspondence: maria.otero@unab.cl
[1] Center for Integrative Medicine and Innovative Science, Facultad de Medicina, Universidad Andrés Bello, Santiago, Chile
Full list of author information is available at the end of the article

Due to their small size, increased stability and sustained drug release properties, biodegradable polymeric nanocarriers display several advantages, being more effective for cancer treatment than other nanoparticles (e.g. metallic ones) [2]. Nanoparticles are being developed for in vivo tumor imaging, targeted drug delivery and biomolecular profiling of cancer biomarkers. Biodegradable polymeric nanoparticles (NPs) have been shown promissory as controlled drug delivery systems, showing high therapeutic potential [3]. Currently, delivery technologies using cell-targeting [4, 5] or specific targeting of organelles inside a cell [6] are becoming increasingly important as an area of scientific investigation. In cancer, polymeric NPs can be used to deliver chemotherapeutics towards tumor cells, with high efficiency and reduced cytotoxicity on healthy tissues [6–8]. The activity of the drug carried by the NPs depends on how and where the NPs are disassembled. The mechanism of NPs endocytosis and trafficking is still a subject of controversy [9–11].

Among polymeric NPs, (3-hydroxybutyric acid-co-3 hydroxyvaleric acid) (PHBV) constitutes a promissory alternative, more accessible and cheaper than similar polymers such as poly (lactic-co-glycolic acid) (PLGA); the last one has been widely used in drug delivery in different cancer types and have an FDA approval [12, 13]. PHBV is a biodegradable and biocompatible polyester, composed of two monomers: hydroxybutyrate and hydroxyvalerate and, since it is produced by bacteria or eukaryotic cells, it can be synthesized in large scales [14]. However, at present, little is known about its uptake and distribution into mammalian cells [14].

In the case of hydroxybutyrate-composed homopolymers (PHB), it has been described that is primarily found complexed with other molecules, and can be found in several subcellular organelles of eukaryotic organisms [15] or associated with proteins [16]. A previous study has suggested that PHB is associated in granules that accumulate in the cytoplasm, similar to the behavior in bacteria, which might constitute an 'energy reservoir'. In the same study, the colocalization of PHV granules with organelles was inspected in U87 cells, showing no remarkable colocalization with lysosomes, mitochondria, or endoplasmic reticulum [17].

PHBV copolymer reduces the chain packing and decreases crystalline melting point, due to the ethyl chain of the valerate, which increases flexibility, impact strength, and ductility in comparison to hydroxybutyrate homopolymer [18]. Furthermore, previous studies have revealed the usefulness of PHBV nanoparticles in the encapsulation of paclitaxel drugs, protecting this anticancer agent against premature degradation, by allowing 48 h of toxicity protection. Other applications of PHBV in medicine include a formulation loaded with antibiotics for veterinary uses, which allows protecting the drug for inactivation [19].

Previous work demonstrates that most NPs, including those formed from biodegradable polymers such as PLGA, are taken up by an endocytic process, and their uptake is concentration- and time-dependent [12, 20, 21]. PLGA NPs for delivery of therapeutics are of particular interest due to their biocompatibility, biodegradability and ability to maintain therapeutic drug levels for sustained periods of time [22]. However, further investigation is required to determine the mechanisms and therapeutic potential of intracellular targeting of nanodelivery systems in vivo for the goal of an anticancer vaccine [14].

Other studies on PLGA uses for targeting have shown that the intracellular fate of nanoparticles may be altered by their surface decoration with a targeting molecule. In the case of PLGA, previous studies in smooth muscle cells have revealed that these NPs can undergo exocytosis in about 65%; meaning that the NPs are effectively internalized, but they escape from the cells before releasing their content [23]. The fraction of NPs that escapes from the endosomes to the cytosolic compartments was shown to be retained inside cells, delivering the encapsulated agent slowly [24]. The polymer matrix prevents the degradation of the drug and the duration and levels of drug released from the NPs can be easily modulated by altering the formulation. Meanwhile, in hepatoma cells, it has been described that PLGA NPs are taken up by the classical clathrin pathway and that they effectively escape from lysosomes and contribute to enhancing the efficiency of intracellular delivery and tumor therapy [25].

In summary, discovering the intracellular fate of NPs and the investigation of their endocytic mechanism is particularly significant with the aim of using them as drug carriers, since in some cases, they may be taken back to the extracellular media (via recycling endosomes), or they may be degraded in lysosomes or be trapped in an organelle, without releasing its content at the desired site [12, 26, 27]. Furthermore, escaping from the endosomes to reach other organelles is another important issue to be addressed.

The present manuscript aims to provide insights about the possible routes of internalization and escaping from the endosomal sorting pathway of PHBV nanoparticles, in two different cell lines: in HeLa (cervix cancer) and SKOV-3 (ovarian cancer). HeLa cells was chosen because its endocytic machinery is very well characterized, and SKOV-3 cells because these cells are challenging due to their common chemoresistance [28] in order to investigate how a possible drug treatment (encapsulated in a nanoparticle) may stay inside the cell, where normally drugs are pumped out.

Several compartments, such as early endosomes (Transferrin Conjugate or EEA1 antibody), late endosomes (Lysotracker or Rab7 antibody) and lysosomes (Lysotracker or LAMP1 antibody), were chosen for this investigation, as they are involved in transport, destination, release, and degradation of soluble and membrane-bound macromolecules [29]. Endosome and lysosome formation require microtubules to form a network in the cytoplasm, whereas actin generates forces to induce membrane invaginations [30, 31], thereby microtubules and actin cytoskeleton were also analyzed.

These methods may contribute to the development of PHBV NPs as drug carriers, enabling to enhance the delivery of chemotherapeutics inside the cells.

Methods

Materials

Poly (3-hydroxybutyric acid-co-hydroxyvaleric acid) (PHBV) with 12 wt% PHV and polyvinyl alcohol (PVA) (average mol wt. 30,000–70,000) and glutathione were obtained from Sigma-Aldrich (St. Louis, MO, USA). Nile Red 552/636 was purchased from Invitrogen (Carlsbad, CA, USA). Dichloromethane (DCM), dimethyl sulfoxide (DMSO) and methanol were purchased from Merck (Darmstadt, Germany). The primary antibody Anti-EEA1 (Early Endosome Antigen 1) was purchased in Santa Cruz Biotechnology (Santa Cruz, CA, USA), the anti-Caveolin-1 and anti-LAMP1 (Lysosomal-associated membrane protein 1) were obtained from Abcam (Cambridge, USA). The secondary antibodies Alexa Fluor 488 Donkey Anti-Mouse IgG (H + L), and Alexa Fluor 488 Goat Anti-Rabbit IgG (H + L) were purchased in Molecular Probes by Life Technologies (Carlsbad, CA, USA). Additionally, membrane glycoproteins marker Wheat Germ Agglutinin, Alexa Fluor 555 Conjugate (WGA), actin (F-actin) Alexa Fluor 488 phalloidin, LysoTracker® Red DND-99 and Hoechst 33342 were obtained from Molecular Probes by Life Technologies (Carlsbad, CA, USA). Fluoromount G was purchased in Electron Microscopy Sciences (Hatfield, PA, USA), and EZ-Link® NHS-SS-Biotin was obtained from Pierce Biotechnology (Rockford, IL, USA).

Preparation of PHBV nanoparticles

PHBV nanoparticles (NPs) were formulated via a modification of the double emulsion (w1/o1/w2) solvent-evaporation method [19]. Briefly, 400 µL of distilled water were added to 1 mL of a solution of 3 mg/mL of PHBV (Sigma-Aldrich, Co., St. Louis, MO, USA) in dichloromethane (Merck KGaA, Darmstadt, Germany). The first emulsion (w1/o1) was prepared by sonication in an ultrasonic processor equipped with a microtip probe for 40 s at 100% in an ice bath. The water-in-oil emulsion was further emulsified by sonication under the same conditions in 4 mL of an aqueous solution of 5 mg/mL PVA (w2). This w1/o1/w2 emulsion was immediately poured into a beaker containing 20 mL of a 0.5 mg/mL PVA solution. The mixture was stirred with an overhead propeller for 12 h under a flow hood, and the solvent was allowed to evaporate. The remaining organic solvent and free molecules were removed by passing the particle solution 3 times through an Amicon Ultra-4 centrifugal filter. Finally, NPs were concentrated by centrifugation and resuspended in 500 µL of phosphate buffered saline (PBS, pH 7.4) for further use. A modification of the technique described was used for the synthesis of functionalized nanoparticles with a marker. The same procedure was followed, except the aqueous phase (W1) was replaced by 100 µL of a solution of fluorescein isothiocyanate (FITC) 1 mg/mL; 100 µL of a solution of Nile Red (RN) 1 mg/mL; or 5 µL of the LysoTracker® Red DND-99 fluorescent probe.

PHBV nanoparticles characterization by dynamic light scattering (DLS)

Each preparation was suspended in 1 mL of PBS pH 7.4 and nanoparticles diameter (nm), polydispersity coefficient (PdI) and zeta potential (mV) were determined by light scattering technique using a Zetasizer Nano-ZS (Malvern Instruments Ltd., UK).

Transmission electron microscopy (TEM)

NPs structure was also characterized using transmission electron microscopy. One drop of the NP sample was placed onto an ultra-thin Lacey carbon-coated 400-mesh copper grid and allowed to dry at room temperature for 10 min prior to image acquisition, ensuring no more than 1 min of electron beam exposure to the sample. TEM images were acquired using an LVEM5 electron microscope (Delong Instrument, Montreal, Quebec, Canada) at a nominal operating voltage of 5 kV. The small volume of the vacuum chamber in the LVEM5 microscope facilitates rapid sample visualization within 3 min before observation. The low voltage used delivers high contrast in soft materials (up to 20-fold) compared with high-voltage electron microscopes, which use accelerating voltages of approximately 100 kV; this procedure facilitates the emission of staining procedures and allows the direct visualization of biological samples. Digital images were captured using a Retiga 4000R camera (QImaging, Inc., USA) at its maximal resolution.

Cell culture

The Human cervical adenocarcinoma HeLa and the human ovarian adenocarcinoma SKOV-3 were obtained from the American Type Culture Collection (ATCC® CCL-2™, and ATCC® HTB-77™ respectively). The cells were maintained in Dulbecco's High Glucose Modified Eagle's Medium (DMEM) supplemented with 10% v/v fetal bovine serum, 1 mM sodium pyruvate and 1% v/v penicillin–streptomycin (Hyclone™ Laboratories, Inc., South Logan, UT, USA), and incubated at 37 °C and 5% CO_2. Peripheral blood mononuclear cell (PBMC) were obtained using a Ficoll® density gradient, and cultured in Roswell Park Memorial Institute Medium (RPMI) 1640, which was supplemented with 10% v/v fetal bovine serum and 1% penicillin–streptomycin (Hyclone™ Laboratories, Inc., South Logan, UT, USA) and incubated at 37 °C and 5% CO_2.

Flow cytometry

HeLa and SKOV-3 cells (2×10^5 cells/well) were seeded separately in 6-well plates and incubated for 48 h.

Subsequently, to evaluate time-dependent intracellular incorporation, a solution of PHBV-RN 100 μg/mL was added in culture medium for 5, 15, 30 min, 1 and 2 h. To assess the dependence of the concentration, different concentrations (1, 10, 100, 500 and 1000 μg/mL) of a PHBV-RN solution were added and incubated for 2 h. Additionally, to evaluate the dependence of the energy, cells were incubated at 4 °C or with a sodium azide (AS) solution 1 mg/mL (Sigma-Aldrich, Co., St. Louis, MO, USA) for 1 h respectively. Then, a PHBV-RN solution (100 μg/mL) was added and incubated for 2 additional hours. After the incubation time, cells were washed 3 times with cold PBS, detached from the culture plate with PBS-EDTA 0.2% (w/v), incubated for 15 min at 37 °C, then washed twice with buffer FACS (PBS-2% FBS) and resuspended in 500 μL of PBS. Finally, the fluorescence intensity of cells containing fluorescent nanoparticles was analyzed using a flow cytometer BD Accuri C6™ and v1.0 software (BD™, Franklin Lakes, NJ, USA).

Immunofluorescence

Cell lines were plated at a confluence of 70–80% in 24-well plates with coverslips. A PHBV-FITC or PHBV-RN nanoparticles solution was added and incubated for different times. Then, cells were washed 2 times with PBS, fixed with paraformaldehyde (PFA) 4% for 10 min, washed 3 times with PBS and then permeabilized with Triton X-100 0.1% by 10 min. Then, cells were washed 2 times with PBS and then cells were incubated with a primary antibody (1:100) for 1 h at RT. Then, cells were washed 5 times again with PBS and incubated with a secondary antibody (1:500) for 1 h at RT. Nucleic acid marker Hoechst 33342 (1:1000) was added for 15 min and finally cells were washed 5 times with PBS. Coverslips were mounted on a slide using Fluoromount G mounting medium for subsequent microscopic observation. Slides were observed using a BX61 fluorescence microscope (Olympus Corp., Tokyo, Japan), coupled to an ORCA-R2 (Hamamatsu Photonics KK, Japan) camera. Images were analyzed by Dimension cellSens v1.7.1 software (Olympus Corp., Tokyo, Japan). Colocalization percentages were based on the merge between the two channels red and green. For this, we use Image J v1.48 software using the plugin Colocalization Threshold which creates a threshold for pixels of the red and green channels. Pixels below this threshold are ignored for the quantification of colocalization.

Reducible biotin assay

To analyze the involvement of proteins involved in endocytosis of nanoparticles, we performed an experiment using reducible biotin. When biotinilated proteins are internalized, they become resistant to extracellular reducing solution, allowing distinction between endocytosed and not endocytosed proteins. Experiment were performed at 4 °C. Cells were grown to 70–80% confluence in 6-well plates and washed 3 times with cold PBS. Biotin (0.5 mg/mL) were incubated for 20 min two times with gentle shaking. Free biotin was blocked adding a solution of 50 mM NH_4Cl for 10 min, and then cells were washed 3 times with PBS. For the reduction, a solution containing glutathione 50 mM, 90 mM NaCl, 10% FBS, 1 mM $MgCl_2$, 0.1 mM $CaCl_2$ and 60 mM NaOH to a final pH of 7.2–7.4 was used. Cells were incubated with the solution two times for 30 min at 4 °C with gentle shaking. Then cells were incubated with lysis buffer (containing antiproteases) with gentle agitation for 30 min at 4 °C. The obtained extract was centrifuged for 30 min at 14,000 rpm at 4 °C and the pellet was discarded. Proteins were quantified using Qubit® 2.0 Fluorometer. 30 μg of protein were denatured in protein loading buffer and incubated at 95 °C for 5 min. Samples were loaded onto a polyacrylamide gel 12% acrylamide-bis.

Western blot

Electrophoresis was performed at 120 V constant current in running buffer in a vertical electrophoresis chamber model Mini PAGE System. After electrophoresis, proteins were transferred to a preactivated PVDF membrane in 100% methanol. The transfer was carried out in a transfer model system Trans-Blot® Turbo™. The PVDF membrane was blocked with gentle agitation for 2 h at room temperature in PBS-Tween 20 0.5, 10% glycerol, glucose 1 M BSA 3% skim milk and 1%. After blocking, the membrane was washed 4 times for 15 min with PBS-0.05% Tween 20 and subsequently incubated for 2 h at room temperature with streptavidin horseradish peroxidase conjugated (1:1000) in PBS with 0.5% Tween 20, 10% glycerol, 1 M glucose, 0.3% BSA. Finally, the membrane is washed 4 times for 15 min with PBS-0.05% Tween 20. For the development of the membrane substrate SuperSignal West Femto Chemiluminescent substrate (Thermo Fisher Scientific Inc., Rockford, IL, USA) was used according to manufacturer's instructions. Development was carried out in the PHOTO/Analyst® Luminary/FX® Systems equipment and using the FOTO/Analyst® PC Image v5.0 (Fotodyne Inc., Hartland, WI, USA) software.

MTT assay

Toxicity of the nanoparticles was determined using the 3-(4,5-dimethylthiazol-2-yl)-2,5-diphenyltetrazolium bromide (MTT) cell viability assay. MTT is a yellow compound that when reduced by functioning mitochondria, produces purple formazan crystals that can be measured spectrophotometrically. For this purpose, HeLa cells were incubated with different concentrations of PHBV solution for 24 h at 37 °C and 5% CO_2. After incubation time, MTT (Sigma-Aldrich) was dissolved in phosphate buffered saline (PBS) to a concentration of 5 mg/mL and further diluted in culture medium (1:11). Cells were incubated with this MTT-solution for 3 h under normal culture conditions. Afterwards 155 μL of the solution were rejected and 90 μL of DMSO were added. To completely dissolve the formazan salts plates were incubated for 10 min on a shaker and afterwards quantified by measuring the absorbance at 535 nm with a ELISA microplate reader. Cell viability was calculated as percentage of surviving cells compared to untreated control cells.

Results and discussion

Characterization of PHBV nanoparticles

Double emulsion solvent evaporation method was used to formulate four preparations of PHBV nanoparticles: PHBV-empty (control), FITC-loaded PHBV NPs (PHBV-FITC), Nile Red-loaded PHBV NPs (PHBV-RN), and Lysotracker-loaded PHBV NPs (PHBV-Lysotracker®). This w/o/w method was selected due of the broad experience of our group in the formulation of NPs, and the incorporation of dyes in the aqueous or organic phase was according to its solubility. The characterization of size and morphology of NPs was performed using dynamic light scattering (DLS) and by transmission electron microscopy (TEM). Table 1 describes the diameter (nm), polydispersity index (PdI), and the Zeta Potential (mV) obtained of NPs, and Additional file 1: Figure S1 exhibits a graphical representation of the data analyzed statistically. The size presented in all formulations of NPs was homogenous with a diameter of ~200 nm, stable in time (Additional file 2: Figure S2), and similar to the size of paclitaxel-loaded PHBV nanoparticles synthesized previously by our group [32].

Studies have revealed a direct relationship between the size of the NPs and the endocytic pathway. Particles with a size less than 200 nm were internalized into non-phagocytic, murine melanome cells B16-F10 via clathrin-mediated endocytosis. The size of clathrin-coated pits varies between different cell types within the same species [33]. The size of clathrin-coated vesicles depends on the size of its cargo, having an upper limit of 200 nm external diameter in the case of virus uptake [34].

Table 1 Nanoparticle physicochemical characterization: size, zeta potential and polydispersity index (PdI) (n = 5)

	Size (nm)	PdI	Zeta potential (mV)
PHBV	200.2 ± 3.66	0.133 ± 0.020	−11.7 ± 1.97
PHBV-FITC	201.7 ± 10.4	0.100 ± 0.023	−9.60 ± 1.25
PHBV-RN	206.0 ± 3.37	0.204 ± 0.043	−7.96 ± 0.41
PHBV-Lysotracker®	201.5 ± 7.19	0.206 ± 0.021	−5.16 ± 1.99

In a similar way, 50 and 120 nm folate-decorated poly(ethylene glycol)-polycaprolactone nanoparticles were found to be internalized via both clathrin- and caveolae-mediated endocytosis in ARPE-19 cells. However, 250 nm nanoparticles, were only internalized via caveolae-mediated pathway [35], despite some articles have described an upper limit size of ~150 nm for passage through caveolae vesicles [36].

In a more recent article, glycopeptide engineered PLGA nanoparticles, closely related to the PHBV nanoparticles studied here, were investigated to determine their endocytic mechanism. As demonstrated by confocal microscopy analysis, these nanoparticles, 170 nm in diameter, strongly colocalized with clathrin related-but not with caveolin related-routes [37].

Figure 1 shows a representative micrograph of control preparation, which was obtained through transmission electron microscopy (TEM). In this figure, the spherical morphology and uniform nanoparticle size is evidenced. It has been reported that nanoparticle morphology may influence its internalization into the cells. In previous reports, spherical-shaped gold NPs displayed higher and faster rate of endocytosis than disk or rod-shaped nanoparticles [38]. In contrast, other researchers have suggested a preferential internalization of polyethylene glycol nanoparticles having a rod or cylindrical shape [39]. All these records indicate that spherical morphology of PHBV nanoparticles may not be decisive in entering the cells, but the setting of other features—such as the chemical nature, size, and charge—are also implied in how they are taken up.

Surface charge of nanoparticles is another important parameter to characterize because the surface of the nanoparticle interacts directly with the biological environment. Due to the chemical nature conferred by the phospholipids, the plasma membrane has a negative charge. For this reason, positively-charged drug delivery systems exhibit better adhesion and a higher internalization rate. An example of this phenomenon has been demonstrated by Harush-Frenkel et al., who revealed in HeLa cells, that polymeric nanoparticles of PLA-PEG coated with cationic lipid stearylamine (~ +35 mV) evidence greater incorporation than the same NPs but negatively charged (~ −35 mV) [40]. However, in the case of possible use in vivo of these nanoplatforms, the interaction with plasma proteins is reduced if the NPs are negatively charged, decreasing the probabilities of activation of immune response by monocytes and macrophages in the bloodstream [41].

Uptake measurement of PHBV nanoparticles

To characterize how PHBV nanoparticles are taken up by the cells, we quantified their incorporation through flow

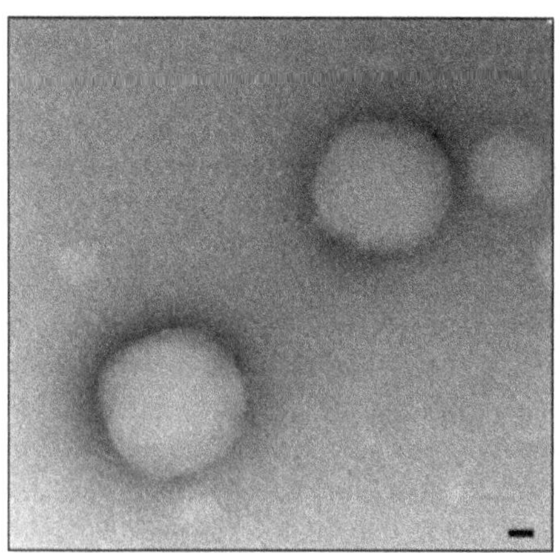

Fig. 1 PHBV nanoparticle micrograph. Representative image of PHBV nanoparticles (naked) by transmission electron microscopy (TEM). Objective ×49.000 magnification. *Bar size* 35 nm

cytometry. This measurement was performed under different conditions, including the use of different formulations of NPs, variation of the incubation time, type of cells, and modifying the energy conditions of the cell culture. The latter condition was evaluated because endocytosis is an active cellular process, which requires energy (ATP).

The time required for the incorporation of NPs in a cell system is critical to evaluate its use for in vivo applications. While NPs interacts with the plasma membrane, cell will package this "foreign agent" with different surface molecules, allowing their uptake [42]. Generally, by increasing the time which cells are exposed to NPs, their uptake increases. However, this process will be finally determined by the balance between entry and excretion of NPs. Time dependence in NPs incorporation was evaluated in PMBC, HeLa and SKOV-3. As shown in Fig. 2, HeLa (a) and SKOV-3 (b) show time-dependence, as evidenced by a high mean fluorescence intensity (MFI) given by the encapsulated fluorophore, showing a similar behavior obtained by Liu et al. [25]. Nevertheless, the same phenomenon was not observed in PBMC. After 5 min of incubation, a high fluorescence intensity was observed, which did not vary under longer incubation, reaching stable values. Most cellular models used so far to investigate nanoparticle cellular uptake do not take into account cellular heterogeneity in the blood stream, for instance, the presence of macrophages or neutrophils. Differences in NPs uptake and/or NPs phagocytosis can be found even between macrophages mouse strains dependent on their preference for either Th1- or Th2-responses [43]. The mentioned phenomena may explain why in our experiments performed with PBMC, we observed a faster NPs internalization compared to the other cell lines (Fig. 2c, d).

NPs concentration is another important feature to be analyzed because it can impact uptake efficiency and cytotoxicity induction. Thus, we evaluated concentration dependence in the uptake of PHBV nanoparticles. HeLa, SKOV-3 and PBMC cells were incubated for 2 h at 37 °C, 5% CO_2 with a solution of PHBV-RN at different concentrations: 1, 10, 100, 500 and 1000 μg/mL. Figure 3a–d shows that in all the three cell types, concentration-dependence is observed, with an apparent saturation at 500 μg/mL. Experiments performed in PBMC indicate that cells need a specific concentration to internalize NPs, although they can internalize less content compared to the other cell lines.

PBMC showed faster cellular uptake in time (comparing with the other cells at 5 min). This phenomenon was not observed in the other cell lines probably because PBMC contains macrophages and neutrophils which endocytosis mechanism is most via phagocytosis [43]. Nevertheless, uptake after 5 min seems to be inefficient probably because PBMC are smaller cells compared to the other cell lines (HeLa and SKOV-3 cells are relative equal in size). This idea is supported by Fig. 2d, which shows that PBMC uptake with different concentrations reach a plateau, absorbing less NPS than the other cell lines. On the other hand, cell lines (which are derived from a tumor) display a high metabolic rate, process that can respond to a higher cellular uptake without reaching a plateau. This effect was more evident in HeLa cells than SKOV-3 cells.

We have estimated an approximate number of NPs at a given NP mass concentration depicted in Fig. 3, following the previous data from Vilos et al. [32]. In this way, the fluctuation in the number of NPs at 500 μg/mL compared to 1000 μg/mL represents only a twofold increase in picomoles/mL. It must be noted that an increase in the concentration of the solution might not be "saturating" the cells, because the NPs could enter progressively. Probably, at a high solution concentration such as 1000 μg/mL, NPs can enter in similar extent than at 500 μg/mL to the cells, as Fig. 3 shows, but probably afterwards, every NP can eventually enter the cells.

Internalization and processing of nanoparticles into the cells is an active process, which requires energy to be performed [44]. Dependence of energy (i.e. ATP) was assessed in HeLa and SKOV-3 cells. Both cell lines were incubated at 4 °C, after that the uptake of nanoparticles was quantified. As seen in Fig. 4, when both cell lines were incubated at 4 °C a decrease of NPs internalization was observed, suggesting endocytosis inhibition. Other groups also have reported this behavior. Qaddoumi et al. [20] reported a decrease of about 90% in the uptake of PLGA nanoparticles in rabbit conjunctival epithelial cells (RCEC), while Liu et al. [25] also observed a decrease in

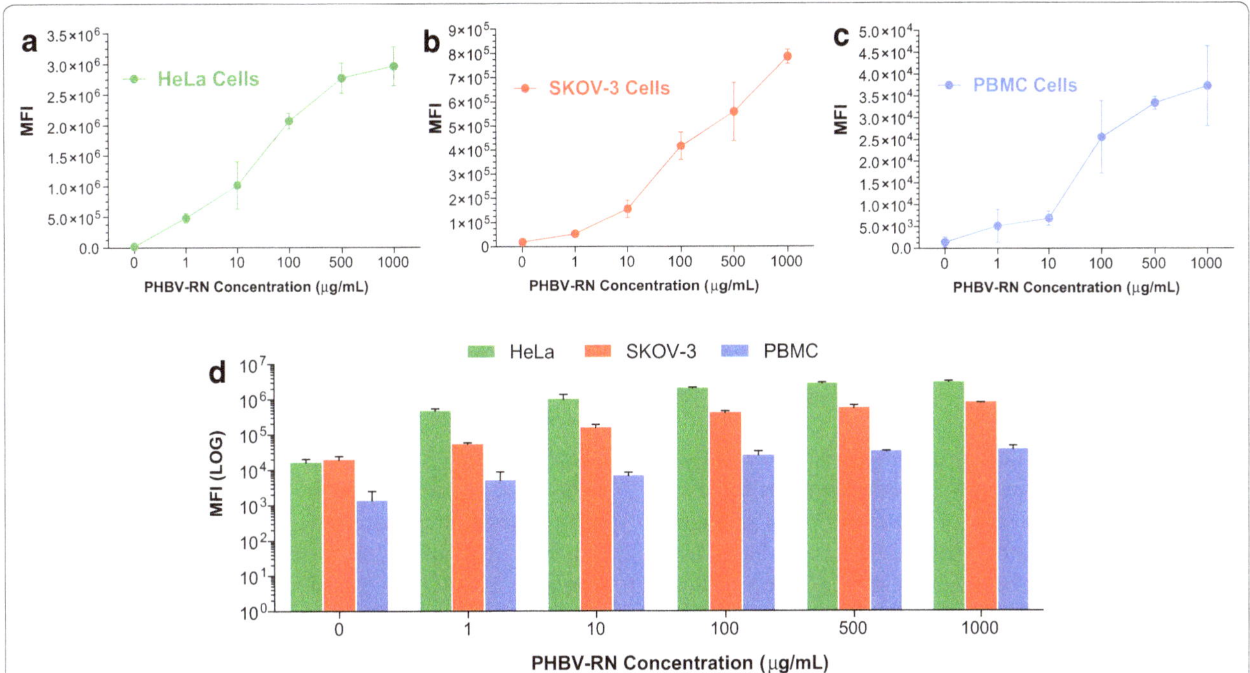

Fig. 2 Quantification of the cellular uptake of PHBV nanoparticles: time dependency. To evaluate time dependency in the cellular uptake of NPs, HeLa (**a**), SKOV-3 (**b**) and PBMC (**c**) cells were incubated with a 100 μg/mL PHBV-RN at several times. **d** Mean fluorescence intensity analysis of the three different cells. Mean fluorescence intensity was determined by flow cytometry (n = 3)

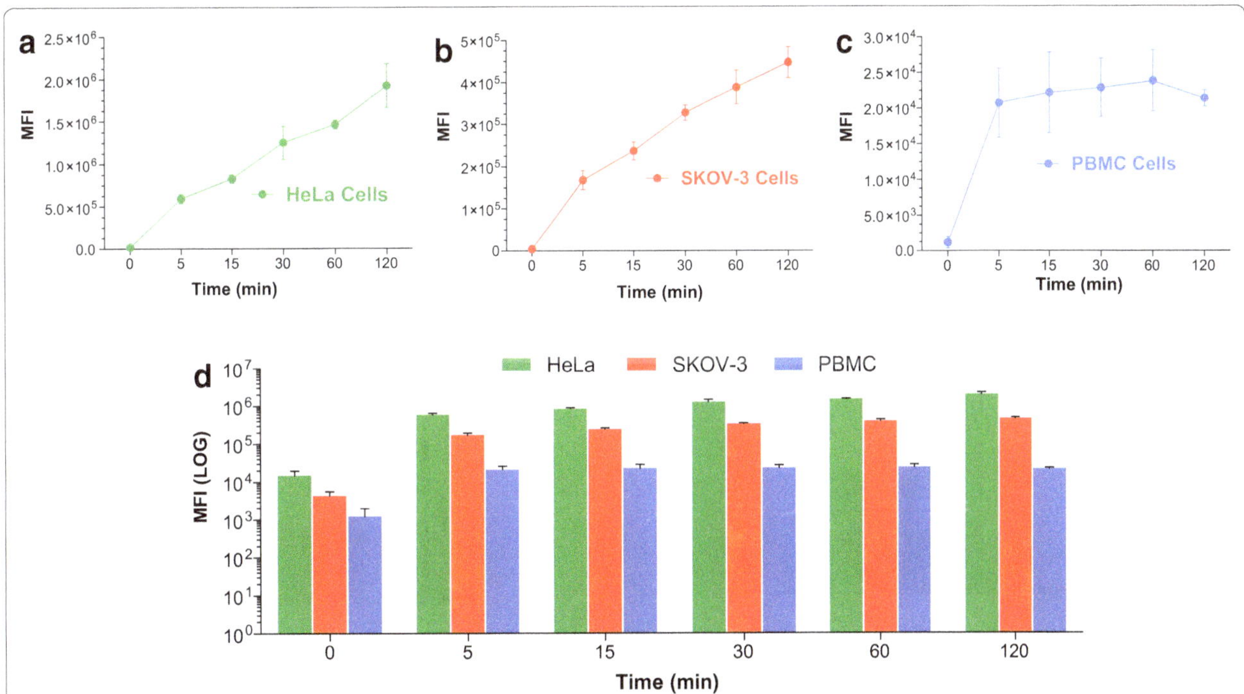

Fig. 3 Quantification of the cellular uptake of PHBV nanoparticles. Concentration dependency. To evaluate concentration dependency in the cellular uptake of NPs, HeLa (**a**), SKOV-3 (**b**) and PBMC (**c**) were incubated with a PHBV-RN solution at different concentrations (1, 10, 100, 500 and 1000 μg/mL) for 2 h. **d** Mean fluorescence intensity analysis of the three different cells. Mean fluorescence intensity was determined through flow cytometry (n = 3)

uptake of polymeric nanoparticles in human cell lines Hep-G2, HuH-7, and PLC. By the other hand, treatment with sodium azide (SA) solution, which blocks electron transport chain, showed a decrease of the entry of NPs, however, this decay is less significant than the observed at 4 °C. One possible explanation for this phenomenon is that SA inhibits synthesis of ATP, then cells may use exogenous ATP for operation, allowing entry of PHBV nanoparticles, as been suggested by Gratton et al. [39]. Other researchers have evaluated inhibition of endocytosis adding sodium azide, obtaining similar results [45]. All these results suggest that uptake of PHBV nanoparticles is a time, concentration, and energy-dependent process.

To study the effect of NPs on cell viability, we analyzed cell viability with annexin V by flow cytometry and caspase-3 by immunofluorescence. Results show no significant levels of cytotoxicity induced by the NPs in HeLa and SKOV-3, which was demonstrated using different assays and different concentrations and incubation times (Additional file 3: Figure S3).

Endocytosis mechanism of PHBV nanoparticles

To determine endocytosis mechanism of PHBV nanoparticles, we used fluorescent marker WGA, which selectively recognizes plasma membrane structures. HeLa and SKOV-3 cell lines were incubated for 5 min with a PHBV-FITC solution and subsequently fixed for analysis. In Fig. 5 is observed colocalization of nanoparticles with WGA marker (arrowheads), demonstrating that uptake is performed by a plasma membrane coated compartment in both cancer cell lines (HeLa 99.98% and SKOV-3 99.95% of colocalization).

As a next step, we assessed whether entry of PHBV NPs is done through classical endocytosis ways. For this purpose, EEA-1 (early endosomes marker) (Fig. 6) and CAV-1 (caveolae-mediated endocytosis marker) (Fig. 7) were used. Both cell lines were exposed for 5 and 15 min to PHBV-RN solution and then fixed. As shown in Fig. 6, HeLa cell line shows clear colocalization of NPs with early endosomes showing 99.42% of colocalization at 5 min and 99.39% of colocalization at 15 min, while SKOV-3 cells colocalization is strongly lesser showing 21.17% of colocalization at 5 min and 23.08% of colocalization at 15 min. Furthermore, Fig. 7 shows in HeLa cells some colocalization of PHBV nanoparticles with endosomes coated with caveolae (41.46% of colocalization at 5 min and 30.62% of colocalization at 15 min). While in SKOV-3 cells, colocalization with CAV-1 was not observed, evidencing that PHBV NPs do not enter the cells through caveolae-coated vesicles, although SKOV-3 have been described as one of the few ovarian carcinoma cells expressing caveolin-1 protein [46]. Experiments conducted by Ekkapongpisi et al., showed that negative-coated silica nanoparticles were uptaken by a caveolin-independent mechanism by SKOV-3 cells, which was similar to neutral PHBV nanoparticles described with mesoporous silica and polystyrene nanoparticles [47].

In contrast to clathrin-mediated, classic route, caveolar uptake has been suggested to avoid lysosomes. Caveolae

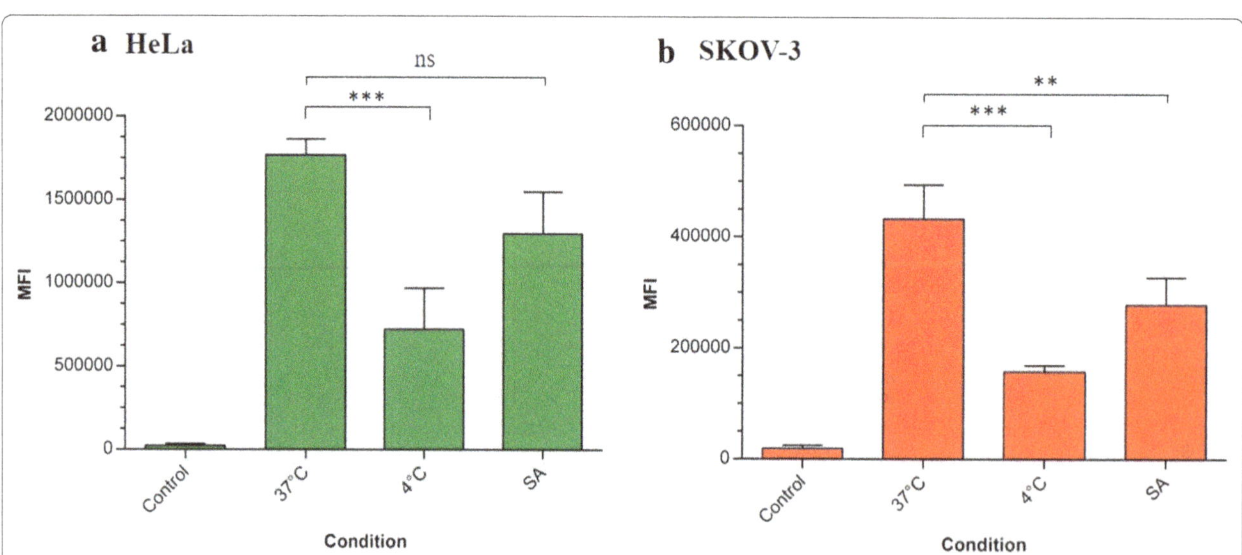

Fig. 4 Quantification of the cellular uptake of PHBV nanoparticles: energy dependency. To evaluate energy dependency in the cellular uptake of NPs, HeLa (**a**) and SKOV-3 cells (**b**) were incubated with 100 µg/mL PHBV-RN under different conditions (37, 4 °C, sodium azide). Mean fluorescence intensity was determined through flow cytometry (n = 3). One-way ANOVA and Bonferroni statistical test were used (** $p < 0.01$, *** $p < 0.001$, **** $p < 0.0001$)

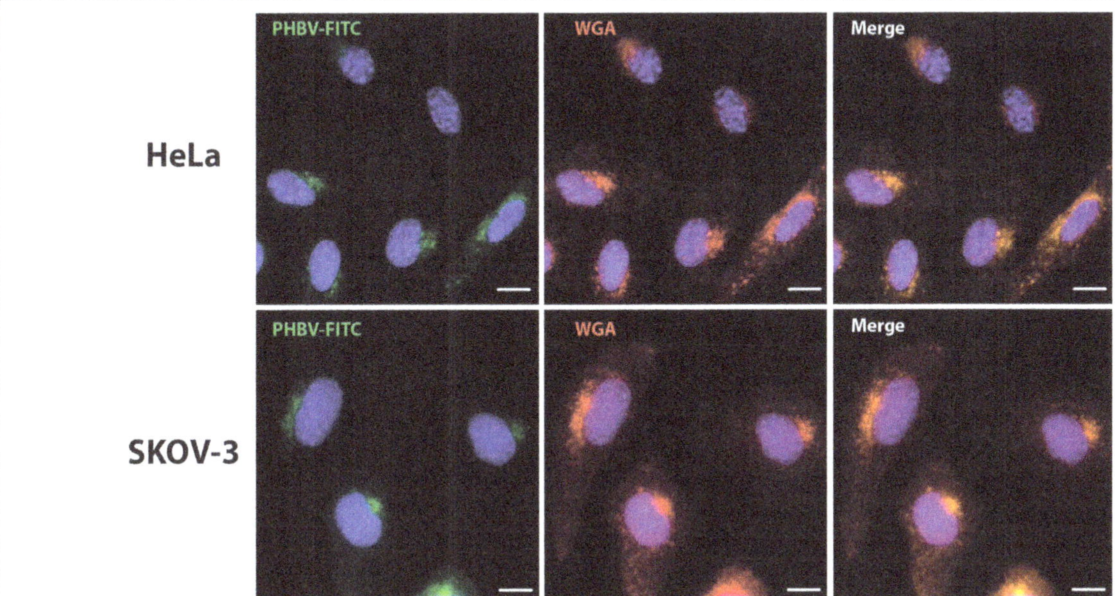

Fig. 5 Characterization of the endocytosis mechanism using WGA marker. HeLa and SKOV-3 cells were incubated with PHBV-FITC for 5 min at 37 °C, in the presence of Alexa Fluor® 555-conjugated WGA. Later, cells were fixed and observed by fluorescence microscopy. Hoechst 33342 was used as a nuclear stain. Objective: ×60 magnification. *Bar size* 10 μm

vesicles fuse with caveosomes, delivering their content to other cellular compartments. By avoiding acidic routes, a caveolar pathway may be advantageous for drug delivery. In this manner, ligands can be developed to target caveolar domains and reaching a more efficient delivery [48].

Macropinocytosis mechanism was also indirectly evaluated. This pathway is characterized by the rearrangement of actin filaments in the cytoplasm when a particle is incorporated. As a marker of actin monomers, Phalloidin, which is a mycotoxin that binds irreversibly to actin monomers, was conjugated to Alexa Fluor® 488 fluorophore. HeLa and SKOV-3 cell lines were incubated for 15 and 30 min with a PHBV-RN solution. Figure 8 shows that HeLa exhibits stress fibers in the presence of PHBV nanoparticles (arrowheads). These fibers are characteristic of rearrangement of actin filaments in cellular processes such as macropinocytosis. It is important to note that in SKOV-3, these types of fibers are not seen, showing the same form as control treatment.

In summary, NPs showed different cell entry ways depending on the cell type; specifically, in SKOV-3, NPs appear to enter through an independent way from the traditional routes, unlike the behavior observed in HeLa cells (summary of the colocalization analysis is given in Table 2).

To determine the participation of the proteins from the cell membrane in the uptake of NPs, a reducible biotin assay was performed. By using a reducing agent such as glutathione, this method allows discriminating surface proteins that enter into the cell under a certain condition. As shown in Fig. 9, apparently there were no specific proteins involved in the incorporation of nanoparticles, because no differences were observed in the protein pattern obtained with NPs compared to the control condition without treatment. However, it is important to remark that the protein membrane pattern evaluated by western blot is completely different between both cell lines, suggesting that endocytosis machinery and plasma membrane proteins together with its dynamics may be different among them. This observation could explain our immunofluorescence results that showed that actin filaments, which are the major cytoskeletal proteins, are not acting in the NPs entry to SKOV-3 cells, differently to Hela cells.

Furthermore, to determine the final fate of the NPs in both cell lines, colocalization with LAMP-1 (Lysosome-Associated Membrane Protein 1) antibody was performed by immunofluorescence (Figs. 10, 11), showing that in both cases NPs do colocalize with lysosomes but at different times: in HeLa cells, transit towards lysosomes was faster than in SKOV-3, probably due to that in Hela, the classical entry pathway is preferred by the NPs, allowing more expedite transit to the lysosomes, unlike SKOV-3. Moreover, in experiments performed with HL-60, a human promyelocytic leukemia cell line, after 4 h of incubation with NPs containing the fluorophore FITC, its content was observed in the cytoplasm meaning that somehow NP content may be released by

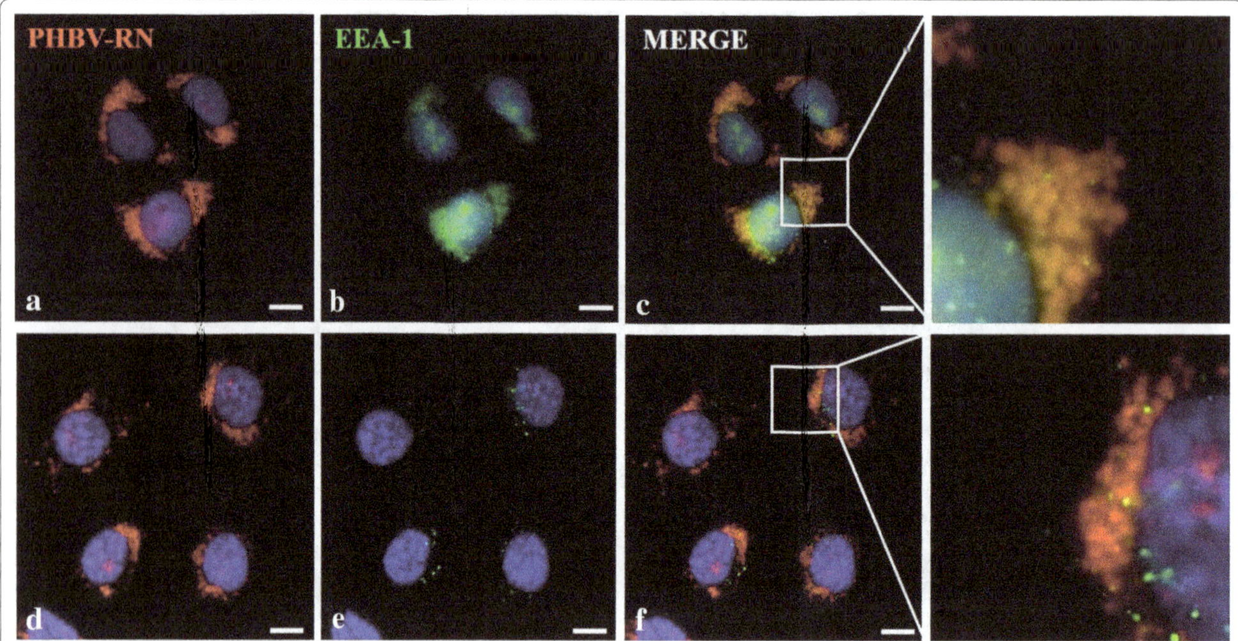

Fig. 6 Characterization of the endocytosis mechanism using EEA-1 marker. HeLa (**a–c**) and SKOV-3 (**d–f**) cells were incubated with PHBV-RN nanoparticles for 15 min at 37 °C; then cells were fixed and permeabilized. Later, an immunofluorescence was performed using anti-EEA1 (1:100) and anti-mouse Alexa Fluor® 488 (1:500) as primary and secondary antibody respectively. Finally, cells were observed through fluorescence microscopy. Hoechst 33342 was used as nuclear stain. Objective: ×60 magnification. *Bar size* 10 μm

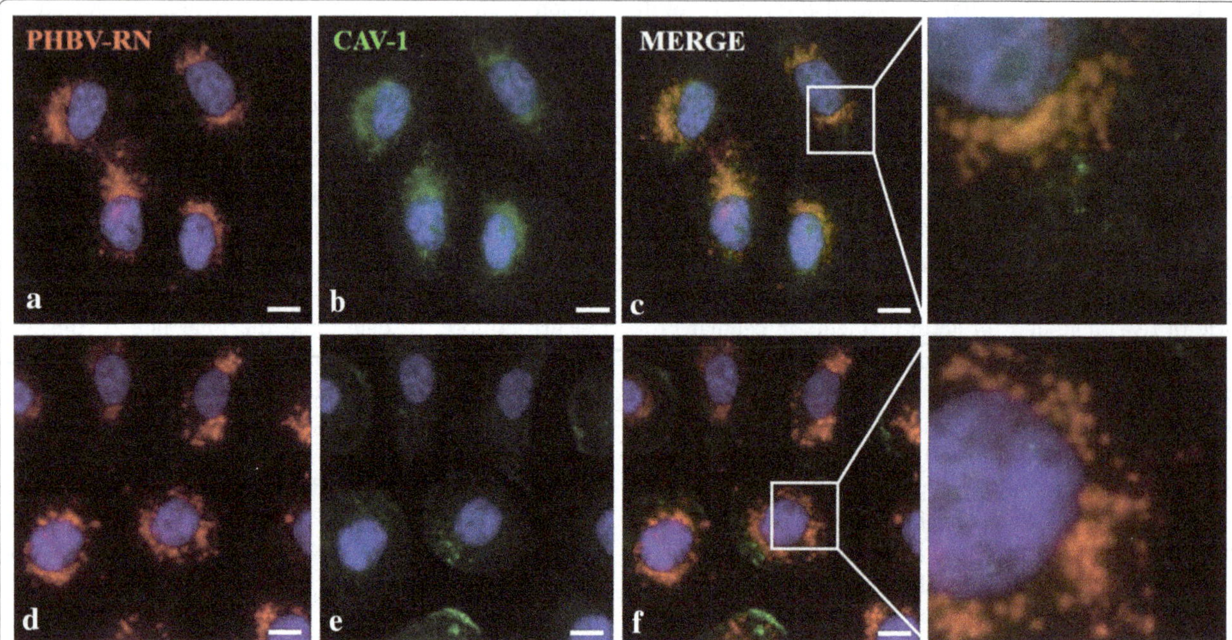

Fig. 7 Characterization of the endocytosis mechanism using CAV-1 marker. HeLa (**a–c**) and SKOV-3 (**d–f**) cells were incubated with PHBV-RN nanoparticles for 15 min at 37 °C, then cells were fixed and permeabilized. Later, an immunofluorescence was performed using anti-CAV1 (1:100) and anti-mouse Alexa Fluor® 488 (1:500) as primary and secondary antibody respectively. Finally, cells were observed through fluorescence microscopy. Hoechst 33342 was used as nuclear stain. Objective: ×60 magnification. *Bar size* 10 μm

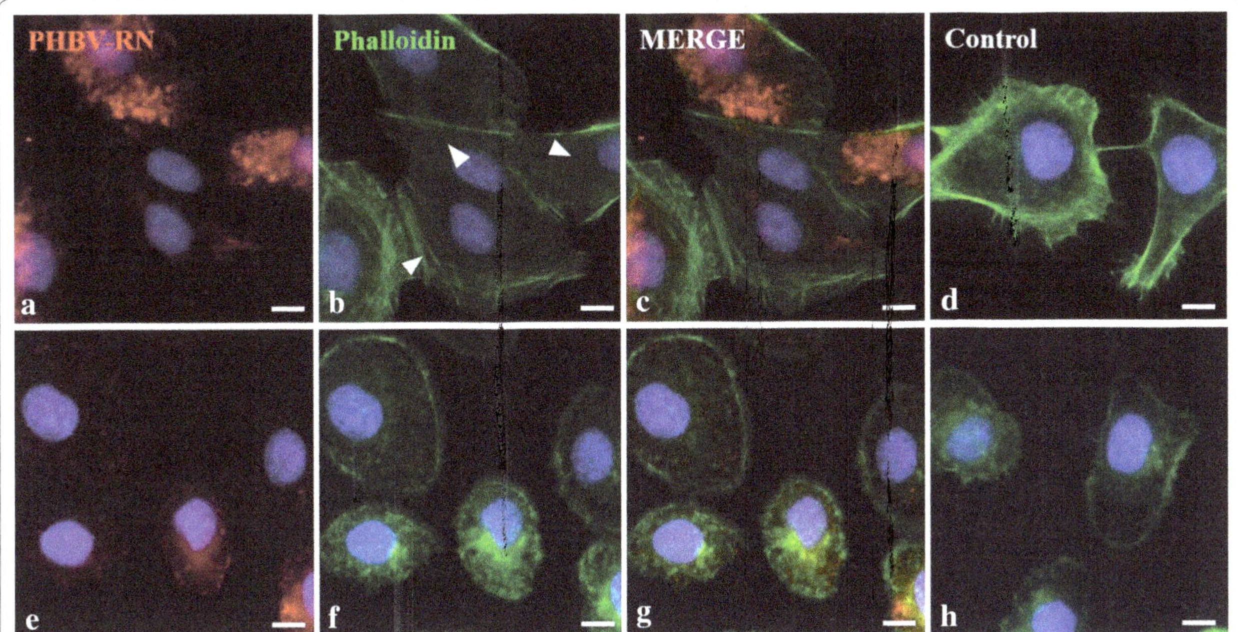

Fig. 8 Evaluation of actin filaments rearrangement in the presence of PHBV nanoparticles. HeLa (**a**–**d**) and SKOV-3 (**e**–**h**) cells were incubated with PHBV-RN for 15 and 30 min at 37 °C; then cells were permeabilized and incubated with Phalloidin Alexa Fluor® 488 conjugated. Hoechst 33242 was used as nuclear stain. Objective: ×60 magnification. *Bar size* 10 µm

Table 2 Extent of colocalization of PHBV nanoparticles with endocytosis markers in HeLa and SKOV-3 cells

	CAV-1		EEA-1		WGA
	5 min	15 min	5 min	15 min	5 min
HeLa (%)	41.46	30.62	99.42	99.39	99.98
SKOV-3 (%)	0	0	21.17	23.08	99.95

the lysosomes where NPs may be degraded (data not shown).

To evaluate NPs final fate through a different approach, we performed an experiment with NPs encapsulating LysoTracker® probe, which consist on a fluorophore linked to a weak base that is only partially protonated at neutral pH [49]. This probe is highly selective for acidic organelles. If the NPs release their content in the lysosome, it may be detected by fluorescence microscopy. In this way, fluorescence is not detected when LysoTracker® is trafficking through the endosomes. Our results in Fig. 12 shows that after 4 h, fluorescence can be detected, demonstrating that NPs are degraded in the lysosome, releasing their content. In vitro degradation of PHBV has been shown very slow [50]. At extreme conditions, such as at acid/basic pH, degradation seems to undergo by a decrease in the molecular weight and then, by a weight loss, when the mechanical strength is lost, and the remaining PHBV breaks down into small fragments [51].

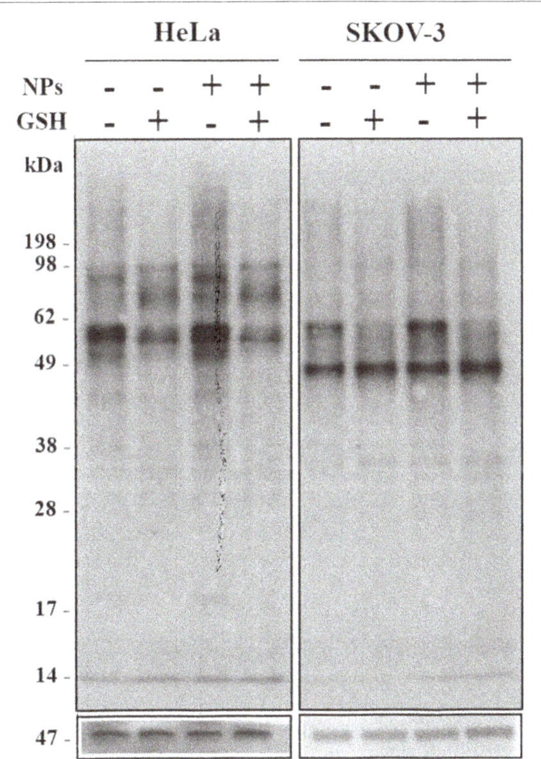

Fig. 9 Analysis of proteins involved in cellular uptake of PHBV nanoparticles. HeLa and SKOV-3 cells were subjected to reducible biotin assay to determine possible membrane proteins involved in nanoparticles uptake. 40 µg of proteins from cell lysate were resolved through SDS-PAGE, then transferred to PDVF membrane and finally, incubated with streptavidin-HRP (1:1000). β-actin (47 kDa) was used as loading control

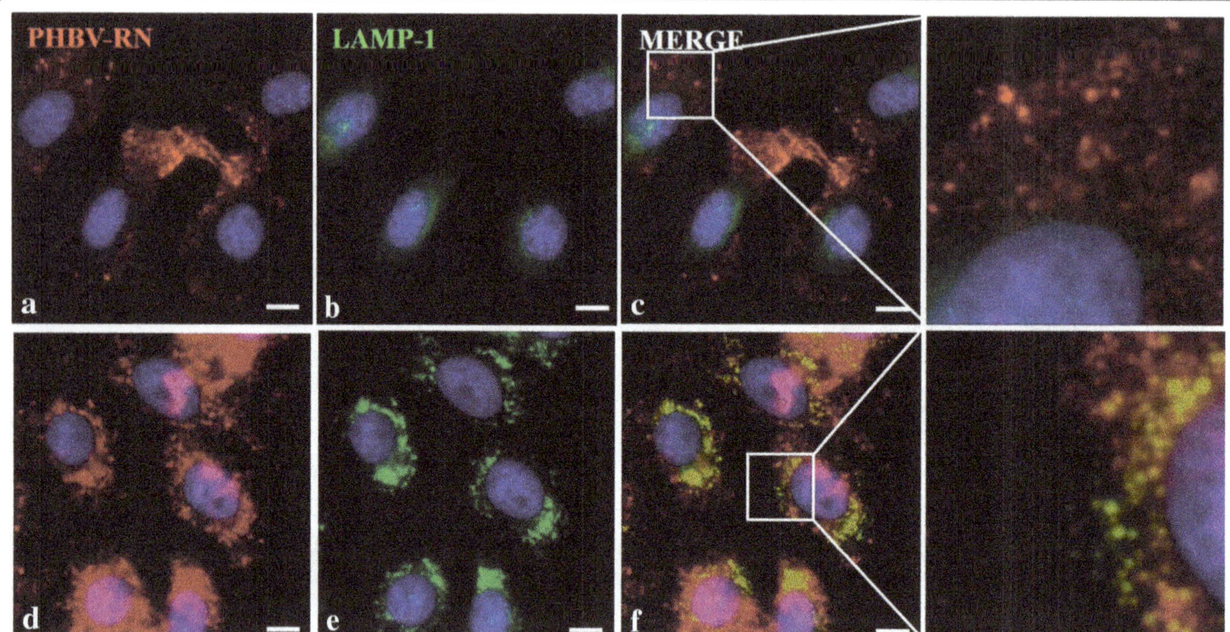

Fig. 10 Determination of final nanoparticle fate in HeLa cells. HeLa cells were incubated with PHBV-RN nanoparticles for 15 min (**a**–**c**) and 1 h (**d**–**f**) at 37 °C; then cells were fixed and permeabilized. Later, an immunofluorescence was performed using anti-LAMP1 (1:100) and anti-mouse Alexa Fluor® 488 (1:500) as primary and secondary antibody respectively. Finally, cells were observed through fluorescence microscopy. Hoechst 33342 was used as nuclear stain. Objective: ×60 magnification. *Bar size* 10 μm

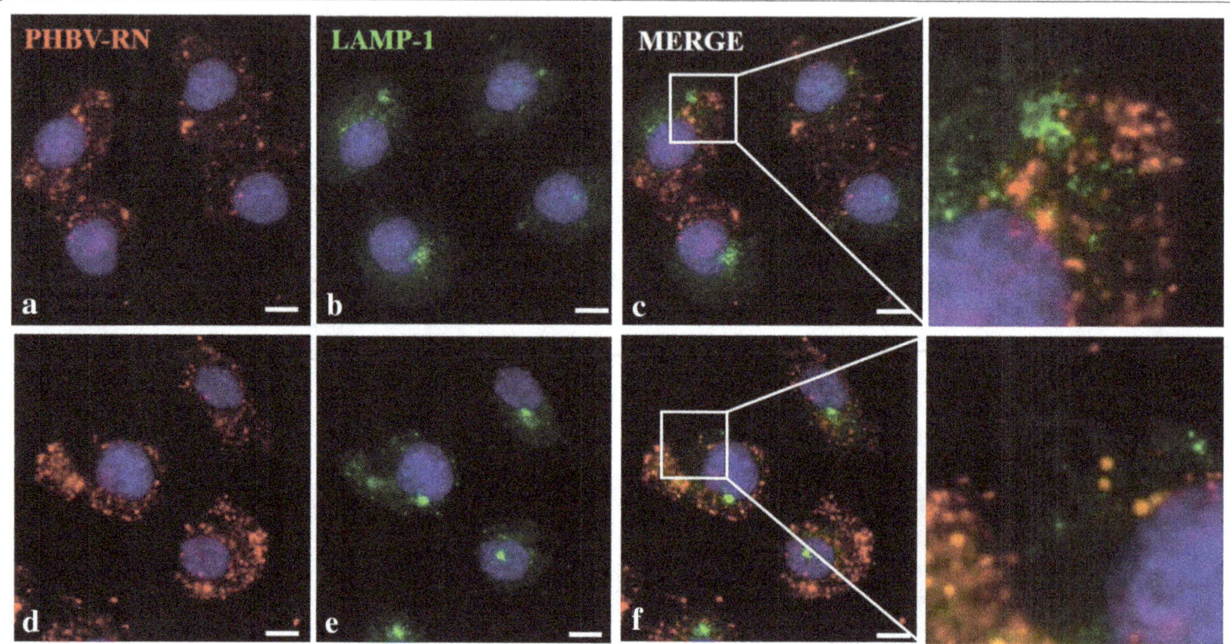

Fig. 11 Determination of final nanoparticle fate in SKOV-3 cells. SKOV-3 cells were incubated with PHBV-RN nanoparticles for 15 min (**a**–**c**) and 1 h (**d**–**f**) at 37 °C, then cells were fixed and permeabilized. Later, an immunofluorescence was performed using anti-LAMP1 (1:100) and anti-mouse Alexa Fluor® 488 (1:500) as primary and secondary antibody respectively. Finally, cells were observed through fluorescence microscopy. Hoechst 33342 was used as nuclear stain. Objective: ×60 magnification. *Bar size* 10 μm

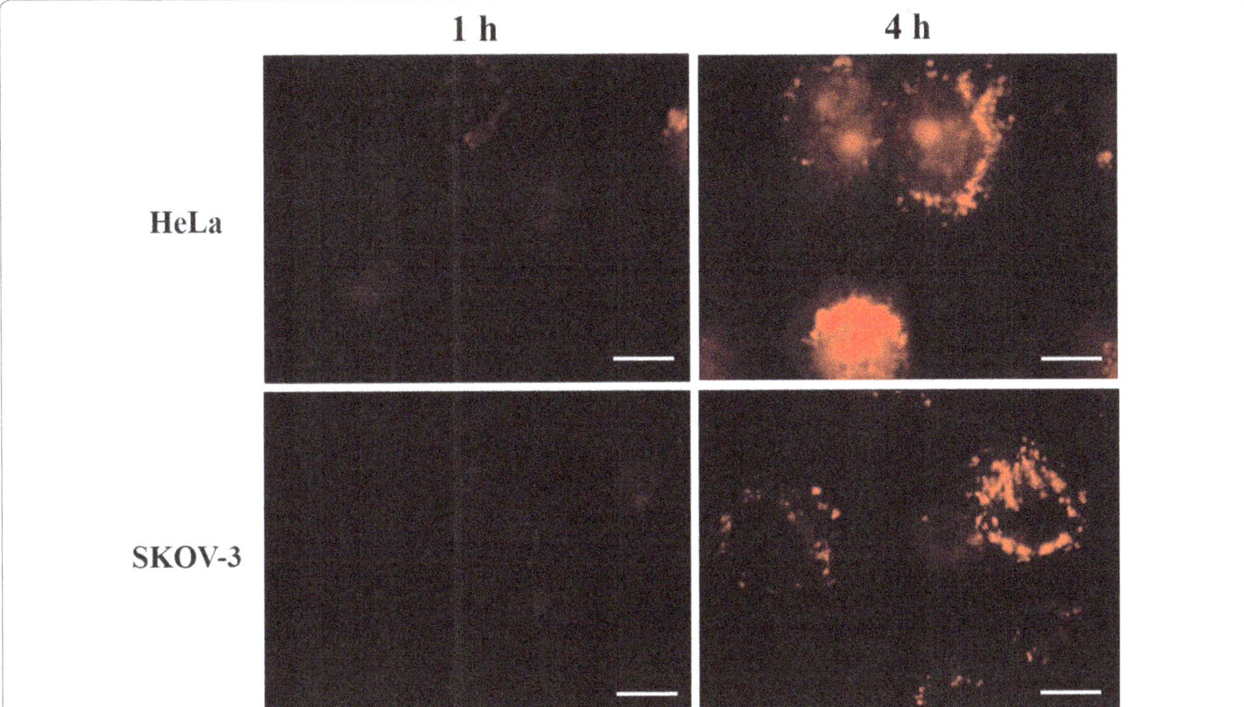

Fig. 12 Evaluation of nanoparticle degradation. HeLa and SKOV-3 cells were incubated with PHBV-Lysotracker® DND-99/nanoparticles for 4 h at 37 °C, then cells were fixed and observed through fluorescence microscopy. Objective: ×60 magnification. *Bar size* 10 μm

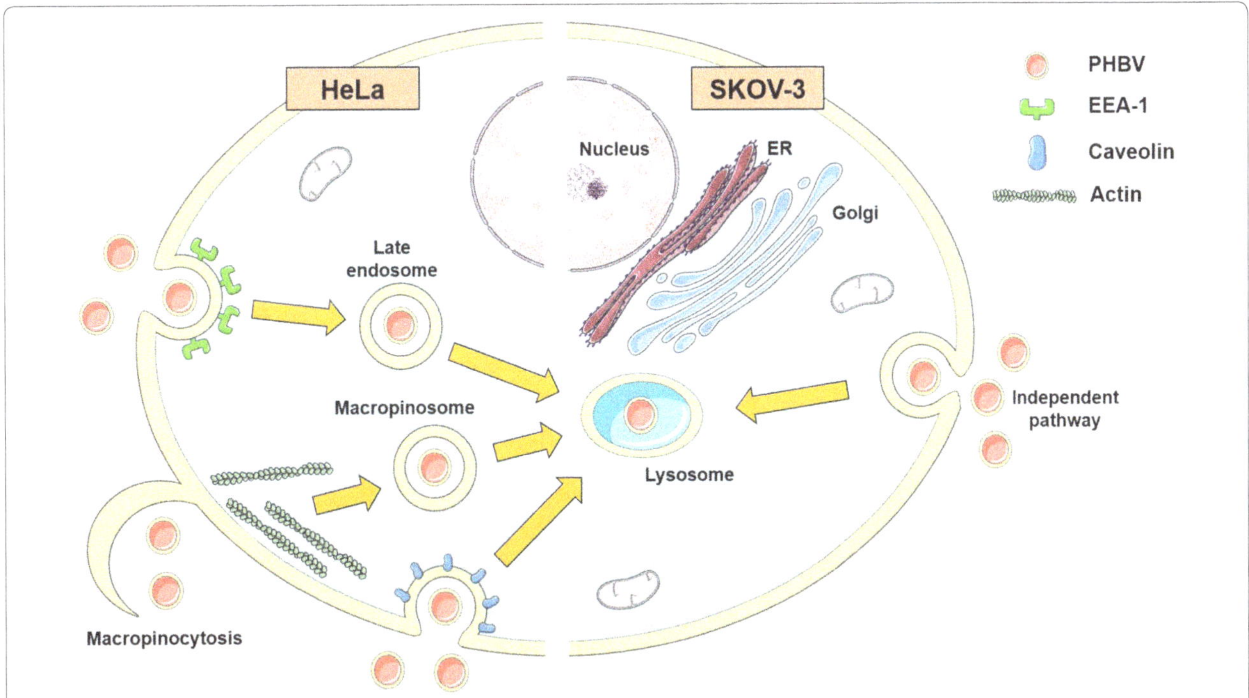

Fig. 13 PHBV nanoparticle intracellular trafficking pathway in HeLa and SKOV-3 cells scheme. Our results showed different endocytosis pathway depending on the cell context but the same final destination

When cells were incubated with PHBV NPs, after trafficking through the endo-lysosome pathway, which finally ends at lysosomes, these organelles enter into a special stage where they fused and get polarized, resembling what normally happens during autophagy and immunological synapsis [52, 53]. To investigate whether NPs/cell incubation may trigger autophagy, we analyze LC3 expression by western blot in the two different cell lines, which showed no LC3 expression (data not shown) suggesting that lysosomal fusion not be triggering by an autophagy event [54]. These findings lead to propose the recruitment of a possible MTOC (Microtubule Organizing Center) towards a particular area, allowing the local regulation of potential exocytosis and endocytosis processes by concentrating at a specific place the required molecular machinery. A scheme of PHBV NPs entry pathway for HeLa and SKOV-3 cell lines suggested in this article is depicted in Fig. 13.

Conclusion

We investigated the cellular uptake mechanism of PHBV for intracellular delivery into HeLa and SKOV3 cells. Our experimental results showed that this process is time, concentration and energy-dependent, and the internalization mechanism depends on the cell line. Results described above, give us a general understanding of the cellular processes required for nanoparticle internalization, which contributes to understanding further targeting properties of PHBV enhancing their targeting efficiency.

Additional files

Additional file 1: Figure S1. Physicochemical characterization of nanoparticles. A) Size, B) Z Potential and C) Polydispersion Index.

Additional file 2: Figure S2. PHBV nanoparticles stability. Synthesized nanoparticles were stored at 4 °C for four week period and then analyzed by DLS (Size and zeta potential). Results are expressed as the mean ± standard deviation of triplicate determinations from three independent experiments. One-way ANOVA with Bonferroni test as statistical analysis was performed. ns = not significant, * $P < 0.05$, ** $P < 0.01$, *** $P < 0.001$.

Additional file 3: Figure S3. Cytotoxicity of PHBV nanoparticles at 1, 10, 100 and 1.000 µg/mL against HeLa cells using the MTT assay. Cells were incubated with the respective concentrations and left untreated to measure cell viability by MTT assay.

Authors' contributions

JPP, CV and CO designed the study; JPP, VMM, MCB, RRR, FML performed experiments; FMB. collected and analysed data; JPP, VMM and CO wrote the manuscript; LAV, JAF, FDGN and MRD gave technical support and conceptual advice. All authors read and approved the final manuscript.

Author details

[1] Center for Integrative Medicine and Innovative Science, Facultad de Medicina, Universidad Andrés Bello, Santiago, Chile. [2] Escuela de Bioquímica, Facultad de Ciencias Biológicas, Universidad Andrés Bello, Santiago, Chile. [3] Center for Bioinformatics and Integrative Biology, Facultad de Ciencias Biológicas, Universidad Andrés Bello, Echaurren #183, 8370071 Santiago, Chile. [4] Departamento de Ciencias Físicas, Facultad de Ciencias Exactas, Universidad Andrés Bello, Santiago, Chile. [5] Laboratorio de Genética y Patógenesis Bacteriana, Facultad de Ciencias Biológicas, Universidad Andrés Bello, Santiago, Chile. [6] Escuela de Química y Farmacia, Facultad de Medicina, Universidad Andrés Bello, Santiago, Chile.

Acknowledgements

This work was supported by UNAB Regular Grant DI-316-13/R. VMM and FDGN thank CONICYT for PhD Scholarship, CONICYT + PAI "Concurso Nacional Tesis de Doctorado en la Empresa" 2014 Folio 781413007, Fraunhofer Chile Research, Innova-Chile CORFO (FCR-CSB 09CEII-6991) and Anillo Científico ACT1107. CV acknowledges support from BASAL Grant FB0807, FONDECYT Regular number 1161438, MECESUP PMI-UAB1301, and UNAB Regular Grant DI-695. C.O acknowledges to Proyecto Núcleo UNAB DI-1419-16/N.

Competing interests

The authors declare that they have no competing interests.

References

1. Wilczewska AZ, Niemirowicz K, Markiewicz KH, Car H. Nanoparticles as drug delivery systems. Pharmacol Rep. 2012;64:1020–37. doi:10.1016/S1734-1140(12)70901-5.
2. De Jong WH. Drug delivery and nanoparticles : applications and hazards. Int J Nanomed. 2008;3:133–49.
3. Chan JM, Valencia PM, Zhang L, Langer R, Farokhzad OC. Polymeric nanoparticles for drug delivery. 2010:163–175. doi:10.1007/978-1-60761-609-2_11.
4. Grabrucker AM, Garner CC, Boeckers TM, Bondioli L, Ruozi B, Forni F, Vandelli MA, Tosi G. Development of novel Zn 2+ loaded nanoparticles designed for cell-type targeted drug release in CNS neurons: in vitro evidences. PLoS ONE. 2011. doi:10.1371/journal.pone.0017851.
5. Wang M, Hu H, Sun Y, Qiu L, Zhang J, Guan G, Zhao X, Qiao M, Cheng L, Cheng L, Chen D. Biomaterials A pH-sensitive gene delivery system based on folic acid-PEG-chitosan e PAMAM-plasmid DNA complexes for cancer cell targeting. Biomaterials. 2013;34:10120–32. doi:10.1016/j.biomaterials.2013.09.006.
6. Sneh-edri H, Stepensky D. Biochemical and Biophysical Research Communications "IntraCell" plugin for assessment of intracellular localization of nano-delivery systems and their targeting to the individual organelles. Biochem Biophys Res Commun. 2011;405:228–33. doi:10.1016/j.bbrc.2011.01.015.
7. Yezhelyev MV, Gao X, Xing Y, Al-hajj A, Nie S, Regan RMO. Emerging use of nanoparticles in diagnosis and treatment of breast cancer. Lancet Oncol. 2006;7:657.
8. Misra R, Sahoo SK. Intracellular trafficking of nuclear localization signal conjugated nanoparticles for cancer therapy. Eur J Pharm Sci. 2010;39:152–63. doi:10.1016/j.ejps.2009.11.010.
9. Xie S, Tao Y, Pan Y, Qu W, Cheng G, Huang L, Chen D, Wang X, Liu Z, Yuan Z. Biodegradable nanoparticles for intracellular delivery of antimicrobial agents. J Control Release. 2014;187:101–17. doi:10.1016/j.jconrel.2014.05.034.
10. Peetla C, Jin S, Weimer J, Elegbede A, Labhasetwar V. Biomechanics and thermodynamics of nanoparticle interactions with plasma and endosomal membrane lipids in cellular uptake and endosomal escape. Langmuir. 2014;30:7522.
11. Vasir JK, Labhasetwar V. Quantification of the force of nanoparticle-cell membrane interactions and its influence on intracellular trafficking of nanoparticles. Biomaterials. 2008;29:4244–52. doi:10.1016/j.biomaterials.2008.07.020.

12. Panyam J, Labhasetwar V. Biodegradable nanoparticles for drug and gene delivery to cells and tissue. Adv Drug Deliv Rev. 2003;55:329–47. doi:10.1016/S0169-409X(02)00228-4.
13. Das S, Das J, Samadder A, Paul A, Khuda-bukhsh AR. Strategic formulation of apigenin-loaded PLGA nanoparticles for intracellular trafficking, DNA targeting and improved therapeutic effects in skin melanoma in vitro. Toxicol Lett. 2013;223:124–38. doi:10.1016/j.toxlet.2013.09.012.
14. Sneh-Edri H, Likhtenshtein D, Stepensky D. Intracellular targeting of PLGA nanoparticles encapsulating antigenic peptide to the endoplasmic reticulum of dendritic cells and its effect on antigen cross-presentation in vitro. Mol Pharm. 2011;8:1266–75. doi:10.1021/mp200198c.
15. Reusch N. Poly-β-hydroxybutyrate/calcium polyphosphate complexes in eukaryotic membranes. Exp Biol Med. 1989;191(1):377–81.
16. Zakharian E, Thyagarajan B, French RJ, Pavlov E, Rohacs T. Inorganic polyphosphate modulates TRPM8 channels. PLoS ONE. 2009. doi:10.1371/journal.pone.0005404.
17. Zakharian E, Pavlov E, Charles S, Building TM, Scotia N. Identification of the polyhydroxybutyrate granules in mammalian cultured. Chem Biodivers. 2012;9:2597–604.
18. Dusek K. Advances in polymer science. Responsive Gels Vol Trans II. 1993. doi:10.1007/BFb0050503.
19. Vilos CA, Constandil L, Herrera N, Solar P, Escobar-Fica J, Velásquez LA. Ceftiofur-loaded PHBV microparticles: a potential formulation for a long-acting antibiotic to treat animal infections. Electron J Biotechnol. 2012;15:1. doi:10.2225/vol15-issue4-fulltext-2.
20. Qaddoumi MG, Ueda H, Yang J, Davda J, Labhasetwar V, Lee VHL. The characteristics and mechanisms of uptake of PLGA nanoparticles in rabbit conjunctival epithelial cell layers. Pharm Res. 2004;21:641–8.
21. Muro S, Cui X, Gajewski C, Murciano J-C, Muzykantov VR, Koval M. Slow intracellular trafficking of catalase nanoparticles targeted to ICAM-1 protects endothelial cells from oxidative stress. Am J Physiol Cell Physiol. 2003;285:C1339–47. doi:10.1152/ajpcell.00099.2003.
22. Cartiera MS, Johnson KM, Rajendran V, Caplan MJ, Saltzman WM. The uptake and intracellular fate of PLGA nanoparticles in epithelial cells. Biomaterials. 2009;30:2790–8. doi:10.1016/j.biomaterials.2009.01.057.
23. Panyam J, Labhasetwar V. Dynamics of endocytosis and exocytosis of poly(d, l-lactide-co-glycolide) nanoparticles in vascular smooth muscle cells. Pharm Res. 2003;20:212–20.
24. Panyam J, Labhasetwar V. Sustained cytoplasmic delivery of drugs with intracellular receptors using biodegradable nanoparticles. Mol Pharm. 2004;1:77–84. doi:10.1021/mp034002c.
25. Liu P, Sun Y, Wang Q, Sun Y, Li H, Duan Y. Intracellular trafficking and cellular uptake mechanism of mPEG-PLGA-PLL and mPEG-PLGA-PLL-Gal nanoparticles for targeted delivery to hepatomas. Biomaterials. 2014;35:760–70. doi:10.1016/j.biomaterials.2013.10.020.
26. des Rieux A, Fievez V, Garinot M, Schneider YJ, Préat V. Nanoparticles as potential oral delivery systems of proteins and vaccines: a mechanistic approach. J Control Release. 2006;116:1–27. doi:10.1016/j.jconrel.2006.08.013.
27. Prego C, Garcia M, Torres D, Alonso MJ. Transmucosal macromolecular drug delivery. J Control Release. 2005;101:151–62. doi:10.1016/j.jconrel.2004.07.030.
28. Morimoto H, Yonehara S, Bonavida B. Overcoming tumor necrosis factor and drug resistance of human tumor cell lines by combination treatment with anti-FAS antibody and drugs or toxins. Cancer Res. 1993;53:2591–6.
29. Mellman I, Fuchs R, Helenius A. Acidification of the endocytic and exocytic pathways. Annu Rev Biochem. 1986;55:663–700. doi:10.1146/annurev.bi.55.070186.003311.
30. Cole NB, Lippincott-Schwartz J. Organization of organelles and membrane traffic by microtubules. Curr Opin Cell Biol. 1995;7:55–64. doi:10.1016/0955-0674(95)80045-X.
31. Schafer DA, Weed SA, Binns D, Karginov AV, Parsons JT, Cooper JA. Dynamin2 and cortactin regulate actin assembly and filament organization. Curr Biol. 2002;12:1852–7. doi:10.1016/S0960-9822(02)01228-9.
32. Vilos C, Morales FA, Solar PA, Herrera NS, Gonzalez-Nilo FD, Aguayo DA, Mendoza HL, Comer J, Bravo ML, Gonzalez PA, Kato S, Cuello MA, Alonso C, Bravo EJ, Bustamante EI, Owen GI, Velasquez LA. Paclitaxel-PHBV nanoparticles and their toxicity to endometrial and primary ovarian cancer cells. Biomaterials. 2013;34:4098–108. doi:10.1016/j.biomaterials.2013.02.034.
33. Ehrlich M, Boll W, Van Oijen A, Hariharan R, Chandran K, Nibert ML, Kirchhausen T. Endocytosis by random initiation and stabilization of clathrin-coated pits. 2004;118:591–605.
34. Cureton DK, Massol RH, Saffarian S, Kirchhausen TL, Sean PJ. Vesicular stomatitis virus enters cells through vesicles incompletely coated with clathrin that depend upon actin for internalization. PLoS Pathog. 2009. doi:10.1371/journal.ppat.1000394.
35. Suen WL, Chau Y. Size-dependent internalisation of folate-decorated nanoparticles via the pathways of clathrin and caveolae-mediated endocytosis in ARPE-19 cells. J Pharm Pharmacol. 2013;66:564–73. doi:10.1111/jphp.12134.
36. Debbage P, Jaschke W. Molecular imaging with nanoparticles: giant roles for dwarf actors. Histochem Cell Biol. 2008;130:845–75. doi:10.1007/s00418-008-0511-y.
37. Vilella A, Tosi G, Grabrucker AM, Ruozi B, Belletti D, Angela M, Boeckers TM, Forni F, Zoli M. Insight on the fate of CNS-targeted nanoparticles. Part I: Rab5-dependent cell-specific uptake and distribution. J Control Release. 2014;174:195–201. doi:10.1016/j.jconrel.2013.11.023.
38. Chithrani BD, Ghazani AA, Chan WCW. Size and shape dependence of nanoparticles on cellular uptake. NANO. 2006;668:662–8. doi:10.1021/nl052396o.
39. Gratton SE, Ropp PA, Pohlhaus PD, Luft JC, Madden VJ, Napier ME, DeSimone JM. The effect of particle design on cellular internalization pathways. Proc Natl Acad Sci USA. 2008;105:11613–8. doi:10.1073/pnas.0801763105.
40. Harush-Frenkel O, Debotton N, Benita S, Altschuler Y. Targeting of nanoparticles to the clathrin-mediated endocytic pathway. Biochem Biophys Res Commun. 2007;353:26–32. doi:10.1016/j.bbrc.2006.11.135.
41. Nel AE, Madler L, Velegol D, Xia T, Hoek EMV, Somasundaran P, Klaessig F, Castranova V, Thompson M. Understanding biophysicochemical interactions at the nano-bio interface. Nat Mater. 2009;8:543–57. doi:10.1038/nmat2442.
42. Sahay G, Alakhova DY, Kabanov AV. Endocytosis of nanomedicines. J Control Release. 2010;145:182–95. doi:10.1016/j.jconrel.2010.01.036.
43. Jones SW, Roberts RA, Robbins GR, Perry JL, Kai MP, Chen K, Bo T, Napier ME, Ting JPY, DeSimone JM, Bear JE. Nanoparticle clearance is governed by Th1/Th2 immunity and strain background. J Clin Invest. 2013;123:3061–73. doi:10.1172/JCI66895.
44. Vácha R, Martinez-Veracoechea FJ, Frenkel D. Receptor-mediated endocytosis of nanoparticles of various shapes. Nano Lett. 2011;11:5391–5. doi:10.1021/nl2030213.
45. Bozavikov P, Rajshankar D, Lee W, McCulloch CA. Particle size influences fibronectin internalization and degradation by fibroblasts. Exp Cell Res. 2014;328:172–85. doi:10.1016/j.yexcr.2014.06.018.
46. Wiechen K, Diatchenko L, Agoulnik A, Scharff KM, Schober H, Arlt K, Zhumabayeva B, Siebert PD, Dietel M, Scha R, Sers C. Caveolin-1 is down-regulated in human ovarian. Am J Pathol. 2001;159:1635–43. doi:10.1016/S0002-9440(10)63010-6.
47. Ekkapongpisit M, Giovia A, Follo C, Caputo G, Isidoro C. Biocompatibility, endocytosis, and intracellular trafficking of mesoporous silica and polystyrene nanoparticles in ovarian cancer cells: effects of size and surface charge groups. Int J Nanomedicine. 2012;7:4147–58. doi:10.2147/IJN.S33803.
48. Bathori G, Cervenak L, Karadi I. Caveolae-an alternative endocytotic pathway for targeted drug delivery. Crit Rev Ther Drug Carrier Syst. 2004;21:30. doi:10.1615/CritRevTherDrugCarrierSyst.v21.i2.10.
49. Soulet D, Gagnon B, Rivest S, Audette M, Poulin R. A fluorescent probe of polyamine transport accumulates into intracellular acidic vesicles via a two-step mechanism. J Biol Chem. 2004;279:49355–66. doi:10.1074/jbc.M401287200.
50. Kanesawa Y, Doi Y. Hydrolytic degradation of microbial poly(3-hydroxybutyrate-co-3-hydroxyvalerate) fibers. Die Makromolekulare Chemie Rapid Commun. 1990;682:679–82.
51. M. Hakkarainen, Aliphatic Polyesters: Abiotic and Biotic Degradation and Degradation Products, in: Degrad. Aliphatic Polyesters, Springer Berlin Heidelberg, Berlin, Heidelberg, n.d.: pp. 113–138. doi:10.1007/3-540-45734-8_4.
52. Todde V, Veenhuis M, van der Klei IJ. Autophagy: principles and significance in health and disease. Biochim Biophys Acta Mol Basis Dis. 2009;1792:3–13. doi:10.1016/j.bbadis.2008.10.016.
53. Griffiths GM, Tsun A, Stinchcombe JC. The immunological synapse: a focal point for endocytosis and exocytosis. J Cell Biol. 2010;189:399–406. doi:10.1083/jcb.201002027.
54. Tanida I, Ueno T, Kominami E. LC3 and autophagy. Autophagosome Phagosome. 2008. doi:10.1007/978-1-59745-157-4_4.

6

Effect of nanoparticles on red clover and its symbiotic microorganisms

Janine Moll[1,2], Alexander Gogos[1], Thomas D. Bucheli[1], Franco Widmer[1] and Marcel G. A. van der Heijden[1,2,3*]

Abstract

Background: Nanoparticles are produced and used worldwide and are released to the environment, e.g., into soil systems. Titanium dioxide (TiO_2) nanoparticles (NPs), carbon nanotubes (CNTs) and cerium dioxide (CeO_2) NPs are among the ten most produced NPs and it is therefore important to test, whether these NPs affect plants and symbiotic microorganisms that help plants to acquire nutrients. In this part of a joint companion study, we spiked an agricultural soil with TiO_2 NPs, multi walled CNTs (MWCNTs), and CeO_2 NPs and we examined effects of these NP on red clover, biological nitrogen fixation by rhizobia and on root colonization of arbuscular mycorrhizal fungi (AMF). We also tested whether effects depended on the concentrations of the applied NPs.

Results: Plant biomass and AMF root colonization were not negatively affected by NP exposure. The number of flowers was statistically lower in pots treated with 3 mg kg^{-1} MWCNT, and nitrogen fixation slightly increased at 3000 mg kg^{-1} MWCNT.

Conclusions: This study revealed that red clover was more sensitive to MWCNTs than TiO_2 and CeO_2 NPs. Further studies are necessary for finding general patterns and investigating mechanisms behind the effects of NPs on plants and plant symbionts.

Keywords: Nanomaterials, Agriculture, Crop, Beneficial soil microbes, Ecosystem services

Background

Titanium dioxide (TiO_2) nanoparticles (NPs), carbon nanotubes (CNTs) and cerium dioxide (CeO_2) NPs are among the ten most produced NPs worldwide [1]. The production and use of these NPs leads to increasing concentrations in the soil system. Estimated material-flow in sludge treated soils for Europe are 2380 t^{-1} y^{-1} and 0.771 t y^{-1} for TiO_2 and CNTs, respectively [2]. For CeO_2 1400 t y^{-1} are assumed to end up in sludge treated soils worldwide [1]. Thus, all of these three NP types are unintentionally released into the soil ecosystem. One NP type that needs special attention regarding risk assessment in soils is TiO_2 because these NPs are listed in patents and publications targeted as additives of plant protection products [3, 4]. Thus, if such products were released to the market and applied in the fields, higher concentrations of TiO_2 NPs would be expected in soils. Due to the potential for increasing amounts of NPs that enter the soil system, it is important to test, whether these NPs affect plants and beneficial soil microorganisms that associate with plant roots and assist plants to acquire nutrients.

Several studies investigated effects of TiO_2 NPs, CNTs and CeO_2 NPs on either plants or microorganisms with variable results. For TiO_2 NPs, contrasting results were found and plant biomass was either decreased or not affected when grown in soil with enhanced TiO_2 NP concentrations [5–7]. Soil microbial community structures were shown to be altered when treated with TiO_2 NPs [7–9]. Also CNTs affected plants and soil microbial community structures: the number of flowers and fruits of tomatoes increased, and bacterial community structure changed [10]. In contrast, in another study with much higher CNT concentrations, soil microbial community structure was not affected [11]. Most often, ecotoxicological tests with NPs (TiO_2, CeO_2 and CNTs) in soil systems are either performed with plants, or with

*Correspondence: marcel.vanderheijden@agroscope.admin.ch
[1] Agroscope, Institute for Sustainability Sciences ISS, 8046 Zurich, Switzerland
Full list of author information is available at the end of the article

microorganisms, but the symbiosis of plants and soil microorganisms has rarely been investigated. Plant symbionts provide important ecosystem functions as e.g., nitrogen-fixation by rhizobia in legumes or phosphorus acquisition by arbuscular mycorrhizal fungi (AMF) [12]. One example is red clover which is used for animal feeding and as green manure. Red clover associates with nitrogen-fixing rhizobia bacteria (rhizobia) [13, 14]. Up to 373 kg N ha^{-1} y^{-1} can be fixed by these bacteria in root nodules of red clover plants [15]. Additionally, red clover performs a second symbiosis with AMF [12, 16–18]. These fungi provide plants with soil nutrients, especially immobile nutrients such as phosphorus. Up to 90 % of plant phosphorus is provided by AMF [18]. The two microbial symbionts, AMF and rhizobia, conduct important ecosystem functions [12], and thus it is important to assess whether nitrogen fixation and root colonization by AMF are affected by NPs.

Earlier studies showed that NPs had adverse effect on the legume-rhizobia symbiosis. For soybeans it has been reported that CeO_2 NPs diminished nitrogen fixation [19], and no effects of TiO_2 and Fe_3O_4 NPs on nodule colonization were found [20]. For barrel clover it has been reported that the number of nodules was decreased and gene expression altered when exposed to biosolids containing Ag, ZnO and TiO_2 NPs [21, 22]. Peas revealed a delayed nitrogen fixation when exposed to TiO_2 and ZnO in hydroponic systems [23, 24], and for faba beans, nodulation and nitrogenase activity were delayed by Ag NPs [25]. AMF root colonization has been reported to not being affected in soybeans exposed to TiO_2 and Fe_3O_4 NPs [20], while colonization of white clover roots was increased by Ag and FeO NPs [26]. Because of these effects on legume-rhizobia and AMF systems, it is important to assess whether root colonization by AMF and nitrogen fixation in soil-grown red clover are affected by NPs, e.g. TiO_2, CeO_2 and CNTs, because these effects might be species and NP dependent. To our best knowledge, there are no studies available on the effects of CNTs on legume-rhizobia-AMF systems.

In the present study, we investigated the effects of three different NP types, i.e., TiO_2 NPs, multi-walled CNTs (MWCNTs) and CeO_2 NPs, on red clover growth, biological nitrogen fixation with rhizobia and on root colonization of AMF in a soil system. We investigated if these NPs affect (1) plant growth, (2) biological nitrogen fixation in plants, (3) AMF root colonization, and (4) phosphorus uptake by red clover. As positive control we chose $ZnSO_4{\cdot}7H_2O$ because Zn^{2+} was reported to decrease plant growth and affect nitrogen fixation of legumes [27]. Effective soil elemental titanium and MWCNT (black carbon) concentrations, their vertical translocation and plant uptake were investigated in detail in a companion paper [28].

Results

Red clover plants were exposed for 14 weeks to agricultural soil spiked with different concentrations of NPs, i.e., TiO_2 NPs (P25), a bigger non-nanomaterial [29] TiO_2 particle (NNM-TiO_2, 20 % particles <100 nm), MWCNTs, CeO_2 NPs and a $ZnSO_4$ treatment. The biomass of red clover plants did not differ between NP spiked substrate and controls without NP addition, both for root and shoot dry weight separately and for total plant dry weight (Fig. 1; Additional file 1: Table S1). Total plant dry weight and effective titanium content per pot were correlated explaining 20 % of variance (Pearson's correlation: $p = 0.041$, $r = 0.45$). The root-shoot ratio was 0.49 ± 0.04 on average, and was also not affected by the presence of NPs ($p > 0.05$). The number of flowers decreased in the 3 mg MWCNT kg^{-1} soil treatment by 34 % ($p = 0.049$, Fig. 1; Additional file 1: Table S1). The higher concentration of 3000 mg MWCNT kg^{-1} exhibited a similar decrease in mean number of flowers (33 %), but the variation was higher and therefore the number of flowers was not significantly different from the control plants ($p = 0.160$).

In addition to plant performance, the interaction of red clover with rhizobia was investigated. All harvested red clover plants contained root nodules and the root nodules had a reddish color which indicates that they fixed nitrogen [14]. In addition, the percentage of fixed nitrogen was assessed based on the ^{15}N concentrations of clover and a reference plant (rye grass; see formula 1 in the "Methods" section). The percentages of fixed nitrogen of control red clover plants and NP treated plants were compared, and confirmed that biological nitrogen fixation took place (Fig. 2). All of the treated red clover plants fixed nitrogen and NP application did not affect nitrogen fixation levels in most of the treatments. Only in the 3000 mg MWCNT kg^{-1} treatment, biological nitrogen fixation was increased by 8 % ($p = 0.016$). Pearson's correlation revealed a correlation of nitrogen fixation and total biomass of $r = 0.28$ ($p = 0.012$).

The second symbiotic partner of red clover, AMF, was assessed by determining root colonization by staining fungal tissue and counting fungal structures by microscopy [30, 31]. In addition the phosphorus content of red clover shoots was assessed, as AMF can contribute significantly to plant P nutrition. Total root colonization by AMF, i.e., % arbuscules, vesicles and hyphae per investigated root intersection, was similar in all treatments (on average 51 ± 4 %; Additional file 1: Figure S1). Also the arbuscular and vesicular colonization revealed no differences between the control and NP treatments (average 23 ± 3 and 6 ± 2 %, respectively;

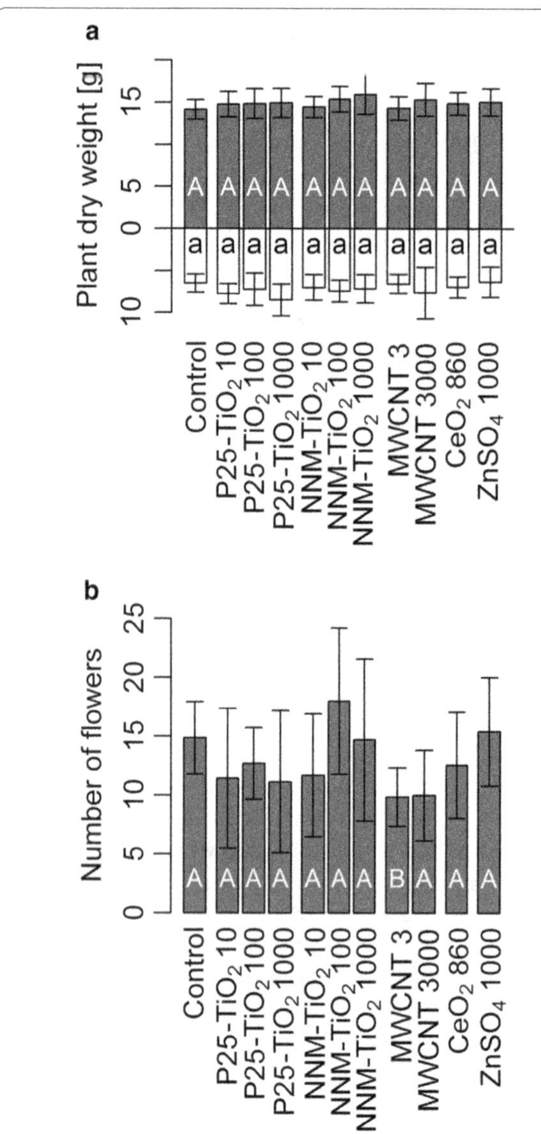

Fig. 1 Plant weight and flowers. **a** Red clover plant dry weight divided in shoot (*grey*) and root (*white*), and **b** number of flowers per pot at the end of the 3 month exposure for control, TiO_2 (P25, non-nanomaterial NNM), MWCNT, CeO_2 NPs, and $ZnSO_4 \cdot 7H_2O$. The number behind the treatment name is the nominal concentration in mg kg^{-1}. *Error bars* show the standard deviations (n = 7). *Capital letters* show significant differences for shoot biomass and number of flowers, and *small letters* for root biomass compared to the control plants ($p \leq 0.05$). The two blocks of starting time were included in the statistical model

Table 1). Phosphorus concentrations of the red clover shoots were not affected in any of the treatments (Additional file 1: Figure S1b, Table S1). Plant phosphorus content and total root colonization by AMF were not correlated (Pearson correlation coefficient: $p = 0.199$; $r = 0.15$).

Discussion

In the present study effects of different NPs, i.e., TiO_2 NPs, MWCNTs and CeO_2 NPs, on red clover and its symbiosis with rhizobia and AMF were assessed in a soil system. Both tested TiO_2 treatments (i.e. P25 and NNM-TiO_2) in all concentrations did not affect plant biomass in our experiment. The absence of effects of TiO_2 NPs on plant biomass are in agreement with other studies, using different plant species. For example plant growth was not affected when soybeans and corn were exposed to 200 mg TiO_2 NP kg^{-1} [7] and when tomatoes were exposed to concentrations between 1000 and 5000 mg P25 TiO_2 NP kg^{-1} [6]. However, in wheat 90 mg TiO_2 NPs kg^{-1} was shown to decrease plant biomass by 13 % [5]. MWCNTs did not affect red clover biomass in our experiment. Contrary to our findings, MWCNTs have been reported to increase biomass of tomatoes exposed to 50 and 200 µg ml^{-1} MWCNTs per plant [10]. In our experiment red clover biomass did not respond to the CeO_2 NP treatment, which is in agreement to a study using CeO_2 NPs at concentrations between 0.1 and 1 g kg^{-1} in an experiment with soybeans [19]. Thus, effects on plant biomass might be influenced by plant species (as shown for the TiO_2 NPs and MWCNTs) as well as by NP type. All of the above cited studies used different soils. Depending on soil properties, NPs might be differently bound to soil particles [32] which could influence the exposure and the effects of NPs on plants.

The number of flower heads was not affected in both TiO_2 and CeO_2 NP treatments at all tested concentrations. However, MWCNTs decreased number of flowers by 34 % ($p = 0.049$) at the lower concentration (3 mg kg^{-1}). The higher MWCNT concentration showed a similar decrease of flower number (33 %), but the variance between the samples was higher and there was no statistically significant difference ($p = 0.16$). Our results indicate that the number of flowers is sensitive to MWCNTs. Khodakovskaya et al. showed that the number of flower increased significantly, when watered weekly with 50 ml of 50 and 200 µg ml^{-1} MWCNTs per pot for 9 weeks [10]. The direction of the effect was in contrast to our observations. Nevertheless, the number of flowers was affected and further research is needed to determine the mechanism responsible for the effects of MWCNT on flowering.

To test effects of NPs on biological nitrogen fixation, the natural abundance of ^{15}N was determined in the red clover shoots and in a reference plant (rye grass) and subsequently the fraction of biological fixed nitrogen in red clover was assessed (see "Methods" section). No nitrogen was added to the pots because increasing the availability of mineral nitrogen has been reported to decline nitrogen fixation rate progressively [33]. The percentage of fixed

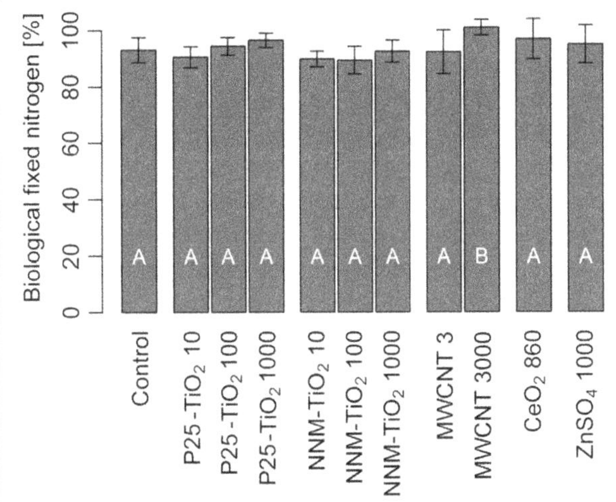

Fig. 2 Biological nitrogen fixation. Percentage of atmospheric nitrogen derived from biological nitrogen fixation in red clover shoots for the control, P25 and NNM-TiO_2, MWCNTs, CeO_2 NPs, and $ZnSO_4 \cdot 7H_2O$. The number behind the treatment name is the nominal concentration in mg kg^{-1}. Rye grass was used as non-nitrogen fixing plant and the B value was assumed to be zero (see text). *Error bars* show the standard deviations (n = 7). *Capital letters* show significant differences compared to the control plants ($p \leq 0.05$)

Table 1 Mean values and standard deviation of the arbuscular and vesicular root colonization

	Arbuscular colonization (%)		Vesicular colonization (%)	
	Mean	SD	Mean	SD
Control	23	8	4	2
P25-TiO_2 10 mg kg^{-1}	22	7	3	3
P25-TiO_2 100 mg kg^{-1}	22	8	6	9
P25-TiO_2 1000 mg kg^{-1}	21	10	8	7
NNM-TiO_2 10 mg kg^{-1}	19	10	7	5
NNM-TiO_2 100 mg kg^{-1}	24	11	6	3
NNM-TiO_2 1000 mg kg^{-1}	25	11	7	8
CNT 3 mg kg^{-1}	20	9	2	2
CNT 3000 mg kg^{-1}	29	7	5	5
CeO_2 860 mg kg^{-1}	24	9	8	5
$ZnSO_4 \cdot 7H_2O$ 1000 mg kg^{-1}	27	8	6	4

nitrogen was high and ranged between 89 and 100 % and was not affected by the TiO_2 NPs in our experiment. These results contrast those of another study performed in a hydroponic system using pea and rhizobia [23]. This study showed that nodulation was negatively affected and that the nitrogen fixation was delayed when TiO_2 NPs were present. However, it needs to be tested whether the results from hydroponic systems can be directly extrapolated to soil systems. In soils, TiO_2 NPs interact with soil particles and are probably heteroaggregated with soil particles such as clay minerals [32]. Thus, the plant roots in soils might be less exposed to the NPs than in hydroponic systems and therefore roots and nodules might be less affected in soils, as indicated by the limited transport of TiO_2 NPs in soils in our experiment [28]. For the higher concentration of MWCNTs (3000 mg kg^{-1}), nitrogen fixation increased by 8 % (p = 0.01) compared to the control and 100 % of the nitrogen content in the shoots originated from nitrogen fixation. Even though the biomass and total nitrogen content of these MWCNT treated plants were not different from those in the control treatment, correlation between biologically fixed nitrogen and total biomass over all treatments was significant but only 8 % of the variation could be explained ($R^2 = 0.08$; p = 0.012). This indicates that enhanced nitrogen fixation had only a small beneficial effect on plant growth. In our experiment, nitrogen fixation was not affected by CeO_2 NPs. For soybeans however, the CeO_2 NPs have been reported to decrease nitrogen fixation potential up to 80 % [19]. This reference investigated a different plant species and effects of NPs might be plant and rhizobia species specific [19]. Also the use of different soils with different soil characteristics might influence the results. Further experiments are needed to consolidate our understanding of the mechanisms of how NPs affect nitrogen fixation.

Total arbuscular, as well as vesicular root colonization of red clover by AMF were not affected in any of the treatments. In support of this finding, but again with another plant species, Burke et al. [20] reported no effects of TiO_2 NPs on AMF root colonization in soybeans using a DNA based approach instead of counting the root colonization. AMF provide plants with nutrient, such as phosphorus [17, 34]. Therefore we assessed phosphorus content in red clover shoots at the harvest. Phosphorus content of red clover shoots was not affected in any of the treatments and there was no correlation between plant phosphorus content and total AMF root colonization (p = 0.2). Again, for TiO_2 NPs this is in agreement with Burke et al. who did not find differences in phosphorus content of soybean leaves [20]. Even though root colonization was not affected by the tested NPs in our experiments, community structure of AMFs in soils might change as shown in Burke et al. [7].

Contrary to our expectations, the $ZnSO_4$ control did not affect any of the measured endpoints. It is known that Zn^{2+} availability is limited at high soil pH conditions [35]. Soil pH was 7.7 [28] and the concentration added was probably not high enough to release enough free Zn^{2+} to cause harmful effects.

The amount of NPs applied to the soil was high and partly outside the exposure range expected in the field.

They were chosen to represent a potential agricultural application scenario, where fluxes between several micrograms to grams of NPs per kilogram of soil are estimated [3]. The highest concentration also simulates accidental spill during transport or pollution in industrial areas or in the field. In our experiment also lower concentrations, i.e. 10 and 100 mg kg^{-1} soil, were tested. This approach ensures that potential negative effects can be detected before a NP is widely used and applied. This approach also facilitates the detection of potential harmful NPs in comparison to non-toxic or less harmful NP. Moreover, in order to be able to detect and measure concentrations of some NPs in the environment (e.g. titanium oxides for this study), high amounts have to be added because element like titanium occur naturally in the soil and the concentrations added need to be higher as natural background levels. For instance, for TiO_2 NPs the lowest concentration of 10 mg kg^{-1} is realistic in comparison with estimations for soils treated with NP containing plant protection products, while the highest tested concentration (1000 mg kg^{-1}) rather represents a worst case scenario [3]. For MWCNTs, yearly increases of estimated environmental concentrations are estimated to range from 5 to 990 ng kg y^{-1} [2]. Hence, both tested concentrations in our experiment are above natural values and represent an upper limit. The addition of these high concentrations was necessary to distinguish the added MWCNTs from the black carbon background of the soil [28, 36]. New methods are currently being developed to distinguish NPs from natural backgrounds as reviewed by others [37, 38]. Further research is needed to measure and characterize NPs in soils at predicted environmental concentrations, both for fate and behavior studies, and to accompany environmentally relevant ecotoxicological tests.

Conclusions

The investigated TiO_2 NPs and CeO_2 NPs did not affect red clover growth, biological nitrogen fixation and AMF root colonization. Opposite to other studies with TiO_2 and CeO_2 that observed effects on N fixing legumes, no effects were observed here with red clover. Further research is needed to search for general patterns and investigate the mechanisms behind such effects. MWCNTs increased nitrogen fixation and decreased the number of flowers compared to the control treatment, which might affect fitness of red clover. However, these effects occurred at concentrations much higher than expected in the environment.

Methods

NPs used for the experiment

P25 (Sigma Aldrich, USA, art. No. 718467) with a particle size of 29 ± 9 nm [28] was used as representative for TiO_2 NPs. In addition, NNM-TiO_2 (Sigma Aldrich, USA, Art. No. 232033) with an average particle size of 145 ± 46 nm [28] was used as non-nano-material, i.e. less than 50 % NPs [29]. MWCNTs were purchased from Cheap Tubes Inc. (USA). They had a length of 10–30 μm, outer diameter of 20–30 nm, a purity of >95 % and an elemental carbon content of >98 % (Additional file 1: Table S2) [28]. CeO_2 NPs (Sigma Aldrich, USA, art. No. 700290) had a diameter of less than 50 nm with cubic crystal structure according to the manufacturer's specifications.

Mixing NPs into the soil

For preparing the substrate, soil classified as brown earth with a sandy loamy to loamy fine fraction was collected from an agricultural field at Agroscope Institute for Sustainability Sciences in Zurich, Switzerland (coordinates N47° 25′ 39.564″ E8° 31′ 20.04″). For this, the top 5 cm were removed and the underlying 15 cm soil were collected and sieved (<0.5 cm). The soil was mixed with quartz sand (50 % v/v) and then characterized as described by Gogos et al. (Additional file 1: Table S3) [28]. Nutrient contents in the mixture were 37.6 mg kg^{-1} phosphorus and 85.3 mg kg^{-1} potassium determined by ammonium acetate EDTA extraction [39]. Soil pH was 7.7. Each of the different NPs was premixed in 300 g substrate (soil and sand) on an overhead mixer (Turbula T2F, Switzerland) in 500 ml Schott bottles by adding 0.3, 3 and 30 g of P25 or NNM-TiO_2, 90 mg and 88 g MWCNTs, 25 g CeO_2 NPs and 30 g $ZnSO_4{\cdot}7H_2O$ (Sigma Aldrich, USA, art. No. Z0251), respectively. P25 (30 g) and MWCNTs (88 g) revealed a too large volume for the 500 ml Schott bottles, necessitating the division of the soil and additives into several bottles (300 g of substrate for each bottle). For P25 15 g were added to two Schott bottles, and for MWCNTs 22 g were added to four bottles. Each of these pre-mixtures was diluted with substrate to a total volume of 30 kg and mixed in a cement mixer for 6 h.

Experimental setup

Pots were prepared by gluing PVC-sewer pipes (15 cm diameter, 20 cm long) on a plastic board with a ball valve as draining device (Fig. 3). A plastic mesh (Propyltex 500 μm, Sefar, Switzerland) was placed on the top of the valve to prevent blockage of the valve by the substrate. Pots were filled with a 500 g quartz sand layer as drainage and 3.3 kg spiked substrate or control substrate. Seven replications per treatment were prepared, i.e., control, P25, NNM-TiO_2, MWCNT, CeO_2 NPs, and $ZnSO_4{\cdot}7H_2O$. Total elemental titanium, black carbon (BC, for MWCNT treatments) and elemental cerium concentrations were determined in the substrate as described in the accompanying study [28]. Average total elemental titanium concentration of the

Fig. 3 Experimental setup. Sketch of the experimental setup of the pots and picture of a part of the pots in the greenhouse 12 weeks after the start of the experiment. All of the pots were randomly arranged in the greenhouse

highest tested concentrations was determined at the end of the experiment using X-ray fluorescence (XRF) and was 1332 ± 100 for the control treatment without titanium, 2059 ± 105 for 1000 mg kg^{-1} (nominal) P25 and 2007 ± 79 mg kg^{-1} for the NNM-TiO_2 treated soils, respectively [28]. For MWCNT the background of BC in control soils was on average 0.50 ± 0.06 mg g^{-1} and BC concentration in MWCNT 3000 mg kg^{-1} treated soil was 2400 ± 100 mg kg^{-1} as quantified by chemothermal oxidation [28]. Average elemental cerium concentration in the 830 mg kg^{-1} CeO_2 treatment was 416 ± 19 mg kg^{-1} determined with XRF at the end of the experiment.

Cultivation of red clover in NP spiked substrate

Red clover (*Trifolium pratense* var. Merula) was germinated on filter paper for 5 days. Thereafter, seven seedlings of equal size were transferred to the pots with substrate spiked with NPs or control soils in a greenhouse (16 h 25 °C 300 W m^2, and 8 h 16 °C in the dark). In addition seven pots with ryegrass (*Lolium perenne* var. Arolus) were prepared in the same way. These plants were grown because a non-nitrogen-fixing plant was needed to estimate biological fixed nitrogen in red clover (see below). The experiment was started in two blocks (n = 4 and 3, respectively), time-shifted with 1 week difference. All pots were regularly watered to keep the water holding capacity between 60 and 70 % (controlled by weighing and adding every time the same amount of water to all of the pots). Clover was fertilized after 6 and 9 weeks with 10 ml of · KH_2PO_4 (5 mM), $MgSO_4 \cdot 7H_2O$ (1 mM), KCl (50 μM), H_3BO_3 (25 μM), $MnSO_4 \cdot H_2O$ (1.3 μM), $ZnSO_4 \cdot 7H_2O$ (2 μM), $CuSO_4 \cdot 5H_2O$ (0.5 μM), $(NH_4)6Mo_7O_{27} \cdot 4H_2O$ (0.5 μM), and Fe(III) EDTA (20 μM). This is comparable to a phosphorus addition of 1.7 kg P ha^{-1}.

After 14 weeks NP exposure of red clover, the number of flowers (flower heads) was determined and the plant shoots were harvested. Soil cores were taken to assess NP concentration as described in Gogos et al. [28]. Roots were separated from the soil and washed. Then the roots were cut in 1 cm pieces, mixed in water and a randomized root subsample of approximately 2 g was taken for determining the AMF colonization. Roots were padded with a paper towel and weighed. The subsample was weighed separately and then stored at 4 °C in 50 % ethanol in Falcon tubes until the colonization was determined. The remaining roots as well as the red clover and ryegrass shoots were dried at 70 °C until they reached constant dry weight and dry weight of roots, shoots and total biomass (root + shoot weight) were determined. The dry weight of the AMF colonization root sample was calculated using the dry/wet weight ratio of the root sample. This AMF sample dry weight was added to the total root dry weight. Shoots of red clover and ryegrass were ground with a centrifugation mill (0.2 mm sieve, Retsch ZM200, Germany) and 2 mg samples were sent for ^{15}N analysis by isotope ratio mass spectrometry at the stable isotope facility at Saskatchewan University (Canada). Root colonization of AMF was analyzed by microscopy following the protocols of Vierheilig et al. [31] for staining the roots and McGonigle et al. [30] for counting the AMF structures. In short, roots were rinsed with deionized water, and transferred to 10 ml 10 % KOH for 20 min at 80 °C. Roots were rinsed again with water and stained in 5 % (v/v) ink (Parker Quink, black) in vinegar for 15 min at 80 °C. After rinsing the stained roots, they were transferred to 50 % glycerol for storage until root colonization was assessed. For microscopy, the root pieces were aligned in parallel onto a glass slide, covered with 50 % glycerol, and the roots were covered with a cover slip [30]. AMF structures in plant roots, i.e., hyphae, arbuscules, and vesicles, were counted for 100 intersections as described by McGonigle et al. [30]. Phosphorus content of shoots was assessed by ICP-OES using a hydrochloric acid digestion of the ashed residues [40].

Nitrogen fixation [%] was calculated using Eq. 1 where B is the value of $\delta^{15}N$ of shoots of plants, that are fully dependent upon nitrogen fixation [33]. For our experiment, a B value of 0 was assumed which reflects $\delta^{15}N$ of plants that are totally dependent on nitrogen fixation. The reference plant $\delta^{15}N$ was derived from the ryegrass shoots.

$$\%\ \text{Nitrogen fixation} = \frac{\delta^{15}\text{N of reference plant} - \delta^{15}\text{N of N}_2\text{ fixing plant}}{\delta^{15}\text{N of reference plant} - \text{B}} \times \frac{100}{1} \quad (1)$$

Statistics

All statistical analyses were performed with R [41]. A generalized linear model with Gaussian distribution was applied to determine differences of each treatment to the control. Thereby the two blocks of the different starting dates of the pot experiment were included as error term. The model was analyzed for homogeneity (Bartlett test) and normality (Shapiro test). Additionally a Dunnett test was performed (R library SimComp) using adjusted p-values for multiple testing [42] when normality and homogeneity were fulfilled. For non-normal residuals or inhomogeneous data, a Mann–Whitney test was used and p-values were adjusted for multiple testing according to Benjamini and Hochberg [43]. Pearson's correlations were calculated with the R command cor.test.

Abbreviations

AMF: arbuscular mycorrhizal fungi; CeO_2: cerium dioxide; CNT: carbon nanotubes; MWCNT: multiwalled carbon nanotubes; ^{15}N: nitrogen isotope; NNM-TiO_2: non-nanomaterial titanium dioxide; NP: nanoparticle; TiO_2: titanium dioxide.

Authors' contributions

JM carried out the study, analyzed data and wrote the manuscript. AG performed the physical–chemical NP analyses and revised the manuscript. MGAvdH, TDB, and FW participated in the design of the study and revised the manuscript. All authors read and approved the final manuscript.

Author details

[1] Agroscope, Institute for Sustainability Sciences ISS, 8046 Zurich, Switzerland. [2] Plant-Microbe-Interactions, Department of Biology, Utrecht University, 3508 TB Utrecht, The Netherlands. [3] Institute of Evolutionary Biology and Environmental Studies, University of Zurich, Winterthurerstrasse 190, 8057 Zurich, Switzerland.

Acknowledgements

We thank Florian Klingenfuss for providing assistance for harvesting the experiment.

Competing interests

The authors declare that they have no competing interests.

Funding

This work is part of the project "Effects of NANOparticles on beneficial soil MIcrobes and CROPS (NANOMICROPS)", within the Swiss National Research Programme NRP 64 "Opportunities and Risks of Nanomaterials". We thank the Swiss National Science Foundation (SNF) for financial support.

References

1. Keller A, McFerran S, Lazareva A, Suh S. Global life cycle releases of engineered nanomaterials. J Nanopart Res. 2013;15:1–17.
2. Sun TY, Gottschalk F, Hungerbühler K, Nowack B. Comprehensive probabilistic modelling of environmental emissions of engineered nanomaterials. Environ Pollut. 2014;185:69–76.
3. Gogos A, Knauer K, Bucheli TD. Nanomaterials in plant protection and fertilization: current state, foreseen applications, and research priorities. J Agric Food Chem. 2012;60:9781–92.
4. Khot LR, Sankaran S, Maja JM, Ehsani R, Schuster EW. Applications of nanomaterials in agricultural production and crop protection: a review. Crop Prot. 2012;35:64–70.
5. Du WC, Sun YY, Ji R, Zhu JG, Wu JC, Guo HY. TiO_2 and ZnO nanoparticles negatively affect wheat growth and soil enzyme activities in agricultural soil. J Environ Monit. 2011;13:822–8.
6. Song U, Jun H, Waldman B, Roh J, Kim Y, Yi J, Lee EJ. Functional analyses of nanoparticle toxicity: a comparative study of the effects of TiO_2 and Ag on tomatoes (*Lycopersicon esculentum*). Ecotoxicol Environ Saf. 2013;93:60–7.
7. Burke DJ, Zhu S, Pablico-Lansigan MP, Hewins CR, Samia ACS. Titanium oxide nanoparticle effects on composition of soil microbial communities and plant performance. Biol Fertility Soils. 2014;50:1169–73.
8. Ge Y, Schimel JP, Holden PA. Identification of soil bacteria susceptible to TiO_2 and ZnO nanoparticles. Appl Environ Microbiol. 2012;78:6749–58.
9. Ge YG, Schimel JP, Holden PA. Evidence for negative effects of TiO_2 and ZnO nanoparticles on soil bacterial communities. Environ Sci Technol. 2011;45:1659–64.
10. Khodakovskaya MV, Kim B-S, Kim JN, Alimohammadi M, Dervishi E, Mustafa T, Cernigla CE. Carbon nanotubes as plant growth regulators: effects on tomato growth, reproductive system, and soil microbial community. Small. 2013;9:115–23.
11. Shrestha B, Acosta-Martinez V, Cox SB, Green MJ, Li S, Cañas-Carrell JE. An evaluation of the impact of multiwalled carbon nanotubes on soil microbial community structure and functioning. J Hazard Mater. 2013;261:188–97.
12. van der Heijden MGA, Bruin Sd, Luckerhoff L, van Logtestijn RSP, Schlaeppi K. A widespread plant-fungal-bacterial symbiosis promotes plant biodiversity, plant nutrition and seedling recruitment. ISME J. 2016;10(2):389–99.

13. Heidstra R, Bisseling T. Nod factor-induced host responses and mechanisms of Nod factor perception. New Phytol. 1996;133:25–43.
14. Somasegaran P, Hoben HJ. Handbook for Rhizobia. New York: Springer-Verlag; 1994.
15. Carlsson G, Huss-Danell K. Nitrogen fixation in perennial forage legumes in the field. Plant Soil. 2003;253:353–72.
16. Smith FA, Smith SE. What is the significance of the arbuscular mycorrhizal colonisation of many economically important crop plants? Plant Soil. 2011;348:63–79.
17. Smith SE, Read D. 5-Mineral nutrition, toxic element accumulation and water relations of arbuscular mycorrhizal plants. In: Read SES, editor. Mycorrhizal symbiosis. 3rd ed. London: Academic Press; 2008. p. 145–8.
18. van der Heijden MGA, Martin FM, Selosse M-A, Sanders IR. Mycorrhizal ecology and evolution: the past, the present, and the future. New Phytol. 2015;205:1406–23.
19. Priester JH, Ge Y, Mielke RE, Horst AM, Moritz SC, Espinosa K, Gelb J, Walker SL, Nisbet RM, An Y-J, et al. Soybean susceptibility to manufactured nanomaterials with evidence for food quality and soil fertility interruption. Proc Natl Acad Sci USA. 2012;109:E2451–6.
20. Burke DJ, Pietrasiak N, Situ SF, Abenojar EC, Porche M, Kraj P, Lakliang Y, Samia ACS. Iron oxide and titanium dioxide nanoparticle effects on plant performance and root associated microbes. Int J Mol Sci. 2015;16:23630–50.
21. Chen C, Unrine JM, Judy JD, Lewis RW, Guo J, McNear DH Jr, Tsyusko OV. Toxicogenomic responses of the model legume *Medicago truncatula* to aged biosolids containing a mixture of nanomaterials (TiO_2, Ag, and ZnO) from a pilot wastewater treatment plant. Environ Sci Technol. 2015;49:8759–68.
22. Judy JD, McNear DH, Chen C, Lewis RW, Tsyusko OV, Bertsch PM, Rao W, Stegemeier J, Lowry GV, McGrath SP, et al. Nanomaterials in biosolids inhibit nodulation, shift microbial community composition, and result in increased metal uptake relative to bulk/dissolved metals. Environ Sci Technol. 2015;49:8751–8.
23. Fan R, Huang YC, Grusak MA, Huang CP, Sherrier DJ. Effects of nano-TiO_2 on the agronomically-relevant *Rhizobium*-legume symbiosis. Sci Total Environ. 2014;466–467:50–512.
24. Huang YC, Fan R, Grusak MA, Sherrier JD, Huang C. Effects of nano-ZnO on the agronomically relevant Rhizobium–legume symbiosis. Sci Total Environ. 2014;497:78–90.
25. Abd-Alla MH, Nafady NA, Khalaf DM. Assessment of silver nanoparticles contamination on faba bean-*Rhizobium leguminosarum bv. viciae*-Glomus aggregatum symbiosis: implications for induction of autophagy process in root nodule. Agric Ecosyst Environ. 2016;218:163–77.
26. Feng Y, Cui X, He S, Dong G, Chen M, Wang J, Lin X. The role of metal nanoparticles in influencing arbuscular mycorrhizal fungi effects on plant growth. Environ Sci Technol. 2013;47(16):9496–504.
27. Vesper SJ, Weidensaul TC. Effects of cadmium, nickel, copper, and zinc on nitrogen fixation by soybeans. Water Air Soil Poll. 1978;9:413–22.
28. Gogos A, Moll J, Klingenfuss F, van der Heijden M, Irin F, Green M, Zenobi R, Bucheli T. Vertical transport and plant uptake of nanoparticles in a soil mesocosm experiment. J Nanobiotechnol **(in press)**.
29. Commission recommendation on the definition of nanomaterial (http://eur-lex.europa.eu/LexUriServ/LexUriServ.do?uri=OJ:L:2011:275:0038:0040:EN:PDF).
30. McGonigle TP, Miller MH, Evans DG, Fairchild GL, Swan JA. A new method which gives an objective measure of colonization of roots by vesicular arbuscular mycorrhizal fungi. New Phytol. 1990;115:495–501.
31. Vierheilig H, Coughlan AP, Wyss U, Piche Y. Ink and vinegar, a simple staining technique for arbuscular-mycorrhizal fungi. Appl Environ Microbiol. 1998;64:5004–7.
32. Fang J, Shan X-q, Wen B, Lin J-m, Owens G. Stability of titania nanoparticles in soil suspensions and transport in saturated homogeneous soil columns. Environ Pollut. 2009;157:1101–9.
33. Unkovich M, Herridge D, Peoples M, Cadisch G, Boddey B, Giller K, Alves B, Chalk P. Measuring plant-associated nitrogen fixation in agricultural systems. Bruce: ACIAR; 2008.
34. Van Der Heijden MGA, Bardgett RD, Van Straalen NM. The unseen majority: soil microbes as drivers of plant diversity and productivity in terrestrial ecosystems. Ecol Lett. 2008;11:296–310.
35. Lindsay WL. Zinc in soils and plant nutrition. In: Brady NC, editor. Advances in agronomy, vol. 24. Cambridge: Academic Press; 1972. p. 147–86.
36. Sobek A, Bucheli TD. Testing the resistance of single-and multi-walled carbon nanotubes to chemothermal oxidation used to isolate soots from environmental samples. Environ Pollut. 2009;157:1065–71.
37. Farré M, Sanchís J, Barceló D. Analysis and assessment of the occurrence, the fate and the behavior of nanomaterials in the environment. TrAC Trends Anal Chem. 2011;30:517–27.
38. Von der Kammer F, Ferguson PL, Holden PA, Masion A, Rogers KR, Klaine SJ, Koelmans AA, Horne N, Unrine JM. Analysis of engineered nanomaterials in complex matrices (environment and biota): general considerations and conceptual case studies. Environ Toxicol Chem. 2012;31:32–49.
39. Stünzi H. The soil P extraction with ammonium acetate EDTA (AAE10). Agrarforschung. 2006;13:448–93.
40. Bassler R. Ausgewählte Elemente in pflanzlichem Material und Futtermitteln mit ICP-OES. In: VDLUFA-Methodenbuch—Band III—Die chemische Untersuchung von Futtermitteln. Vol 3; 1993.
41. R Core Team. R: a language and environment for statistical computing. Vienna: R Foundation for Statistical Computing; 2014.
42. Hasler M, Hothorn LA. A Dunnett-type procedure for multiple endpoints. Int J Biostat. 2011;7:1–15.
43. Benjamini Y, Hochberg Y. Controlling the false discovery rate: a practical and powerful approach to multiple testing. J Roy Stat Soc Ser B (Stat Method). 1995;57:289–300.

7

Imipenem/cilastatin encapsulated polymeric nanoparticles for destroying carbapenem-resistant bacterial isolates

Mona I. Shaaban[1,2], Mohamed A. Shaker[1,3*] and Fatma M. Mady[1,4]

Abstract

Background: Carbapenem-resistance is an extremely growing medical threat in antibacterial therapy as the incurable resistant strains easily develop a multi-resistance action to other potent antimicrobial agents. Nonetheless, the protective delivery of current antibiotics using nano-carriers opens a tremendous approach in the antimicrobial therapy, allowing the nano-formulated antibiotics to beat these health threat pathogens. Herein, we encapsulated imipenem into biodegradable polymeric nanoparticles to destroy the imipenem-resistant bacteria and overcome the microbial adhesion and dissemination. Imipenem loaded poly ε-caprolactone (PCL) and polylactide-*co*-glycolide (PLGA) nanocapsules were formulated using double emulsion evaporation method. The obtained nanocapsules were characterized for mean particle diameter, morphology, loading efficiency, and in vitro release. The in vitro antimicrobial and anti adhesion activities were evaluated against selected imipenem-resistant *Klebsiella pneumoniae* and *Pseudomonas aeruginosa* clinical isolates.

Results: The obtained results reveal that imipenem loaded PCL nano-formulation enhances the microbial susceptibility and antimicrobial activity of imipenem. The imipenem loaded PCL nanoparticles caused faster microbial killing within 2–3 h compared to the imipenem loaded PLGA and free drug. Successfully, PCL nanocapsules were able to protect imipenem from enzymatic degradation by resistant isolates and prevent the emergence of the resistant colonies, as it lowered the mutation prevention concentration of free imipenem by twofolds. Moreover, the imipenem loaded PCL eliminated bacterial attachment and the biofilm assembly of *P. aeruginosa* and *K. pneumoniae* planktonic bacteria by 74 and 78.4%, respectively.

Conclusions: These promising results indicate that polymeric nanoparticles recover the efficacy of imipenem and can be considered as a new paradigm shift against multidrug-resistant isolates in treating severe bacterial infections.

Keywords: Imipenem, Antibiotic resistance, Biodegradable, PCL, PLGA, Nanoparticles

Background

Over the last decades, the frightening spread of antibiotic-resistant infections all over the world inherently emerges multidrug-resistant/pan-resistant pathogens that evade even powerful antibiotics [1]. Various strategies have been used by the bacteria to develop resistance, including secretion of antibiotic specific/nonspecific inactivation enzyme, active efflux of antibiotics and surfaces adhesion with the formation of a protective biofilm [2, 3]. This biofilm provide an inaccessible barrier to even small molecule antibacterial agents affording the suitable support for the bacterial proliferation/colonization and development of a severe health threatening microbial infections [2]. Consequently, searching for effective and biofilm preventing bactericidal agents is deemed necessary in the clinical prospective of antibacterial therapy [4]. Designing new generation or derivative of antibiotics is incredibly costly investment process and it wastes much time until it is distinguished in the pharmaceutical

*Correspondence: mshaker@mun.ca; mona_ibrahem@mans.edu.eg
[1] Pharmaceutics and Pharmaceutical Technology Department, College of Pharmacy, Taibah University, PO Box 30040, Al Madina, Al Munawara, Saudi Arabia
Full list of author information is available at the end of the article

production pipelines, however, protection through a smart delivery system can potentiate the bactericidal efficacy of existing antibiotics and adequately address a solution to cease the current progression of resistant bacteria [5].

Carbapenems are new broad spectrum beta lactam antibacterial agents with potent activity against serious/complicated bacterial infections, which keeps them as the last reliable hospitalization treatment line for the life threatening microbial infections [6]. Initially, they had demonstrated a great stability against the hydrolysis by the beta-lactamases produced from the resistant pathogens, however, the emergence of carbapenem-resistance has been noticed globally with Gram-negative pathogenic bacteria such as *Enterobacteriaceae* and *Pseudomonas*. The prevalence of such resistance has been associated mainly with the bacterial formation of carbapenemases (carbapenem hydrolyzing enzymes), elimination of carbapenem influx, activation of the multi-drug efflux pump and mutation of the outer membrane protein [7, 8]. Such developed resistance rapidly disseminate between isolates and even among various species via integrons, transposons and exchange of plasmids [8]. Hazardously, carbapenem-resistant strains have been recently found to be associated with resistance to an expansive diversity of antimicrobials such as aminoglycosides, quinolones and cephalosporins [1]. Facing the aforementioned therapeutic challenge, various approaches have been investigated to formulate/deliver carbapenems, trying to accomplish two main goals. The first one is to overcome the stability issue through protecting the carbapenems molecular entity from the degradation bacterial enzymes to circumvent this bacterial blow [2]. The second is to target carbapenems to their site of action through increasing their bacterial penetration/uptake to predominate their therapeutic action on the bacterial renitences. The research efforts to reach these two goals are still ongoing.

Nano-size carriers claim the sufficient chemical protection and the adequate targeting effect needed for the effective delivery of antimicrobial molecules [5]. These nano-carriers include but not limited to liposomes, solid lipid nanoparticles, polymeric nanoparticles, metal nanoparticles, quantum dots and self-assembled micelles [9–12]. An extensive evaluation for the pros and cons of each of those nano-carriers for combatting microbial resistance has recently been reviewed [4, 5, 11–14]. Reader is strongly advised to refer/read those reviews for detailed discussions related to the improvement in the antibacterial activity of each of those nano-carrier delivery systems against various pathogenic microbes [11, 12]. Among the previously mentioned nano-size carriers, using polymer nanoparticles for the delivery of antibiotics has been growing steadily over the last years and getting special attention in the landscape of microbial resistance as well as prevention/eradication of biofilm formation [13, 14]. Polymeric nanoparticles are considered a promising antibiotics delivery vehicle due to the appropriate thermodynamic stability of the prepared self-assembled nano-sized particles. In addition to their ability to increase bacterial uptake and the penetration power of the loaded antibiotics to combat the developed multi-drug resistances. Meantime, they provide more in vivo stability for the loaded antibiotics against biodegradation, increase the circulation time inside the body and decrease the therapeutic dose and administration frequency [15]. Also, polymeric nanoparticles enhance localization of the loaded antibiotics to the infected organ, which is associated with minimizing the side effects accompanying the common systemic administration [16].

The highlighted biodegradable polymers that are widely used in literature as pharmaceutical carriers are either natural (such as alginate and chitosan) [17], or synthetic polyester from alpha hydroxy-acids (such as polylactide, polyglycolide and poly ε-caprolactone) [12] and polyamino acids [18]. Meantime, the only reported study for imipenem delivery was by Fazli et al. who physically entrap imipenem/cilastatin in the nanopores of ZnO nanoparticles and incorporate it with chitosan–polyethylene oxide nanofibrous mat [17]. Successfully they were able to sustain the release of loaded imipenem/cilastatin, however, there are still some unresolved issues and problems surrounding the potential toxicity of using ZnO nanoparticles in pharmaceutical applications [19]. On the other hand, polyhydroxy-acid ester are brilliant biodegradable biomaterials that have been employed as a promising vehicle to antimicrobials delivery in a controlled/sustained categorical form for a distinct period of time [20]. The specification of used polymer (e.g. molecular weight, hydrophilicity/hydrophobicity) easily manipulated to tailor the prepared nanoparticles for a specific delivery purpose [20]. To our knowledge, no research study has evaluated the using of polyhydroxy-acid esters for the protective delivery of carbapenems as explored in this study for the first time. Herein, we selected imipenem as the first candidate drug to represent carbapenems and in combination with cilastatin. Cliastatin has no antibacterial activity but inhibits the enzymatic degradation of imipenem molecules occurs by the renal dehydropeptidase [17]. In this paper, we formulated imipenem/cilastatin as polymeric nanocapsules using polylactide-*co*-glycolide (PLGA) and poly ε-caprolactone (PCL). We also characterized the prepared nanoparticles and evaluated their in vitro antibacterial efficiency against selected imipenem-resistant isolates.

Methods

Bacterial isolates and growth setting

The bacterial isolates were collected from clinical laboratory, Mansoura University Hospitals, Mansura, Egypt. *Klebsiella pneumoniae* isolates KMU5.5, KMU4.5, KMU2.3 were separated from urine samples. *Pseudomonas aeruginosa* isolates PUMU2.3, PWMU2.3 and PSMU8.0 were obtained from urine, wound and sputum samples, respectively. Isolates were identified according to the standard procedures of Bergey's manual, 1989 [21]. The sensitivity of the collected isolates to imipenem was first detected through the standard disc diffusion method (CLSI, 2015) [22]. Non-imipenem resistant standard strain *Klebsiella pneumoniae* ATTCC 4352 and *Pseudomonas aeruginosa* PAO1 were used as negative control. All the collected strains were propagated in Luria–Bertani (LB) medium (Yeast extracts 0.5%, Tryptone 1% and Sodium Chloride 1%, pH 6.8).

Materials

Poly ε-caprolactone (Mn = 45,000), polylactide-*co*-glycolide (50:50 Mw = 7000–17,000) and didodecyl dimethyl ammonium bromide (DMAB) were obtained from Aldrich-Sigma chemical company, USA. Imipenem and cilastatin (pharmaceutical grade) were kindly gifted from Merck Sharp & Dohme Corp., Canada. All the other chemicals were of analytical grade and used as received. Deionised water (Millipore®, 18.2 MΩ cm) was utilized as the water source in the experimental procedures.

Preparation of imipenem/cilastatin encapsulated nanoparticles

Polymeric nanocapsules prepared by double emulsion evaporation technique [23]. Briefly, 20 mg of both imipenem and cilastatin were first dissolved in 4 ml of deionized water. Sixty milligrams of the used polymer (either PCL or PLGA) were dissolved in 8 ml of chloroform. The antibiotic solution was blended with the polymer solution and emulsified by using a probe homogenizer (IKA, Ultra-Turrax) at 20,000 rpm for 15 min, to form the primary w/o emulsion and 0.2 ml of ethyl cellulose solution (10% w/v in chloroform) was added as a stabilizer for the prepared emulsion. Then, 25 ml of 1% w/v DMAB aqueous solution was added directly into the prepared w/o emulsion and emulsified by using a probe homogenizer at 20,000 rpm for 15 min, to produce the multiple-emulsion (w/o/w). The obtained emulsion was sonicated for 2 min and diluted with 20 ml of deionized water. The chloroform was then evaporated under vacuum by stirring for further 2 h at room temperature. The produced nanoparticles were collected by centrifugation for 20 min at 20,000 rpm and then rinsed three times by deionized water. In the following discussion, IMP refers to the physical mixture of equal weight of imipenem and cilastatin, however, IMP/PCL and IMP/PLGA refer to imipenem/cilastatin loaded poly ε-caprolactone and polylactide-*co*-glycolide nanoparticles, respectively.

Size and morphology of the prepared nanoparticles

Laser diffraction was used for particle analysis through measuring the particle size and zeta potential for the prepared nanoparticles (Microtrac, nanotrac wave II Q). Before measurement, all the samples were properly sonicated to avoid any aggregation and diluted with deionized water.

The shape and topology of the nanoparticles were detected by a Bioscope Catalyst-atomic force microscope (Bruker, Santa Barbra, CA, USA) mounted on Lica inverted microscope (Wetzlar, Germany). A single ScanAsyst-Fluid probe with a tip radius of 20 nm (Bruker, Santa Barbra, CA, USA) was used. The deflection sensitivity (nm/V) was determined on a glass and was found to be 45.5 nm/V. The nanoparticles suspension was placed on the silicon wafer and air dried.

Measurement of encapsulation efficiency, drug loading and nanoparticle yield

The encapsulation efficiency (EE) and drug loading (DL) were simply measured by quantizing the imipenem and cilastatin encapsulated inside the obtained nanoparticles after being disassembled and dissolved with methanol. The imipenem and cilastatin concentrations were determined by measurement of their UV absorbance at 318 and 243 nm, respectively, using a UV/visible spectroscopy (Evolution 201 UV–visible spectrophotometer, Thermo scientific). Simultaneously, nanoparticles' yield (NY) was determined gravimetrically. The percentage of the EE, DL and NY were calculated as follows:

$$\text{EE\%} = \left(\text{weight of encapsulated drug in nanoparticles}/\text{initial weight of added drug}\right) \times 100,$$

$$\text{DL\%} = \left(\text{weight of encapsulated drug in nanoparticles}/\text{weight of nanoparticles}\right) \times 100,$$

$$\text{NY\%} = \left(\text{weight of nanoparticles}/\text{initial weight of added drug and polymer}\right) \times 100.$$

Fourier transform infrared

Fourier transform infrared (FTIR) spectroscopic analysis was performed by using IRAffinity-1S spectrophotometer (Shimadzu, Japan) supplied with sealed interferometer with auto dryer attenuated total reflection accessory, and dynamic alignment system. All spectra were obtained at room temperature with examining spectrum of 400–4000 cm^{-1}. Each sample scanned for 32 times through a mercury cadmium telluride detector and the resolution of the spectrum was 4 cm^{-1} for all step-scan FTIR measurements.

X-ray powder diffraction analysis

X-ray powder diffraction (X-RPD) analysis were obtained using the Bruker D8-Advance X-ray diffractometer (Bruker, Germany). Scanning was implemented at a voltage of 40 kV and 30 mA using Cu Kα radiation. The scanned angle was set from 5° to 80° with accuracy of 0.02 throughout the measurement range, and the scanned rate was 4°/min.

Antimicrobial effect of the prepared nanoparticles

Primary screening of antimicrobial activity of IMP, IMP/PCL and IMP/PLGA was carried out by agar diffusion assay. The activity of IMP/PCL and IMP/PLGA against imipenem resistant isolates was compared to IMP. The minimal inhibitory concentrations (MICs) of IMP and its nano preparations were quantified throughout the microtiter plate assay method [22]. Muller–Hinton (100 μl) was distributed in a sterile microtitre plate. Twofold serial dilutions of IMP, IMP/PLGA, and IMP/PCL were prepared 1250, 625, 312.5, 156.5, 78.12, 39, 19.5, 9.75, 4.87, 2.4, 1.2 and 0.6 μg/ml in Muller–Hinton broth medium. All dilutions were inoculated with imipenem resistant isolates at final concentrations of 0.5 × 10^5 CFU/ml. Wells for positive and negative controls were included with all experiments. The MICs of plain nanoparticles of PLGA and PCL were also determined under the same conditions. The plates were incubated at 37 °C for 24 h and the minimum concentration that inhibit the growth of bacteria was determined.

Mutation prevention concentration (MPC)

The MPC was determined for IMP, IMP/PCL and IMP/PLGA according to the agar plate dilution procedures as reported previously by Credito et al. [24]. Briefly, twofold serial dilutions of the tested preparations (0.5–16 × MIC) were incorporated into Mueller–Hinton agar plates. Each plate was inoculated with 50 μl of a concentrated bacterial suspension containing 10^{10} CFU/ml. The plates were incubated for 24 h at 37 °C and each dilution was tested in triplicate. The isolates growing on the plates supplemented with antibiotic concentration ≥MIC were transferred on antibiotic-free medium, followed by a redetermination of their MICs to assess the development of microbial resistance. MPC was estimated as the concentration of the first plate that showed no bacterial growth.

Effect of carbapenemase on IMP, IMP/PCL and IMP/PLGA

The ability of the polymeric nano-formulation to protect imipenem from degrading enzymes was studied. *Pseudomonas aeruginosa* and *Klebsiella pneumoniae* isolates were first tested for carbapenemase production by using Modified Hodge Test (MHT) [22]. Carbapenemase enzymes were then isolated from carbapenemases producing strains by applying the published method by Masuda et al. with some modifications [3]. The isolates were propagated by overnight incubation at 37 °C in 40 ml LB medium. Bacterial pellets were collected by centrifugation at 4000 rpm for 10 min. Cells were suspended in equal volume of phosphate buffered saline (PBS) and disrupted by five cycles of sonication (30 s each) on ice. Crude enzymes in the supernatant were separated through centrifugation for 10 min at 8000 rpm. The presence of carbapenemase in the cell extracts was tested microbiologically by monitoring the hydrolysis of imipenem.

Finally, the stability of the nano-preparations was evaluated in the presence/absence of degrading enzymes compared to IMP. In this study, IMP, IMP/PCL and IMP/PLGA (1 mg/ml) were mixed with 100 μl of the enzyme extract and incubated for 1 h at 37 °C. Mueller–Hinton agar plates were inoculated with *Escherichia coli* (ATCC 25922) with adjusted turbidity to 0.5 McFarland standard. Wells were performed in the plates for IMP, IMP/PCL and IMP/PLGA either treated or untreated with the degrading enzyme and each sample was applied to the corresponding wells. The plates were incubated at 37 °C overnight for detection of growth inhibition zone.

Antibiotic kill test

The rate of killing imipenem-resistant bacteria by IMP/PCL and IMP/PLGA was determined and compared to IMP free antibiotic. KMU5.5 was propagated till bacterial count 5 × 10^6 CFU/ml. The culture was mixed with 5 μg/ml of each IMP or IMP/PCL or IMP/PLGA. Samples were taken at zero, 1, 2, 3, 4, 6, 10, and 24 h and each sample was serially diluted 1:10 to determine the number of the viable bacteria. In the time kill assay of PUMU2.3, 10 μg/ml of each preparation were used. Bacterial growth without IMP or IMP formulated particles was also tested. The number of bacteria recovered overtime post treatment was determined by surface drop method in triplicates [25]. The viable count (CFU/ml) = dilution factor × (Average number of colonies/drop)/volume of drop (0.01 ml). The count of the recovered cells post antibiotic treatment was plotted as the CFU/ml over time. The experiment was performed in triplicate.

Effect on bacterial attachment

The efficacy of the formulated nano-imipenem on microbial adhesion and biofilm formation was evaluated using microtitre plate method [26]. Growth medium containing 1/4 MICs of IMP, IMP/PCL and IMP/PLGA were prepared and 100 µl of each preparation was distributed into wells in triplicates. Overnight cultures of the tested isolates were diluted and 10 µl of each bacterial suspension was placed in the corresponding well and the plates were incubated at 37 °C for 24 h. Microbial controls containing no antibiotic were performed under the same conditions. At the second day, the wells were aspirated and free cells were washed with saline and attached biofilms were fixed with methyl alcohol. The wells were stained with crystal violet for 15 min and the plates were washed with deionized water. The stained biofilm with crystal violet was dissolved with glacial acetic acid (33% v/v) and measured with an E max microtitre ELISA reader (Microplate reader, Per long Medical Equipment Co., Ltd. China) at OD 490 nm and the mean reading of three wells was calculated [27].

In vitro drug release

Membrane dialysis was used for studying the in vitro release of imipenem and cilastatin. Nanoparticles were first suspended at a concentration of 10 mg/ml in PBS pH = 7.4 and then transferred to a pre-soaked Fischer dialysis tube (MWCO = 12,000–14,000 daltons). The assembled dialysis bags were immersed into 100 ml PBS of pH = 7.4 at 37 °C under shaking (100 rpm/min). At different time intervals, 4 ml of the release buffer was withdrawn and returned back after being measured by UV/vis spectrophotometer (Evolution 201 UV–visible spectro-photometer, Thermo scientific) to calculate the released amount of both imipenem and cilastatin.

Statistical analysis

Statistical analysis of the data was manipulated by using two-way ANOVA followed by the Bonferroni post-tests, using GraphPad Prism version 5.02 (GraphPad Software, Inc. La Jolla, USA). Data represented as mean ± standard deviation (SD). Statistical differences between the groups were considered significant if the p value was <0.05.

Results and discussion

Imipenem-resistant pathogens account for serious public and hospital-gained infections [28]. Restoring the antibacterial properties of imipenem modifies this health care concern and polymeric nano-encapsulation shall be the approach to catch that desire. Herein, we prepared polymeric nanocapsules by double emulsion formation followed by organic solvent evaporation [23]. The principles to generate core–shell structures by the two-step emulsification process "emulsions of emulsions". In the primary w/o emulsion, the aqueous phase represents the core to carry the antibiotic (imipenem and cilastatin). However, chloroform was used as a water immiscible organic phase, containing the shell-forming polymer (PCL or PLGA) and emulsion stabilizing agent (ethyl cellulose) [29]. The emulsifying agent (DMAB) was then added to the system to obtain the multiple-emulsion (w/o/w) and to achieve full diffusion of solvents. Subsequently, the solvent vaporization step was performed for nanoparticle solidification that occurred by solvent dispersion in association with polymer nanoprecipitation principles [23].

Characterization of the prepared nanoparticles

The measured particle sizes and zeta potentials using laser diffraction particle analyzer are listed in Table 1. The prepared IMP/PCL nanoparticles showed smaller size and higher zeta potential compared to that of IMP/PLGA nanoparticles. However, both exhibited high encapsulation efficiency, drug loading and yield value. Meanwhile, Fig. 1 shows the atomic forced microscopy image for IMP/PCL nanoparticles. The contour photo views that the shape of PCL nanoparticles is well defined and have a distinctive spherical structure and a smooth surface. Furthermore, an acceptable match was observed in the mean diameter as measured by both laser diffraction particle sizer (Fig. 1a) and atomic forced microscopy (Fig. 1c).

FTIR spectrum for IMP, PCL nanoparticles, IMP/PCL, PLGA nanoparticles and IMP/PLGA are shown in Fig. 2. As it observed the imipenem/cliastatin had characteristic peaks/bands (cm^{-1}) that were attributed to vibration of OH stretching (3400 and 3250), N–H stretching

Table 1 Physical characterization of imipenem/cilastatin loaded polycaprolactone (IMP/PCL), and imipenem/cilastatin loaded polylactide-*co*-glycolide (IMP/PLGA) nanoparticles

Used polymer	Size (nm)	Zeta potential (mv)	Encapsulation efficiency (EE) %		Drug loading (DL) %		Nanoparticle yield (NY) %
			Imipenem	Cilastatin	Imipenem	Cliastatin	
PCL	132 ± 20	17 ± 1.6	83 ± 2.2	81 ± 1.2	17.7 ± 0.6	17.3 ± 0.7	93 ± 2.7
PLGA	348 ± 65	15 ± 0.6	76 ±1.2	77 ± 3.8	17.2 ± 1.3	17.6 ± 0.9	88 ± 1.9

PCL polycaprolactone, *PLGA* polylactide-*co*-glycolide

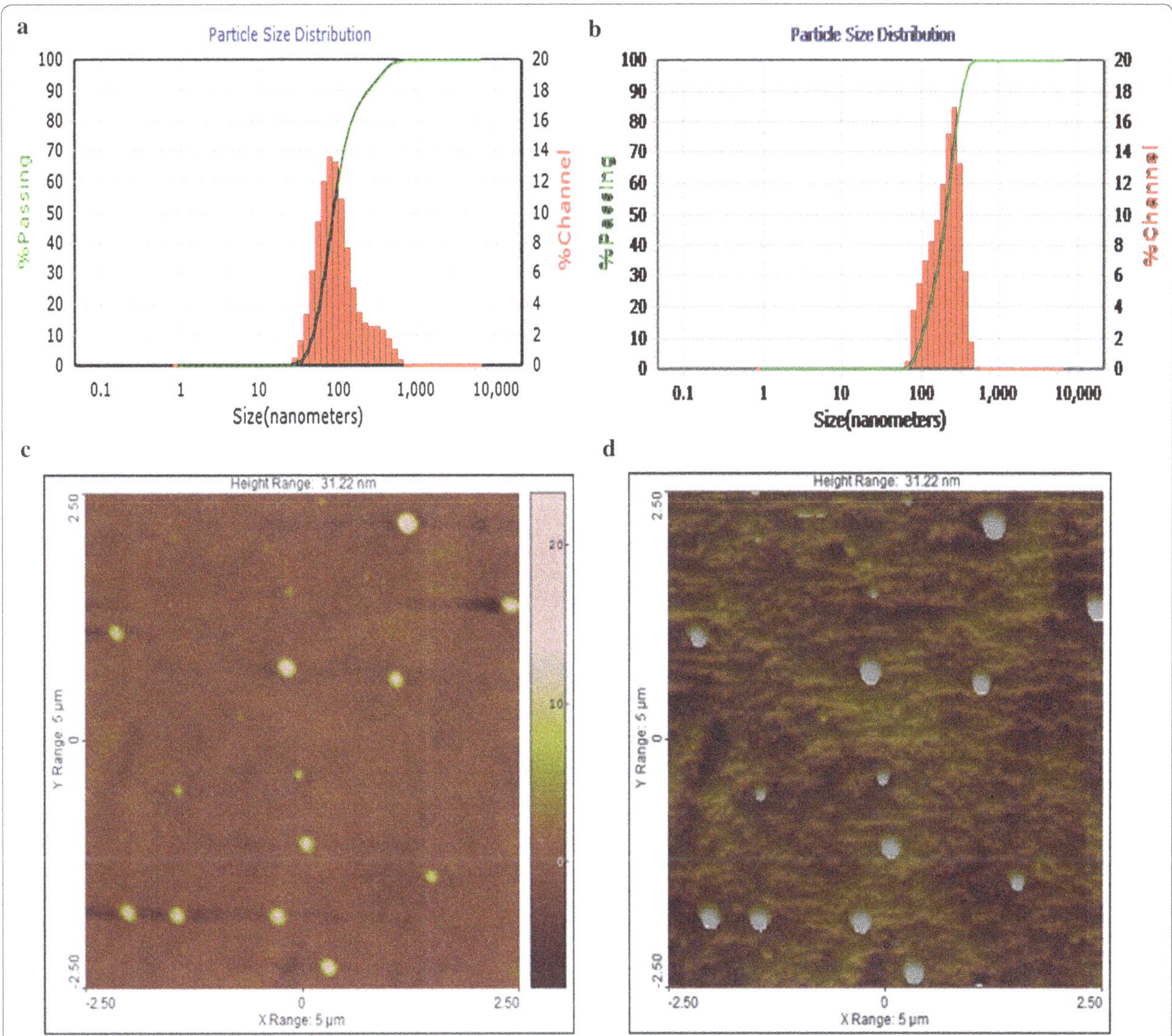

Fig. 1 **a** Particle size distribution for imipenem/cilastatin loaded polycaprolactone nanoparticles as measured by laser diffraction particle analyzer. **b** Particle size distribution for imipenem/cilastatin loaded polylactide-co-glycolide nanoparticles as measured by laser differaction particle analyzer. **c** Image for imipenem/cilastatin loaded polycaprolactone nanoparticles as measured by atomic forced microscopy (AFM), a 2-D view and **d** the image in 3-D

(2950), C–H stretching (2860), C=O stretching (1776 and 1737), amide bands (1655 and 1580), C=N stretching (1460 and 1440), C=C stretching (1395 and 1230), N–C–S vibration (1070/1030), C=C bending (991 and 952), C–S stretching (892 and 809) C–N–H bending (727), C–S stretching (697) and C–O–H bending (666). The characteristic peaks/bands for PCL appear in both PCL nanoparticles and IMP/PCL nanoparticles which are OH stretching (3600–3200), C–H stretching (2950 and 2860), C=O stretching (1725), C–H symmetric/asymmetric deformation (1474, 1425, 1400 and 1370), C–O–C stretching (1297, 1246, 1194), C–O stretching (1108, 1070 and 1050), C–H bending (965 and 935), and C–C stretching (842 and 733). Likewise, the characteristic peaks/bands for PLGA appear with both PLGA nanoparticles and IMP/PLGA nanoparticles which are OH starching (3650–3250), C–H stretching (2980, 2930 and 2878), C=O stretching (1760), C–H symmetric/asymmetric deformation (1460 and 1425), C–O–C stretching (1395, 1275, 1175 and 1095), CH_3 asymmetric rocking (1070/1030), C–O stretching (1054), and C–C stretching (872 and 751).

The encapsulation of IMP in the IMP/PCL nanoparticles was confirmed by the association of amide I band (1655) and amide II band (1580) in the spectrum. Also, the association of IMP in the IMP/PLGA nanoparticles

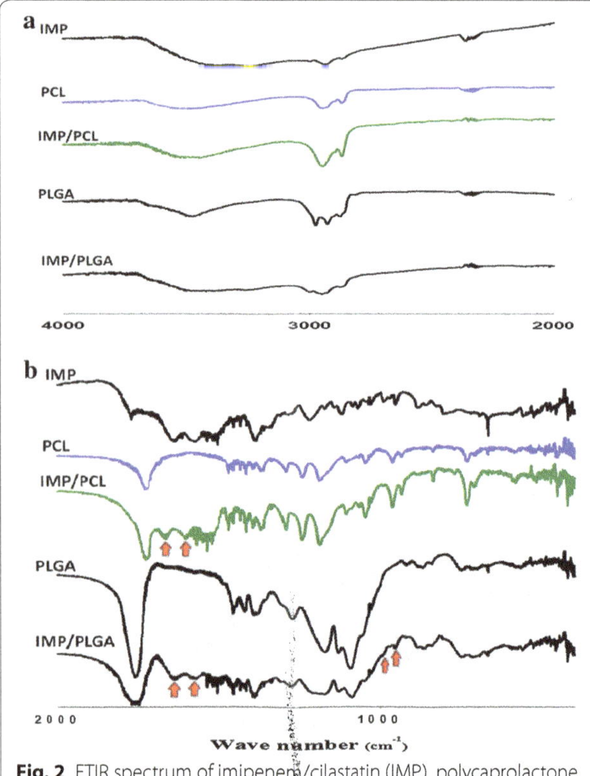

Fig. 2 FTIR spectrum of imipenem/cilastatin (IMP), polycaprolactone nanoparticles (PCL), imipenem/cilastatin loaded polycaprolactone nanoparticles (IMP/PCL), polylactide-*co*-glycolide nanoparticles (PLGA), and imipenem/cilastatin loaded polylactide-*co*-glycolide nanoparticles (IMP/PLGA). **a** The spectrum from 4000 to 2000 cm^{-1}. **b** The spectrum from 2000 to 400 cm^{-1}

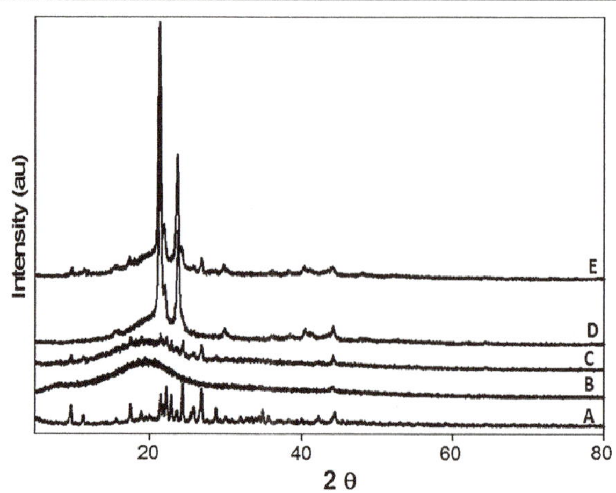

Fig. 3 X-ray diffractograms for imipenem/cilastatin (*A*), polylactide-*co*-glycolide nanoparticles (*B*), imipenem/cilastatin loaded polylactide-*co*-glycolide nanoparticles (*C*), polycaprolactone nanoparticles (*D*) and imipenem/cilastatin loaded polycaprolactone nanoparticles (*E*)

was confirmed by the amide I band (1655), amide II band (1580) and the C=C bending (991 and 952). Assignment of the imipenem peaks in both IMP/PCL and IMP/PLGA nanoparticles confirms the encapsulation of imipenem and demonstrates the thermodynamic stability for the chemical structure of encapsulated imipenem [17].

X-ray diffractographs for IMP, PLGA, IMP/PLGA, PCL and IMP/PCL nanoparticles are shown in Fig. 3. The IMP showed its crystalline peaks at diffraction angles (2θ) of 9.6°, 11.3°, 17.5°, 18.9°, 21.3°, 22.2°, 22.8°, 23.5°, 24.3°, 25.7°, 26.7°, 28.6°, 34.8°, 42.1° and 44.4°. The prepared PLGA nanoparticles were amorphous and did not show any crystalline peaks, nonetheless, IMP/PLGA nanoparticles showed the IMP characteristic peaks but at low intensity. The PCL nanoparticles showed sharp and intense crystalline peaks at 21.4°, 21.9° and 23.6°, meantime, all these peaks in addition to IMP characteristic peaks were detected in IMP/PCL. The presence of IMP peaks within the pattern of both IMP/PLGA and IMP/PCL confirmed the crystalline pattern of the encapsulated IMP within the prepared nanoparticles and this is also in agreement with the previously published reports [17].

Antibacterial activity

The initial screening for the prepared nanoparticles indicated that clinical isolates which were non-susceptible to IMP, but IMP/PCL showed marked antimicrobial activity against carbapenem-resistant isolates (Fig. 4). Table 2 displays the MIC values as measured through the broth micro-dilution technique, for IMP/PCL and IMP/PLGA compared to IMP. The obtained results illustrate that IMP inhibited the growth of PSMU8.0 isolate with MIC of 40 µg/ml, while, PWMU2.3 and PUMU2.3 isolated possessed high-level of resistance and showed MIC equals 625 µg/ml. In addition, IMP showed MIC value ranged from 80 to 625 µg/ml against the tested *Klebsiella pneumoniae* isolates. Meantime, IMP/PCL was more effective than IMP as the susceptibility of both *P. aeruginosa* and *K. pneumoniae* significantly increased (Table 2). The MIC of IMP/PCL has decreased by five to seven folds compared to the free IMP. IMP/PLGA formulation was less effective than IMP/PCL and was only able to decrease the MIC of KMU4.5 (two folds) as well as PWMU2.3 and PSMU80 (1-2 folds) from all of the tested isolates (Table 2). At the same time, the control nanoparticles either PLGA or PCL revealed no activity on the bacterial growth, highlighting that the antimicrobial effects were only obtained from the encapsulated drug itself.

These results are also in accordance with reported outcomes that have been published previously. Xiong et al. had been demonstrated that a vancomycin-loaded PCL nanoparticles are more effective in penetrating cells infected with *Staphylococcus aureus* compared to free vancomycin [30]. Also, acrylated penicillin G polymer

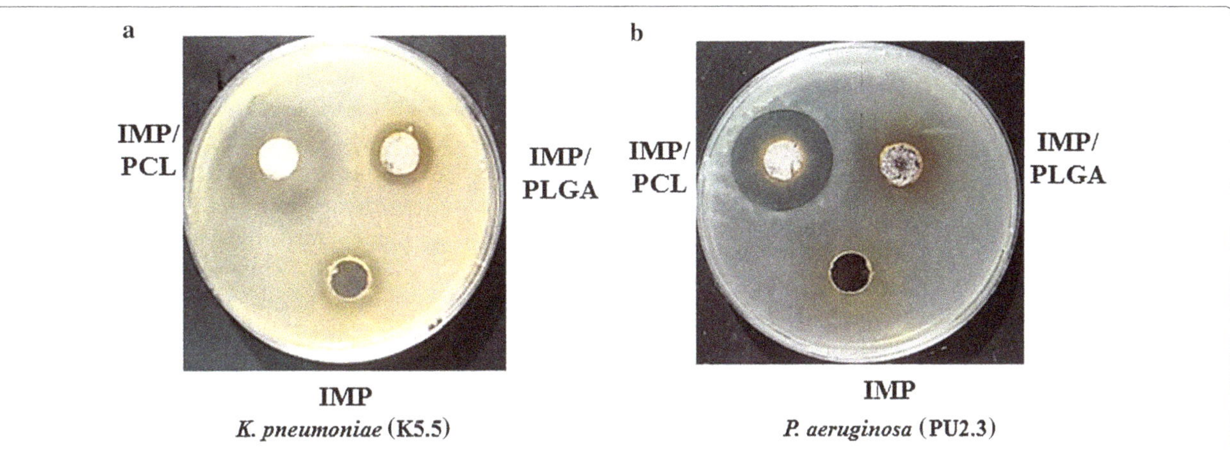

Fig. 4 Antimicrobial screening of the formulated imipenem/cilastatin loaded polycaprolactone (IMP/PCL) and imipenem/cilastatin loaded polylactide-*co*-glycolide (IMP/PLGA) nanoparticles against *Klebsiella pneumoniae*, KMU5.5 (**a**) and *Pseudomonas aeruginosa* PUMU2.3 (**b**) compared to imipenem/cilastatin (IMP). IMP/PCL retained antimicrobial activity against both isolates

Table 2 The minimal inhibitory concentrations (MIC) of imipenem/cilastatin loaded polycaprolactone (IMP/PCL), and imipenem/cilastatin loaded polylactide-*co*-glycolide (IMP/PLGA) compared to imipenem/(IMP) against *Klebsiella pneumoniae* and *Pseudomonas aeruginosa* isolates

	MIC (μg/ml)		
	IMP/PCL	IMP/PLGA	IMP
Klebsiella pneumoniae			
ATCC4352	0.6	1.25	0.6
KMU5.5	5	312.5	312.5
KMU4.5	2.5	156.5	625
KMU2.3	2.5	80	80
Pseudomonas aeruginosa			
PAO1	1.25	1.25	1.25
PWMU2.3	20	312.5	625
PUMU2.3	10	625	625
PSMU8.0	1.25	20	40

exhibits potent antimicrobial activity against MRSA [16]. Nano-penicillin with particle size 70 nm exhibits higher bactericidal activity than the bulk penicillin [31]. As it has been reported, the ability of imipenem to pass through the Gram-negative bacterial membrane is mainly attributed to its low molecular weight and zwitterion nature [32]. Imipenem molecules penetrate the bacteria through membrane transporting proteins (porins) which act as pores/channels for selective/nonselective diffusion of molecules as nutrients, such as outer membrane porins (OMPs) [24, 33]. Some of these porins are nonspecific such as omp F and omp C whereas others are special for certain molecules such as OprD which is specifically expressed in *P. aeruginosa* [6]. Concurrently, it was revealed that gained bacterial resistance to imipenem is associated with alteration in the structure of the imipenem binding and penetration part in porins [6, 34] such a mutational change resulting in specific bacterial identification to imipenem molecules with the loss of imipenem transmembrane entrance ability [34]. Nonetheless, encapsulated imipenem will not be easily identified by bacteria and is able to disguisedly diffuse through the membrane porins. Consequently, the observed enhancement in the antibacterial action of IMP/PCL could be associated with the enhanced membrane permeability associated with nanosize-encapsulated imipenem [14, 15, 30]. Meantime, the bacterial porins act as size exclusion filters and the size of the prepared particles plays that role in penetrability through the membrane, therefore, IMP/PCL nanoparticles (132 $\pm$ 20 nm) were more penetrable than that of the IMP/PLGA (348 $\pm$ 65 nm).

Protection from carbapenemases

Production of carbapenemase is also included as an essential factor in carbapenem-resistance for Gram-negative bacteria [7, 35]. Incorporation of IMP into the nanocapsules could provide protection from bacterial degrading enzymes. In order to address this hypothesis, IMP and IMP formulations were screened against *E. coli* sensitive isolate in the presence and the absence of carbapenem degrading enzymes (Fig. 5). The image shows that IMP, IMP/PCL and IMP/PLGA, had the same antimicrobial effect against the carbapenem-susceptible *E. coli*. When they were treated with the hydrolyzing enzymes, IMP lost all its activity while both nanoparticles retained their inhibitory effects. This could be attributed to the protective capacity of the polymeric nanoparticle

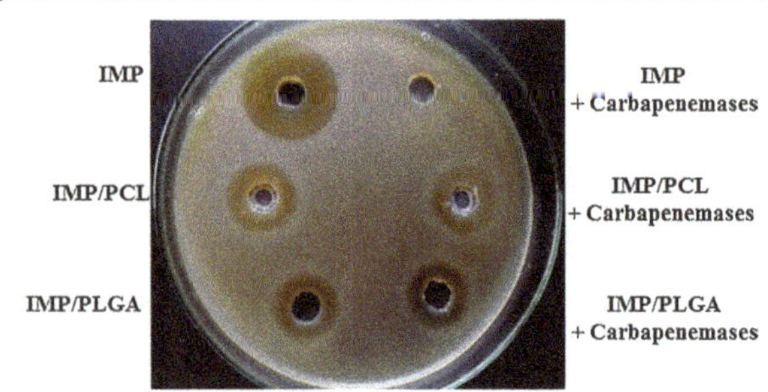

Fig. 5 Effect of carbapenemases on activity of imipenem/cilastatin (IMP), imipenem/cilastatin loaded polycaprolactone nanoparticles (IMP/PCL) and imipenem/cilastatin loaded polylactide-*co*-glycolide nanoparticles (IMP/PLGA) against *Escherichia coli* ATCC 25922

to the incorporated antimicrobial agent from the inactivation by hydrolyzing enzymes without affecting its antibacterial properties. As such loaded imipenem molecules remained effective against resistant Gram-negative isolates. Likewise, coating penicillin G in polyacrylate nanoparticles provides protection to the loaded antibiotic from extracellular penicillinases and retains penicillin efficacy [16].

Mutation inhibition effect

Above the minimal inhibitory concentration, bacteria exhibit a selective emergence of resistant sub populations. The mutation prevention concentration represents the concentration that suppresses the emergence of resistance mutations and equivalent to the MIC of the strains. On that ground, we screened the ability of IMP/PCL to induce antibiotic mutation in comparison to IMP control. IMP induced the resistance in *K. pneumoniae* isolate KMU2.3 and enhanced the phenotypic switching to high-level antibiotic resistance rapidly and irreversibly. The MPC of KMU2.3 emerging mutants was 320 µg/ml with twofolds increase in the MIC of IMP. However, IMP/PCL did not induce mutation and the MPC was equal to the MIC (2.5 µg/ml) in KMU2.3.

Previous studies revealed the ability of imipenem and meropenem to induce mutational changes in Gram-negative bacilli [24, 33]. This mutational change has been attributed to various resistance mechanisms such as changing cell permeability [36], induction of carbapenemase [37] or expression of metallic β-lactamases from an integron [35]. However incorporation of imipenem into polymeric nanoparticle was able to prevent such a mutational changes.

Time kill assay

The rate of microbial killing following antibiotic treatment is so critical for avoiding any future resistance from bacteria. For that reason, we tracked bacterial viability in the presence of IMP, IMP/PLGA and IMP/PCL compared to control untreated cultures. As shown in Fig. 6, treating *K. pneumoniae* KMU5.5 isolate with IMP/PCL caused significant decline ($P < 0.01$) in the bacterial growth within 2 h of incubation and complete bacterial killing after 6 h. However, IMP and IMP/PLGA did not affect bacterial growth (Fig. 6a). Similarly, the count of *Pseudomonas aeruginosa* PUMU2.3 significantly decreased after 3 h of incubation with IMP/PCL ($P < 0.01$), within 6 h the count of PUMU2.3 was reduced from 2×10^8 to 5.5×10^3 CFU/ml and declined to 45 colonies by 10 h. Coating of IMP with PCL increased its killing capacity. Instant microbial killing induced by IMP/PCL could be attributed to bacterial hydrophobic affinity for the PCL

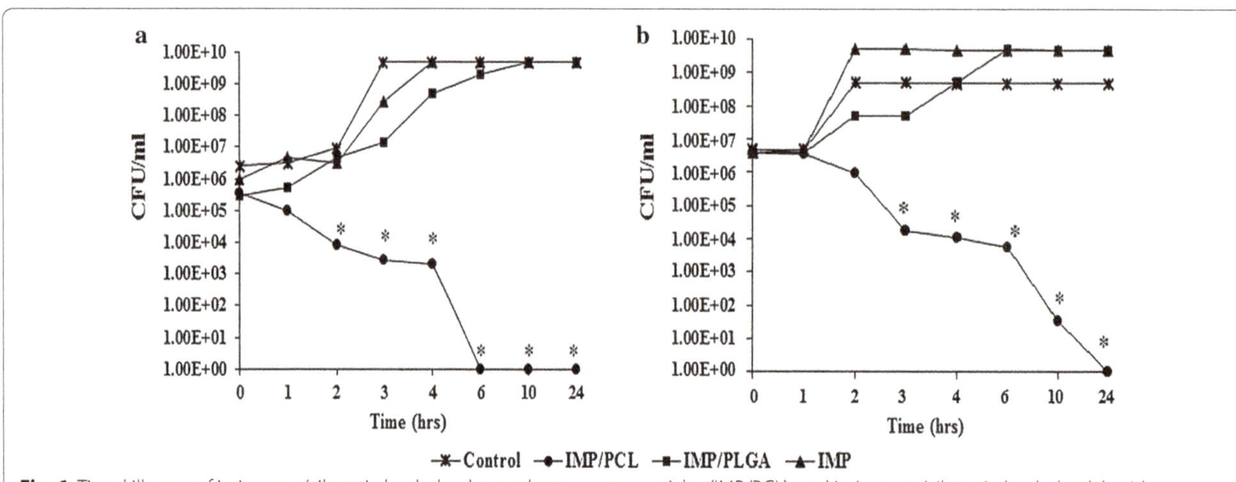

Fig. 6 Time kill assay of imipenem/cilastatin loaded polycaprolactone nanoparticles (IMP/PCL), and imipenem/cilastatin loaded polylactide-*co*-glycolide nanoparticles (IMP/PLGA) against **a** *Klebsiella pneumoniae* KMU5.5 and **b** *Pseudomonas aeruginosa* PUMU2.3. Low concentration of IMP/PCL (5-10 µg/ml) caused rapid and significant decrease in bacterial count within 2–3 h of interaction (*significant difference $P < 0.05$)

polymer and the nano-size/charge of the formulated imipenem, which assist the rapid penetration through the bacterial outer membrane [5].

Anti-adhesion activity

Biofilm is a disseminated microbial growth that is difficult to penetrate and develops resistance to conventional treatment. Various strategies have been investigated to enhance the antimicrobial to microbial biofilms especially those biofilms formed on the implanted medical devices [38]. As demonstrated in Fig. 7, the influence of IMP and nano-formulated imipenem at sub-inhibitory concentrations (1/4 MIC) on the attachment of *K. pneumoniae* and *P. aeruginosa* were evaluated. IMP non-functionalized antibiotic maintained bacterial ability for biofilm synthesis. The adhesion of KMU5.5 and PUMU2.3 in the presence of IMP/PCL was significantly ($P < 0.01$) reduced by 74 and 78.4%, respectively, compared to the control untreated cultures.

Formulation of nano-therapy increases the solubility and decrease the aggregation of the antimicrobial agents hence, improve their penetration and efficacy. Nano-formulation of IMP resulted in reducing the particle size of imipenem and enhanced the penetration of the loaded antibiotic [9]. In addition to the hydrophobic nature of PCL chains that assists the rapid penetration of the antibiotic and rupture of the bacterial cell wall so prevents microbial colonization and block biofilm formation [13].

In vitro drug release

Diffusion-driven release profiles of both imipenem and cilastatin from the prepared nanoparticles are demonstrated in Fig. 8. With PLGA nanoparticles, fast release occurs within the initial phase (the first three days) followed by a much slower release period. The release profiles show that 61.8 and 50.8% of the loaded imipenem and cilastatin, respectively, released within the first 3 days and 67.8 and 59.6%, released at the day 7. Meanwhile, PCL nanoparticles demonstrated a slow diffusion rate all over the release study periods. The release profiles show that only 9.7 and 6.9% of the loaded imipenem and cilastatin, respectively, released within the first 3 days and 17.9 and 13.7% of the loaded imipenem and cilastatin, respectively, released at the day 7. This turns out that PCL nanoparticles are very good at slowly releasing imipenem/cilastatin over extended periods of time.

PLGA is relatively amorphous low molecular weight polymer with reasonable hydrophilicity [39], whereas, PCL is a semi-crystalline high molecular weight polymer with reasonable hydrophobic nature [40]. Hence, hydrophilic drug (imipenem/cilastatin) was incorporated within the polymeric shell of PLGA nanoparticles (poor entrapment) or in the core of PCL nanoparticles (good entrapment). Shell and surface associated drug is adjacent

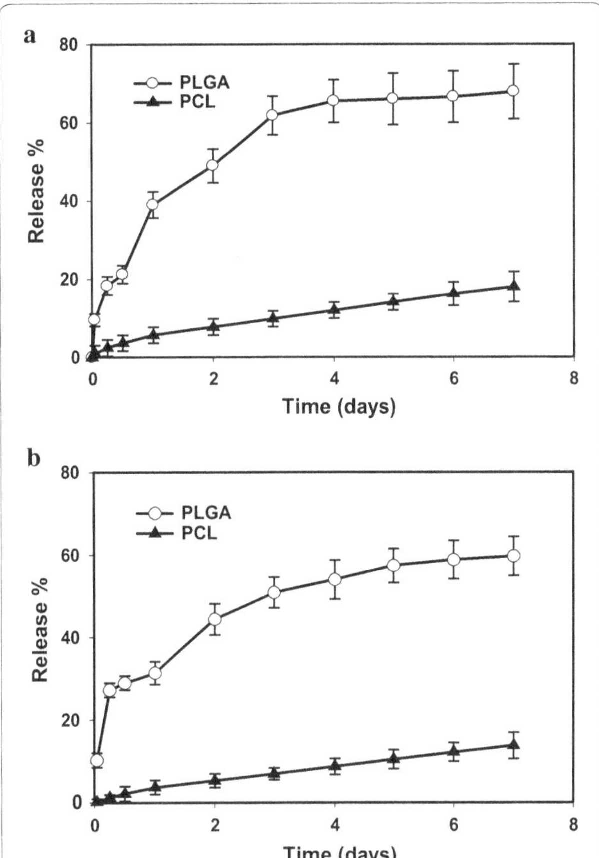

Fig. 8 Cumulative percent of released imipenem (**a**) and cilastatin (**b**) from the prepared polymeric nanoparticles polycaprolactone (PCL) and polylactide-*co*-glycolide (PLGA) in phosphate buffer saline (pH = 7.4) as a release media at 37 °C. *Error bars* represent the standard deviation of the mean of measurements from four samples

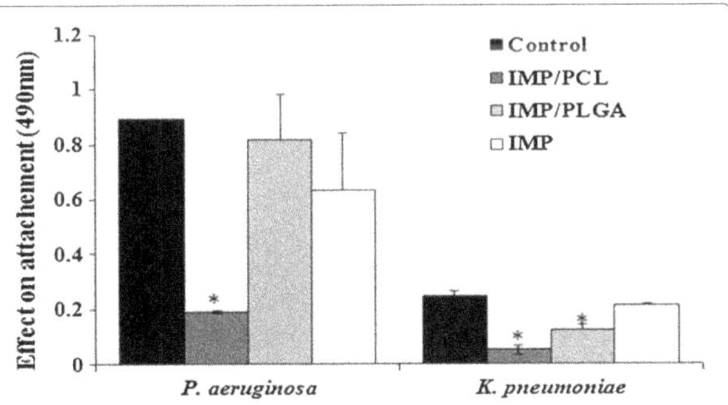

Fig. 7 Effect of imipenem/cilastatin (IMP), imipenem/cilastatin loaded polycaprolactone nanoparticles (IMP/PCL), and imipenem/cilastatin loaded polylactide-*co*-glycolide nanoparticles (IMP/PLGA) on biofilm formation. IMP/PCL significantly eliminated the adhesion of *Pseudomonas aeruginosa* and *Klebsiella pneumoniae* and inhibited biofilm assembly (*significant difference $P < 0.05$)

to the PBS and is easily released, depending on water solubility and water imbibition within the nanoparticles shell (polymeric layer) [41]. Therefore, the initial burst release of imipenem from PLGA is mainly attributed to the diffusion from the outer most layers of the nanoparticles as well as emancipation on the external layers of the nanoparticles. In addition to the PLGA bulk degradation that is associated with a gradual decline of its molecular weight; even without weight loss or soluble monomer formation [39]. Such a fast release phase is associated with a linear slow release phase, where the drug diffusion only was occurring. On the other hand, the slow release rate from PCL nanoparticles can be attributed to the hydrophobic properties of drug depleted shell layers [40].

Conclusion

Imipenem/cilastatin encapsulated polylactide-*co*-glycolide and polycaprolactone nanoparticles were successfully prepared by double emulsion evaporation method. Antibacterial efficacy evaluation results showed that polycaprolactone nanoparticles were more effective than polylactide-*co*-glycolide nanoparticles and free drug, against all of the tested resistant isolates. Hence, the antimicrobial efficacy of carbapenems can be regained to smash the developing resistant isolates if the polymeric nanoencapsulation of the drug is virtually adopted. More effective therapies/clinical outcomes are expected from the strategy of using carbapenem-encapsulated polycaprolactone nanoparticles in comparison with using the current marketed formule administered in the hospitals. It cannot be overlooked that such a suggested application still requires further strong and dependable in vivo testing of antimicrobial activity, for confirming the efficacy and the validity for pharmaceutical production. A highlighted area for future perspective is applying carbapenem-loaded polycaprolactone nanoparticles in vivo in different local/systemic infections using suitable animal models and our study in this regard is still undergoing.

Authors' contributions

Authors MSH, MS and FM contributed equally in the design, analysis and interpretation of data for the work. They also worked on drafting the paper. All authors read and approved the final manuscript.

Author details

[1] Pharmaceutics and Pharmaceutical Technology Department, College of Pharmacy, Taibah University, PO Box 30040, Al Madina, Al Munawara, Saudi Arabia. [2] Microbiology and Immunology Department, Faculty of Pharmacy, Mansoura University, PO Box 35516, Mansoura, Egypt. [3] Pharmaceutics Department, Faculty of Pharmacy, Helwan University, PO Box 11795, Cairo, Egypt. [4] Pharmaceutics Department, Faculty of Pharmacy, El-Minia University, El-Minia, Egypt.

Acknowledgements

The authors would like to thank the Deanship of Scientific Research at Taibah University, Al-Madinah Al-Munawarah, Kingdom of Saudi Arabia. Thanks and appreciation to clinical laboratories at Mansoura University for providing clinical isolates.

Competing interests

The authors declare that they have no competing interests.

Funding

The authors have received no funding for the research or in the preparation of this manuscript.

References

1. Ventola CL. The antibiotic resistance crisis: part 1: causes and threats. Pharm Ther. 2015;40:277–83.
2. Fair RJ, Tor Y. Antibiotics and bacterial resistance in the 21st century. Perspect Med Chem. 2014;6:25–64.
3. Masuda G, Tomioka S, Hasegawa M. Detection of beta-lactamase production by gram-negative bacteria. J Antibiot (Tokyo). 1976;29:662–4.
4. Beyth N, Houri-Haddad Y, Domb A, Khan W, Hazan R. Alternative antimicrobial approach: nano-antimicrobial materials. Evid Based Complement Alternat Med. 2015;2015:246012.
5. Brooks BD, Brooks AE. Therapeutic strategies to combat antibiotic resistance. Adv Drug Deliv Rev. 2014;78:14–27.
6. Kohler T, Michea-Hamzehpour M, Epp SF, Pechere JC. Carbapenem activities against *Pseudomonas aeruginosa*: respective contributions of OprD and efflux systems. Antimicrob Agents Chemother. 1999;43:424–7.
7. Quale J, Bratu S, Gupta J, Landman D. Interplay of efflux system, ampC, and oprD expression in carbapenem resistance of *Pseudomonas aeruginosa* clinical isolates. Antimicrob Agents Chemother. 2006;50:1633–41.
8. Meletis G, Exindari M, Vavatsi N, Sofianou D, Diza E. Mechanisms responsible for the emergence of carbapenem resistance in *Pseudomonas aeruginosa*. Hippokratia. 2012;16:303–7.
9. Brown AN, Smith K, Samuels TA, Lu J, Obare SO, Scott ME. Nanoparticles functionalized with ampicillin destroy multiple-antibiotic-resistant isolates of *Pseudomonas aeruginosa* and *Enterobacter aerogenes* and methicillin-resistant *Staphylococcus aureus*. Appl Environ Microbiol. 2012;78:2768–74.
10. Roe D, Karandikar B, Bonn-Savage N, Gibbins B, Roullet JB. Antimicrobial surface functionalization of plastic catheters by silver nanoparticles. J Antimicrob Chemother. 2008;61:869–76.
11. Zazo H, Colino CI, Lanao JM. Current applications of nanoparticles in infectious diseases. J Control Release. 2016;224:86–102.
12. Zhang Z, Tsai PC, Ramezanli T, Michniak-Kohn BB. Polymeric nanoparticles-based topical delivery systems for the treatment of dermatological diseases. Wiley Interdiscip Rev Nanomed Nanobiotechnol. 2013;5:205–18.
13. Cheow WS, Hadinoto K. Antibiotic polymeric nanoparticles for biofilm-associated infection therapy. Methods Mol Biol. 2014;1147:227–38.
14. Samiei M, Farjami A, Dizaj SM, Lotfipour F. Nanoparticles for antimicrobial purposes in Endodontics: a systematic review of in vitro studies. Mater Sci Eng C Mater Biol Appl. 2016;58:1269–78.
15. Hofmann D, Messerschmidt C, Bannwarth MB, Landfester K, Mailander V. Drug delivery without nanoparticle uptake: delivery by a kiss-and-run mechanism on the cell membrane. Chem Commun (Camb). 2014;50:1369–71.
16. Turos E, Shim JY, Wang Y, Greenhalgh K, Reddy GS, Dickey S, Lim DV. Antibiotic-conjugated polyacrylate nanoparticles: new opportunities for development of anti-MRSA agents. Bioorg Med Chem Lett. 2007;17:53–6.
17. Fazli Y, Shariatinia Z, Kohsari I, Azadmehr A, Pourmortazavi SM. A novel chitosan-polyethylene oxide nanofibrous mat designed for controlled co-release of hydrocortisone and imipenem/cilastatin drugs. Int J Pharm. 2016;513:636–47.
18. Elsabahy M, Wooley KL. Design of polymeric nanoparticles for biomedical delivery applications. Chem Soc Rev. 2012;41:2545–61.
19. Rasmussen JW, Martinez E, Louka P, Wingett DG. Zinc oxide nanoparticles for selective destruction of tumor cells and potential for drug delivery applications. Expert Opin Drug Deliv. 2010;7:1063–77.
20. Forier K, Raemdonck K, De Smedt SC, Demeester J, Coenye T, Braeckmans K. Lipid and polymer nanoparticles for drug delivery to bacterial biofilms. J Control Release. 2014;190:607–23.

21. Bergey DH, Krieg NR, Holt JG. Bergey's manual of systematic bacteriology. Baltimore: Williams and Wilkins; 1989.
22. Institute CLS. Performance standards for antimicrobial susceptibility testing. In: Twenty-fifth informational supplement. Vol M100. Wayne: Clinical and Laboratory Standards Institute CLSI document M100-S25; 2015.
23. Mora-Huertas CE, Fessi H, Elaissari A. Polymer-based nanocapsules for drug delivery. Int J Pharm. 2010;385:113–42.
24. Credito K, Kosowska-Shick K, Appelbaum PC. Mutant prevention concentrations of four carbapenems against gram-negative rods. Antimicrob Agents Chemother. 2010;54:2692–5.
25. Barbosa HR, Rodrigues MFA, Campos CC, Chaves ME, Nunes I, Juliano Y, Novo NF. Counting of viable cluster-forming and non cluster-forming bacteria: a comparison between the drop and the spread methods. J Microbiol Methods. 1995;22:39–50.
26. Thomas R, Nair AP, Kr S, Mathew J, Ek R. Antibacterial activity and synergistic effect of biosynthesized AgNPs with antibiotics against multidrug-resistant biofilm-forming coagulase-negative *staphylococci* isolated from clinical samples. Appl Biochem Biotechnol. 2014;173:449–60.
27. El-Mowafy SA, Shaaban MI, Abd El Galil KH. Sodium ascorbate as a quorum sensing inhibitor of *Pseudomonas aeruginosa*. J Appl Microbiol. 2014;117:1388–99.
28. Meletis G. Carbapenem resistance: overview of the problem and future perspectives. Ther Adv Infect Dis. 2016;3:15–21.
29. Melzer E, Kreuter J, Daniels R. Ethylcellulose: a new type of emulsion stabilizer. Eur J Pharm Biopharm. 2003;56:23–7.
30. Chang HI, Perrie Y, Coombes AG. Delivery of the antibiotic gentamicin sulphate from precipitation castmatrices of polycaprolactone. J Control Release. 2006;110:414–21.
31. Yariv I, Lipovsky A, Gedanken A, Lubart R, Fixler D. Enhanced pharmacological activity of vitamin B12 and penicillin as nanoparticles. Int J Nanomed. 2015;10:3593.
32. Wise R. In vitro and pharmacokinetic properties of the carbapenems. Antimicrob Agents Chemother. 1986;30:343–9.
33. Dahdouh E, Shoucair SH, Salem SE, Daoud Z. Mutant prevention concentrations of imipenem and meropenem against *Pseudomonas aeruginosa* and *Acinetobacter baumannii*. Sci World J. 2014;2014:979648.
34. Li H, Luo YF, Williams BJ, Blackwell TS, Xie CM. Structure and function of OprD protein in *Pseudolmonas aeruginosa*: from antibiotic resistance to novel therapies. Int J Med Microbiol. 2012;302:63–8.
35. Poirel L, Naas T, Nicolas D, Collet L, Bellais S, Cavallo J-D, Nordmann P. Characterization of VIM-2, a carbapenem-hydrolyzing metallo-β-lactamase and its plasmid- and integron-borne gene from a *Pseudomonas aeruginosa* clinical isolate in France. Antimicrob Agents Chemother. 2000;44:891–7.
36. Wang XD, Cai JC, Zhou HW, Zhang R, Chen GX. Reduced susceptibility to carbapenems in *Klebsiella pneumoniae* clinical isolates associated with plasmid-mediated beta-lactamase production and OmpK36 porin deficiency. J Med Microbiol. 2009;58:1196–202.
37. Nordmann P, Poirel L. The difficult-to-control spread of carbapenemase producers among Enterobacteriaceae worldwide. Clin Microbiol Infect. 2014;20:821–30.
38. Kasimanickam RK, Ranjan A, Asokan GV, Kasimanickam VR, Kastelic JP. Prevention and treatment of biofilms by hybrid- and nanotechnologies. Int J Nanomed. 2013;8:2809–19.
39. Hines DJ, Kaplan DL. Poly(lactic-*co*-glycolic) acid-controlled-release systems: experimental and modeling insights. Crit Rev Ther Drug Carrier Syst. 2013;30:257–76.
40. Dash TK, Konkimalla VB. Poly-ε-caprolactone based formulations for drug delivery and tissue engineering: a review. J Control Release. 2012;158:15–33.
41. Khoee S, Yaghoobian M. An investigation into the role of surfactants in controlling particle size of polymeric nanocapsules containing penicillin-G in double emulsion. Eur J Med Chem. 2009;44:2392–9.

8

The impact of species and cell type on the nanosafety profile of iron oxide nanoparticles in neural cells

Freya Joris[1], Daniel Valdepérez[2], Beatriz Pelaz[2], Stefaan J. Soenen[3], Bella B. Manshian[3], Wolfgang J. Parak[2], Stefaan C. De Smedt[1*†] and Koen Raemdonck[1†]

Abstract

Background: While nanotechnology is advancing rapidly, nanosafety tends to lag behind since general mechanistic insights into cell-nanoparticle (NP) interactions remain rare. To tackle this issue, standardization of nanosafety assessment is imperative. In this regard, we believe that the cell type selection should not be overlooked since the applicability of cell lines could be questioned given their altered phenotype. Hence, we evaluated the impact of the cell type on in vitro nanosafety evaluations in a human and murine neuroblastoma cell line, neural progenitor cell line and in neural stem cells. Acute toxicity was evaluated for gold, silver and iron oxide (IO)NPs, and the latter were additionally subjected to a multiparametric analysis to assess sublethal effects.

Results: The stem cells and murine neuroblastoma cell line respectively showed most and least acute cytotoxicity. Using high content imaging, we observed cell type- and species-specific responses to the IONPs on the level of reactive oxygen species production, calcium homeostasis, mitochondrial integrity and cell morphology, indicating that cellular homeostasis is impaired in distinct ways.

Conclusions: Our data reveal cell type-specific toxicity profiles and demonstrate that a single cell line or toxicity end point will not provide sufficient information on in vitro nanosafety. We propose to identify a set of standard cell lines for screening purposes and to select cell types for detailed nanosafety studies based on the intended application and/or expected exposure.

Keywords: Nanosafety, High content imaging, Inorganic nanoparticles, Iron oxide nanoparticles, Stem cells, Multiparametric analysis

Background

In recent years, many inorganic nanoparticles (NPs) have made their way to the market as they are being incorporated into various consumer products [1]. Moreover, their unique properties are being extensively explored for various biomedical applications. For instance, gold NPs (AuNPs) and iron oxide NPs (IONPs) hold great promise as theranostic agents for cancer treatment through hyperthermia combined with tumour detection via respectively photoacoustic or magnetic resonance imaging [2]. Additionally, silver NPs (AgNPs) are good candidates for wound dressings and antibacterial coatings of medical devices due to their enhanced antimicrobial properties [3]. However, to date only a few nano-enabled products were successfully translated into the clinic. Besides general targeting issues, this can primarily be attributed to their elusive safety profiles [4]. Despite extensive efforts, a general paradigm on how inorganic NPs are able to affect homeostasis on the level of the cell, organ or organism and to which physicochemical NP properties this can be attributed, is largely lacking [5].

*Correspondence: stefaan.desmedt@ugent.be
†Stefaan C. De Smedt and Koen Raemdonck contributed equally to this work
[1] Lab of General Biochemistry and Physical Pharmacy, Department of Pharmaceutics, Faculty of Pharmaceutical Sciences, Ghent University, Ottergemsesteenweg 460, 9000 Ghent, Belgium
Full list of author information is available at the end of the article

In general, nanosafety evaluations struggle with two important obstacles. The first is the fast pace at which nanotechnology keeps advancing, leading to the development of a plethora of NPs with distinct physicochemical properties, which should ideally undergo safety evaluation prior to their (biomedical) implementation. The second is the lack of standardization of in vitro nanosafety studies, as various groups apply different assays on various cell types. This results in low inter-study comparability and the publication of conflicting data, which complicates the elucidation of general paradigms on NP-cell interactions [6, 7].

The first hurdle can be overcome by implementing high throughput or high content techniques in order to speed up in vitro nanosafety testing [8, 9]. Secondly, much effort is being put into the standardization of various factors of in vitro nanosafety studies [10, 11]. In this regard, we believe that the cell type selection should receive equal attention. In most studies a cell line is selected since they are in general more readily accessible, less expensive and easier to cultivate when compared to primary cells [7, 12]. However, cancer cell lines have a disturbed anti-apoptotic balance as well as an altered metabolism to sustain their high proliferation rate [13]. The phenotype expressed by immortalized cells is in turn not entirely stable and might undergo changes due to the extensive in vitro manipulation or the initial immortalization [14]. Hence, a shift towards the use of primary or stem cells as well as more complex cell culture models for in vitro nanosafety testing strategies could be noted recently. In contrast, primary cells can suffer from clonal variations and have a limited lifespan in vitro, making rational cell type selection a balancing act [7].

Subsequent to the realization that the cell type could be of substantial importance, several groups have shown that NP-induced effects vary in cell lines retrieved from different tissues or species [15–18]. On the contrary, only a few studies compared NP effects in a cancer or immortalized cell line versus primary cells representing the same tissue and species [19, 20]. Unfortunately, available data contrast one another wherefore no unambiguous conclusions could yet be formulated on whether cell lines can generally be applied as a reliable model for in vitro nanosafety studies. In addition, many of the abovementioned reports choose to either focus on interspecies variations or cell-type related differences in NP-evoked effects and do not address both factors in a single study.

Here, we present a side-by-side comparison of NP-evoked effects in six related neural cell types thereby evaluating the extent of both species and cell type related variations in NP-induced cytotoxicity. We selected a neuroblastoma cell line, neural progenitor cell line and neural stem cells derived from either humans or mice (Table 1) and purposely applied the optimal culture conditions for each cell type. These cell types were selected as potential models to assess the safety of neural stem cell labeling with nanosized contrast agents prior to transplantation in the context of regenerative medicine [21–23]. In turn, the synthesized AuNPs, AgNPs and IONPs had a diameter below 10 nm, making them good candidates for the proposed application [24]. First, we surveyed the acute toxicity of AuNPs, AgNPs and IONPs in all cell types. Subsequently we selected the IONPs for further evaluation given the minor acute toxicity. Hereto we applied a validated multiparametric approach, using automated imaging, to evaluate the effect of sublethal doses on the production of reactive oxygen species (ROS), the calcium (Ca^{2+}) homeostasis, mitochondrial health and cell morphology [25]. Importantly, our data reveal distinct and cell type specific toxicity profiles that warrant careful selection of appropriate cell models for future nanosafety studies, taking both species and target tissue into account, and caution misinterpretation of experimental results based on a single cell type and/or toxicity end point.

Table 1 Cell types applied in this study

	Stem cells	Progenitor cell line	Cancer cell line
Human	hNSC [26]	ReNcell [27]	LA-N-2 [28]
Mouse	mNSC [26]	C17.2 [29]	Neuro-2a [30]

Results and discussion

Synthesized inorganic NPs display similar physicochemical characteristics

AuNP, AgNP and IONP synthesis was initiated with the aim of obtaining a similar core diameter. All NPs had a mean core diameter around 3.8 nm, as measured by transmission electron microscopy (Additional file 1: Figure S4). Subsequently all NPs were coated with poly(isobutylene-*alt*-maleic anhydride) grafted with dodecylamine (PMA), which was selected as it ensures colloidal stability over a wide pH range and a uniform coating of the different core materials [31]. Dynamic light scattering measurements in water showed a hydrodynamic diameter of 9.0, 8.9 and 12.3 nm and a negative zeta-potential around −45, −35, and −54 mV for the coated AuNPs, AgNPs and IONPs respectively. All obtained values correspond well to data reported on the characterization of NPs synthesized via similar protocols [32, 33]. The NPs were synthesized with the intention of obtaining similar physicochemical properties so that discrepancies in cell responses could be related to variations between the cell types. Additional characterization data on the plasmon resonance peaks, molecular extinction coefficients, initial NP dispersion concentrations, and

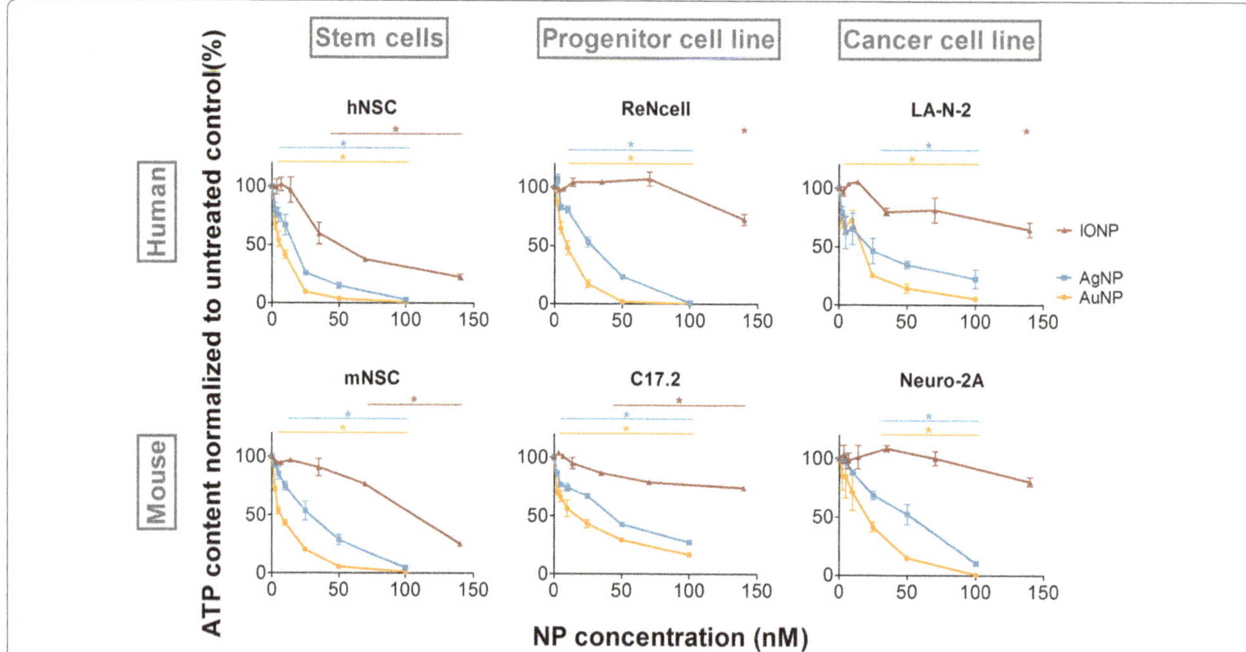

Fig. 1 A concentration-dependent decrease in ATP content, as measured via the CellTiter GLO® assay, is observed for every NP-cell type combination tested. Results for the AuNPs (*yellow*), AgNPs (*blue*) and IONPs (*red*) are represented as mean ± standard error of the mean (SEM, n = 3,). Statistical significance is indicated when appropriate for each type of NP in the corresponding color of the graphs [*$p < 0.05$, AuNPs (*yellow*), AgNPs (*blue*) and IONPs (*red*)]

electrophoretic mobility can be found in Additional file 1: Figures S5, S6 and Tables S1, S2.

Acute toxicity depends on both the NP core material and the cell type

In initial cell experiments, we evaluated cell viability following 24 h NP exposure with the CellTiter GLO® assay. In Fig. 1, a general concentration-dependent decrease in ATP signal can be observed for every evaluated NP—cell type combination. Although the extent of this decrease clearly varies, the onset of this downward trend depended on both the applied NP and the cell type. In all cell types, the most severe effect was observed following AuNP treatment, while the cells were least affected by the IONPs. The toxicity observed for the AgNPs can likely in part be explained in terms of Ag^{+}-ion leaching [34]. In turn, the severe acute cytotoxicity induced by the AuNPs could possibly be attributed to genotoxicity due to direct interactions between the 3.8 nm diameter AuNPs and DNA [35]. In addition, note that determining NP concentrations is not straightforward, as various methods/models need to be applied for different NP materials (Additional file 1). This may affect the comparison of absolute concentrations of NPs of different materials and may additionally explain the severe toxicity observed here for the AuNPs. Given the limited loss of cell viability observed for the IONPs, the latter were selected for further evaluation of sublethal effects.

Independent of the core material, the Neuro-2a cells were least susceptible to NP exposure whereas the hNSC, followed by the mNSC, were most sensitive. The susceptibility ranking for the other cell types varied with the NP core material. This greater sensitivity of the NSC, as found under the conditions reported here, is dissimilar to several studies where cell lines were found to be more susceptible to NP-induced acute cell injury [18, 19]. However, our data correlate well with previous work from Bregoli et al. [14] who did not observe any toxic effects in several hematopoietic cell lines, while primary bone marrow cells were clearly affected. Similar observations were recorded by Wilkinson et al. and Schlinkert et al. who independently found normal bronchial epithelial cells to experience more acute toxicity than the A549 cancer cell line [36, 37].

ROS induction is observed in two out of six cell types

Since we found the IONPs to induce the least acute cell damage, it was decided to probe for sublethal effects caused by these NPs using a multiparametric methodology. The evaluation of effects on cell function has become crucial, as it is generally recognized that nanosafety evaluations should go beyond live/dead scoring in order to establish a more predictive paradigm [15, 38]. Subtle changes in cell function might indeed be more predictive towards in vivo adverse outcomes. For instance, NP-promoted reactive oxygen species (ROS) production in pulmonary cells has been linked to acute inflammation in the lung [39, 40]. ROS induction is also stated to be the main mechanism via which metallic NPs induce cell stress. Persisting ROS induction can subsequently lead to oxidative stress and damage cellular components such as DNA, proteins and membrane lipids [6].

Upon IONP treatment, we observed an increased ROS production in two out of six cell types, namely the mNSC and human ReNcells (Fig. 2). In all other four cell types, ROS production was significantly reduced. Notably, for both the reduced or increased ROS levels, the effect was most outspoken in the NSC. Again the murine neuroblastoma cell line (Neuro-2a) was least affected in terms of ROS. Given the variable effects, no general statements can be made on whether the human or murine cell types were more severely affected than their counterparts.

Although ROS induction by IONPs is often observed [19, 39, 41], it has been shown that NP-induced cytotoxicity cannot always be attributed to an increased ROS production [42]. Interestingly, Harris et al. also witnessed reduced ROS levels in their high content analysis of IONP-induced effects on a mammalian fibroblast cell line [6]. Additionally, IONPs can exhibit an intrinsic peroxidase-like activity in mesenchymal stem cells and thus reduce the cellular ROS content, especially of H_2O_2 [43, 44]. As this effect was only witnessed when IONPs remained intact, IONP biocompatibility is presumably to a large extent affected by the intracellular location and the way the cell processes the IONPs. In confirmation, Sabella et al. [45] found greater cell perturbation by metallic NPs when they were trafficked to the acidic lysosomes in comparison to the same NPs present in the cytoplasm, due to the enhanced degradation in the acidic compartments. Indeed, this degradation will be accountable for an increased amount of free iron ions, which may in turn enhance ROS production via for instance Fenton chemistry [6, 46]. A final factor that could clarify our observation is the intrinsically different anti-oxidative capacity of the various cell types [15, 17]. Thus, the cell itself likely determines NP biocompatibility to a large extent.

IONP exposure perturbs cellular calcium homeostasis

Subsequently, we evaluated the effect of IONP exposure on the Ca^{2+} homeostasis. The intracellular free Ca^{2+} concentration ($[Ca^{2+}]_c$) is a valuable toxicity marker since Ca^{2+} is involved in a plethora of processes such as cell proliferation, mitochondrial function and gene

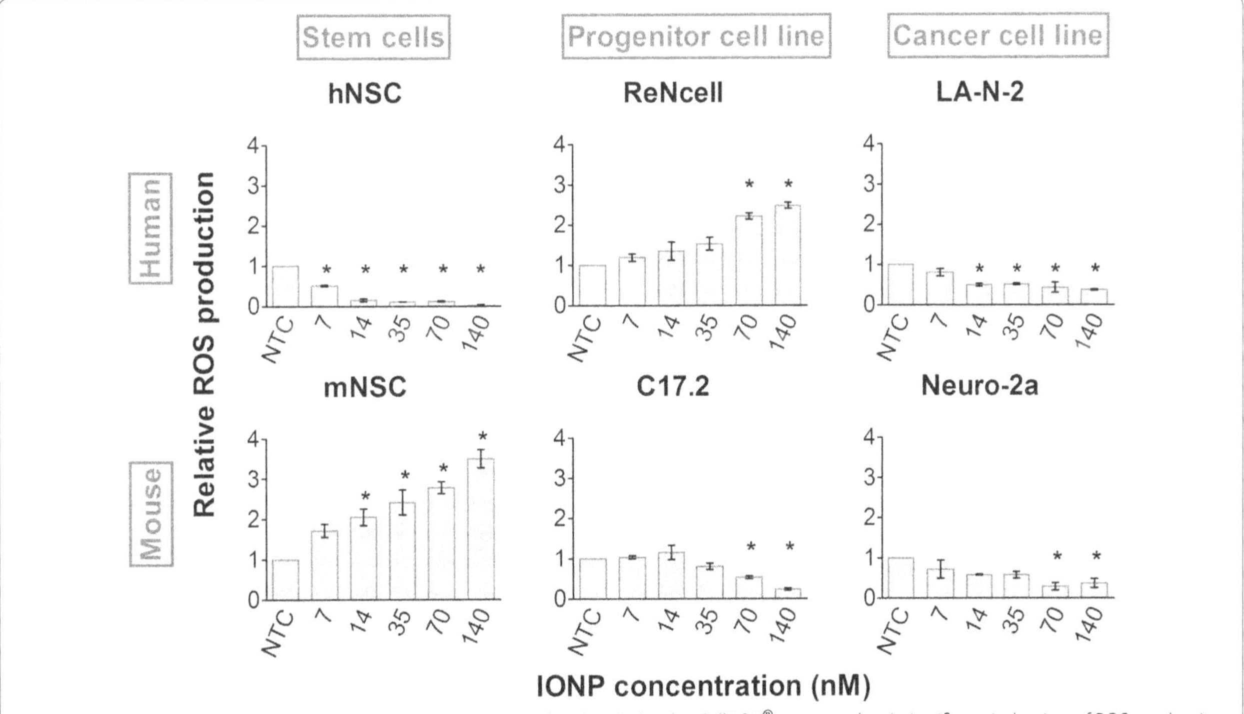

Fig. 2 Effects on ROS production following IONP exposure visualized with the the CellROX® green probe. A significant induction of ROS production was observed in the mNSC and human ReNcells. In the other four cell types a significant reduction was observed. Statistical significance is indicated when appropriate (*$p < 0.05$). *NTC* not treated control

expression [47, 48]. Ca^{2+} is furthermore of ultimate importance for proper cell function in neural cells, as it is required for neurotransmitter release and cellular excitability [8, 49]. Additionally, Ca^{2+} is since long known to be an important regulator of cell death, where a significant increase in $[Ca^{2+}]_c$ is noted [47]. A mild reduction can on the contrary be correlated with an impaired cell function due to enhanced intracellular Ca^{2+} storage or efflux in an effort to retain cell homeostasis, while cell lysis is correlated to a more severe decrease [48, 50].

On the one hand, a significant concentration-dependent increase in $[Ca^{2+}]_c$ was observed in the hNSC, ReNcells, and C17.2 cells (Fig. 3). The effect was more severe in the progenitor cell lines compared to the hNSC and the ReNcells showed the highest $[Ca^{2+}]_c$. On the other hand, a decline of the $[Ca^{2+}]_c$ was detected in the mNSC, LA-N-2 and Neuro-2a cells. In contrast to previous parameters, the Neuro-2a cells showed more severe effects in terms of the perturbation of the calcium homeostasis. Again, no unambiguous conclusions could be drawn on whether human or murine cell types are more sensitive towards NP exposure.

Multiple studies investigating the influence of NP exposure on the Ca^{2+} homeostasis also found $[Ca^{2+}]_c$ to be augmented [51]. Since this response could be interpreted as a cell death signal, this outcome could be correlated to the initially observed acute toxicity (Fig. 1) [47, 52]. Although we would have expected to observe a greater increase in $[Ca^{2+}]_c$ in the hNSC when compared to the ReNcells based on the acute toxicity data, the opposite was true. In line with the observed decline in $[Ca^{2+}]_c$ in three out of six cell types, Haase et al. [3] documented diminished Ca^{2+} responses at cytotoxic NP doses in mNSC. This observation could on the one hand be explained in terms of cell lysis. On the other hand, stressed cells can maintain their Ca^{2+} homeostasis by elevating Ca^{2+} efflux via the plasma membrane Ca^{2+} ATPase pump [50]. We hypothesized that this occurred in the neuroblastoma cell lines where ATP levels were to a minor extent reduced and thus still allowed sufficient pump function.

In general, two groups could be distinguished based on the elevation or diminution of $[Ca^{2+}]_c$. Even though similar trends were retrieved in each group, it is clear that the extent of the perturbation of cellular Ca^{2+} homeostasis varied with the cell type. Since Ca^{2+} homeostasis is significantly altered upon cell transformation or immortalization in favour of cell proliferation [53], it was not surprising that NP exposure variably altered the $[Ca^{2+}]_c$. Notably, cell type specific toxicity profiles started to emerge as various combinations of the thus far evaluated effects were obtained.

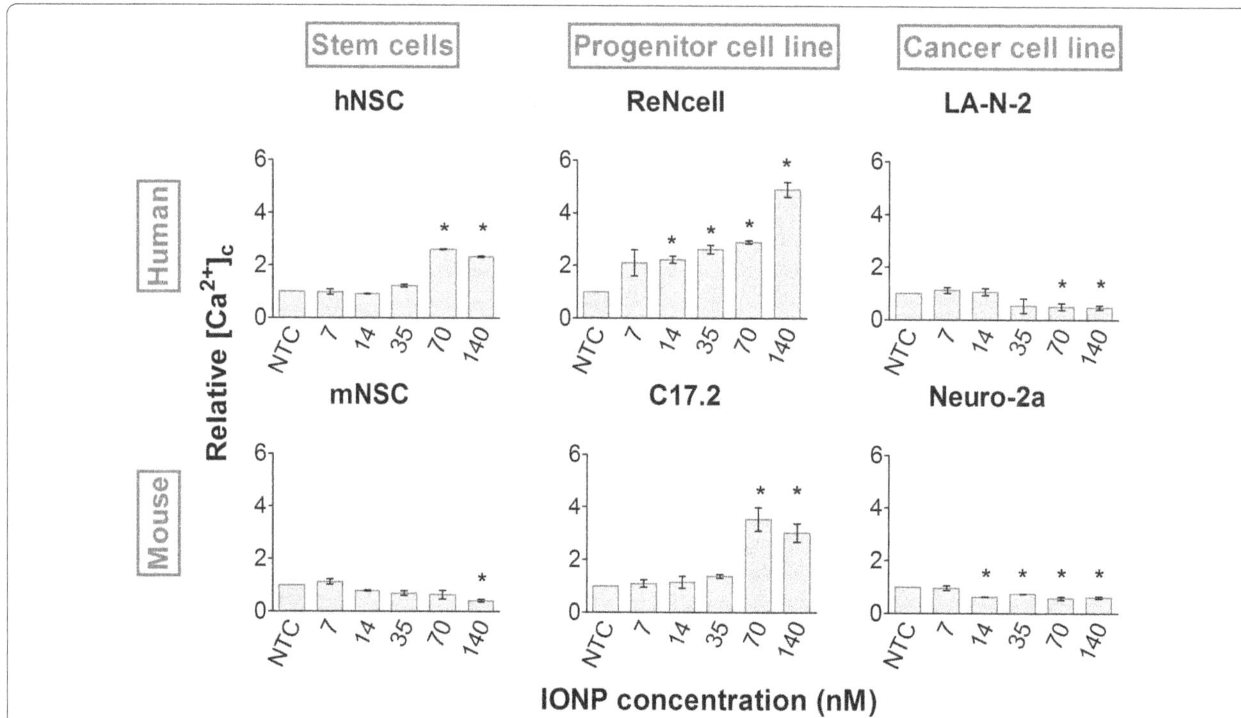

Fig. 3 Effect on $[Ca^{2+}]_c$ as determined following labelling with Rhod-2 AM. A significant increase in $[Ca^{2+}]_c$ was observed in the hNSC and both progenitor cell lines whereas a significant reduction was observed in the remaining three cell types ($p < 0.05$). Statistical significance is indicated when appropriate (*$p < 0.05$). *NTC* not treated control

Mitochondria are affected by IONP loading

Next, the effect of IONP exposure on mitochondrial homeostasis was evaluated. The mitochondria are interesting organelles as they are the cell's main energy suppliers, involved in programmed cell death, an important source of ROS and to a large extent regulated by Ca^{2+} [52, 54]. This Ca^{2+}-mediated regulation is furthermore influenced by external stimuli: in combination with a stress inducer Ca^{2+} promotes ROS production and possibly cell death, whereas under physiological conditions Ca^{2+} stimulates the oxidative respiration in the mitochondria and thus ATP production [52]. Interestingly, the importance of oxidative respiration for overall cellular ATP production varies with the cell type: both cancer cells and stem cells rather rely on cytosolic glycosylation for their ATP production [54, 55]. Hence, it is conceivable that mitochondria will not only be differentially affected, but also that the impact of mitochondrial perturbation on overall cell homeostasis will vary in the different cell types.

To visualize the mitochondria, we selected a probe that specifically labels the organelles based on their membrane potential ($\Delta\Psi_m$). Loss of this potential, as a result of mitochondrial membrane permeabilization, will render the organelle undetectable and has been associated with cytochrome C release and cell death initiation [52, 56]. During data analysis, such events could be detected as a reduction of the relative mitochondrial area. Figure 4 shows that all cell types, except the Neuro-2a cells, showed significant mitochondrial damage. Accordingly, the loss of $\Delta\Psi_m$ following NP exposure has already been described in multiple studies for several NPs in cell types from various lineages and species [8, 9, 42]. In the NSC all IONP doses caused a decreased signal area, though the effect was only significant starting from 7 nM. In contrast, the affected cell lines (ReNcell, C17.2 and LA-N-2) were significantly affected by all IONP doses. The effects were most outspoken in the ReNcells, closely followed by the hNSC and mNSC. The mitochondria in the C17.2 and LA-N-2 cell lines were perturbed to a lesser extent. Notably, the human cell types were more severely affected than the murine counterpart. In addition, the neuroblastoma cell lines were most resilient on the mitochondrial level. In correspondence, Heerdt et al. [57] have previously found mitochondria in transformed cells to be less sensitive to perturbation due to an intrinsically lower mitochondrial activity and higher $\Delta\Psi_m$.

IONP loading affects cell morphology

Lastly, we examined alterations in cell morphology following IONP exposure. Cell morphology is a convenient

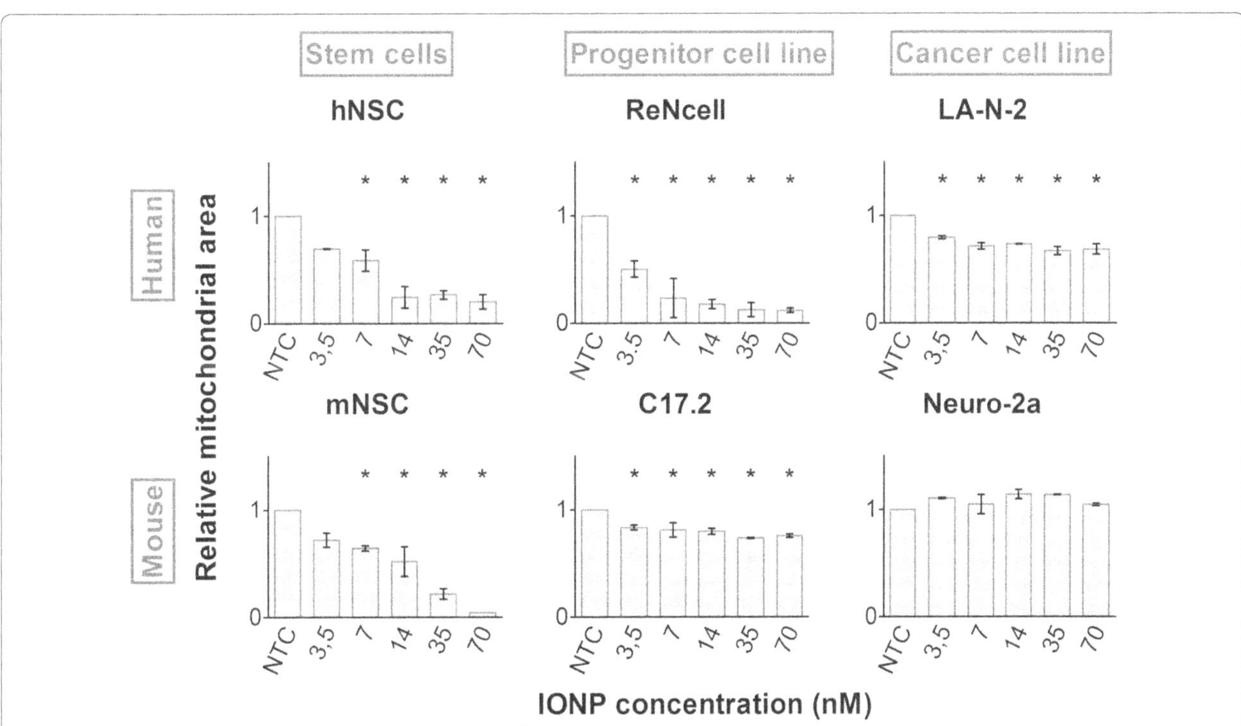

Fig. 4 Effects on the mitochondria labelled with Mitotracker® CMX-ROS in terms of the relative signal area representing the size of the mitochondrial compartment relative to the total cell area. Except for the Neuro-2a cell line, all cell types showed a significant decrease in mitochondrial area. Statistical significance is indicated when appropriate (*$p < 0.05$). *NTC* not treated control

parameter, especially for neural cells given their intricate architecture [8]. Moreover, numerous NPs have been shown to alter cell morphology as a secondary effect of ROS induction or via direct interactions with elements of the cytoskeleton [58, 59]. In addition to the changes in the morphological appearance, certain cell functions that require signaling via these components can subsequently be impaired [59, 60]. Thus, subtle effects on cell morphology can indirectly herald perturbation of cell function whereas severe morphological alterations, i.e. cell rounding and shrinking, can be interpreted in terms of cell death [47].

After staining the entire cell cytoplasm, the impact of IONP exposure on cell morphology was quantified via two parameters: cell area and cell circularity. The latter is applied as a measure of cell spreading and is a value between zero and one, where one represents a perfect sphere (Additional file 1). Although the extent of neurite outgrowth is often applied to evaluate the morphology of neural cells [61], this parameter was not selected for this work, as several cell types are not capable of forming neurites.

While only a significantly decreased cell area was noted for the C17.2 cell line, both a reduced cell area and an increase in circularity were observed in the NSC, ReNcells and Neuro-2a cells (Fig. 5; Additional file 1: Figure S8). Thus, the cells became both smaller and more spherical in a concentration-dependent fashion (Fig. 6), which was most outspoken in the ReNcells. Such loss of specific morphological features and cell shrinking has already been described in numerous studies for multiple NPs and cell types [3, 8, 22, 42]. Since it is known that cell transformation or immortalization affects cell morphology, it is not surprising that morphology was also differentially affected in the various cell types. For instance the mNSCs were more strongly affected in terms of morphology whereas only minor effects were observed in the C17.2 or Neuro-2a cell line. Since stem cells have a more intricate architecture in comparison to most cell lines, it was not surprising that the morphology of the former was impaired more extensively. Finally, as the LA-N-2 cells tend to grow in clusters we evaluated effects on cell morphology in terms of the total cluster area and number of cells per cluster, which both showed a similar concentration-dependent decrease starting from 3.5 nM IONPs. Since the decrease in cluster area was slightly more severe than the number of cells per cluster, we concluded that the cell area also decreased with every dose tested.

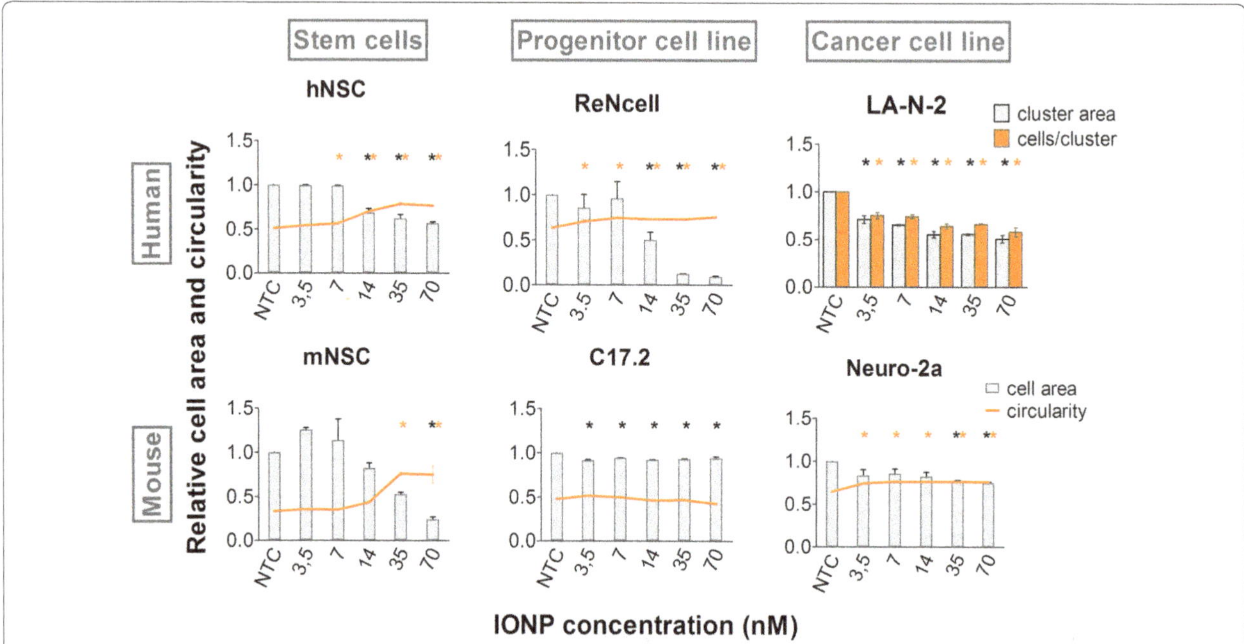

Fig. 5 IONP-induced alterations in cell area (*grey bars*) and cell circularity (*orange lines*) visualized after labelling of the cytoplasm with the CellMask™ Blue probe for the NSC, progenitor cell lines and murine neuroblastoma cell line. Cell circularity is a measure of cell spreading and is a value between zero and one, where one represents a perfect sphere. LA-N-2 cell morphology was analysed in terms of cluster area (*grey bars*) and number of cells per cluster (*orange bars*). A decreased cell area and increased cell circularity were detected in the NSC, ReNcell and Neuro-2a cell line. For the C17.2 cells only a diminution in cell area was detected. In the LA-N-2 cell line a reduction in cells per cluster and cluster size were observed. Statistical significance is indicated when appropriate (*$p < 0.05$), in *black* for the cell area and *orange* in case of the cell circularity (respectively cluster area and cells per cluster in case of the LA-N-2 cell line). *NTC* not treated control

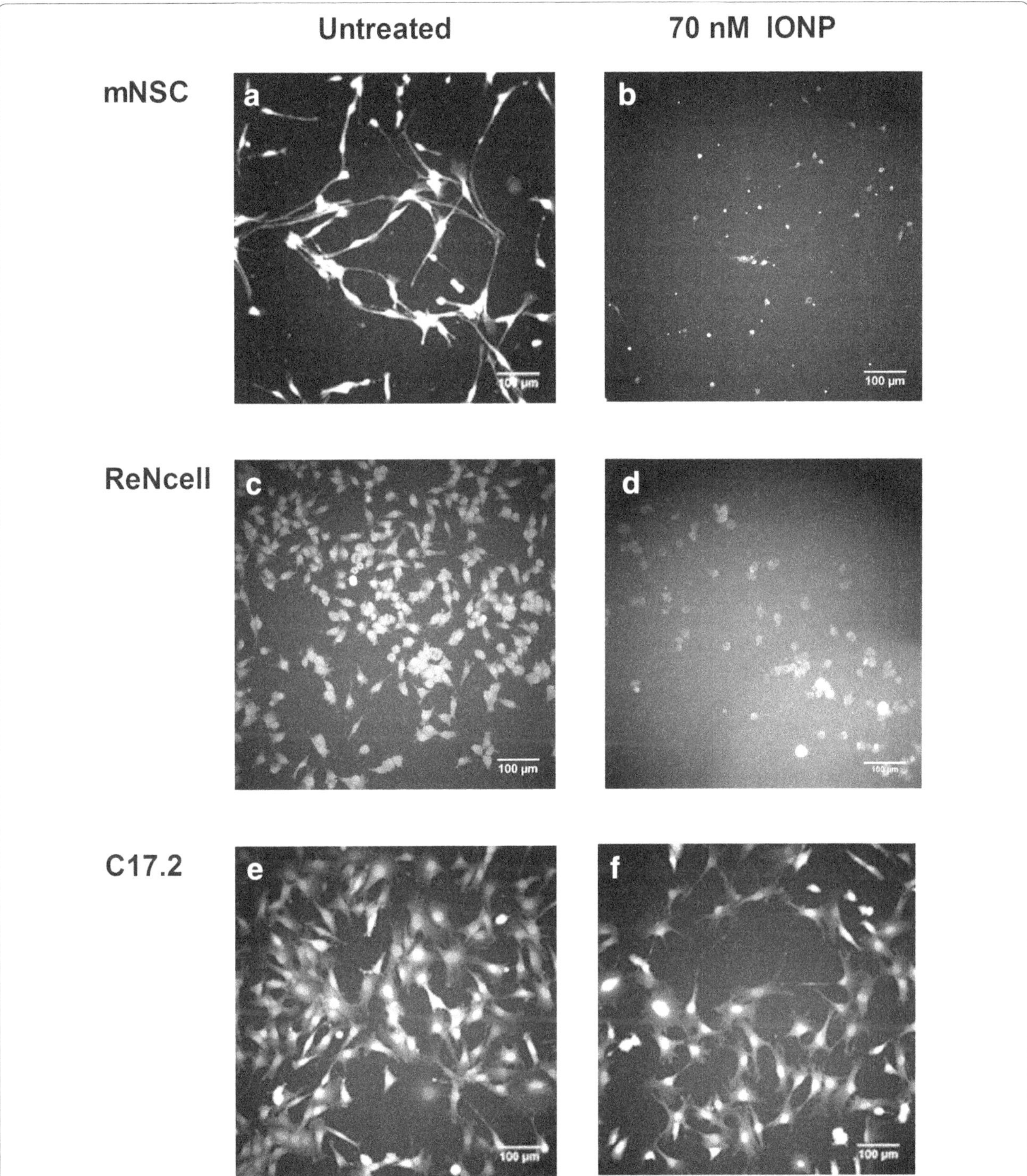

Fig. 6 Representative images of untreated mNSCs (**a**), ReNcells (**c**) and C17.2 cells (**e**) as well exposed to 70 nM IONP (**b**, **d**, **f**). The mNSCs are affected in terms of cell area and circularity. The altered circularity in the ReNcells is less outspoken as initial morphology is less complex. Only the cell area is affected in the C17.2 cells

Overall, we observed similar effects on cell morphology (cell rounding and shrinking) in the various cell types in contrast to previously evaluated parameters. However, the exact trends and extent of the responses clearly differed. Importantly, these variations could not unequivocally be linked to one or a specific combination of responses observed for the other toxicity parameters investigated in this study, underscoring the cell type specific nature of the recorded toxicity profiles.

Multiparametric analysis reveals cell type-specific toxicity profiles

In general, our data set reveals that each cell type reacted in a specific way to IONP exposure in terms of both extent and nature of the responses (Table 2). This could not have been deduced from the acute toxicity assessment (Fig. 1) but became increasingly clear with every additionally evaluated parameter. Furthermore, the obtained profiles would likely become increasingly complex with the addition of supplementary end points such as the influence on autophagy, induction of endoplasmic reticulum stress or genotoxicity. Note that it was not the primary objective of this study to unravel the underlying toxicity mechanisms. Hereto, additional experiments, for instance on the type of cell death or gene expression, should be performed. Instead, the aim was to clearly show the impact of both the species and the cell type, under its optimal cell culture conditions on the nanotoxicity profile within one single study. We show that for 3 different, though related neural cell types (stem cells, immortalized cells and cancer cells) the effects in the human cells were often more outspoken than the murine alternative. In addition, we found the NSC from each species to be more sensitive to IONP exposure than the cell lines.

The observed variations in cell responses can be explained in several possible ways. One may argue that variations in NP uptake in the various cell types will be an important factor. In this regard, dose heterogeneity at single cell level due to variations in NP uptake in the same population will also lead to response heterogeneity [62]. In addition, NP uptake is related to the colloidal stability in the applied cell culture media. Although we did not evaluate the abovementioned parameters in detail, it was previously shown that PMA-coated NPs show good stability in biological media and that they are taken up well by various cell types [25, 63]. Besides the

Table 2 Cell type-specific nanotoxicity profiles induced by 24 h exposure to 70 nM IONPs

	ROS	Ca^{2+}	Mitochondria	Cell morphology	
				Area	Circularity
hNSC	↓	↑	↓	↓	↑
mNSC	↑	↓	↓	↓	↑
ReNcell	↑	↑	↓	↓	↑
C17.2	↓	↑	↓	↓	=
LA-N-2	↓	↓	↓	/	/
Neuro-2a	↓	↓	=	↓	↑

extent of NP uptake, we believe that the cellular response is strongly related to the intracellular NP processing. This will in part depend on the uptake pathway since the latter will co-determine the intracellular trafficking route and the ultimate intracellular location. Indeed, as previously mentioned when NPs are present in the acidic and degrading environment of the endo-lysosomes, stronger cytotoxicity is observed than when the NPs reside in the cytosol [45]. In addition, the variations in intrinsic cell properties, such as the anti-oxidative capacity, metabolic rate (e.g. Ca^{2+} homeostasis) and mitochondrial activity, are to a large extent accountable for the revealed divergent toxicity profiles. Combined, these elements advocate an in vitro toxicity profiling that takes intrinsic cell properties and variations in the studied cell population into account. Indeed, to understand the intrinsic cellular capacity to traffic and handle exogenous materials could be of key importance to anticipate NP-evoked effects.

Furthermore, our data indicate that it is imperative to apply multiparametric methods that look beyond live/dead scoring. Notably, even when only minor variations could be detected in the cell viability, as for instance for the Neuro-2a and ReNcells, cellular homeostasis was distinctly altered. In addition, minor cell viability alterations for the ReNcells did not imply that the cell homeostasis was not impaired. Accordingly, Ge et al. [64] found IONPs to evoke important effects on cell function without affecting cell viability. Also, toxicity endpoints included in nanosafety screens should be carefully selected as some are more sensitive or indicative of the induced damage. An example of the latter is the use of cell area and circularity as parameters to describe alterations in cell morphology. Although effects on cell circularity occurred sooner, the impact on cell area was more outspoken and illustrative for the extent of the actual damage in cell types without a complex architecture. Finally, the safety of the coating should be investigated in further detail to determine its possible contribution to some of the observed effects.

Notably, we found that none of the cell types included in this work would be a suitable substitute for any other tested. In contrast, other groups did succeed in identifying a cell line alternative for primary cells based on similar cellular responses to NP exposure [12]. In such cases the use of those cell lines should be encouraged. However, the generalized use of cell lines should be approached with caution, especially when performing a detailed toxicity profiling to elucidate the mechanisms via which NPs alter cell homeostasis. Indeed, cell lines are not always ideal candidates for the analysis of cell function and may not be representative in terms of discrete cell perturbation [19]. Thus, it would be fitting to select a cell type based on the expected exposure and/or intended application of the NPs. We also propose to cautiously apply non-human cell types since we, as well as several other groups, have observed notable interspecies variations [15, 16].

For screening purposes the selection of a proper cell type is a balancing act. Indeed, primary cells can suffer from several drawbacks like an often limited availability, specific cultivation requirements, a limited life-span, and possible inter-batch and individual variations, which possibly limit the throughput [7, 12]. Hence, cell lines are still the preferred candidates when performing a large-scale screening of numerous NPs. For this reason and because it is highly unlikely that one single cell type will emerge as a universal model, we strongly believe that the definition of a set of standard cell lines would constitute a definite asset in standardizing nanosafety assessments. Additionally, the use of multiple cell types should be encouraged as it was shown to enhance the predictive power of in vitro nanosafety assessments [65]. The selected cell types would preferably be known to mimic responses observed in primary cells and would ideally be thoroughly characterized in terms of their intrinsic properties in order to enhance our understanding of the NP-induced effects.

Conclusions

In this work, we investigated the effect of both species and cell type related variations on NP-evoked responses in six related neural cell types via a multiparametric approach. Interestingly, the observed impact on cellular health varied widely in each cell type in terms of both the nature and extent of the analyzed effects and cell type-specific nanotoxicity profiles were obtained. Hence, conclusions on the safety of a NP should preferably not be based on the evaluation of a single toxicity end point in a single cell type. We propose to rationally select a cell model based on the envisioned (biomedical) application and/or exposure scenario, especially when performing an extensive in vitro toxicity assessment with the aim of unveiling mechanisms via which the NPs inflict cell injury. Finally, with regard to standardization of in vitro nanosafety evaluations, we strongly believe that for the safety screening of large sets of nanomaterials the selection of a set of standard cell types, representing relevant target tissues, would contribute to the generation of more consistent nanosafety data.

Methods

NP synthesis and characterization

AuNPs, AgNPs and IONPs were synthesized and coated with the polymer poly(isobutylene-*alt*-maleic anhydride) grafted with dodecylamine (PMA), as described in Additional file 1. Following synthesis, the core diameter was measured using transmission electron microscopy. UV/Vis spectroscopy was applied to evaluate the spectral characteristics of the NPs. With the combination of UV/Vis spectroscopy and inductively coupled plasma mass spectrometry the concentrations of the dispersions were determined. Finally the hydrodynamic diameter and zeta-potential were measured using a Zetasizer Nano ZS (Malvern Instruments). Detailed information on the characterization procedures is provided in Additional file 1.

Cell culture

All assays were performed on six neural cell types (Table 1): human and murine neural stem cells (hNSC and mNSC, Invitrogen and Millipore, Belgium), a human and mouse-derived progenitor cell line, respectively ReNcell (Millipore, Belgium) and C17.2 (Sigma, Belgium), and finally a human neuroblastoma cell line (LA-N-2, European Collection of Cell Cultures) as well as a murine counterpart (Neuro-2a, Sigma, Belgium). All cell types were cultured according to the supplier's guidelines. Detailed information on the applied coatings and culture media compositions can be found in Additional file 1.

The cells were cultured at 37 °C in a humidified atmosphere completed with 5 % CO_2. Cell medium was renewed every other day and cells were split after reaching 80 % confluency. Hereto, the cells were dissociated with 0.05 % trypsin–EDTA (Invitrogen, Belgium), after which the cells were centrifuged (4 min, 300 g), resuspended in fresh culture medium and seeded at appropriate densities.

Acute toxicity

All cell types were seeded at 25,000 cells per well in opaque 96-well plates and were allowed to settle overnight. Thereafter the cells were incubated with 2.5, 5, 10, 25, 50 and 100 nM of the AuNPs and AgNPs and 3.5, 7, 14, 35, 70 and 140 nM of the IONPs during 24 h at 37 °C (5 % CO_2). After 24 h NP incubation, the CellTiter-GLO® assay (Promega, Belgium) was performed according to the manufacturer's instructions. In short, 100 μL of the assay buffer was added to each well. Plates were shaken during 2 min after which a 10-min incubation period was respected. Finally, the signal was measured using a GloMax® 96 Microplate Luminometer (Promega, Belgium). Experiments were performed in triplicate and the data are represented as the mean ± the standard error to the mean (SEM).

High content imaging

For the multiparametric analysis, cells were seeded in 24-well plates and were allowed to attach overnight. Optimal seeding cell densities were identified for each cell type individually. The optimal seeding density was

Table 3 Seeding densities and incubation volumes per well applied in the multiparametric analysis

	hNSC	mNSC	ReNcell	C17.2	LA-N-2	Neuro-2a
Cell density	35,000	17,500	17,500	15,000	50,000	15,000
Volume (μL)	700	350	350	300	1000	300

defined as the density that would result in an 80 % confluent cell layer in the untreated control at the end point of the assay. In order to preserve the cell density/cell medium volume ratio for all cell types, we varied the latter according to the optimal cell seeding density (Table 3).

For the evaluation of effects on ROS production and $[Ca^{2+}]_c$ 7, 14, 35, 70 and 140 nM IONP dispersions were applied, whereas for the effects on cell morphology and the mitochondria 3.5, 7, 14, 35 and 70 nM were tested as effects on cell function were expected to occur starting from lower NP doses. As the volume of cell medium used for incubation was adjusted according to the cell density, the NP number/volume cell medium/cell number remained equal in all high content experiments. Similar to acute toxicity experiments, the cells were incubated with the IONPs during 24 h at 37 °C in an atmosphere containing 5 % CO_2 after which staining and analysis were performed. This set of data is presented as mean ± SEM from to independent replicates.

Reactive oxygen species and cytoplasmic calcium levels

To allow detection of reactive oxygen species (ROS) the general ROS marker CellROX® green probe (Molecular Probes, Invitrogen, Belgium) was selected. The latter was combined with the Rhod-2 AM (Molecular Probes, Invitrogen, Belgium), which becomes strongly fluorescent upon interaction with free Ca^{2+} in the cytoplasm. Following 24 h IONP incubation, the cells were labelled with both probes as described in Additional file 1.

Effect on mitochondrial health and cell morphology

The mitochondria were labelled with Mitotracker® CMX-ROS Red (Molecular Probes, Invitrogen, Belgium), which specifically accumulates in the mitochondria based on its membrane potential. To allow evaluation of cell morphology the HCS CellMask™ Blue probe (Molecular Probes, Invitrogen, Belgium) was applied. Again, cells were labeled following 24 h of IONP exposure as explained in Additional file 1.

Statistics

Acute toxicity data are expressed as mean ± SEM (n = 3). IN Cell data are presented as mean values normalized against the untreated control ± SEM (n = 2). Statistical analysis was performed using one-way ANOVA combined with post hoc Dunnett test.

Abbreviations

$\Delta\Psi_m$: mitochondrial membrane potential; $[Ca^{2+}]_c$: cytosolic free calcium concentration; AgNPs: silver nanoparticles; ATP: adenosine triphosphate; AuNPs: gold nanoparticles; Ca^{2+}: calcium; d_c: core diameter; HCS: high content screening; hNSC: human neural stem cell; IONPs: iron oxide nanoparticles; mNSC: murine neural stem cell; NPs: nanoparticles; NTC: not treated control; PMA: poly(isobutylene-*alt*-maleic anhydride); ROS: reactive oxygen species; SEM: standard error of the mean; TiO_2 NPs: titanium dioxide nanoparticles.

Authors' contributions

FJ, SJS and KR designed the study and the in vitro assays were performed by FJ. DV was responsible for NP synthesis and characterization under the supervision of WJP and BP. SJS and BBM aided with the initiation of the IN Cell experiments. SCDS and KR provided supervision and guidance throughout this work. The manuscript was written through contributions of all authors and all authors approved the final version of the manuscript. All authors read and approved the final manuscript.

Author details

[1] Lab of General Biochemistry and Physical Pharmacy, Department of Pharmaceutics, Faculty of Pharmaceutical Sciences, Ghent University, Ottergemsesteenweg 460, 9000 Ghent, Belgium. [2] Department of Physics, Philipps University of Marburg, Renthof 7, 35037 Marburg, Germany. [3] Biomedical MRI Unit/MoSAIC, Department of Medicine, KULeuven, Herestraat 49, 3000 Louvain, Belgium.

Acknowledgements

We would like to thank Shareen Doak from the DNA damage group, Institute of Life Sciences, Swansea University, UK for the use of the IN Cell Analyzer 2000. Additionally we wish to thank Sebastian Munck and Nicky Corthout from the VIB Centre for Biology of Disease, Belgium for the technical guidance in the use of the IN Cell Analyzer 2000 and IN Cell Developer Toolbox software. We would also like to thank Tianqiang Wang for support in nanoparticle characterization and Karsten Kantner for the ICP measurements.

Competing interests

The authors declare that they have no competing interests.

Funding

FJ is a doctoral fellow of the Agency for Innovation by Science and Technology in Flanders (IWT). KR and SJS are postdoctoral fellows of the Research Foundation-Flanders (FWO). Funding of both agencies was applied to design, perform and analyze all cell-based experiments. The DFG Germany (GRK 1782 to WJP), the European Commission (grant FutureNanoNeeds to WJP) and Alexander von Humboldt Foundation (postdoctoral fellowship of BP) supported the NP synthesis and characterization part of this work.

References

1. Stark WJ, Stoessel PR, Wohlleben W, Hafner A. Industrial applications of nanoparticles. Chem Soc Rev. 2015;44:5793–805.
2. Lim EK, Kim T, Paik S, Haam S, Huh YM, Lee K. Nanomaterials for theranostics: recent advances and future challenges. Chem Rev. 2015;115:327–94.
3. Haase A, Rott S, Mantion A, Graf P, Plendl J, Thunemann AF, Meier WP, Taubert A, Luch A, Reiser G. Effects of silver nanoparticles on primary mixed neural cell cultures: uptake, oxidative stress and acute calcium responses. Toxicol Sci. 2012;126:457–68.
4. Min Y, Caster JM, Eblan MJ, Wang AZ. Clinical translation of nanomedicine. Chem Rev. 2015;115:11147–90.
5. Rivera-Gil P, De Aberasturi DJ, Wulf V, Pelaz B, Del Pino P, Zhao YY, De La Fuente JM, De Larramendi IR, Rojo T, Liang XJ, Parak WJ. The challenge to relate the physicochemical properties of colloidal nanoparticles to their cytotoxicity. Acc Chem Res. 2013;46:743–9.
6. Nel A, Xia T, Madler L, Li N. Toxic potential of materials at the nanolevel. Science. 2006;311:622–7.
7. Joris F, Manshian BB, Peynshaert K, De Smedt SC, Braeckmans K, Soenen SJ. Assessing nanoparticle toxicity in cell-based assays: influence of cell culture parameters and optimized models for bridging the in vitro-in vivo gap. Chem Soc Rev. 2013;42:8339–59.
8. Jan E, Byrne SJ, Cuddihy M, Davies AM, Volkov Y, Gun'ko YK, Kotov NA. High-content screening as a universal tool for fingerprinting of cytotoxicity of nanoparticles. ACS Nano. 2008;2:928–38.
9. George S, Pokhrel S, Xia T, Gilbert B, Ji Z, Schowalter M, Rosenauer A, Damoiseaux R, Bradley KA, Madler L, Nel AE. Use of a rapid cytotoxicity screening approach to engineer a safer zinc oxide nanoparticle through iron doping. ACS Nano. 2010;4:15–29.
10. Ong KJ, MacCormack TJ, Clark RJ, Ede JD, Ortega VA, Felix LC, Dang MK, Ma G, Fenniri H, Veinot JG, Goss GG. Widespread nanoparticle-assay interference: implications for nanotoxicity testing. PLoS ONE. 2014;9:e90650.
11. Feliu N, Pelaz B, Zhang Q, del Pino P, Nyström A, Parak WJ. Nanoparticle dosage—a nontrivial task of utmost importance for quantitative nanotoxicology. WIREs Nanomed Nanobiotechnol. 2015.
12. Kermanizadeh A, Lohr M, Roursgaard M, Messner S, Gunness P, Kelm JM, Moller P, Stone V, Loft S. Hepatic toxicology following single and multiple exposure of engineered nanomaterials utilising a novel primary human 3D liver microtissue model. Part Fibre Toxicol. 2014;11:56.
13. Wang J, Fang X, Liang W. Pegylated phospholipid micelles induce endoplasmic reticulum-dependent apoptosis of cancer cells but not normal cells. ACS Nano. 2012;6:5018–30.
14. Bregoli L, Chiarini F, Gambarelli A, Sighinolfi G, Gatti AM, Santi P, Martelli AM, Cocco L. Toxicity of antimony trioxide nanoparticles on human hematopoietic progenitor cells and comparison to cell lines. Toxicology. 2009;262:121–9.
15. Zhang H, Wang X, Wang M, Li L, Chang CH, Ji Z, Xia T, Nel AE. Mammalian cells exhibit a range of sensitivities to silver nanoparticles that are partially explicable by variations in antioxidant defense and metallothionein expression. Small. 2015;11:3797–805.
16. Luengo Y, Nardecchia S, Morales MP, Serrano MC. Different cell responses induced by exposure to maghemite nanoparticles. Nanoscale. 2013;5:11428–37.
17. Mukherjee SG, O'Claonadh N, Casey A, Chambers G. Comparative in vitro cytotoxicity study of silver nanoparticle on two mammalian cell lines. Toxicol In Vitro. 2012;26:238–51.
18. Wang Y, Aker WG, Hwang HM, Yedjou CG, Yu H, Tchounwou PB. A study of the mechanism of in vitro cytotoxicity of metal oxide nanoparticles using catfish primary hepatocytes and human HepG2 cells. Sci Total Environ. 2011;409:4753–62.
19. Ekstrand-Hammarstrom B, Akfur CM, Andersson PO, Lejon C, Osterlund L, Bucht A. Human primary bronchial epithelial cells respond differently to titanium dioxide nanoparticles than the lung epithelial cell lines A549 and BEAS-2B. Nanotoxicology. 2012;6:623–34.
20. Kermanizadeh A, Gaiser BK, Ward MB, Stone V. Primary human hepatocytes versus hepatic cell line: assessing their suitability for in vitro nanotoxicology. Nanotoxicology. 2013;7:1255–71.
21. Shen WB, Vaccaro DE, Fishman PS, Groman EV, Yarowsky P. SIRB, sans iron oxide rhodamine B, a novel cross-linked dextran nanoparticle, labels human neuroprogenitor and SH-SY5Y neuroblastoma cells and serves as a USPIO cell labeling control. Contrast Media Mol Imaging. 2016;11:222–8.
22. Soenen SJ, Himmelreich U, Nuytten N, De Cuyper M. Cytotoxic effects of iron oxide nanoparticles and implications for safety in cell labelling. Biomaterials. 2011;32:195–205.
23. Lei H, Nan X, Wang Z, Gao L, Xie L, Zou C, Wan Q, Pan D, Beauchamp N, Yang X, et al. Stem cell labeling with superparamagnetic iron oxide nanoparticles using focused ultrasound and magnetic resonance imaging tracking. J Nanosci Nanotechnol. 2015;15:2605–12.
24. Hahn MA, Singh AK, Sharma P, Brown SC, Moudgil BM. Nanoparticles as contrast agents for in vivo bioimaging: current status and future perspectives. Anal Bioanal Chem. 2011;399:3–27.
25. Manshian BB, Moyano DF, Corthout N, Munck S, Himmelreich U, Rotello VM, Soenen SJ. High-content imaging and gene expression analysis to study cell-nanomaterial interactions: the effect of surface hydrophobicity. Biomaterials. 2014;35:9941–50.
26. Wu YY, Mujtaba T, Rao MS. Isolation of stem and precursor cells from fetal tissue. Methods Mol Biol. 2002;198:29–40.
27. Donato R, Miljan EA, Hines SJ, Aouabdi S, Pollock K, Patel S, Edwards FA, Sinden JD. Differential development of neuronal physiological responsiveness in two human neural stem cell lines. BMC Neurosci. 2007;8:36.
28. Seeger RC, Rayner SA, Banerjee A, Chung H, Laug WE, Neustein HB, Benedict WF. Morphology, growth, chromosomal pattern, and fibrinolytic-activity of 2 new human neuroblastoma cell lines. Cancer Res. 1977;37:1364–71.
29. Ryder EF, Snyder EY, Cepko CL. Establishment and characterization of multipotent neural cell-lines using retrovirus vector-mediated oncogene transfer. J Neurobiol. 1990;21:356–75.
30. Klebe RJ, Ruddle RH. Neuroblastoma—cell culture analysis of a differentiating stem cell system. J Cell Biol. 1969;43:A69.
31. Lin CA, Sperling RA, Li JK, Yang TY, Li PY, Zanella M, Chang WH, Parak WJ. Design of an amphiphilic polymer for nanoparticle coating and functionalization. Small. 2008;4:334–41.
32. Sun S, Zeng H, Robinson DB, Raoux S, Rice PM, Wang SX, Li G. Monodisperse MFe_2O_4 (M = Fe Co, Mn) nanoparticles. J Am Chem Soc. 2004;126:273–9.
33. Caballero-Diaz E, Pfeiffer C, Kastl L, Rivera-Gil P, Simonet B, Valcarcel M, Jimenez-Lamana J, Laborda F, Parak WJ. The toxicity of silver nanoparticles depends on their uptake by cells and thus on their surface chemistry. Part Part Syst Charact. 2013;30:1079–85.
34. Soenen SJ, Parak WJ, Rejman J, Manshian B. (Intra)cellular stability of inorganic nanoparticles: effects on cytotoxicity, particle functionality, and biomedical applications. Chem Rev. 2015;115:2109–35.
35. Soenen SJ, Rivera-Gil P, Montenegro JM, Parak WJ, De Smedt SC, Braeckmans K. Cellular toxicity of inorganic nanoparticles: common aspects and guidelines for improved nanotoxicity evaluation. Nano Today. 2011;6:446–65.
36. Wilkinson KE, Palmberg L, Witasp E, Kupczyk M, Feliu N, Gerde P, Seisenbaeva GA, Fadeel B, Dahlen SE, Kessler VG. Solution-engineered palladium nanoparticles: model for health effect studies of automotive particulate pollution. ACS Nano. 2011;5:5312–24.
37. Schlinkert P, Casals E, Boyles M, Tischler U, Hornig E, Tran N, Zhao JY, Himly M, Riediker M, Oostingh GJ, et al. The oxidative potential of differently charged silver and gold nanoparticles on three human lung epithelial cell types. J Nanobiotechnol. 2015;13:1.
38. George S, Xia T, Rallo R, Zhao Y, Ji Z, Lin S, Wang X, Zhang H, France B, Schoenfeld D, et al. Use of a high-throughput screening approach coupled with in vivo zebrafish embryo screening to develop hazard ranking for engineered nanomaterials. ACS Nano. 2011;5:1805–17.
39. Park EJ, Park K. Oxidative stress and pro-inflammatory responses induced by silica nanoparticles in vivo and in vitro. Toxicol Lett. 2009;184:18–25.
40. Zhang H, Pokhrel S, Ji Z, Meng H, Wang X, Lin S, Chang CH, Li L, Li R, Sun B, et al. PdO doping tunes band-gap energy levels as well as oxidative stress responses to a Co(3)O(4) p-type semiconductor in cells and the lung. J Am Chem Soc. 2014;136:6406–20.
41. Sharma G, Kodali V, Gaffrey M, Wang W, Minard KR, Karin NJ, Teeguarden JG, Thrall BD. Iron oxide nanoparticle agglomeration influences dose rates and modulates oxidative stress-mediated dose-response profiles in vitro. Nanotoxicology. 2014;8:663–75.
42. Fujioka K, Hanada S, Inoue Y, Sato K, Hirakuri K, Shiraishi K, Kanaya F, Ikeda K, Usui R, Yamamoto K, et al. Effects of silica and titanium oxide particles on a human neural stem cell line: morphology, mitochondrial

activity, and gene expression of differentiation markers. Int J Mol Sci. 2014;15:11742–59.

43. Huang DM, Hsiao JK, Chen YC, Chien LY, Yao M, Chen YK, Ko BS, Hsu SC, Tai LA, Cheng HY, et al. The promotion of human mesenchymal stem cell proliferation by superparamagnetic iron oxide nanoparticles. Biomaterials. 2009;30:3645–51.
44. Gao L, Zhuang J, Nie L, Zhang J, Zhang Y, Gu N, Wang T, Feng J, Yang D, Perrett S, Yan X. Intrinsic peroxidase-like activity of ferromagnetic nanoparticles. Nat Nanotechnol. 2007;2:577–83.
45. Sabella S, Carney RP, Brunetti V, Malvindi MA, Al-Juffali N, Vecchio G, Janes SM, Bakr OM, Cingolani R, Stellacci F, Pompa PP. A general mechanism for intracellular toxicity of metal-containing nanoparticles. Nanoscale. 2014;6:7052–61.
46. Petters C, Thiel K, Dringen R. Lysosomal iron liberation is responsible for the vulnerability of brain microglial cells to iron oxide nanoparticles: comparison with neurons and astrocytes. Nanotoxicology. 2016;10:332–42.
47. Zhivotovsky B, Orrenius S. Calcium and cell death mechanisms: a perspective from the cell death community. Cell Calcium. 2011;50:211–21.
48. Clapham DE. Calcium signaling. Cell. 2007;131:1047–58.
49. Ariano P, Zamburlin P, Gilardino A, Mortera R, Onida B, Tomatis M, Ghiazza M, Fubini B, Lovisolo D. Interaction of spherical silica nanoparticles with neuronal cells: size-dependent toxicity and perturbation of calcium homeostasis. Small. 2011;7:766–74.
50. Tseng YC, Yang A, Huang L. How does the cell overcome lcp nanoparticle-induced calcium toxicity? Mol Pharm. 2013;10:4391–5.
51. Anguissola S, Garry D, Salvati A, O'Brien PJ, Dawson KA. High content analysis provides mechanistic insights on the pathways of toxicity induced by amine-modified polystyrene nanoparticles. PLoS ONE. 2014;9:e108025.
52. Brookes PS, Yoon Y, Robotham JL, Anders MW, Sheu SS. Calcium, ATP, and ROS: a mitochondrial love-hate triangle. Am J Physiol Cell Physiol. 2004;287:C817–33.
53. Capiod T, Shuba Y, Skryma R, Prevarskaya N. Calcium signalling and cancer cell growth. Subcell Biochem. 2007;45:405–27.
54. Fulda S, Galluzzi L, Kroemer G. Targeting mitochondria for cancer therapy. Nat Rev Drug Discov. 2010;9:447–64.
55. Ito K, Suda T. Metabolic requirements for the maintenance of self-renewing stem cells. Nat Rev Mol Cell Biol. 2014;15:243–56.
56. Gottlieb E, Armour SM, Harris MH, Thompson CB. Mitochondrial membrane potential regulates matrix configuration and cytochrome c release during apoptosis. Cell Death Differ. 2003;10:709–17.
57. Heerdt BG, Houston MA, Wilson AJ, Augenlicht LH. The intrinsic mitochondrial membrane potential (Deltapsim) is associated with steady-state mitochondrial activity and the extent to which colonic epithelial cells undergo butyrate-mediated growth arrest and apoptosis. Cancer Res. 2003;63:6311–9.
58. Buyukhatipoglu K, Clyne AM. Superparamagnetic iron oxide nanoparticles change endothelial cell morphology and mechanics via reactive oxygen species formation. J Biomed Mater Res A. 2011;96:186–95.
59. Tay CY, Cai P, Setyawati MI, Fang W, Tan LP, Hong CH, Chen X, Leong DT. Nanoparticles strengthen intracellular tension and retard cellular migration. Nano Lett. 2014;14:83–8.
60. Wu YL, Putcha N, Ng KW, Leong DT, Lim CT, Loo SCJ, Chen XD. Biophysical responses upon the interaction of nanomaterials with cellular interfaces. Acc Chem Res. 2013;46:782–91.
61. Soenen SJ, Manshian B, Montenegro JM, Amin F, Meermann B, Thiron T, Cornelissen M, Vanhaecke F, Doak S, Parak WJ, et al. Cytotoxic effects of gold nanoparticles: a multiparametric study. ACS Nano. 2012;6:5767–83.
62. Ware MJ, Godin B, Singh N, Majithia R, Shamsudeen S, Serda RE, Meissner KE, Rees P, Summers HD. Analysis of the influence of cell heterogeneity on nanoparticle dose response. ACS Nano. 2014;8:6693–700.
63. Kirchner C, Liedl T, Kudera S, Pellegrino T, Javier AM, Gaub HE, Stolzle S, Fertig N, Parak WJ. Cytotoxicity of colloidal CdSe and CdSe/ZnS nanoparticles. Nano Lett. 2005;5:331–8.
64. Ge GY, Wu HF, Xiong F, Zhang Y, Guo ZR, Bian ZP, Xu JD, Gu CR, Gu N, Chen XJ, Yang D. The cytotoxicity evaluation of magnetic iron oxide nanoparticles on human aortic endothelial cells. Nanoscale Res Lett. 2013;8:1.
65. Shaw SY, Westly EC, Pittet MJ, Subramanian A, Schreiber SL, Weissleder R. Perturbational profiling of nanomaterial biologic activity. Proc Natl Acad Sci USA. 2008;105:7387–92.

9

Nanoparticles based on essential metals and their phytotoxicity

Branislav Ruttkay-Nedecky[1,2], Olga Krystofova[1,2], Lukas Nejdl[1,2] and Vojtech Adam[1,2*]

Abstract

Nanomaterials in agriculture are becoming popular due to the impressive advantages of these particles. However, their bioavailability and toxicity are key features for their massive employment. Herein, we comprehensively summarize the latest findings on the phytotoxicity of nanomaterial products based on essential metals used in plant protection. The metal nanoparticles (NPs) synthesized from essential metals belong to the most commonly manufactured types of nanomaterials since they have unique physical and chemical properties and are used in agricultural and biotechnological applications, which are discussed. The paper discusses the interactions of nanomaterials and vascular plants, which are the subject of intensive research because plants closely interact with soil, water, and atmosphere; they are also part of the food chain. Regarding the accumulation of NPs in the plant body, their quantification and localization is still very unclear and further research in this area is necessary.

Keywords: Agriculture, Fertilizers, Nanomaterials, Essential metal nanoparticles, Nanoparticles uptake, Phytotoxicity

Background

The main issues, of which agriculture worldwide have been facing to, are loss of fertile land due to pollution, desertification and climate changes. Due to unique and outstanding properties of nanomaterials it is not surprising that an effort to improve the agrarian sector using nanotechnology and nanomaterials has been developing [1–11]. Particularly, the use of various types of nanomaterials made of metal oxides, ceramics, silicates, magnetic materials, semiconductor quantum dots (QDs), lipids, polymers, dendrimers, and emulsions [12–15] aims to reduce the applied amount of plant protection products (PPP), to minimize the loss of nutrients during fertilization, and increase revenues through optimized nutrient management in agriculture [3, 4, 16–18].

Greater utilization of nanoparticles (NPs) in agriculture depends on several factors including well known effects, monitored fate as well as their potential toxicity and levels of overdosing. NPs may interact with their environment and plants are a fundamental part of all ecosystems. It can therefore be assumed that NPs will interact with plants and these interactions, such as income and their accumulation in plant biomass, will affect their fate and transport in the environment. NPs may also adhere to the roots of the plants and cause physical or chemical toxicity to plants. Interaction with microorganisms in the soil cannot be excluded because they can positively interact with plants [19–22]. Based on these fact it is clear that there is an ability of nanomaterials to penetrate live plant tissues, but it has ramifications for their accumulation in the food chain and for their utility as smart delivery systems in living plants. Our ability to evaluate these impacts requires an understanding of how NPs are transported within a plant. It is important to understand whether intact NPs can be taken up by plants and transported to other plant tissues. In this area, it was found that NPs can enter plant tissues through either the root tissues or the aboveground organs and tissues (e.g., cuticles, trichomes, stomata, stigma, and hydathodes), as well as through wounds and root junctions (Fig. 1). Only several studies have reported 'direct' uptake, translocation, and localization of NPs in plants using various insoluble NPs including mesoporous silica NPs [23], silica NPs (SNPs) [24], carbon nanotubes [25], fullerenes (C70) [26], QDs [27],

*Correspondence: vojtech.adam@mendelu.cz
[1] Department of Chemistry and Biochemistry, Mendel University in Brno, Zemedelska 1, 613 00 Brno, Czech Republic
Full list of author information is available at the end of the article

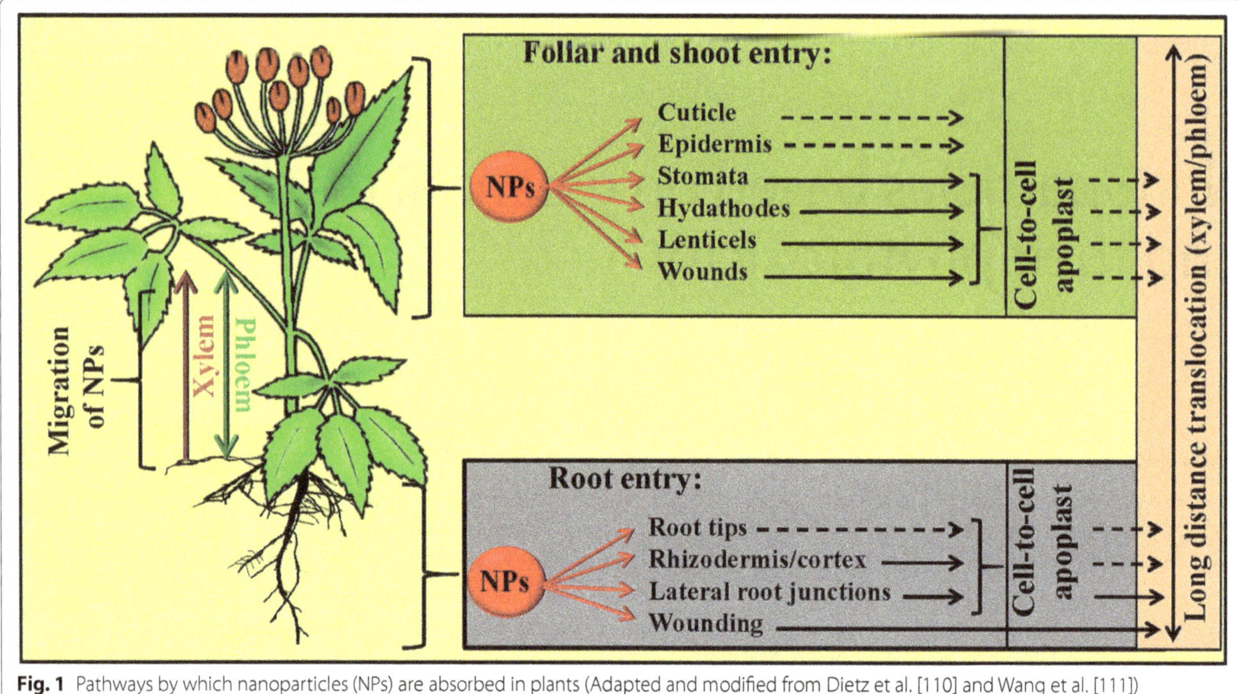

Fig. 1 Pathways by which nanoparticles (NPs) are absorbed in plants (Adapted and modified from Dietz et al. [110] and Wang et al. [111])

Au-NPs [28], titanium dioxide NPs (TiO_2 NPs) [29, 30], iron (II, III) oxide (Fe_3O_4) NPs [31, 32], and virus-based NPs [33].

The toxicity of NPs in plants has been discussed several times [8, 34–37]. The conclusions showed that not all plants treated with nanoparticles exhibit toxic effects; substantially more studies showed positive or no consequential effects on plants. Nanotoxicity mechanisms remain unknown, however, it can be assumed to be closely related to the chemical composition, chemical structure, size, and surface area of nanoparticles. The presence of nanoparticles on the root surface can change the surface chemistry of the roots and consequently affect the uptake of nutrients into the plant root [21, 22], thus, these have to be taken into consideration too. Generally, the toxicity of nanoparticles is attributable to two different steps: (1) chemical toxicity based on chemical composition, for example the release of (toxic) ions, and (2) stress stimuli caused by surface, size, or shape of the particles.

In the review, first we describe the basic methods for phytotoxicity testing, then we provide an overview of the most common techniques for detecting and imaging nanoparticles in plants, and then we will focus on the benefits of using essential metal nanoparticles (Zn, Cu, Fe, Mn, and their oxides) in agriculture and current knowledge on their potential phytotoxicity.

Methods for testing the phytotoxicity of metal nanoparticles

Phytotoxicity tests

There are no specific test guidelines for nanotoxicity so EPA48 or OECD49 directives from the US for chemical testing are currently used [38]. Phytotoxicity tests generally use plants recommended by these guidelines. These are mostly species of crops, and include both monocotyledonae and dicotyledonae [38]. Species that are recommended most are bean (*Phaseolus vulgaris*), cabbage (*Brassica oleracea*), carrot (*Daucus carota* subsp. *sativus*), cucumber (*Cucumis sativus*), lettuce (*Lactuca sativa*), maize (*Zea mays* subsp. *mays*), oat (*Avena sativa*), onion (*Allium cepa*), radish (*Raphanus sativus*), rice (*Oryza sativa*), ryegrass (*Lolium perenne* L.), soybean (*Glycine max*), tomato (*Solanum lycopersicum*), and wheat (*Triticum aestivum*). Recently, research model species such as the well characterized thale cress (*Arabidopsis thaliana*) were also included [39].

Phytotoxicity tests are carried out in two stages of plant development: (1) during germination, when the germination percentage is measured, where the seeds must be exposed to the test solution for the duration of germination (preferably at least 4 days) [38], and (2) during seedling growth, in which root/shoot elongation and dry weight are frequently used variables to assess the effects of plant exposure to harmful substances [40]. The aforementioned protocols have been applying for testing the effects of nanoparticles in water, wastewater, sediment, and slurry.

For phytotoxicity testing different media for the growth of plants are used. The simplest medium is water (Fig. 2). Other applications include soft gels or agars, which better represent the soil, and finally soil itself is also often used [39]. Nanoparticles tend to adsorb to soil matrix and aggregate in the natural environment which reduces their mobility and bioavailability (Fig. 2). Also, the study of the interactions of nanoparticles with plants in alternative substrates does not take into account the potential interaction of nanoparticles with soil and the associated water phase [41]. For example, nanoparticles in soil can influence the growth of soil bacteria, which may then indirectly affect the plant growth [42].

Rico et al. [36] and Peralta-Videa et al. [43] are the pioneers of studies dealing with the effects of nanoparticles on vascular plants. The most common monitored parameters include the germination rate and root/stem growth rate. Recently, the number of leaves [44] and chlorophyll content [45, 46] of exposed plants were included as new monitored parameters for phytotoxicity tests. In addition, the cytotoxicity and genotoxicity of nanoparticles are assessed [39, 47], which is also indicated in Fig. 2.

Nanomaterials based on essential metals and their use in agriculture

Essential metal nanoparticles are chosen because they are essential metals for plants, and are nontoxic in wide concentration range. This group of metals involves nanoparticles based on Zn, Cu, Fe, Mn, and their oxides. From these, zinc oxide (ZnO) and copper oxide (CuO) nanoparticles (NPs) are used in numerous commercial applications including antimicrobial formulations. Recent studies suggest the use of these NPs as fungicides in agriculture and in the food industry, whereas treatments with ZnO or CuO NPs inhibit the growth of fungal plant pathogens such as *Botrytis cinerea, Penicillium expansum, Fusarium graminearum,* and *Phytophthora infestans* [48–50]. Consequently, although NPs may be formulated for use in agriculture as crop protectants, their impact on nontarget soil microbes is not fully known. Both CuO and ZnO NPs cause bacterial cell death at doses that vary with the microorganism [42, 51, 52].

Nanoscale zerovalent iron (Fe^0) and bimetallic Ni^0–Fe^0 nanoparticles have emerged as effective redox media for the detoxification of organic and inorganic pollutants in aqueous solutions. These nanomaterials (10–100 nm) have larger surface areas and reactivity than bulk Fe^0 particles [53–55]. Mn deficiency has been widely reported all over the world, especially in soils with higher pH (>6.0) or in calcareous, sandy, peat, or muck soils [56]. Therefore, Mn fertilization is very important to improve

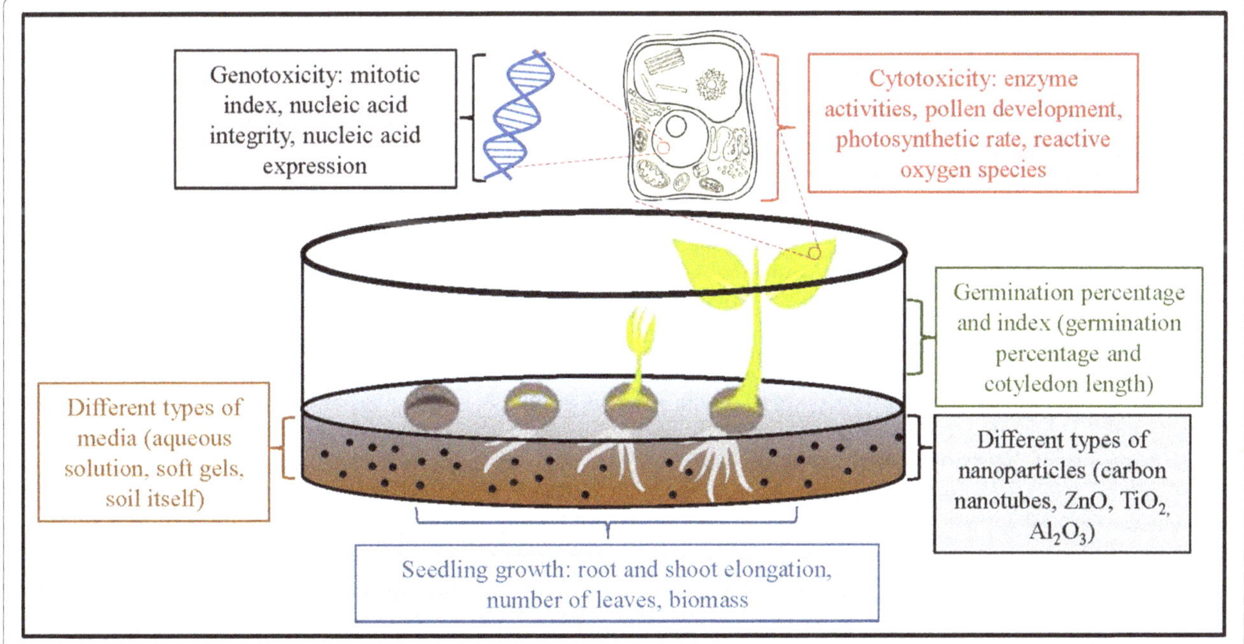

Fig. 2 Important considerations when designing phytotoxicity studies and endpoints in phytotoxicity studies (Adapted and modified from Miralles et al. [39])

agronomic production [56]. Manganese nanoparticles (MnNPs) have been also proposed as a suitable alternative to commercially used manganese salts $MnSO_4$ (MS) for nanobiotechnology based crop management studies [57].

Nanomaterials based on Zn

Zinc nanoparticles (NPs) are spherical or polished metal particles with a high specific surface area. The applications of zinc nanocrystals include antimicrobial, antibiotic and antifungal agents, which are a part of painting buildings, dressing materials, nanofibers, plastics, and textiles [58]. Moreover, ZnO NPs are widely used in personal care products such as sunscreens, cosmetics, textiles, lipsticks, and hair dyes. In industrial products they are used in floor coatings, solar cells, as an antibacterial agent, and with optical and electronic materials [59–61]. These sprays are one of the direct routes of ZnO NPs into the environment. ZnO NPs are also present in agricultural spraying as a protecting material against UV radiation [62], where ZnO contributes together with an organic filter for the protection of photosensitive pesticides, is used directly for crop protection against UV radiation [63]. In addition, ZnO NPs have been also studied as a nutrient to increase the efficiency of plant fertilization, but the larger surface area of nanoparticles do not ensure improved solubility or even higher availability of Zn^{2+} for plants [64]. One may suggest that the solubility of Zn^{2+} in Zn fertilizers plays an important role in the agronomic effectiveness of the fertilizer. On the basis of thermodynamics, ZnO NPs should dissolve faster and to a greater extent than bulk ZnO particles (equivalent spherical diameter >100 nm). These novel solubility features of ZnO NPs might be exploited to improve the efficiency of Zn fertilizers. In this field, coated monoammonium phosphate granules show greater Zn solubility and faster dissolution rates in sand columns compared to coated urea granules, which may be related to pH differences in the solution surrounding the fertilizer granules. The kinetics of Zn dissolution was not affected by the size of the ZnO NPs applied for coating of either fertilizer type, possibly because solubility was controlled by the formation of the same compounds irrespective of the size of the original ZnO NPs used for coating [64]. In another study ZnO NPs were investigated for their use as a Zn supplement. Seeds of several plants (*Z. mays, G. max, Cajanus cajan* and *Abelmoschus esculentum*) were coated with ZnO NPs. The germination test carried out with coated and uncoated seeds indicated a better germination percentage (93–100%) due to the ZnO coating when compared to uncoated seeds (80%). A pot culture experiment was conducted with coated seeds and this also revealed that the crop growth with ZnO coated seeds was similar to that observed with soluble Zn treatment applied as zinc sulfate heptahydrate [47].

Besides supplementation by Zn, ZnO NPs, synthesized by soil fungi in a concentration 10 mg L^{-1}, has been shown to enhance the mobilization of native phosphorus in the mung bean (*Vigna radiata*) rhizosphere. The analyses made by authors showed that they synthesized ZnO NPs with average diameter as 22.4 nm as they claimed to have stable nanoparticles due to in situ corona formation by fungal extracellular protein used in the synthesis procedure. Zn acts as a cofactor for phosphorus-solubilizing enzymes such as phosphatase and phytase, and ZnO NPs increased their activity. The level of resultant phosphorus uptake in *V. radiata* increased by 10.8%. In addition, biosynthesized ZnO NPs also improved plant plienology such as stem height, root volume, and biochemical indicators such as leaf protein and chlorophyll contents [65]. Not only the effect but also the way of application was considered by authors as they choose foliar application on 2-week-old mung bean plants. The concentration of the ZnO suspension was 10 mg L^{-1}, where a total of 25 mL of suspension ZnO NPs was sprayed on each plant by an atomizer generating droplets. In spite of the foliar application, the aforementioned positive effects have been evidenced.

Next, the fungicide activity of ZnO NPs against *F. graminearum* was investigated, too. Wheat plants were inoculated with *F. graminearum* and treated with ZnO NPs (100 mM). When the wheat plants reached maturation, the grains were harvested and evaluated for *Fusarium* (number of colonies, CFU g^{-1}). ZnO NPs showed a reduction in number of CFU of *F. graminearum* when compared to the control [66].

In another work, ZnO NPs were shown to have interactive effects on *Pseudomonas chlororaphis* O6 (PcO6) to inhibit the plant pathogen *F. graminearum.* ZnO NPs were commercial ones with diameter less than 100 nm. Growth of *F. graminearum* was significantly ($p = 0.05$) inhibited by the inclusion of ZnO NPs in a mung bean both in mung bean agar and in sand tested in the presence and also in no presence of PcO6. The treatment itself lasted for 7 days. The ZnO NPs were significantly more inhibitory to fungal growth than micro-sized particles of ZnO, although both types of particles released similar levels of soluble Zn, indicating size-dependent toxicity of the particles [49].

Thus, one can say that low concentrations of ZnO NPs are beneficial to plants. Positive effects of ZnO NPs are manifested in promoting germination, stem and root growth, increase in phosphorus mobilizing enzymes, phosphorus uptake, and antifungal properties. The observed positive effects of ZnO NPs on plants are summarized in Table 1.

Table 1 The observed positive effects of ZnO NPs on plants

Plant	Particle size (nm)	Particle concentration	Comment	Observed effect	References
Vigna radiata	22.4 ± 1.8	10 mg L^{-1}	NPs were synthesized by soil fungi	Increase in stem height and root length. Increase in phosphorus mobilizing enzymes and phosphorus uptake by 10.8%	[65]
Zea mays, Glycine max, Cajanas cajan, Abelmoschus esculentus	<100	25 or 50 mg Zn g^{-1} seed	Seeds were coated with ZnO NPs	Improved germination Coated seeds 93–100% Uncoated seeds 80%	[47]
Triticum aestivum	30	100 mM Zn	Wheat plants were inoculated with *Fusarium graminearum*	Reduction in number of CFU of *F. graminearum*	[66]

Nanomaterials based on copper

Cu/CuO NPs are used in optoelectronics, catalysis, solar cells, as semiconductors, as they are also used as pigments, and fungicides [50, 67, 68]. Copper as fungicide is especially used in vineyards and in organic farming [62]. The ability of copper ions to prevent spore germination of fungi has been known for a long time, but to achieve this effect it is necessary to apply a large amount of copper (500–1500 g ha^{-1}). Of note, it is certainly worth a patent of the BASF company [69]. Subject to the patent is the nanoparticulate amorphous Cu^{2+} salt, which forms by a reaction with polymer CuNPs within the size from 1 to 200 nm. Compared with commonly used non-nano product containing cupric hydroxide (Cuprozin, Spiess Urania Chemicals), the same dose of copper in the form of nanoparticles improves efficiency by 8% against a phytopathogenic fungus on vines [62]. This is an example of how the nanoparticle form can reduce the amount of Cu discharged into the environment. Recently, one study demonstrated that CuNPs absorbed in chitosan hydrogel had positive effects on tomato growth and quality [70]. During this process, the activity of some enzymes can increase such as catalase, or decrease in the case of ascorbate peroxidase [71]. It is believed that the stimulatory effects of CuNPs are related to the induction of antioxidant activity [72]. The positive effect of CuNPs on *S. lycopersicum* is shown in Table 2 in the same way as in the case of Zn NPS.

Nanomaterials based on iron

Iron nanoparticles (INPs) represent a new generation of environmental remediation technologies that could provide cost-effective solutions to some of the most challenging environmental issues. Because of large surface areas and high surface reactivity [73], INPs have found their main application in remediation [71]. This method is relatively cheap and uses both free (soil application, where INPs can penetrate ground water) and into matrix fixed nanoparticles (cleaning water or air) [54]. In the greatest extent INPs are used to decompose substances such as chlorinated hydrocarbons (e.g. trichloroethylene), organochlorine pesticides, and polychlorinated biphenyls [54]. Besides decomposing, INPs can be further applied to bind, for example, to a significant pollutant, arsenic ions [74]. Materials composed of nanoscaled iron particles exhibit high absorbency and a second advantage is their response to external magnetic fields by which they can be, even with bound arsenic compounds, removed. The mentioned procedure can also be used for other metals such as mercury or lead [75].

In agriculture, Fe_2O_3 NPs may be used instead of Fe fertilizers [76]. Rui et al. evaluated the effectiveness of iron oxide nanoparticles (IONPs; Fe_2O_3 NPs) as a fertilizer to replace traditional Fe fertilizers [77]. The effects of the Fe_2O_3 NPs and a chelated-Fe fertilizer (ethylenediaminetetraacetic acid-Fe; EDTA-Fe) on the growth and development of peanut (*Arachis hypogaea*), a crop that is very sensitive to Fe deficiency, were studied in a pot experiment. The results showed that Fe_2O_3 NPs increased root length, plant height, biomass, and soil plant analysis development (SPAD) values of peanut plants. The Fe_2O_3 NPs promoted the growth of peanuts by regulating phytohormone contents and antioxidant enzyme activity. The Fe contents in peanut plants with Fe_2O_3 NPs and EDTA-Fe treatments were higher than the control group. The next study was conducted to examine the effect FeNPs (prepared by reduction with a gum kondagogu) on the growth of a mung bean (*V. radiata*). The radical length and biomass was increased in seeds exposed to FeNPs in comparison with the ions [78]. In the following study, the uptake of iron oxide (Fe_2O_3) nanoparticles by spinach (*Spinacea oleracea*) via hydroponics was demonstrated and its effects on the growth rate and productivity of the spinach plant were examined. The experimental studies such as plant growth (stem and root length) and biomass analysis revealed a dose and time dependent increase due to the uptake of Fe_2O_3 [19]. In the next study, Trujillo-Reyes et al. showed that iron NPs, unlike CuNPs, did not affect the chlorophyll content, plant growth, catalase (CAT), and ascorbate peroxidase (APX) activities of lettuce (*L. sativa*) [71].

In another work, INPs after foliar application had significant effect on yield, leaf Fe content, stem Mg content, plasma membrane stability, and chlorophyll content of *Vigna unguiculata* [79]. In the following study, Alidoust et al. investigated the effect of 6-nm IONPs and citrate-coated IONPs (IONPs-Cit) on photosynthetic characteristics and root elongation during germination of a soybean (*G. max* L.) [20]. Plant physiological performance was assessed after foliar and soil IONPs fertilization. No adverse impacts at any growth stage of the soybeans were observed after the application of IONPs.

Table 2 The observed positive effects of CuNPs on *Solanum lycopersicum*

Plant	Particle size (nm)	Particle concentration	Comment	Observed effect	Reference
Solanum lycopersicum	<100	15, 30, 60, 150 mg L^{-1}	CuNPs were adsorbed on chitosan hydrogels	Application of chitosan hydrogels with CuNPs was favorable to tomato growth and quality	[70]

The Fe_2O_3 nanoparticles produced a significant positive effect on root elongation, particularly when compared to the bulk counterpart (IOBKs) suspensions of concentrations greater than 500 mg L^{-1}. In the next study of Ghafariyan et al. [80] seed germination of a soybean exposed to superparamagnetic iron oxide nanoparticles (SPIONs) was investigated. It was found that SPIONs, which were entered and translocated in the soybean, increased chlorophyll levels with no trace of toxicity. Furthermore, it was found that physicochemical characteristics of the SPIONs had a crucial role in the enhancement of chlorophyll content in subapical leaves of soybeans. The equivalent ratio of chlorophyll *a* to *b* in all treatments with conventional growth, medium iron chelate, and SPIONs (as iron source) indicated no significant difference on the photosynthesis efficiency. An overview of the positive effects of INPs and iron oxide nanoparticles (IONPs) on plants is shown in Table 3.

Nanomaterials based on manganese

Manganese (Mn) is an essential micronutrient for growth regulation and the development of plants [81]. It plays a pivotal role in oxygenic photosynthesis both directly and indirectly. The major drawbacks associated with Mn deficiency are plant nutritional disorders [81]. To circumvent this nutritional disorder of plants, nanoparticle mediated crop management has of late found potential applications [57].

In a study by Pradhan et al. the effect of manganese nanoparticles (MnNPs) on nitrogen uptake in mung bean plants (*V. radiata*) was investigated [82]. The objective of this study was to determine the response of manganese nanoparticles (MnNP) in nitrate uptake, assimilation, and metabolism compared with the commercially used manganese salt, manganese sulfate (MS). MnNPs were modulated to affect the assimilatory process by enhancing the net flux of nitrogen assimilation through NR-NiR and GS-GOGAT pathways. This study was associated with toxicological investigation on in vitro and in vivo systems to promote MnNPs as a nanofertilizer and can be used as an alternative to MS.

In another study from the same research group [57] MnNP-treated chloroplasts showed greater photophosphorylation, oxygen evolution with respect to control, and $MnSO_4$-treated chloroplasts as determined by biophysical and biochemical techniques. Positive effects on root and shoot elongation was observed. MnNP-treated plants did not trigger oxidative stress.

In the next study, Liu et al. [83] investigated the effects of laboratory-prepared MnOx NPs on the germination of lettuce (*L. sativa*) seeds in a water medium. MnOx NPs only slightly reduced the germination percentage from 84% (control) to 63% even at a high concentration of 50 mg L^{-1} and was not significantly different from that of the control. Furthermore, MnOx NPs specifically improved the growth of lettuce seedlings by enhancing root elongation. For example, the 5-day root length of the seedlings increased by 68%. Similarly, 10- and 5-mg L^{-1} NPs also significantly increased the elongations by 41.6 and 53.9%, respectively. An overview of the positive effects of MnNPs and manganese oxide nanoparticles (MnOx NPs) in plants is shown in Table 4.

Phytotoxicity of ZnO, Cu (CuO), and iron oxide nanoparticles

A good understanding of the mechanisms of the nanoparticle phytotoxicity is important for the targeted application of nanoparticles [84]. Essential metal NPs can cause phytotoxicity via the dissolution and release of higher concentration of essential ions [85, 86] such as Zn^{2+} and Cu^{2+} or the production of excess reactive oxygen species (ROS) through redox cycling and the Fe^{2+}-mediated Fenton reaction [87].

Phytotoxicity of nanoparticles based on ZnO

Ecotoxicity studies on ZnO NPs are most abundant in bacteria and are relatively lacking in other species [88]. These studies suggest relative high acute toxicity of ZnO NPs (in the low mg L^{-1} levels) to environmental species, although this toxicity is highly dependent on test species, physicochemical properties of the material, and test methods. Particle dissolution to ionic zinc and particle-induced generation of ROS represent the primary modes of action for ZnO NPs toxicity across all species tested, and photo-induced toxicity associated with its photocatalytic property may be another important mechanism of toxicity under environmentally relevant UV radiation [85].

ZnO NPs have been shown to induce oxidative stress in soybean (*G. max*) seedlings in a concentration of 500 mg L^{-1}. Plant growth, rigidity of roots, and root cell viability were markedly affected by ZnO NPs stress. Oxidation–reduction cascade related genes, such as GDSL motif lipase 5, SKU5 similar 4, galactose oxidase, and quinone reductase were down-regulated in ZnO NPs treatment [89].

In the next study, Mukherjee et al. [90] investigated the impact of different zinc oxide (ZnO) NPs on green pea plants (*Pisum sativum* L.). Pea plants were grown for 65 days in soil amended with commercially available bare ZnO NPs (10 nm), 2 wt% alumina doped Al_2O_3/ZnO NPs (15 nm), or 1 wt% aminopropyltriethoxysilane coated KH550/ZnO NPs (20 nm) at 250 and 1000 mg $NPs.kg^{-1}$ soil inside a greenhouse. Although all treated plants showed higher tissue Zn content than controls, those exposed to Al_2O_3/ZnO NPs at 1000 mg kg^{-1} had greater

Table 3 The observed positive effects of iron/iron oxide NPs on plants

Plant	Particle size (nm)	Particle concentration	Comment	Observed effect	References
Arachis hypogaea	γ-Fe_2O_3, 20 nm	2, 10, 50, 250, 1000 mg kg^{-1} of soil	Fe_2O_3 NPs were applied into soil and compared with a chelated-Fe fertilizer	Fe_2O_3 NPs increased root length, plant height, biomass, and SPAD values of peanut plants. Fe_2O_3 NPs adsorbed onto sandy soil and improved the availability of Fe to the plants. Fe_2O_3 NPs can replace traditional Fe fertilizers in the cultivation of peanut plants	[77]
Vigna radiata	FeNPs 2–6 nm +0.2% gum, +0.4% gum	1 mM Fe^{2+}ions	Natural biopolymer gum kondagogu as reducing and capping agent was used	The radical length and biomass was increased in seeds exposed to Fe NPs in comparison to Fe^{2+} ions. The α-amylase activity was increased in the seeds exposed to Fe NPs	[78]
Spinacea oleracea	α-Fe_2O_3 50 nm	100, 150, 200 mg kg^{-1} of soil	Experiments were performed in a solid hydroponic medium consisting of sawdust and coco peat and adequate amounts of water	Positive effects on spinach plant due to uptake of Fe_2O_3 nanoparticles such as increase in stem and root lengths, biomass production and magnetic properties were observed	[19]
Lactuca sativa	Core–shell NPs Fe/Fe_3O_4 13/9 nm	10, 20 mg L^{-1}	15-days treatment of hydroponically grown lettuce	The nano-Fe/Fe_3O_4 at 10 and 20 mg L^{-1} and $FeSO_4 \cdot 7H_2O$ at 10 mg L^{-1} did not affect lettuce growth and chlorophyll content	[71]
Vigna unguiculata	<100 nm	25, 500 mg L^{-1}	The elements were applied 56 and 72 days after sowing over the leaves, and data was collected after day 85	Iron had significant effect on yield, leaf Fe content, stem Mg content, plasma membrane stability, and chlorophyll content, probably as a result of more efficient photosynthesis	[79]
Glycine max	γ-Fe_2O_3 (IONPS) and citrate coated IONPs 6 nm	500, 1000 mg L^{-1}	Plant physiological performance was assessed after foliar and soil IONPs fertilization	IONPs produced a significant positive effect on root elongation. IONPs-Cit significantly enhanced photosynthetic parameters when sprayed foliarly. More pronounced positive effects of IONPs via foliar application than by soil treatment was observed	[20]
Glycine max	Superparamagnetic iron oxide NPs (SPIONs) 8–12 nm	200, 400, 1000 and 2000 mg L^{-1}	Seed germination of soybean exposed to SPIONs was investigated	SPIONs, which were entered and translocated in the soybean, increased chlorophyll levels, with no trace of toxicity	[80]

Table 4 The observed positive effects of MnNPs on plants

Plant	Particle size (nm)	Particle concentration	Comment	Observed effect	References
Vigna radiata	MnNPs	50, 100, 500, 1000 mg L^{-1}	Leaf and root enzyme extract was analyzed for use as nanofertilizer	Nitrogen uptake, its assimilation, and metabolism was increased after MnNPs soil application	[82]
Vigna radiata	MnNPs	50, 100, 500, 1000 mg L^{-1}	Leaf and root enzyme extract was analyzed. Chloroplasts from leaves were isolated and analyzed for their level of photophosphorylation and oxygen evolution	MnNP-treated chloroplasts showed greater photophosphorylation, oxygen evolution with respect to control and MnSO4-treated chloroplasts. Positive effects on root and shoot elongation was observed	[57]
Lactuca sativa	MnOx NPs 5–15 nm	0.25, 0.5, 5, 10, mg L^{-1}	Overall, the data suggests that MnOx NPs can be used as an Mn fertilizer (better than their soluble or bulk solid counterparts) for crop growth improvement	MnOx NPs specifically improved the growth of lettuce seedlings by enhancing root elongation	[83]

Zn concentration in roots and seeds, compared to bulk Zn and the other NPs treatments. In leaves, Al_2O_3/ZnO NPs at 250 mg kg^{-1} significantly increased chlorophyll-a and carotenoid concentrations relative to the bulk, ionic, and other NPs treatments. The protein and carbohydrate profiles remained largely unaltered across all treatments with the exception of Al_2O_3/ZnO NPs at 1000 mg kg^{-1} where the sucrose concentration of green peas increased significantly, which is likely a biomarker of stress. Most importantly, these findings demonstrate that lattice and surface modification can significantly alter the fate and phytotoxic effects of ZnO NPs in food crops and seed nutritional quality.

In another study, ZnO NPs at concentrations of 2000 mg L^{-1} have been shown to inhibit root elongation (50.45% for maize and 66.75% for rice) of two crop plants [91]. Similarly, Xiang et al. [92] observed that ZnO NPs did not affect germination rates at concentrations of 1–80 mg L^{-1} but significantly inhibited the root and shoot elongation of Chinese cabbage seedlings, with the roots being more sensitive. Both the production of free hydroxyl groups and the Zn bioaccumulation in roots or shoots resulted in toxicity of ZnO NPs to Chinese cabbage seedlings. In another work, the impact of ZnO NPs on rhizobium-legume symbiosis was studied with the garden pea (*P. sativum*) and its compatible bacterial partner *Rhizobium leguminosarum*. Exposure of peas to ZnO NPs (500–1000 mg L^{-1}) had no impact on germination, but significantly affected root length. Chronic exposure of the plant to ZnO NPs impacted its development by decreasing the number of the first- and the second-order lateral roots, stem length, leaf surface area, and transpiration. Exposure of *R. leguminosarum by. viciae 3841* to ZnO NPs brought about morphological changes by rendering the microbial cells toward round shapes and damaging the bacterial surface. Furthermore, the presence of ZnO NPs in the rhizosphere affected root nodulation, delayed the onset of nitrogen fixation, and caused early senescence of nodules. The attachment of NPs on the root surface and dissolution of Zn^{2+} are important factors affecting the phytotoxicity of ZnO NPs, hence, the presence of ZnO NPs in the environment is potentially hazardous to the rhizobium-legume symbiosis system [93].

Wang et al. used synchrotron-based techniques to examine the uptake and transformation of Zn in various tissues of cowpea [*V. unguiculata* (L.) Walp.] exposed to ZnO NPs or $ZnCl_2$ following growth in either a solution or soil culture. In the solution culture, soluble Zn ($ZnCl_2$) was more toxic than the ZnO NPs, although there was a substantial accumulation of ZnO NPs on the root surface. When grown in soil, however, there was no significant difference in plant growth and accumulation or speciation of Zn between soluble Zn and ZnO NPs treatments, indicating that the added ZnO NPs underwent rapid dissolution following their entry into the soil [94].

In next study, the effect of exposure to 100 mg L^{-1} ZnO NPs on gene expression in *A. thaliana* roots was studied using microarrays. The genes induced by ZnO NPs include mainly ontology groups annotated as stress responsive, including both abiotic (oxidative, salt, water deprivation) and biotic (wounding and defense to pathogens) stimuli. The down-regulated genes upon ZnO NPs exposure were involved in cell organization and biogenesis, including translation, nucleosome assembly, and microtubule based process [95].

In another study, soybean plants (*G. max*) were grown through the seed production stage in soil amended with ZnO NPs (0, 50, 100 or 500 mg kg^{-1}). Although ZnO NPs slightly stimulated plant growth, most striking was the degree to which Zn bioaccumulated in all tissues and especially in the leaves. Zn that translocated aboveground in the present study may have been substantially dissolved from the ZnO NPs added to the soil. This study shows that two high productions of ZnO NPs are able to change soybean agriculture, and demonstrates the importance of managing waste streams to control such exposures [96]. In the following work, the effects of ZnO NPs on the soil plant interactive system were estimated. The growth of plant seedlings in the presence of different concentrations of ZnO NPs within microcosm soil (M) and natural soil (NS) was compared. Changes in dehydrogenase activity (DHA) and soil bacterial community diversity were estimated based on the microcosm with plants (M + P) and microcosm without plants (M − P) in different concentrations of ZnO NPs treatment. The shoot growth of M + P and NS + P was significantly inhibited by 24 and 31.5% relative to the control at a ZnO NPs concentration of 1000 mg kg^{-1}. The DHA levels decreased following increased ZnO NPs concentration. Specifically, these levels were significantly reduced from 100 mg kg^{-1} in M − P and only 1000 mg kg^{-1} in M + P [21].

Dimkpa et al. [67] investigated the impact of commercial ZnO (<100 nm) NPs on wheat (*T. aestivum*) grown in a solid matrix, sand. Solubilization of metals occurred in the sand at similar rates from ZnO NPs as their bulk equivalents. Amendment of the sand with Zn (500 mg kg^{-1}) from the ZnO NPs significantly ($p = 0.05$) reduced root growth, growth reduction was less with the bulk amendment. Bioaccumulation of Zn as Zn-phosphate was detected in the shoots of ZnO NP-challenged plants. Oxidative stress in the ZnO NPs-treated plants was evidenced by increased lipid peroxidation and oxidized glutathione in roots and decreased chlorophyll content in shoots; higher peroxidase and catalase activities were present in roots. These findings correlate with the ZnO NPs causing increased production of ROS.

The next study was carried out to examine the phytotoxicity and oxidative stress by ZnO NPs in cucumber (*Cucumis sativus*). Kim et al. [97] estimated the bioaccumulation of ZnO NPs in plants, reactive oxygen species enzyme [superoxide dismutase (SOD), catalase (CAT), and peroxidase (POD)] activities in plant root tissue, and observed ZnO NPs with transmission electron microscopy. They found that the seedling biomass significantly decreased to 35% of that of control at 1000 mg L^{-1} of ZnO NPs. The median inhibition concentrations of ZnO NPs were 215 mg L^{-1}. In transmission electron microscopy, ZnO NPs greatly adhered to the root cell wall and some of the ZnO NPs were observed in the root cells. Another finding indicated that ZnO NPs caused statistically significant increases in SOD, CAT, and POD activities and a significant increase already at 100 mg L^{-1} concentration levels. These results indicate that ZnO NPs altered both phytotoxicity and oxidative stress in plant assays.

In the following study the effects of zinc oxide NPs (ZnO NPs) on the root growth, root apical meristem mitosis, and mitotic aberrations of garlic (*Allium sativum* L.) were investigated. ZnO NPs caused a concentration-dependent inhibition of root length. When treated with 50 mg L^{-1} ZnO NPs for 24 h, the root growth of garlic was completely blocked. The 50% inhibitory concentration (IC50) was estimated to be 15 mg L^{-1}. ZnO NPs also induced several kinds of mitotic aberrations, mainly consisting of chromosome stickiness, bridges, breakages, and laggings. The total percentage of abnormal cells increased with the increase of ZnO NPs concentration and the prolonging of treatment time. The investigation provided new information for the possible genotoxic effects of ZnO NPs on plants [98].

In the next study, transport of ZnO NPs in a sandy loam soil and their uptake by corn plants (*Z. mays*) was investigated. Results showed that ZnO NPs exhibit low mobility in a soil column at various ionic strengths. By using an electron microprobe, Zn/ZnO NPs aggregates were visualized associating them with soil clay minerals. The uptake (mg kg^{-1}) of Zn by 1-month old corn plants varied from 69 to 409 in roots and from 100 to 350 in shoots, respectively, in soils contaminated with different concentrations of ZnO NPs (from 100 to 800 mg kg^{-1} soil). Confocal microscope images showed that ZnO NPs aggregates penetrated the root epidermis and cortex through the apoplastic pathway, however, the presence of a few NP aggregates in xylem vessels suggests that the aggregates passed the endodermis through the symplastic pathway. Most of the aggregates, however, remained around the endodermis border [99]. An overview of the phytotoxicity of nanoparticles based on ZnO is shown in Table 5.

Phytotoxicity of nanoparticles based on Cu/CuO

CuNPs toxicity mechanisms have been extensively studied in animal/human systems [100, 101]. In plants, toxicity of Cu and Cu ions was thoroughly investigated [102] but not so Cu/CuO NPs phytotoxicity [103]. Zhao et al. [104] investigated the response of cucumber plants in hydroponic culture at early development to two concentrations of CuNPs (10 and 20 mg L^{-1}). Results showed that CuNPs interferes with the uptake of a number of micro- and macronutrients such as Na, P, S, Mo, Zn, and Fe. Metabolomics data revealed that CuNPs at both levels triggered significant metabolic changes in cucumber leaves and root exudates. The root exudate metabolic changes revealed an active defense mechanism against CuNPs stress: up-regulation of amino acids to sequester/exclude Cu/CuNPs, down regulation of citric acid to reduce the mobilization of Cu ions, ascorbic acid up-regulation to combat reactive oxygen species, and up-regulation of phenolic compounds to improve antioxidant system. It also observed a decrease in root length, reduction of root biomass, and bioaccumulation of Cu mainly in roots.

The following study was conducted to assess the effects of laboratory-prepared CuNPs in low concentrations (<50 ppm) on the germination of lettuce (*L. sativa*) seeds in a water medium. The data showed that CuO NPs were slightly more toxic than Cu^{2+} ions and a reduction of seed germination and root elongation was observed [83]. In the next study 18-day-old hydroponically grown lettuce (*L. sativa*) seedlings were treated for 15 days with core–shell Cu/CuO NPs at two concentrations (10 and 20 mg L^{-1}). The results showed that Cu^{2+} ions or Cu/CuO NPs reduced water content, root length, and dry biomass of the lettuce plants. ICP-OES results showed that Cu/CuO NPs treatments produced significant accumulations of Cu in roots compared to the Cu^{2+} ions. In roots, all Cu treatments increased CAT activity but decreased APX activity. In addition, relative to the control, nano-Cu/CuO altered the nutritional quality of the lettuce, since the treated plants had significantly more Cu, Al, and S, but less Mn, P, Ca, and Mg [71].

In another study Atha et al. reported that copper oxide nanoparticles induced DNA damage in agricultural and grassland plants. Significant accumulation of oxidatively modified, mutagenic DNA lesions (7,8-dihydro-8-oxoguanine; 2,6-diamino-4-hydroxy-5-formamidopyrimidine; 4,6-diamino-5-formamidopyrimidine) and strong plant growth inhibition was observed for radish (*R. sativus*), perennial ryegrass (*L. perenne*), and annual ryegrass

Table 5 Phytotoxicity of nanoparticles based on ZnO

Plant	Type of nanoparticle, particle size (nm)	Particle concentration	Comment	Observed effect	References
Glycine max	ZnO NPs <50 nm	500 ppm	Effect of ZnO NPs on soybean seedlings was studied	Decrease in root growth (length and weight), loss of root cell viability, accumulation of superoxide and decrease in leaf weight, down regulation of oxidative cascade related genes	[89]
Pisum sativum	Bare ZnO NPs 10 nm, Al_2O_3/ZnONPs 15 nm, KH550/ZnO NPs 20 nm	250, 1000 mg L^{-1} of soil	Pea plants were grown for 65 days in soil amended with three types of ZnO NPs	Al_2O_3/ZnO NPs at 250 mg kg^{-1} significantly increased chlorophyll-a and carotenoid concentrations. Al_2O_3/ZnO NPs at 1000 mg kg^{-1} significantly increased sucrose concentration of green peas	[90]
Zea mays, Oryza sativa	ZnO NPs <50 nm	500, 1000, 2000 mg L^{-1}	Seed germination was investigated	ZnO NPs inhibited root elongation at 2000 mg L^{-1} (50.45% for maize and 66.75% for rice) of two crop plants	[91]
Brassica pekinensis	Spheric ZnO NPs 30 nm, spheric ZnO NPs 50 nm, columnar ZnO NPs 90 nm, hexagon rod-like ZnO NPs 150 nm	1, 5, 10, 20, 40, 80 mg L^{-1}	There were no significant differences in observed effects between different NPs	ZnO NPs inhibited the root and shoot elongation of Chinese cabbage seedlings. The highest inhibition of root elongation at 80 mg L^{-1} was observed	[92]
Pisum sativum	ZnO NPs <50 nm	250, 500, 750, 1000 mg L^{-1}	No impact on germination	ZnO NPs (500–1000 mg L^{-1}) significantly inhibited root elongation	[93]
Vigna unguiculata	ZnO NPs <100 nm	25 mg L^{-1}	More pronounced effects were observed with $ZnCl_2$ than with ZnO NPs	Significant decrease in biomass production of roots and leaves observed in solution culture, but not observed in soil culture	[94]
Arabidopsis thaliana	ZnO NPs <100 nm	100 mg L^{-1}	Effect of ZnO NPs on gene expression in plant roots were studied	Induction of stress responsive genes, down regulation of genes involved in cell organization and biogenesis	[94]
(*Glycine max*)	ZnO NPs 10 nm	50, 100, and 500 mg kg^{-1} of soil	ZnO NPs were added to the soil	Zn bioaccumulated in all tissues and especially in the leaves	[96]
(*Fagopyrum esculentum*)	ZnO NPs <50 nm	10, 100, and 1000 mg kg^{-1} of soil	ZnO NPs were added to the soil and growth of plant seedlings were observed	Inhibition of shoot growth	[21]
Triticum aestivum	ZnO NPs <100 nm	500 mg kg^{-1} sand	ZnO NPs were added to the sand	Reduced root growth, increased lipid peroxidation and oxidized glutathione in roots. Bioaccumulation of Zn and decreased chlorophyll content in the shoots	[67]
Cucumis sativus	ZnO NPs 50 nm	10, 50, 100, 500, 1000 mg L^{-1}	Hydroponic experiments	Decrease in seedling biomass. ZnO NPs adhered to the root cell wall, and some of them were observed in the root cells	[97]
Allium sativum	ZnO NPs 3–5 nm	10, 20, 30, 40, 50 mg L^{-1}	Hydroponic experiments	Concentration-dependent inhibition of root length, observed mitotic aberrations	[98]
Zea mays	ZnO NPs 370–410 nm	20 mg L^{-1}	ZnO NPs were added to the sandy loam soil or to the water	ZnO NPs aggregates penetrated the root epidermis and cortex. Some of the NPs aggregates were also present in xylem vessels	[112]

(*Lolium rigidum*) under controlled laboratory conditions [105].

The next study investigated the phytotoxicity and accumulation of copper oxide (CuO) NPs to *Elsholtzia splendens* (a Cu-tolerant plant) under hydroponic conditions. The Cu content in the shoots treated with 1000 mg L^{-1} CuO NPs was much higher than those exposed to the comparable 0.5 mg L^{-1} soluble Cu and CuO bulk particles. CuO NPs-like deposits were found in the root cells and leaf cells. Cu K-edge X-ray absorption near-edge structure analysis further revealed that the accumulated Cu species existed predominantly as CuO NPs in the plant tissues. All these results suggested that CuO NPs can be absorbed by the roots and translocated to the shoots in *E. splendens*.

In another study, phytotoxicity of CuO NPs was assessed in two crop plants, maize (*Z. mays*) and rice (*O. sativa*). The results showed that seed germination was not affected by CuO NPs at any of the investigated concentrations. However, at the concentration of 2000 mg L^{-1}, the root elongation was significantly inhibited by CuO NPs (95.73% for maize and 97.28% for rice), and the shoot length of maize was reduced by 30.98%. An overview of phytotoxicity of nanoparticles based on Cu/CuO is shown in Table 6.

Phytotoxicity of nanoparticles based on iron oxides

Most of the available studies on the phytotoxicity of iron nanomaterials have focused mainly on their advantages while relatively few have examined the mechanisms of phytotoxicity, uptake, translocation, and bioaccumulation [73]. Martinez-Fernandez et al. [106] investigated if water uptake by the roots could be affected by the adsorption of γ-Fe_2O_3 nanoparticles (50, 100 mg L^{-1}) on the root surface of *Helianthus annuus*. The main effect was related to the reduction of the root hydraulic conductivity (Lo) and the nutrient uptake. The concentrations of the macronutrients Ca, K, Mg, and S in the shoot were reduced relative to the control plants, which resulted in lower contents of chlorophyll pigments. In the next study, the same group of authors [73] investigated the effects of nano zerovalent Fe (nZVI) and maghemite NPs (γ-Fe_2O_3) on the nutritional status of *S. lycopersicum*, through distinct effects on root functionality. A hydroponic experiment together with an incubation experiment helped to relate the reduction of the root water uptake with the potential blockage of root nutrient uptake by each nanomaterial. The treatment with 100 mg L^{-1} of γ-Fe_2O_3 inhibited 40% of the root hydraulic conductivity (Lo) of tomato plants (*S. lycopersicum* L.), which could explain the reduction in the Mo and Zn concentrations in their shoots. On the other hand, compared to γ-Fe_2O_3, nZVI seems to be less harmful since no effects on Lo were detected for the exposed roots, or regarding the shoot nutrient composition.

Liu et al. [83] investigated the effects of laboratory-prepared FeOx NPs (probably γ-Fe_2O_3) on the germination of lettuce (*L. sativa*) seeds in a water medium. NPs were not only less toxic than their ionic counterparts but also significantly stimulated the growth of root elongation by 12–26% in a concentration range (5–20 mg L^{-1}). Conversely, at 50 mg L^{-1} root elongation was inhibited by 12%.

Gui et al. [107] performed a glasshouse study to quantify the uptake of γ-Fe_2O_3 NPs on transgenic and non-transgenic rice *O. sativa* plants. Nutrient concentrations, biomass, enzyme activity, and the concentration of two phytohormones, abscisic acid (ABA) and indole-3-acetic acid (IAA), and malondialdehyde (MDA) was measured. Root phytohormone inhibition was positively correlated with γ-Fe_2O_3 NP concentrations, indicating that Fe_2O_3 had a significant influence on the production of these hormones. The activities of antioxidant enzymes were significantly higher as a factor of low γ-Fe_2O_3 NP treatment concentration and significantly lower at high NPs concentrations, but only among transgenic plants. An overview of phytotoxicity of nanoparticles based on iron oxides is shown in Table 7.

Conclusions

Phytotoxicity of any nanoparticle is largely influenced by its shape, size, chemical composition, and coating material composition. Sometimes, the phytotoxicity of nanoparticles may be as a result of the toxicity of substances, which were used for its preparation. Further, phytotoxicity may depend on the environment and on the physical and chemical nature of the plant species. The nanoparticles may have potentiating or inhibitory effects on plant growth in different developmental stages. Some nanoparticles are taken up by plant roots and transported to the aboveground parts of the plant through the vascular system, depending on the composition, shape, size of nanoparticle, and anatomy of the plant. Some nanoparticles remain adhered to the plant roots. In the discussed studies, sometimes nanoparticles have not been properly characterized and/or their composition vs. their shapes have not been considered, which is one the biggest obstacles need to be overcome for further planning of nanoparticles-plant research [108, 109]. Moreover, we have mentioned some papers and studies, where some metal based particles were both beneficial and toxic, but the right reason for these misleading findings lies in very high doses used together with a number of artifacts and misinterpretations especially regarding description of nanoparticles uptake. Despite the fact that a lot of knowledge has been acquired through many previous studies,

Table 6 Phytotoxicity of nanoparticles based on Cu/CuO

Plant	Type of nanoparticle, particle size (nm)	Particle concentration	Comment	Observed effect	References
Cucumis sativus	CuNPs 40 nm	10, 20 mg L^{-1}	Analysis of plants and root exudates	Decrease in root length, reduction of root biomass, bioaccumulation mainly in roots, a little in stems	[104]
Lactuca sativa	CuO NPs 5–15 nm	0.02, 0.04, 0.4, 4, 8 mg L^{-1}	5-day seed germination test	Reduction of seed germination and root elongation	[58]
Lactuca sativa	Core–shell NPs Cu/CuO 13/9 nm	10, 20 mg L^{-1}	15-days treatment of hydroponically grown lettuce	Reduction of water content, root length, and dry biomass of the plant, alteration of the nutritional quality of lettuce	[71]
Raphanus sativus, Lolium perenne, Lolium rigidum	CuO NPs 6 nm, CuO bulk particles 200 nm	10, 100, 500, 1000 mg L^{-1}	The seeds were allowed to germinate for 6 days	Oxidative damage to plant DNA, inhibition of seedling growth (root and shoot growth)	[105]
Elsholtzia splendens	CuO NPs 34–52 nm, CuO bulk particles ˃1000 nm	100, 200, 500, 1000, 2000 mg L^{-1}	Cu—tolerant plant, the seeds were allowed to germinate for 5 days, hydroponic experiments	No effect on seed germination, reduction of root length, accumulation of CuO NPs in root and leaf cells	[113]
Zea mays, Oryza sativa	CuO NPs 40–80 nm	500, 1000, 2000 mg L^{-1}	Seed germination was investigated	CuO NPs inhibited root elongation at 2000 mg L^{-1} (95.73% for maize and 97.28% for rice) of two crop plants and reduced shoot length of maize by 30.98%	[91]

Table 7 Phytotoxicity of nanoparticles based on iron oxides

Plant	Type of nanoparticle, particle size (nm)	Particle concentration	Comment	Observed effect	References
Helianthus annuus	γ-Fe_2O_3 20–100 nm	50, 100 mg L^{-1}	Effect on the root functionality was investigated	The treatment with 50 mg L^{-1} FeNPs significantly reduced the root hydraulic conductivity (*Lo*) by up to 26% at 100 mg L^{-1} FeNPs, but it had no effect on plant biomass production, on shoot or root elongation, and it did not induce oxidative stress in the plant	[106]
Solanum lycopersicum	nZVI < 50 nm γ-Fe_2O_3 20–100 nm	50, 100 mg L^{-1}	Effect on the root functionality was investigated	The treatment with 100 mg L^{-1} of Fe_2O_3 NPs inhibited 40% of the root hydraulic conductivity (*Lo*), with nZVI no effect on *Lo* was observed	[73]
Lactuca sativa	FeOx NPs <50 nm	1, 5, 10, 20, 50 mg L^{-1}	A 5-day seed germination test was used to test how different FeOx NPs affected the plant growth in comparison with their respective ionic or solid counterparts	FeOx NPs significantly enhanced root elongation of lettuce seedlings by 12%–26%, indicating that FeOx NPs could be used as an Fe fertilizer as well at low application rates (5–20 mg L^{-1}). At a concentration of 50 mg L^{-1}, FeOx NPs decreased root elongation of lettuce seedlings by 20%	[83]
Oryza sativa	γ-Fe_2O_3 7–13 nm	2, 20, 200 mg L^{-1}	A 7-day seed germination test was used	Root phytohormone inhibition abscisic acid (ABA) and indole-3-acetic acid (IAA) was positively correlated with Fe_2O_3 NPs concentrations, indicating that Fe_2O_3 had a significant influence on the production of these hormones	[107]

many questions still remain unanswered such as the fate and behavior of nanoparticles in plant systems, or the role of surface area or activity of nanoparticles on phytotoxicity, and the role of plant cell walls in the internalization of nanoparticles.

In a study of phytotoxicity nanoparticles, the most urgent need is to build a connection between the characteristics of nanoparticles (surface area, particle size, surface tension) and phytotoxicity. Equally important is the need to understand the role of plant species and composition of the nanoparticles phytotoxicity. Finally, most studies on phytotoxicity and uptake of nanoparticles plants were performed in a hydroponic setup. Hydroponic studies do not reflect the interaction of nanoparticles with soil and soil microorganisms.

Finally, it can be concluded that the nanoparticles prepared from essential heavy metals and their oxides have proven to be suitable for use in the agriculture. The least phytotoxic of these appear to be nanoparticles made of iron oxides and manganese oxides.

Abbreviations

ABA: abscisic acid; APX: ascorbate peroxidase; CAT: catalase; DHA: dehydrogenase activity; GDSL: motif consensus amino acid sequence of Gly, Asp, Ser, and Leu around the active site Ser; GDSL motif lipase 5: serine esterase and lipase with GDSL sequence motif; GS-GOGAT pathway: glutamine synthetase-glutamate synthase pathway; IAA: indole-3-acetic acid; ICP-OES: inductively coupled plasma optical emission spectrometry; MDA: malondialdehyde; NPs: nanoparticles; NR-NiR pathway: nitrate reductase-nitrite reductase pathway; POD: peroxidase; ROS: reactive oxygen species; SKU5 similar 4: multi-copper oxidase type I family protein expressed in plant roots; SOD: superoxide dismutase; SPIONs: superparamagnetic iron oxide nanoparticles.

Authors' contributions

BRN and LN reviewed the literature, drafted and wrote significant portions of the manuscript. OK and VA created the reviews' concept and edited the manuscript. All authors critically reviewed the manuscript. All authors read and approved the final manuscript.

Author details

[1] Department of Chemistry and Biochemistry, Mendel University in Brno, Zemedelska 1, 613 00 Brno, Czech Republic. [2] Central European Institute of Technology, Brno University of Technology, Technicka 3058/10, 616 00 Brno, Czech Republic.

Acknowledgements

Not applicable.

Competing interests

The authors declare that they have no competing interests. The authors are entirely responsible for the content of the review of the opinions contained within it.

Funding

This research has been financially supported by the Ministry of Education, Youth and Sports of the Czech Republic under the project CEITEC 2020 (LQ1601) and by IGA_MENDELU_Tym003.

References

1. Knauer K, Bucheli TD. Nano-materials: research needs in agriculture. Rev Suisse Agric. 2009;41:341–5.
2. Nair R, Varghese SH, Nair BG, Maekawa T, Yoshida Y, Kumar DS. Nanoparticulate material delivery to plants. Plant Sci. 2010;179:154–63.
3. Ghormade V, Deshpande MV, Paknikar KM. Perspectives for nano-biotechnology enabled protection and nutrition of plants. Biotechnol Adv. 2011;29:792–803.
4. Perez-de-Luque A, Rubiales D. Nanotechnology for parasitic plant control. Pest Manag Sci. 2009;65:540–5.
5. Aitken RJ, Chaudhry MQ, Boxall ABA, Hull M. Manufacture and use of nanomaterials: current status in the UK and global trends. Occup Med. 2006;56:300–6.
6. Grillo R, Rosa AH, Fraceto LF. Engineered nanoparticles and organic matter: a review of the state-of-the-art. Chemosphere. 2015;119:608–19.
7. Handford CE, Dean M, Henchion M, Spence M, Elliott CT, Campbell K. Implications of nanotechnology for the agri-food industry: opportunities, benefits and risks. Trends Food Sci Technol. 2014;40:226–41.
8. Husen A, Siddiqi KS. Phytosynthesis of nanoparticles: concept, controversy and application. Nanoscale Res Lett. 2014;9:1–24.
9. Jampilek J, Kral'ova K. Application of nanotechnology in agriculture and food industry, its prospects and risks. Ecol Chem Eng S. 2015;22:321–61.
10. Khot LR, Sankaran S, Maja JM, Ehsani R, Schuster EW. Applications of nanomaterials in agricultural production and crop protection: a review. Crop Prot. 2012;35:64–70.
11. Ju-Nam Y, Lead JR. Manufactured nanoparticles: an overview of their chemistry, interactions and potential environmental implications. Sci Total Environ. 2008;400:396–414.
12. Parisi C, Vigani M, Rodriguez-Cerezo E. Agricultural Nanotechnologies: what are the current possibilities? Nano Today. 2015;10:124–7.
13. Bhagat Y, Gangadhara K, Rabinal C, Chaudhari G, Ugale P. Nanotechnology in agriculture: a review. J Pure App Microbiol. 2015;9:737–47.
14. Bindraban PS, Dimkpa C, Nagarajan L, Roy A, Rabbinge R. Revisiting fertilisers and fertilisation strategies for improved nutrient uptake by plants. Biol Fertil Soils. 2015;51:897–911.
15. Dasgupta N, Ranjan S, Mundekkad D, Ramalingam C, Shanker R, Kumar A. Nanotechnology in agro-food: from field to plate. Food Res Int. 2015;69:381–400.
16. Garcia M, Forbe T, Gonzalez E. Potential applications of nanotechnology in the agro-food sector. Ciencia Tecnol Aliment. 2010;30:573–81.
17. Savage N, Diallo MS. Nanomaterials and water purification: opportunities and challenges. J Nanopart Res. 2005;7:331–42.
18. Servin A, Elmer W, Mukherjee A, De la Torre-Roche R, Hamdi H, White JC, Bindraban P, Dimkpa C. A review of the use of engineered nanomaterials to suppress plant disease and enhance crop yield. J Nanopart Res. 2015;17:1–21.
19. Jeyasubramanian K, Thoppey UUG, Hikku GS, Selvakumar N, Subramania A, Krishnamoorthy K. Enhancement in growth rate and productivity of spinach grown in hydroponics with iron oxide nanoparticles. RSC Adv. 2016;6:15451–9.
20. Alidoust D, Isoda A. Effect of gamma Fe_2O_3 nanoparticles on photosynthetic characteristic of soybean (*Glycine max* (L.) Merr.): foliar spray versus soil amendment. Acta Physiol Plant. 2013;35:3365–75.
21. Lee S, Kim S, Lee I. Effects of soil-plant interactive system on response to exposure to ZnO nanoparticles. J Microbiol Biotechnol. 2012;22:1264–70.
22. Mirzajani F, Askari H, Hamzelou S, Farzaneh M, Ghassempour A. Effect of silver nanoparticles on *Oryza sativa* L. and its rhizosphere bacteria. Ecotox Environ Safe. 2013;88:48–54.
23. Hussain HI, Yi ZF, Rookes JE, Kong LXX, Cahill DM. Mesoporous silica nanoparticles as a biomolecule delivery vehicle in plants. J Nanopart Res. 2013;15:1–15.
24. Slomberg DL, Schoenfisch MH. Silica nanoparticle phytotoxicity to *Arabidopsis thaliana*. Environ Sci Technol. 2012;46:10247–54.

25. Liu QL, Chen B, Wang QL, Shi XL, Xiao ZY, Lin JX, Fang XH. Carbon nanotubes as molecular transporters for walled plant cells. Nano Lett. 2009;9:1007–10.
26. Lin SJ, Reppert J, Hu Q, Hudson JS, Reid ML, Ratnikova TA, Rao AM, Luo H, Ke PC. Uptake, translocation, and transmission of carbon nanomaterials in rice plants. Small. 2009;5:1128–32.
27. Koo Y, Wang J, Zhang QB, Zhu HG, Chehab EW, Colvin VL, Alvarez PJJ, Braam J. Fluorescence reports intact quantum dot uptake into roots and trans location to leaves of *Arabidopsis thaliana* and subsequent ingestion by insect herbivores. Environ Sci Technol. 2015;49:626–32.
28. Zhu ZJ, Wang HH, Yan B, Zheng H, Jiang Y, Miranda OR, Rotello VM, Xing BS, Vachet RW. Effect of surface charge on the uptake and distribution of gold nanoparticles in four plant species. Environ Sci Technol. 2012;46:12391–8.
29. Kurepa J, Paunesku T, Vogt S, Arora H, Rabatic BM, Lu JJ, Wanzer MB, Woloschak GE, Smalle JA. Uptake and distribution of ultrasmall anatase TiO2 alizarin red S nanoconjugates in *Arabidopsis thaliana*. Nano Lett. 2010;10:2296–302.
30. Larue C, Laurette J, Herlin-Boime N, Khodja H, Fayard B, Flank AM, Brisset F, Carriere M. Accumulation, translocation and impact of TiO2 nanoparticles in wheat (*Triticum aestivum* spp.): influence of diameter and crystal phase. Sci Total Environ. 2012;431:197–208.
31. Gonzalez-Melendi P, Fernandez-Pacheco R, Coronado MJ, Corredor E, Testillano PS, Risueno MC, Marquina C, Ibarra MR, Rubiales D, Perez-De-Luque A. Nanoparticles as smart treatment-delivery systems in plants: assessment of different techniques of microscopy for their visualization in plant tissues. Ann Bot. 2008;101:187–95.
32. Zhu H, Han J, Xiao JQ, Jin Y. Uptake, translocation, and accumulation of manufactured iron oxide nanoparticles by pumpkin plants. J Environ Monit. 2008;10:713–7.
33. Huang XL, Stein BD, Cheng H, Malyutin A, Tsvetkova IB, Baxter DV, Remmes NB, Verchot J, Kao C, Bronstein LM, Dragnea B. Magnetic virus-like nanoparticles in *N. benthamiana* plants: a new paradigm for environmental and agronomic biotechnological research. ACS Nano. 2011;5:4037–45.
34. Ruffini Castiglione M, Cremonini R. Nanoparticles and higher plants. Caryologia. 2009;62:161–5.
35. Ma XM, Geiser-Lee J, Deng Y, Kolmakov A. Interactions between engineered nanoparticles (ENPs) and plants: phytotoxicity, uptake and accumulation. Sci Total Environ. 2010;408:3053–61.
36. Rico CM, Majumdar S, Duarte-Gardea M, Peralta-Videa JR, Gardea-Torresdey JL. Interaction of nanoparticles with edible plants and their possible implications in the food chain. J Agric Food Chem. 2011;59:3485–98.
37. Arruda SCC, Silva ALD, Galazzi RM, Azevedo RA, Arruda MAZ. Nanoparticles applied to plant science: a review. Talanta. 2015;131:693–705.
38. Wang WC, Freemark K. The use of plants for environmental monitoring and assessment. Ecotox Environ Safe. 1995;30:289–301.
39. Miralles P, Church TL, Harris AT. Toxicity, uptake, and translocation of engineered nanomaterials in vascular plants. Environ Sci Technol. 2012;46:9224–39.
40. Wang WC. Literature-review on higher-plants for toxicity testing. Water Air Soil Pollut. 1991;59:381–400.
41. Petersen EJ, Zhang LW, Mattison NT, O'Carroll DM, Whelton AJ, Uddin N, Nguyen T, Huang QG, Henry TB, Holbrook RD, Chen KL. Potential release pathways, environmental fate, and ecological risks of carbon nanotubes. Environ Sci Technol. 2011;45:9837–56.
42. Dimkpa CO, Calder A, Britt DW, McLean JE, Anderson AJ. Responses of a soil bacterium, *Pseudomonas chlororaphis* O6 to commercial metal oxide nanoparticles compared with responses to metal ions. Environ Pollut. 2011;159:1749–56.
43. Peralta-Videa JR, Zhao LJ, Lopez-Moreno ML, de la Rosa G, Hong J, Gardea-Torresdey JL. Nanomaterials and the environment: a review for the biennium 2008–2010. J Hazard Mater. 2011;186:1–15.
44. Lee CW, Mahendra S, Zodrow K, Li D, Tsai YC, Braam J, Alvarez PJJ. Developmental phytotoxicity of metal oxide nanoparticles to *Arabidopsis thaliana* (vol 29, pg 669, 2010). Environ Toxicol Chem. 2010;29:1399.
45. Racuciu M, Creanga DE. TMA-OH coated magnetic nanoparticles internalized in vegetal tissue. Rom J Phys. 2007;52:395–402.
46. Parsons JG, Lopez ML, Gonzalez CM, Peralta-Videa JR, Gardea-Torresdey JL. Toxicity and biotransformation of uncoated and coated nickel hydroxide nanoparticles on mesquite plants. Environ Toxicol Chem. 2010;29:1146–54.
47. Adhikari T, Sarkar D, Mashayekhi H, Xing BS. Growth and enzymatic activity of maize (*Zea mays* L.) plant: solution culture test for copper dioxide nano particles. J Plant Nutr. 2016;39:102–18.
48. He LL, Liu Y, Mustapha A, Lin MS. Antifungal activity of zinc oxide nanoparticles against *Botrytis cinerea* and *Penicillium expansum*. Microbiol Res. 2011;166:207–15.
49. Dimkpa CO, McLean JE, Britt DW, Anderson AJ. Antifungal activity of ZnO nanoparticles and their interactive effect with a biocontrol bacterium on growth antagonism of the plant pathogen *Fusarium graminearum*. Biometals. 2013;26:913–24.
50. Giannousi K, Avramidis I, Dendrinou-Samara C. Synthesis, characterization and evaluation of copper based nanoparticles as agrochemicals against *Phytophthora infestans*. RSC Adv. 2013;3:21743–52.
51. Gajjar P, Pettee B, Britt DW, Huang WJ, Johnson WP, Anderson AJ: Antimicrobial activity of commercial nanoparticles. In: Hendy SC, Brown IWM, editors. Advanced materials and nanotechnology, proceedings. volume 1151 (AIP conference proceedings). Melville: American Institute of Physics; 2009. p. 130–2.
52. Jones N, Ray B, Ranjit KT, Manna AC. Antibacterial activity of ZnO nanoparticle suspensions on a broad spectrum of microorganisms. FEMS Microbiol Lett. 2008;279:71–6.
53. Schrick B, Blough JL, Jones AD, Mallouk TE. Hydrodechlorination of trichloroethylene to hydrocarbons using bimetallic nickel–iron nanoparticles. Chem Mat. 2002;14:5140–7.
54. Zhang WX. Nanoscale iron particles for environmental remediation: an overview. J Nanopart Res. 2003;5:323–32.
55. Nurmi JT, Tratnyek PG, Sarathy V, Baer DR, Amonette JE, Pecher K, Wang CM, Linehan JC, Matson DW, Penn RL, Driessen MD. Characterization and properties of metallic iron nanoparticles: spectroscopy, electrochemistry, and kinetics. Environ Sci Technol. 2005;39:1221–30.
56. Fageria NK. Manganese. In: Fageria NK, editor. The use of nutrients in crop plants. London: CRC Press; 2008. p. 333–58.
57. Pradhan S, Patra P, Das S, Chandra S, Mitra S, Dey KK, Akbar S, Palit P, Goswami A. Photochemical modulation of biosafe manganese nanoparticles on *Vigna radiata*: a detailed molecular, biochemical, and biophysical study. Environ Sci Technol. 2013;47:13122–31.
58. Liu RQ, Lal R. Potentials of engineered nanoparticles as fertilizers for increasing agronomic productions. Sci Total Environ. 2015;514:131–9.
59. Singh S, Thiyagarajan P, Kant KM, Anita D, Thirupathiah S, Rama N, Tiwari B, Kottaisamy M, Rao MSR. Structure, microstructure and physical properties of ZnO based materials in various forms: bulk, thin film and nano. J Phys D Appl Phys. 2007;40:6312–27.
60. Huang ZB, Zheng X, Yan DH, Yin GF, Liao XM, Kang YQ, Yao YD, Huang D, Hao BQ. Toxicological effect of ZnO nanoparticles based on bacteria. Langmuir. 2008;24:4140–4.
61. Zhou J, Xu NS, Wang ZL. Dissolving behavior and stability of ZnO wires in biofluids: a study on biodegradability and biocompatibility of ZnO nanostructures. Adv Mater. 2006;18:2432–5.
62. Gogos A, Knauer K, Bucheli TD. Nanomaterials in plant protection and fertilization: current state, foreseen applications, and research priorities. J Agric Food Chem. 2012;60:9781–92.
63. Ishaque M, Schnabel G, Anspaugh DD: Agrochemical formulation comprising a pesticide, an organic uv photoprotective filter and coated metal oxide nanoparticles. In: Patentscope, vol. WO/2009/153231; 2009.
64. Milani N, McLaughlin MJ, Stacey SP, Kirby JK, Hettiarachchi GM, Beak DG, Cornelis G. Dissolution kinetics of macronutrient fertilizers coated with manufactured zinc oxide nanoparticles. J Agric Food Chem. 2012;60:3991–8.
65. Raliya R, Tarafdar JC, Biswas P. Enhancing the mobilization of native phosphorus in the mung bean rhizosphere using ZnO nanoparticles synthesized by soil fungi. J Agric Food Chem. 2016;64:3111–8.
66. Savi GD, Piacentini KC, de Souza SR, Costa MEB, Santos CMR, Scussel VM. Efficacy of zinc compounds in controlling Fusarium head blight and deoxynivalenol formation in wheat (*Triticum aestivum* L.). Int J Food Microbiol. 2015;205:98–104.
67. Dimkpa CO, McLean JE, Latta DE, Manangon E, Britt DW, Johnson WP, Boyanov MI, Anderson AJ. CuO and ZnO nanoparticles: phytotoxicity, metal speciation, and induction of oxidative stress in sand-grown wheat. J Nanopart Res. 2012;14:1–15.

68. Martineau N, McLean JE, Dimkpa CO, Britt DW, Anderson AJ. Components from wheat roots modify the bioactivity of ZnO and CuO nanoparticles in a soil bacterium. Environ Pollut. 2014;187:65–72.
69. Schneider KH, Karpov A, Voss H, Dunker S, Merk M, Kopf A, Kondo S. Method for treating phytopathogenic microorganisms using surface-modified nanoparticulate copper salts. In: Patentscope, vol. WO/2011/067186; 2011.
70. Juarez-Maldonado A, Ortega-Ortiz H, Perez-Labrada F, Cadenas-Pliego G, Benavides-Mendoza A. Cu nanoparticles absorbed on chitosan hydrogels positively alter morphological, production, and quality characteristics of tomato. J Appl Bot Food Qual. 2016;89:183–9.
71. Trujillo-Reyes J, Majumdar S, Botez CE, Peralta-Videa JR, Gardea-Torresdey JL. Exposure studies of core-shell Fe/Fe_3O_4 and Cu/CuO NPs to lettuce (*Lactuca sativa*) plants: are they a potential physiological and nutritional hazard? J Hazard Mater. 2014;267:255–63.
72. Fu PP, Xia QS, Hwang HM, Ray PC, Yu HT. Mechanisms of nanotoxicity: generation of reactive oxygen species. J Food Drug Anal. 2014;22:64–75.
73. Martinez-Fernandez D, Komarek M. Comparative effects of nanoscale zero-valent iron (nZVI) and Fe_2O_3 nanoparticles on root hydraulic conductivity of *Solanum lycopersicum* L. Environ Exp Bot. 2016;131:128–36.
74. Kanel SR, Manning B, Charlet L, Choi H. Removal of arsenic(III) from groundwater by nanoscale zero-valent iron. Environ Sci Technol. 2005;39:1291–8.
75. Feng LY, Cao MH, Ma XY, Zhu YS, Hu CW. Superparamagnetic high-surface-area Fe_3O_4 nanoparticles as adsorbents for arsenic removal. J Hazard Mater. 2012;217:439–46.
76. Ali A, Zafar H, Zia M, Haq IU, Phull AR, Ali JS, Hussain A. Synthesis, characterization, applications, and challenges of iron oxide nanoparticles. Nanotechnol Sci Appl. 2016;9:49–67.
77. Rui MM, Ma CX, Hao Y, Guo J, Rui YK, Tang XL, Zhao Q, Fan X, Zhang ZT, Hou TQ, Zhu SY. Iron oxide nanoparticles as a potential iron fertilizer for peanut (*Arachis hypogaea*). Front Plant Sci. 2016;7:1–10.
78. Raju D, Mehta UJ, Beedu SR. Biogenic green synthesis of monodispersed gum kondagogu (*Cochlospermum gossypium*) iron nanocomposite material and its application in germination and growth of mung bean (*Vigna radiata*) as a plant model. IET Nanobiotechnol. 2016;10:141–6.
79. Delfani M, Firouzabadi MB, Farrokhi N, Makarian H. Some physiological responses of black-eyed pea to iron and magnesium nanofertilizers. Commun Soil Sci Plant Anal. 2014;45:530–40.
80. Ghafariyan MH, Malakouti MJ, Dadpour MR, Stroeve P, Mahmoudi M. Effects of magnetite nanoparticles on soybean chlorophyll. Environ Sci Technol. 2013;47:10645–52.
81. Mukhopadhyay MJ, Sharma A. Manganese in cell-metabolism of higher-plants. Bot Rev. 1991;57:117–49.
82. Pradhan S, Patra P, Mitra S, Dey KK, Jain S, Sarkar S, Roy S, Palit P, Goswami A. Manganese nanoparticles: impact on non-nodulated plant as a potent enhancer in nitrogen metabolism and toxicity study both in vivo and in vitro. J Agric Food Chem. 2014;62:8777–85.
83. Liu RQ, Zhang HY, Lal R. Effects of stabilized nanoparticles of copper, zinc, manganese, and iron oxides in low concentrations on lettuce (*Lactuca sativa*) seed germination: nanotoxicants or nanonutrients? Water Air Soil Pollut. 2016;227:1–14.
84. Rai M, Ingle A, Gupta I, Gaikwad S, Gade A, Rubilar O, Duran N. Cyto-, geno-, and ecotoxicity of copper nanoparticles. In: Duran N, Guterres SS, Alves OL, editors. Nanotoxicology: materials, methodologies, and assessments. Berlin: Springer; 2014. p. 325–45.
85. Ma HB, Williams PL, Diamond SA. Ecotoxicity of manufactured ZnO nanoparticles—a review. Environ Pollut. 2013;172:76–85.
86. Anjum NA, Adam V, Kizek R, Duarte AC, Pereira E, Iqbal M, Lukatkin AS, Ahmad I. Nanoscale copper in the soil-plant system—toxicity and underlying potential mechanisms. Environ Res. 2015;138:306–25.
87. Letelier ME, Sanchez-Jofre S, Peredo-Silva L, Cortes-Troncoso J, Aracena-Parks P. Mechanisms underlying iron and copper ions toxicity in biological systems: pro-oxidant activity and protein-binding effects. Chem-Biol Interact. 2010;188:220–7.
88. Jin T, Sun D, Su JY, Zhang H, Sue HJ. Antimicrobial efficacy of zinc oxide quantum dots against *Listeria monocytogenes*, *Salmonella enteritidis*, and *Escherichia coli* O157:H7. J Food Sci. 2009;74:M46–52.
89. Hossain Z, Mustafa G, Sakata K, Komatsu S. Insights into the proteomic response of soybean towards Al2O3, ZnO, and Ag nanoparticles stress. J Hazard Mater. 2016;304:291–305.
90. Mukherjee A, Sun YP, Morelius E, Tamez C, Bandyopadhyay S, Niu GH, White JC, Peralta-Videa JR, Gardea-Torresdey JL. Differential toxicity of bare and hybrid ZnO nanoparticles in green pea (*Pisum sativum* L.): a life cycle study. *Front*. Plant Sci. 2016;6:1–13.
91. Yang ZZ, Chen J, Dou RZ, Gao X, Mao CB, Wang L. Assessment of the phytotoxicity of metal oxide nanoparticles on two crop plants, maize (*Zea mays* L.) and rice (*Oryza sativa* L.). Int J Environ Res Public Health. 2015;12:15100–9.
92. Xiang L, Zhao HM, Li YW, Huang XP, Wu XL, Zhai T, Yuan Y, Cai QY, Mo CH. Effects of the size and morphology of zinc oxide nanoparticles on the germination of Chinese cabbage seeds. Environ Sci Pollut Res. 2015;22:10452–62.
93. Huang YC, Fan R, Grusak MA, Sherrier JD, Huang CP. Effects of nano-ZnO on the agronomically relevant Rhizobium-legume symbiosis. Sci Total Environ. 2014;497:78–90.
94. Wang P, Menzies NW, Lombi E, McKenna BA, Johannessen B, Glover CJ, Kappen P, Kopittke PM. Fate of ZnO nanoparticles in soils and cowpea (*Vigna unguiculata*). Environ Sci Technol. 2013;47:13822–30.
95. Landa P, Vankova R, Andrlova J, Hodek J, Marsik P, Storchova H, White JC, Vanek T. Nanoparticle-specific changes in *Arabidopsis thaliana* gene expression after exposure to ZnO, TiO2, and fullerene soot. J Hazard Mater. 2012;241:55–62.
96. Priester JH, Ge Y, Mielke RE, Horst AM, Moritz SC, Espinosa K, Gelb J, Walker SL, Nisbet RM, An YJ, et al. Soybean susceptibility to manufactured nanomaterials with evidence for food quality and soil fertility interruption. Proc Natl Acad Sci USA. 2012;109:E2451–6.
97. Kim S, Lee S, Lee I. Alteration of phytotoxicity and oxidant stress potential by metal oxide nanoparticles in *Cucumis sativus*. Water Air Soil Pollut. 2012;223:2799–806.
98. Shaymurat T, Gu JX, Xu CS, Yang ZK, Zhao Q, Liu YX, Liu YC. Phytotoxic and genotoxic effects of ZnO nanoparticles on garlic (*Allium sativum* L.): a morphological study. Nanotoxicology. 2012;6:241–8.
99. Zhao LJ, Sun YP, Hernandez-Viezcas JA, Servin AD, Hong J, Niu GH, Peralta-Videa JR, Duarte-Gardea M, Gardea-Torresdey JL. Influence of CeO2 and ZnO nanoparticles on cucumber physiological markers and bioaccumulation of Ce and Zn: a life cycle study. J Agric Food Chem. 2013;61:11945–51.
100. Minocha S, Mumper RJ. Effect of carbon coating on the physico-chemical properties and toxicity of copper and nickel nanoparticles. Small. 2012;8:3289–99.
101. Piret JP, Mejia J, Lucas S, Zouboulis CC, Saout C, Toussaint O. Sonicated and stirred copper oxide nanoparticles induce similar toxicity and pro-inflammatory response in N-hTERT keratinocytes and SZ95 sebocytes. J Nanopart Res. 2014;16:1–18.
102. Yruela I. Copper in plants: acquisition, transport and interactions. Funct Plant Biol. 2009;36:409–30.
103. Dimkpa CO, Latta DE, McLean JE, Britt DW, Boyanov MI, Anderson AJ. Fate of CuO and ZnO nano- and microparticles in the plant environment. Environ Sci Technol. 2013;47:4734–42.
104. Zhao LJ, Huang YX, Hu J, Zhou HJ, Adeleye AS, Keller AA. H-1 NMR and GC-MS based metabolomics reveal defense and detoxification mechanism of cucumber plant under nano-Cu stress. Environ Sci Technol. 2016;50:2000–10.
105. Atha DH, Wang HH, Petersen EJ, Cleveland D, Holbrook RD, Jaruga P, Dizdaroglu M, Xing BS, Nelson BC. Copper oxide nanoparticle mediated DNA damage in terrestrial plant models. Environ Sci Technol. 2012;46:1819–27.
106. Martinez-Fernandez D, Barroso D, Komarek M. Root water transport of *Helianthus annuus* L. under iron oxide nanoparticle exposure. Environ Sci Pollut Res. 2016;23:1732–41.
107. Gui X, Deng YQ, Rui YK, Gao BB, Luo WH, Chen SL, Nhan LV, Li XG, Liu ST, Han YN, et al. Response difference of transgenic and conventional rice (*Oryza sativa*) to nanoparticles (gamma Fe_2O_3). Environ Sci Pollut Res. 2015;22:17716–23.
108. Krug HF. Nanosafety research-are we on the right track? Angew Chem Int Edit. 2014;53:12304–19.
109. Nau K, Bohmer N, Kuhnel D, Marquardt C, Paul F, Steinbach C, Krug HF. The DaNa(2.0) knowledge base on nanomaterials—communicating

current nanosafety research based on evaluated literature data. J Mater Educ. 2016;38:93–108.
110. Dietz KJ, Herth S. Plant nanotoxicology. Trends Plant Sci. 2011;16:582–9.
111. Wang P, Lombi E, Zhao FJ, Kopittke PM. Nanotechnology: a new opportunity in plant sciences. Trends Plant Sci. 2016;21:699–712.
112. Zhao LJ, Peralta-Videa JR, Ren MH, Varela-Ramirez A, Li CQ, Hernandez-Viezcas JA, Aguilera RJ, Gardea-Torresdey JL. Transport of Zn in a sandy loam soil treated with ZnO NPs and uptake by corn plants: electron microprobe and confocal microscopy studies. Chem Eng J. 2012;184:1–8.
113. Shi JY, Peng C, Yang YQ, Yang JJ, Zhang H, Yuan XF, Chen YX, Hu TD. Phytotoxicity and accumulation of copper oxide nanoparticles to the Cu-tolerant plant *Elsholtzia splendens*. Nanotoxicology. 2014;8:179–88.

10

Enhanced detection with spectral imaging fluorescence microscopy reveals tissue- and cell-type-specific compartmentalization of surface-modified polystyrene nanoparticles

Kata Kenesei[1,2*], Kumarasamy Murali[1,2], Árpád Czéh[3], Jordi Piella[4], Victor Puntes[4] and Emília Madarász[2]

Abstract

Background: Precisely targeted nanoparticle delivery is critically important for therapeutic applications. However, our knowledge on how the distinct physical and chemical properties of nanoparticles determine tissue penetration through physiological barriers, accumulation in specific cells and tissues, and clearance from selected organs has remained rather limited. In the recent study, spectral imaging fluorescence microscopy was exploited for precise and rapid monitoring of tissue- and cell-type-specific distribution of fluorescent polystyrene nanoparticles with chemically distinct surface compositions.

Methods: Fluorescent polystyrene nanoparticles with 50–90 nm diameter and with carboxylated- or polyethylene glycol-modified (PEGylated) surfaces were delivered into adult male and pregnant female mice with a single intravenous injection. The precise anatomical distribution of the particles was investigated by confocal microscopy after a short-term (5 min) or long-term (4 days) distribution period. In order to distinguish particle-fluorescence from tissue autofluorescence and to enhance the detection-efficiency, fluorescence spectral detection was applied during image acquisition and a post hoc full spectrum analysis was performed on the final images.

Results: Spectral imaging fluorescence microscopy allowed distinguishing particle-fluorescence from tissue-fluorescence in all examined organs (brain, kidney, liver, spleen and placenta) in NP-treated slice preparations. In short-time distribution following in vivo NP-administration, all organs contained carboxylated-nanoparticles, while PEGylated-nanoparticles were not detected in the brain and the placenta. Importantly, nanoparticles were not found in any embryonic tissues or in the barrier-protected brain parenchyma. Four days after the administration, particles were completely cleared from both the brain and the placenta, while PEGylated-, but not carboxylated-nanoparticles, were stuck in the kidney glomerular interstitium. In the spleen, macrophages accumulated large amount of carboxylated and PEGylated nanoparticles, with detectable redistribution from the marginal zone to the white pulp during the 4-day survival period.

Conclusions: Spectral imaging fluorescence microscopy allowed detecting the tissue- and cell-type-specific accumulation and barrier-penetration of polystyrene nanoparticles with equal size but chemically distinct surfaces. The data revealed that polystyrene nanoparticles are retained by the reticuloendothelial system regardless of surface functionalization. Taken together with the increasing production and use of nanoparticles, the results highlight the necessity of long-term distribution studies to estimate the potential health-risks implanted by tissue-specific nanoparticle accumulation and clearance.

*Correspondence: kenesei.kata@koki.mta.hu
[2] Institute of Experimental Medicine, Hungarian Academy of Sciences, Szigony Street 43, Budapest 1083, Hungary
Full list of author information is available at the end of the article

Keywords: Spectral imaging fluorescence microscopy, Polystyrene nanoparticle, Nanoparticle surface, Toxicity, Macrophage, In vivo distribution

Background

Nanoparticles (NPs) are increasingly popular tools with widespread industrial, medical and every-day applications. While the continuous progress in NP production is appealing, the escalation of environmental NP pollution raises serious health concerns. Despite of world-wide efforts, our understanding on the penetration, distribution, potential accumulation and clearance of various NPs in the living body is far from complete.

Several imaging approaches have already been used to visualize nanoparticles in vitro or in vivo [1]. Some studies investigated NP distribution at the whole-body level by using magnetic resonance imaging, computed tomography, positron emission tomography, or radiolabeling techniques [2–4]; whereas other reports focused on the subcellular localization of NPs by exploiting transmission or scanning electron microscopy [5, 6]. Relatively few studies attempted to follow the in vivo distribution of distinct types of NPs at the tissue and cellular levels [3, 7, 8]. This approach would be important from a medical perspective, because specific tissues and cells may be differentially involved in pathophysiological responses to nanoparticle exposure. Moreover, NP-aided drug delivery seeks to target certain cell types in selected organs, while must avoid loading others in order to reduce unwanted side-effects [9–11]. To achieve targeted NP distribution, it is pivotal to understand the impact of the physical and chemical parameters of NPs on their tissue- and cell-type-specific accumulation. Yet the availability of high-throughput imaging modalities to compare the distribution of different NPs, is rather limited.

Fluorescence microscopy is generally the method-of-choice to monitor the cellular and tissue-level distribution of biologically-relevant fluorescent materials, including nanoparticles [12]. However, visualization of fluorescent NPs even with high-resolution confocal microscopy has been notoriously difficult, because of their size (typically being between 1 and 100 nm) is below Abbe's diffraction limit of ~250 nm. Therefore, fluorochrome-functionalized NPs can only be detected if they passively aggregate in body fluids or at biological interfaces (e.g. on the blood vessel wall or on the cell surface) [13], or if they are actively taken up by cells and are concentrated within endocytotic vesicles or lysosomes [14–16]. Moreover, the detection of fluorescent NPs is further hindered by the high autofluorescence of biological samples [12], which does not allow visualizing small aggregates due to the low signal-to-noise ratio. Increasing fluorescent dye concentration on the NP surface may lead to enhanced cytotoxicity, whereas encapsulation (in "core–shell" nanoparticles) limits dye concentration per particle, thereby reducing NP fluorescence intensity [17, 18]. These constraints stressed the importance of applying new imaging approaches in the studies on tissue- and cell-type-specific NP distribution, and prompted us to exploit spectral detection and spectral analysis with the aim to overcome the limitations caused by low NP fluorescence *versus* high tissue autofluorescence.

The in vivo distribution of NPs is influenced by several physical and chemical parameters including size, shape, core material and surface composition [19]. Importantly, NPs readily adsorb various chemical substances from their environments due to the highly reactive surface [20, 21]. The composition and thickness of adsorbed layers (the so-called corona) depends on the chemical properties of both the NP surface and the environment [22–24]. Because the corona governs the interaction of NPs with biological structures, it plays a decisive role in the tissue- and cell-type-specific NP distribution [25, 26]. Moreover, as a result of chemical exchange reactions, the corona is expected to change with time even within the same tissue environment [27, 28]. While NP surfaces are ultimately functionalized by the actual environment, this process can be regulated by changing the surface charge of NPs or by coating the NP with chemically less reactive, hydrophilic polymers [29]. Polyethylene glycol (PEG) polymers with different oligomer-numbers and linear or branching chains have been widely used to reduce the chemical reactivity of surfaces [30]. Accordingly, protein adsorption by NPs could be reduced by PEGylation, and PEG-coating was shown to inhibit the cellular uptake of NPs [31–33], as well. In vivo studies demonstrated that PEGylated nanoparticles remained longer in the circulation due to their reduced attachment to vessel walls and cell surfaces [34, 35]. These findings together suggested that NPs displaying different adsorption-characteristics will show different tissue-, and cell-type-specific integration.

To investigate the impact of molecular surface characteristics on the in vivo tissue penetration and accumulation of otherwise identical NPs, we followed the fate of non-toxic polystyrene NPs with carboxylated or PEGylated surfaces by spectral imaging fluorescence microscopy. Spectral imaging has been used for localization of quantum dots before [1, 36]. The object of the study was to show that spectral imaging is a valuable tool to study the biodistribution and subcellular localization of fluorescently labeled NPs with broader emission bandwidths as well.

Results

In vitro characterization of polystyrene nanoparticles

Polystyrene nanoparticles core-labelled with fluorescent dyes and surface coated with either carboxyl groups (PS-COOH) or PEG (PS-PEG) were used throughout this study. The physical and chemical parameters of particles including size, aggregation properties, and protein adsorption were thoroughly analyzed. These parameters were determined in distinct inorganic or biological environments, including solutions used during particle handling and solutions that mimic the characteristics of body fluids.

Dynamic light scattering (DLS) measurements verified the similar size of PS-COOH and PS-PEG NPs (Fig. 1a, b): 70.81 ± 21.09 and 68.69 ± 18.68 nm for PS-COOH and PS-PEG, respectively; and showed no aggregation of particles in distilled water. Transmission electron microscopic (TEM) images showed slight agglomeration of dried particles (Fig. 1a, b, insert).

The zeta potential of particles measured by DLS in distilled water, showed significant differences: −42.1 ± 0.9 mV for PS-COOH and −28.5 ± 1.8 mV for PS-PEG NPs. According to hydrodynamic particle size distribution, the nanoparticles did not aggregate in distilled water or in phosphate-buffered saline (PBS) during a 96-h assay period (Fig. 1c), indicating that the ionic strength of organic material-free physiological saline did not induce aggregation of PS-COOH and PS-PEG nanoparticles. In contrast, a time-dependent, heavy particle aggregation of both NPs was found in serum-free Dulbecco's modified Eagle's cell culture medium (DMEM), which represents an ionic strength similar to PBS, but contains various organic compounds including glucose, amino acids, vitamins and non-peptide hormones (Fig. 1d). In DMEM, a moderate increase in NP size was observed after 4 h, which elevated robustly by the end of the 96-h incubation [(hydrodynamic diameters for PS-COOH: 66.40 ± 0.82 nm (5 min), 116.90 ± 2.10 nm (4 h), 178.47 ± 17.39 nm (24 h), 851.77 ± 34.27 nm (96 h); for PS-PEG: 68.55 ± 0.45 nm (5 min), 115.93 ± 0.60 nm

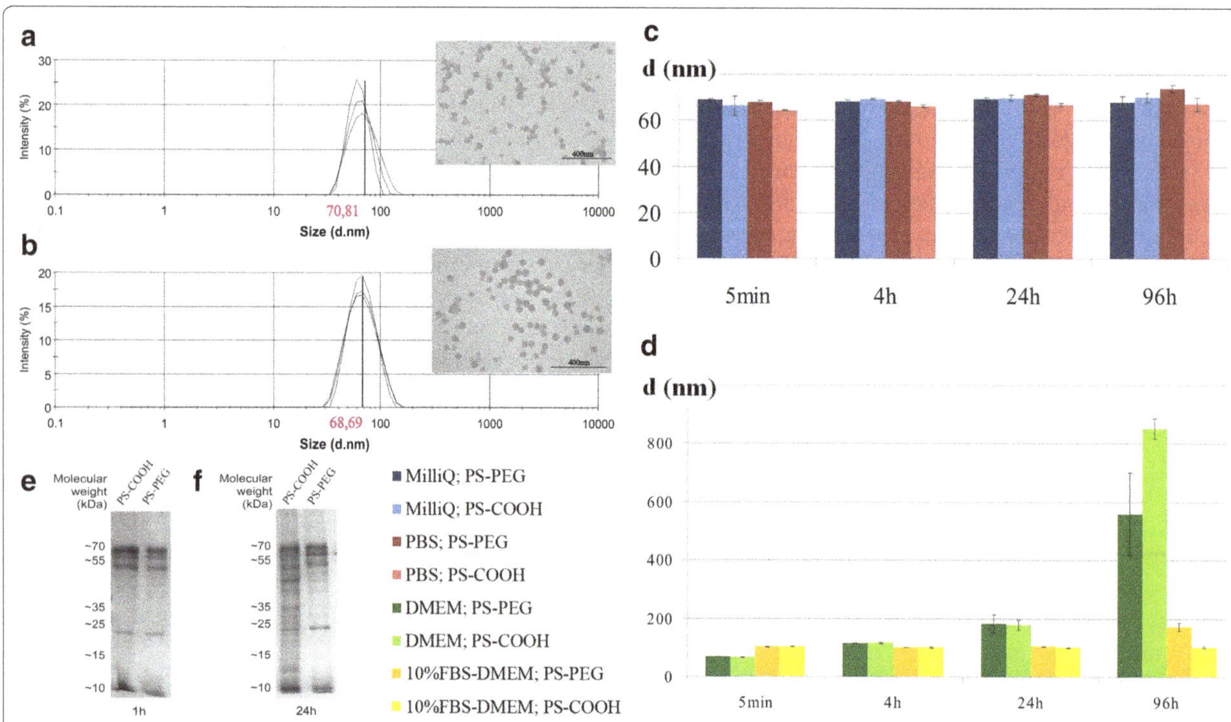

Fig. 1 Physical-chemical characterization of polystyrene NPs. Intensity weighted size distribution of carboxylated (**a**) and PEGylated (**b**) polystyrene nanoparticles measured by dynamic light scattering. Representative TEM images of the particles are shown in the top right panels of each DLS plot, *scale bars* represent 400 nm. **c, d** Aggregation–agglomeration properties and corona thickening of polystyrene nanoparticles was measured by DLS as a change in hydrodynamic size. Particles were incubated in inorganic (**c**) or biological solutions (**d**) for 96 h, and the size distribution was monitored. *Color codes* of samples are shown in the figure. Data are presented as mean ± standard deviation (n = 3). Particles showed no aggregation in distilled water or in PBS during the 96-h assay period (**c**). In contrast a time-dependent, heavy particle aggregation was found in serum-free DMEM (d). Incubation of nanoparticles with 10 % FBS also evoked an immediate size increase, but prevented large-scale aggregation (**d**). SDS-PAGE analysis confirmed the adsorption of serum proteins to both PS-COOH and PS-PEG nanoparticles after 1 h incubation in 10 % serum containing MEM (e) and verified reduced protein adsorption of PEG-coated nanoparticles after 24 h incubation (**f**)

(4 h), 182.07 ± 30.53 nm (24 h), 559.67 ± 141.11 nm (96 h)]. The kinetics of particle-size enlargement is consistent with an immediate deposition of material on particle surfaces and a large-scale aggregation thereafter. The data showed that PEG-coating reduced the aggregation, an effect which was evident after long-term incubation (Fig. 1d).

The incubation of nanoparticles with 10 % fetal bovine serum also evoked an immediate size increase, but prevented large-scale NP aggregation during long-term treatment (Fig. 1d). The observation indicated that serum components were immediately adsorbed by particle surfaces, but instead of cross-linking particles, the protein corona could stabilize the suspension of the dispersed particles. Additional electrophoresis data further verified the rapid adsorption of proteins to both PS-COOH and PS-PEG nanoparticles (Fig. 1e). PEG-coated nanoparticles, however, exhibited reduced protein adsorption, which was evident after 24-h incubation (Fig. 1f) suggesting that PEGylation makes nanoparticles less prone to interactions with the environment. (Raw data of characterization experiments is shown in the Additional file 1).

Physicochemical characterization demonstrated that both, protein adsorption on nanoparticle surfaces and NP aggregation are influenced by the chemical milieu and by the type of surface-coating of NPs.

Spectral imaging fluorescence microscopy visualizes in vivo distribution of polystyrene nanoparticles

Previous experiments demonstrated that fresh PS-COOH or PS-PEG nanoparticle preparations did not have toxic effects on several mouse neural cell lines, even when applied at a very high concentration (up to 125 µg/ml) [33]. In accordance with the in vitro observations, single intravenous injection of 33.3 µg/ml PS-COOH or PS-PEG nanoparticles did not cause detectable physiological or behavioral changes.

To monitor the in vivo distribution of PS-COOH and PS-PEG polystyrene nanoparticles, we developed a spectral imaging fluorescence microscopy based method, which could overcome the limitations of nanoparticle detection caused by the high autofluorescence of native tissues, and could allow analyzing larger samples in less time, compared to electron microscopic approaches. Using a multidetector array confocal arrangement [Nikon A1R Confocal Laser Microscope System equipped with a spectral detector unit (Nikon, Shinjuku, Tokyo, Japan)], fluorescence intensity was recorded at defined wavelengths throughout the investigated spectral range, from 468 to 548 nm, with a spectral resolution of 2.5 nm. This acquisition approach combined with post hoc spectrum analysis selectively visualized nanoparticles in various tissues (Fig. 2a). We found that the highest signal-to-noise ratio was achieved by illuminating the tissue-embedded fluorescent nanoparticles at 457 nm excitation wavelength (Additional file 2).

Using the optimized imaging settings, the fluorescence spectra of PS-COOH and PS-PEG NPs were measured in various environments including PBS; PBS complemented with 1 % bovine serum albumin (BSA); mounting medium (Mowiol); and seeded onto fixed tissue slices. The measured spectra of particles showed parity with the spectrum provided by the manufacturer and did not vary substantially under different conditions (Fig. 2b). Furthermore, the different surface coating had no direct effect on the NP fluorescence, PS-COOH and PS-PEG nanoparticles displayed identical fluorescence spectra (Fig. 2c). For spectral analysis, the autofluorescence of particle-free, fixed tissue-sections was used as a negative control (Additional file 3), whereas the fluorescence of NPs seeded on control tissue sections served as positive controls.

Tissue- and cell-type-specific distribution of PS-COOH and PS-PEG nanoparticles

To evaluate short-term and the long-term distribution of NPs, mice were sacrificed either 5 min after a single tail-vein injection, or after a 4-day long survival period. In general, all investigated organs including the brain, kidney, liver, placenta and the spleen showed different amounts of nanoparticles after a single intravenous injection, and displayed distinct clearing after the 4-day post-injection period (Table 1).

Five minutes after nanoparticle injection, high abundance of both PS-COOH and PS-PEG NPs were found in organs involved in the elimination of toxic products from the body, such as the kidney, liver and the spleen. In organs protected by complex physiological barriers, such as the brain and the placenta, high density of PS-COOH NPs was revealed, while PS-PEG NPs were not present (Table 1).

The high-resolution of confocal microscopy combined with spectral imaging enabled more detailed analysis of regional and cellular distribution of nanoparticles in various tissues. Notably, aggregated PS-COOH NPs were concentrated in large vessels and capillaries within the brain, whereas the parenchyma was largely devoid of NPs (Fig. 3a, b; Additional file 4). In the placenta, PS-COOH NPs, but not PS-PEG NPs were seen in the lacunas (Fig. 3c, d), and importantly, neither type of nanoparticles was found in embryonic tissues (Additional file 5), indicating a proper placental barrier function.

In the kidney, both PS-COOH and PS-PEG NPs were present in the glomeruli and also in the interstitium around the tubuli 5 min after nanoparticle administration (Fig. 4a, c). Numerous NPs were identified in the

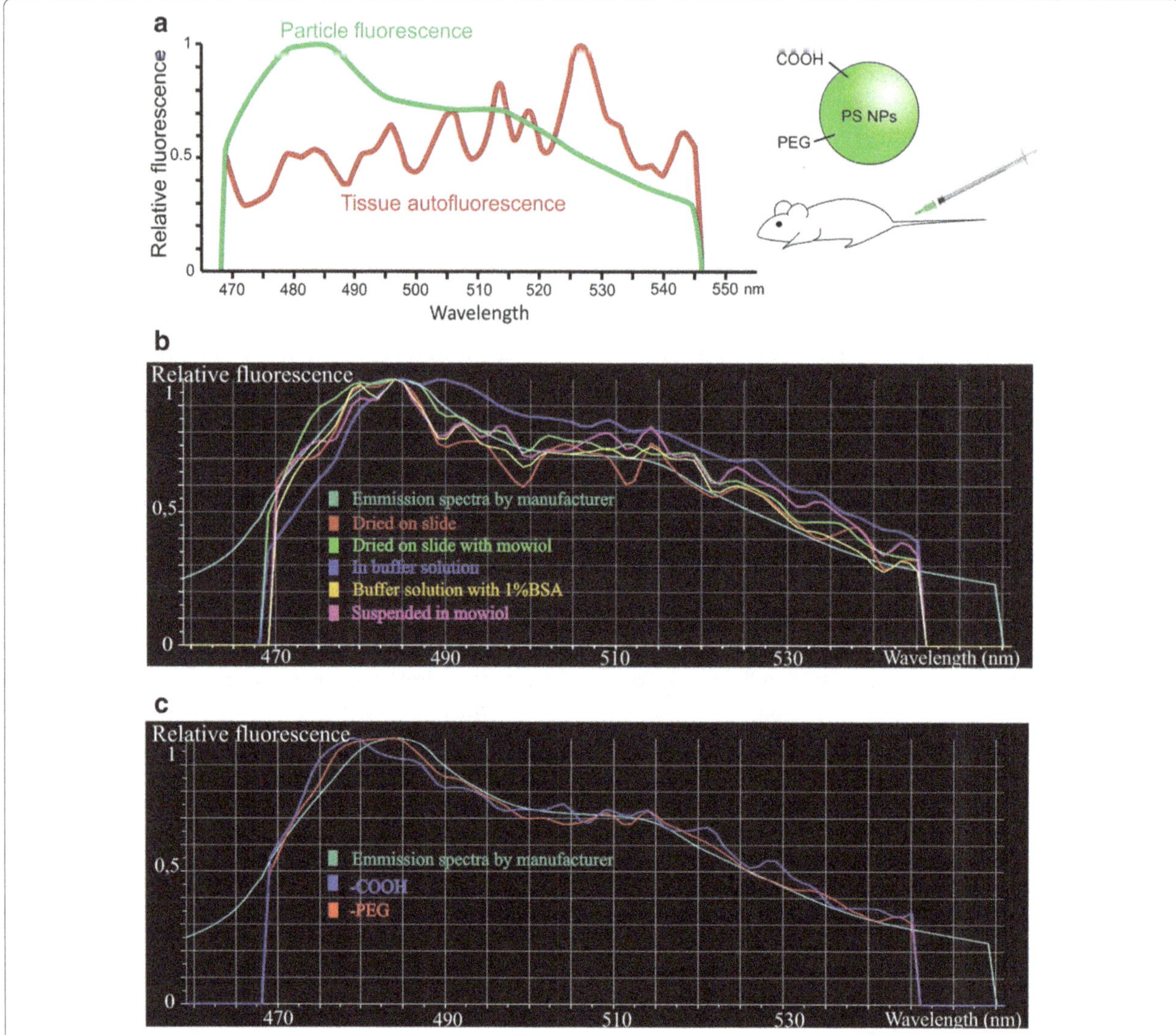

Fig. 2 Spectral imaging fluorescence microscopy and full spectrum analysis visualizes in vivo distribution of polystyrene nanoparticles. **a** Summary of the spectral analysis of particle-fluorescence (*green*) versus tissue-autofluorescene (*red*) after a single injection of COOH- or PEG-coated polystyrene nanoparticles. **b, c** Emission spectra of nanoparticles measured by spectral imaging fluorescence microscopy. Specimens were excited at 457 nm and the emitted light was detected in a wavelength range from 468 to 548 nm, with a spectral resolution of 2.5 nm. Spectra of polystyrene nanoparticles did not change in different conditions (shown as *color-codes* in **b**) or after functionalization (**c**) compared to the native spectrum provided by the manufacturer (represented by the aquamarine *color-code* in the figures)

Table 1 Biodistribution of polystyrene nanoparticles after intravenous injection. Summary of the distribution of differently functionalized polystyrene NPs after intravenous injection into mice. Organs were excised 5 min or 4 days after particles administration

	Brain	Placenta	Kidney	Liver	Spleen	Embryonic tissue
PS-COOH						
5 min	+	+	+	+	+	–
4 days	–	–	–	+	+	–
PS-PEG						
5 min	–	–	+	+	+	–
4 days	–	–	+	+	+	–

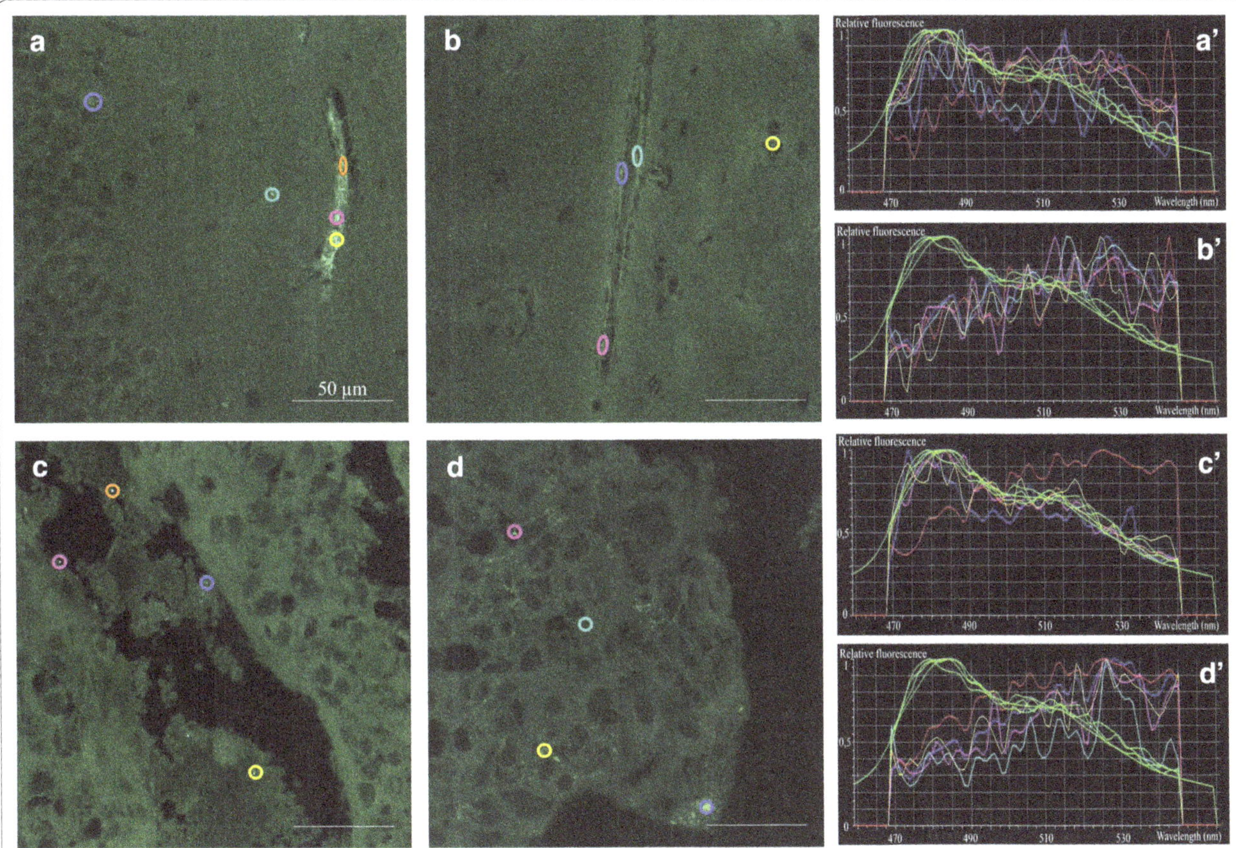

Fig. 3 Distribution of polystyrene NPs in the brain and the placenta 5 min after systemic exposure. Spectral images of tissue sections from the mouse brain (**a**, **b**) and placenta (**c**, **d**) 5 min after injection of PS-COOH (**a**, **c**) or PS-PEG (**b**, **d**) nanoparticles into the tail vein. **a′**–**d′**: spectrum profiles of ROIs in the corresponding images. The spectrum of each ROI is marked with the same *color* as it is delineated in the microscopic image. *Green curves* represent particle fluorescence (positive controls); *red curve* represents tissue autofluorescence (negative control). *Scale bars* 50 µm

liver (Fig. 5b) and in the spleen (Fig. 6a, c), regardless of the type of NP surface functionalization. In the spleen, NP distribution was mainly restricted to the marginal zone, enriched in monocytes/macrophages. Immunocytochemical analysis directly demonstrated the presence of both PS-COOH and PS-PEG NPs within the intracellular vesicles of Iba-1-positive phagocytotic cells (Fig. 7).

Four days after nanoparticle administration, PS-COOH NPs were completely cleared from the brain and the placenta and PS-PEG NPs were not found in these tissues (Additional file 6). Significant reduction of nanoparticle density was also observed in the kidney. While PS-COOH particles were not found in the kidney sections anymore, a few PS-PEG NPs were still deposited within the glomeruli (Fig. 4b, d). In striking contrast, high densities of both NPs were found in the liver (Fig. 5c, d) and in the spleen (Fig. 6b, d). The regional distribution of polystyrene nanoparticles did not change during the 4-day post-loading period in the liver. Characteristic redistribution of NPs was, however, observed within the spleen: 4-days after the exposure, NPs were identified in the white pulp, regardless of surface functionalization (Fig. 7b, d).

Taken together, spectral imaging fluorescence microscopy was instrumental in characterizing the penetration and accumulation of nanoparticle in different organs. The findings also show that differentially-coated nanoparticles exhibit distinct tissue- and cell-type-specific distribution and clearance.

Discussion

To investigate the impact of surface chemical characteristics of NPs in particle-distribution in the living body, fluorescently core-labeled polystyrene nanoparticles, with different surface characteristics were used. Polystyrene as a core material was chosen, because of its stability in biological solutions, and because ions or toxic compounds could not dissolve from it [37]. Polystyrene in bulk material is not toxic [38] and our previous in vitro data demonstrated that 50 nm polystyrene particles were not toxic to several mouse neural cell lines, when used at concentrations up to 125 µg/ml [33]. Particle surfaces were

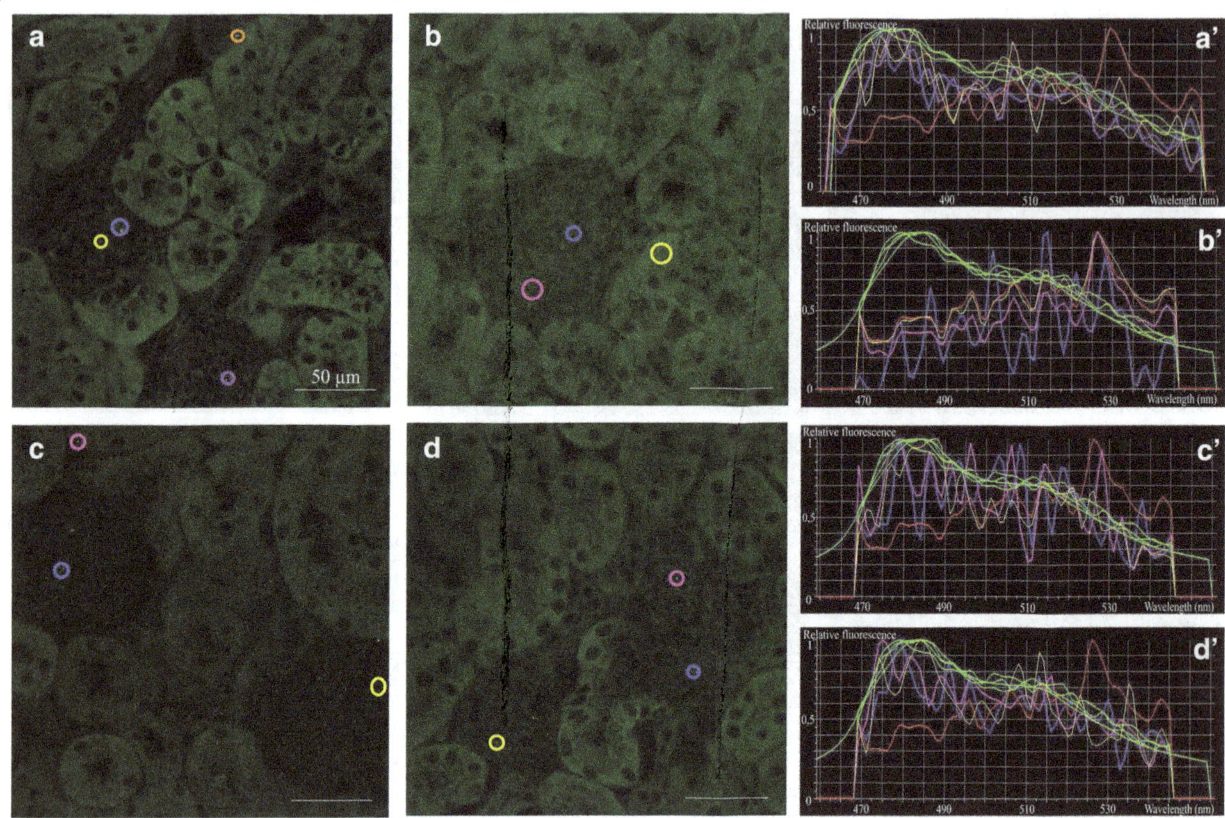

Fig. 4 Distribution and accumulation of polystyrene nanoparticles in the kidney. Spectral images of adult mouse kidney sections 5 min (**a**, **c**) and 4 days (**b**, **d**) after injection of PS-COOH (**a**, **b**) or PS_PEG (**c**, **d**) nanoparticles through the tail vein. **a'**–**d'**: spectrum profiles of ROIs in the corresponding images. The spectrum of each ROI is marked with the *same color* as it is delineated in the microscopic image. *Green curves* represent particle fluorescence (positive controls); *red curve* represents tissue autofluorescence (negative control). *Scale bars* 50 μm

functionalized either with -$COOH^-$ groups presenting strong negative surface charge and mediating ionic interactions or with hydrophilic and charge neutralizing -PEG chains.

Physicochemical characterization of the particles revealed that the applied PS-COOH and PS-PEG particles could be considered identical in their physicochemical aspects, except for the chemical reactivity of their surfaces. At early time points particles did not show significant hydrodynamic size differences in serum containing physiological solutions. Accordingly, PS-COOH and PS-PEG particles were expected to similarly influence the blood flow conditions at the time of intravenous administration. The DLS measurements demonstrated that particles were stabilized in serum-containing fluids, accordingly large-scale aggregate formation was not expected in the circulation, either. The 50–90 nm particle size-range promised several advantages. Particles were comparable in size with natural assemblies of protein complexes, thus were not expected to block circulation in the applied concentration. On the other hand, particles were big enough not to be excreted rapidly by the kidney, which is known to retain proteins and particles larger than 10 nm [4]. Fluorescence spectrum of particles was identical for the two particles regardless of surface functionalization, and was not affected by different environmental conditions, ranging from the dry state to protein containing solutions.

Taken together, their stable spectra, identical hydrodynamic size, and the expected similar effect on flow conditions, PS-COOH and PS-PEG nanoparticles represented ideal tools to investigate the effect of distinct surface characteristics on in vivo accumulation.

Visualization of fluorescent nanoparticles even with high-resolution confocal microscopy is hindered by the size of nanoparticles and by the high autofluorescence of biological samples [12]. The intensity of particle fluorescence depends on the amount of fluorescent dye or on the properties of the core material. In physiological experimentation the use of potentially inert particles labelled with non-toxic signaling compounds is critically important. Therefore, interest

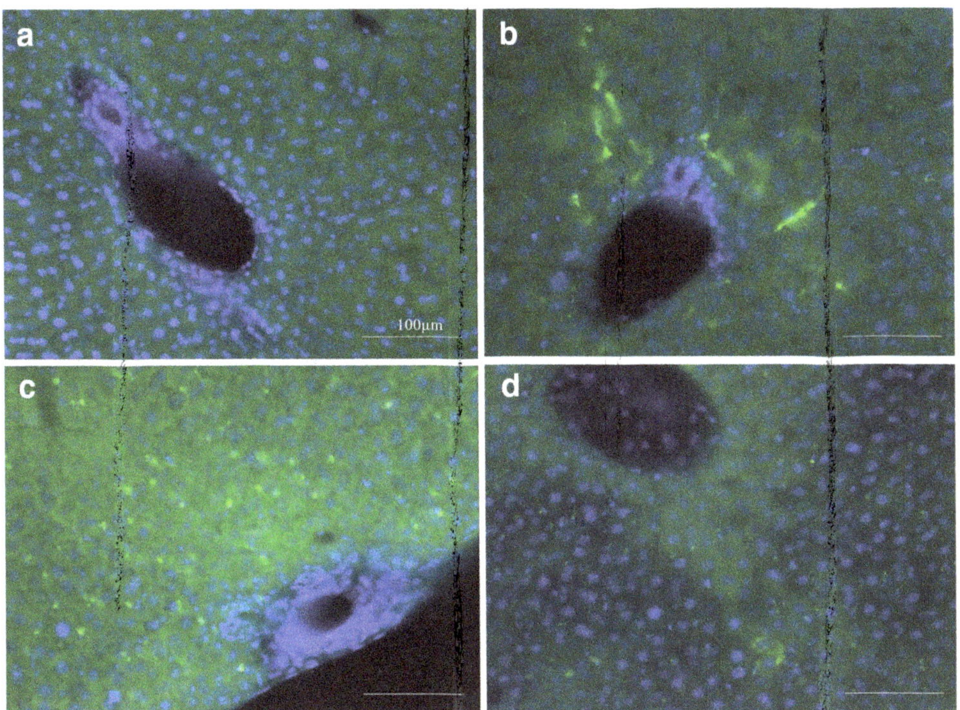

Fig. 5 Accumulation of polystyrene nanoparticles in the liver. Traditional fluorescence microscopic images of sections made from livers of non-treated (**a**) and NP-injected (**b**, **c**, **d**) adult mice. Animals were sacrificed 5 min (**b**) and 4 days (**c**) after intravenous injection of PS-COOH and 4 days after injection of PS-PEG (**d**) polystyrene nanoparticles. Particle fluorescence is shown in *green*, and bisbenzimide nuclear staining in *blue*. *Scale bars* 100 μm

was turned to polymer nanoparticles which encapsulate fluorescent dyes. While the covalent embedding of the dye into the core-material prevents dissolution, the amount of "in-core" dye is limited and hinders the fluorescence detection.

In the present study, spectral imaging fluorescence microscopy was used to overcome the limitations caused by high autofluorescence and low particle fluorescence. Spectral image acquisition combined with spectrum analysis could reliably detect fluorescent polystyrene nanoparticles at the tissue- and cellular level.

The distribution of PS-COOH and PS-PEG nanoparticles was investigated at two time points, 5 min and 4 days after particle loading. Initial distribution of particles could not be revealed by histological methods because of the high heart rate (300–800 beats/min) and 20–45 μl stroke volume of mice resulting in an average 20 ml/min cardiac output [39]. The 5 min exposure allowed investigating short-term tissue distribution of particles after a single injection and comparison to a longer term (4 day) distribution.

Five minutes after NP-injection, significantly more PS-COOH then PS-PEG particles were found at the walls of brain vessels and in the placenta. As it was expected [32, 33, 40], PEGylation inhibited the attachment of particles to biological interfaces, while PS-COOH particles could interact with vessel walls and cell surfaces. On the other hand, PEGylation could not prevent the active uptake of PS-PEG NPs by professional macrophages [41, 42]. We found a remarkable accumulation of both PS-COOH and PS-PEG nanoparticles in the kidney, liver and the spleen, e.g. in organs responsible for elimination of particulate polluting agents from the circulation.

Particles were never seen in embryonic tissues indicating a proper placental barrier function against polystyrene nanoparticles, as it was shown also for human placenta [43].

After 4 days, the particles were completely cleared from the vessels of the brain and the placenta, but were accumulated in the liver and the spleen. The presented study did not aim and allow analyzing the renal clearance of the particles. The persisting presence of PS-PEG particles in the glomeruli, however, suggested that particles were accumulated by intraglomerular mesangial cells known to phagocytose contaminating particles, mesangial matrix material and cell debris [44]. In the spleen, several particles were translocated from the marginal zones into the white pulp during the 4-day post-injection period. Our observations on the storage and translocation of these nanoparticles are worthy of further consideration, since they may suggest a

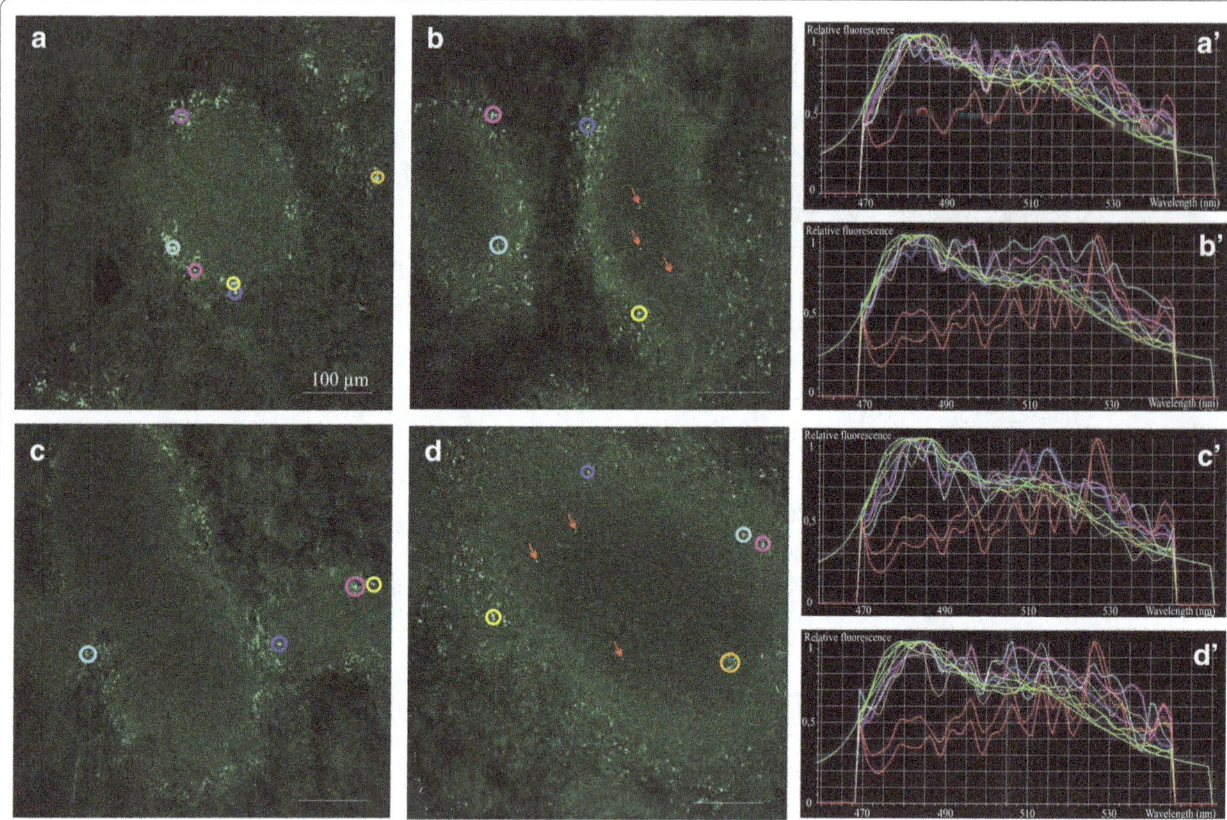

Fig. 6 Distribution and accumulation of PS nanoparticles in the spleen. Spectral images of mouse spleen sections after a single intravenous injection of carboxylated (**a**, **b**) or PEGylated (**c**, **d**) polystyrene nanoparticles, 5 min (**a**, **c**) and 4 days (**b**, **d**) after exposure. *Red arrows* indicate translocated particles to the white pulp during the 4-day after exposure period. **a′**–**d′**: spectrum profiles of ROIs in the corresponding images. The spectrum of each ROI is marked with the *same color* as it is delineated in the microscopic image. *Green curves* represent particle fluorescence (positive controls); *red curves* represent tissue autofluorescence (negative controls). *Scale bars* 100 µm

method through which compounds on particle surfaces can be presented to the lymphocytes of the marginal zone and the white-pulp [45] and thus manipulate the initiation of adaptive immune responses. Regarding the large-scale material adsorption by NP surfaces including LPS contamination [33], the inflammation-initiating effects of "harmless" NPs should not be neglected.

The observations indicate that middle-sized (50–90 nm) non-biodegradable nanoparticles which are captured by macrophages in the spleen and the liver or even in the kidney interstitium, cannot be easily removed from the body. Regarding the fact that material of dying cells will be ingested by neighboring phagocytes, the question can be raised whether such particles can be cleared at all. Long-term accumulation and limited clearance may cause problems if repeated nanoparticle loading is considered, even if the acute single dose of potentially "harmless" particles is low. Considering the continuous growth of the production and use of nanoparticles, together with the consequent increase in environmental nanoparticle pollution, the long-term fate of nanoparticles in exposed organisms needs further thorough studies. These concerns stress the importance of high-throughput imaging modalities which help to perform distribution studies of tissue- and cell-type-specific NP accumulation. Spectral imaging fluorescence microscopy provides technological tools to monitor the penetration, distribution and accumulation of fluorescently labeled NPs in various tissues with high accuracy and in a time-frame which cannot be achieved with electron microscopy.

Conclusions

In this work, the in vivo distribution of 50–90 nm polystyrene nanoparticles with distinct surface characteristics was investigated using spectral imaging fluorescence microscopy. Spectral imaging combined with post hoc spectrum analysis allowed visualizing nanoparticles in various tissues and helped to overcome the limitations caused by the high autofluorescence of native tissues. As an additional advantage, the method allowed analyzing larger samples in less time, compared to electron microscopic approaches.

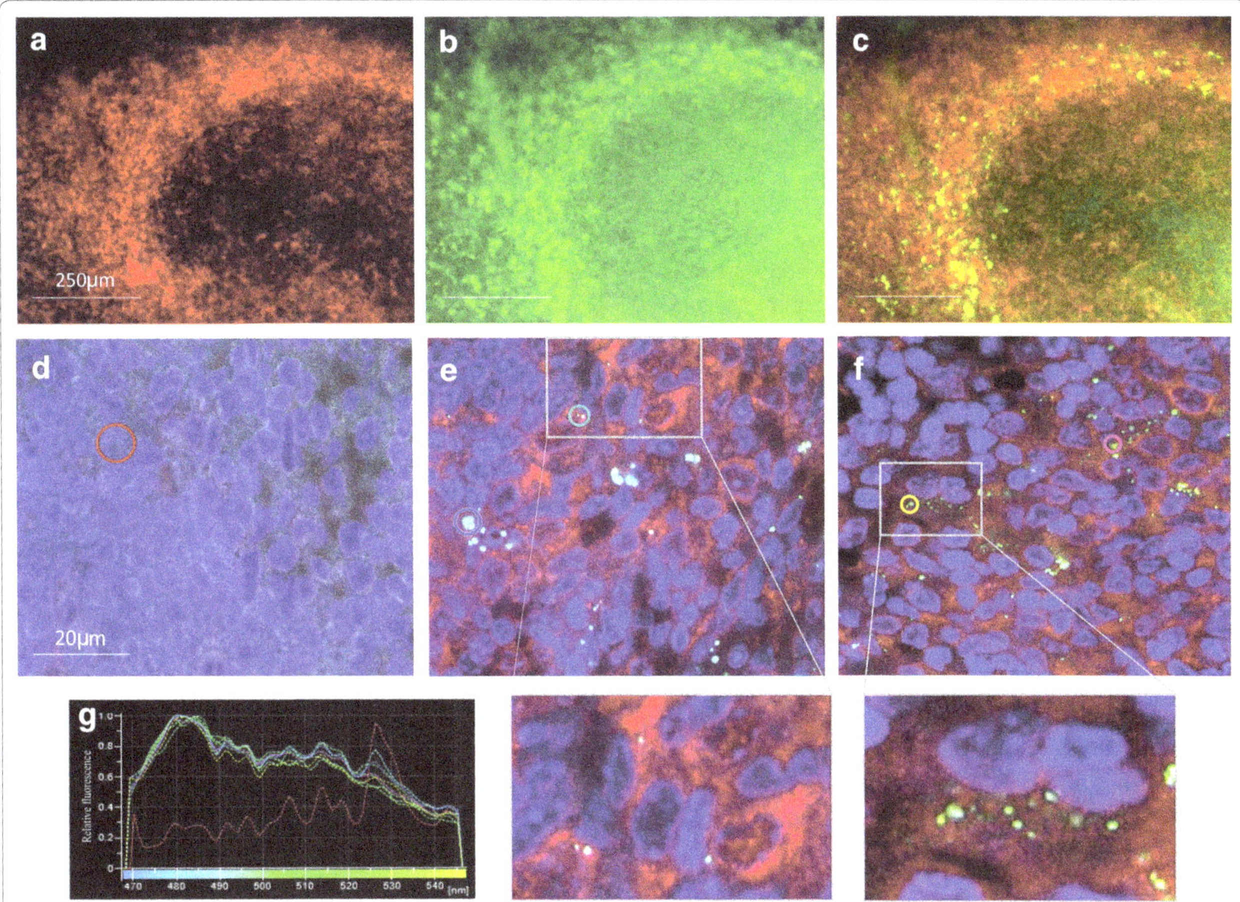

Fig. 7 PS-NPs are associated with the marginal zone monocytes/macrophages, identified by Iba-1 staining. Immunohistochemical staining for the spleen macrophages, stained with anti-Iba-1 antibody (**a** *red*), and carboxylated PS-NPs (**b** *green*) 4 days after a single intravenous injection of nanoparticles. Merged image (**c**) shows that residual nanoparticles (*green*) are co-localized with Iba-positivity of marginal zone macrophages (*red*). *Scale bars* 250 µm. Confocal images of spleen sections from non-treated (**d**) or PS-PEG-injected (**e**, **f**) mice. Samples were collected 5 min (**e**) or 4 days (**f**) after exposure to PS- PEG, and were stained for Iba-1. The sample from non-treated animal (**d**) serves also as staining control. Enlarged areas of the boxed regions in **e** and **f**, show the presence of PS-PEG NPs in Iba-1 positive phagocytotic cells. **g** The spectra of ROIs indicated the presence of PS-PEG NPs on images of both, 5-min and 4-day samples. The spectrum of each ROI is marked with the *same color* as it is delineated in the microscopic image. Fluorescence spectra of ROIs were compared against spectrum of tissue autofluorescence (*red curves* as negative controls) and particle fluorescence (*green curves* as positive controls). *Scale bars* 20 µm

Analyses of the body-distribution of carboxylated or PEGylated polystyrene NPs indicated proper barrier-functions for the blood–brain-barrier and the placenta. While NPs were stuck in the brain vessels and the lacunas of the placenta, barriers completely prevented the penetration of particles into the brain parenchyma and embryonic tissues. Particles with both surface functionalization were accumulated in the reticuloendothelial organs. While PEGylation reduced short-term attachment of particles to vessel walls, it did not prevent accumulation in the liver and the spleen or in the intraglomerular mesangium.

The accumulation and long-term storage of nanoparticles in the reticuloendothelial systems rise important questions on the long-term health-risk even of otherwise non-toxic particles.

Methods

Characterization of nanoparticles

Nominal size of 50–70 nm carboxylated and PEGylated ($Mw_{PEG} = 300$ g/mol) nanoparticles made from "Yellow" fluorochrome-labelled polystyrene, were obtained from Spherotech Inc. (Lake Forest, IL, USA). (Fluorescence spectrum provided by the manufacturer for these "yellow" particles [46]). Concentration of the stock suspensions was 10 mg/ml.

Size and zeta-potential of nanoparticles were measured by ZetasizerNano ZS90 (Horiba Instruments Inc., Irvine,

CA). For dynamic light scattering analysis 0.01 mg/ml nanoparticle suspensions were measured using a 633 nm He–Ne laser. Zeta-potential measurements were carried out at 25 °C using a folded capillary cell (DTS1070, Malvern Instruments, Worcestershire, United Kingdom).

Size and shape of nanoparticles were confirmed by transmission electron microscopy (JEOL JEM 1010, JEOL Ltd., Tokyo, Japan) at 80 keV. For TEM analysis, 3 µl aliquots of 0.01 mg/ml nanoparticle suspensions in distilled water were transferred to and dried on 200 mesh copper grids with carbon film.

Particles were stored in sterile MilliQ water and were used within 2–3 months after synthesis.

Raw data of characterization experiments is shown in the Additional file 1.

Aggregation and protein adsorption of particles in biological solutions

Increase in particle diameter was measured by DLS. Nanoparticle aggregation was monitored in inorganic or biological environments, including solutions used during particle handling and solutions that mimic the characteristics of body fluids. 1 mg/ml carboxylated and PEGylated particle preparations were made by 1:10 dilution of stock suspensions with distilled water. The suspensions were further diluted 1:10 with PBS (pH 7.4), DMEM cell culture medium (Sigma-Aldrich, St. Louis, MO, USA) or DMEM supplemented with 10 % fetal bovine serum (Invitrogen/Gibco, Carlsbad, CA, USA). After 0, 4, 24 or 96-h incubation at 37 °C, particle preparations were diluted in 1:10 with distilled water and the hydrodynamic diameter of particles was measured by DLS.

Proteins adsorbed by particles in 10 % fetal bovine serum containing minimum essential medium (FBS-MEM; Sigma-Aldrich, St. Louis, MO, USA) were analyzed by SDS-PAGE (Sodium dodecyl sulfate polyacrylamide gel electrophoresis). After 1 or 24-h incubation in 10 % FBS-MEM nanoparticles were sedimented by centrifugation (45 min at 20,000×*g*) and were washed with PBS to remove non-bound proteins. Washed NPs were resuspended in Laemmli buffer containing 1 % (w/v) sodium dodecyl sulfate, and loaded onto 10 % polyacrylamide gel. Gel electrophoresis was performed at 130 V for about 60 min. The gels were stained with silver staining 33 kit (Cosmobio Ltd., Tokyo, Japan), according to the manufacturer's instructions.

In vivo experiments

Animal experiments were conducted with the approval of the Animal Care Committee of the Institute of Experimental Medicine of Hungarian Academy of Sciences and according to the official license (No.: 22.1/353/3/2011; exp. date: 4/7/2016) issued by National Food Chain Safety Office (http://www.NEBIH.gov.hu), Hungary.

Male mice (aged 25–30 days) and pregnant female mice on the 10th to 15th post conception days were anesthetized with a mixture of ketamine (CP-Pharma mbH, Burgdorf, Germany) and xylazine (CEVA-PHYLAXIA, Budapest, Hungary), 100 and 10 µg/g bodyweight, respectively. Nanoparticle stock suspensions (10 mg/ml) were diluted 1:30 in PBS and dispersed by sonication. Under proper anesthesia, 7 µl/g bodyweight aliquots of carboxylated ($n_{COOH(male)} = 6$, $n_{COOH(pregnant\ female)} = 6$) or PEGylated ($n_{PEG(male)} = 6$, $n_{PEG(pregnant\ female)} = 6$) nanoparticle suspensions were introduced into the tail vein. Animals were sacrificed by overdose of anesthetics after a 5-min or 4-day exposure to the single-injection loading. Various organs including brain, liver, kidney, spleen as well as placenta and embryos were removed and fixed with paraformaldehyde (8 w/v % in PBS) for 24 h at 4 °C. Organs and embryos were collected from animals not exposed to nanoparticles, as controls ($n_{control(male)} = 3$, $n_{control(pregnant\ female)} = 3$).

Microscopic evaluation

For microscopic evaluation, 30 or 60 µm thick vibratome sections (VT1000S, Leica, Wetzlar, Germany) were made from fixed organs.

For immunocytochemical staining, the sections were permeabilized and non-specific antibody binding was blocked by incubating the sections in PBS containing 10 % FBS and 0.1 % Triton-X for 2 h. Primary antibodies, anti-Iba-1 goat polyclonal antibodies and anti-claudin V rabbit polyclonal antibodies (Abcam; Cambridge, UK) were diluted in 1–1000 with PBS-FBS, and the sections were incubated at 4 °C, overnight. After incubation, the cells were washed three times (15 min each) with PBS and incubated with alexa-594 conjugated secondary antibodies (1:1000; Molecular Probes, Invitrogen, Carlsbad, CA, USA) for 1 h. Slices were mounted with mowiol (Sigma-Aldrich, Budapest, Hungary) containing 10 µg/ml bisbenzimide (Sigma-Aldrich) for nuclear staining.

Sections were analyzed with a Zeiss Axiovert 200 M fluorescence microscope (Zeiss, Jena, Germany) and a Nikon A1R Confocal Laser Microscope System equipped with a spectral detector unit (Nikon, Shinjuku, Tokyo, Japan).

Optimized detection of particles by spectral imaging fluorescence microscopy

Fluorescence spectra of NPs were determined in dry and in buffer-dispersed particle preparations, as well as in contact with fixed tissue sections from control animals by spectral imaging fluorescence microscopy. The extrinsic spectra emitted by the particles in contact with the corresponding tissue section from control animals were used as positive controls. For negative control, the intrinsic autofluorescence spectra of corresponding sections of control organs were used (Additional file 3).

For spectral evaluation 457 nm argon ion laser was used as excitation source, and the emitted light was detected by the spectral detector unit from 468 to 548 nm, with a spectral resolution of 2.5 nm. In order to record continuous spectrum, a 20/80 beam splitter (BS20/80) with continuous transmission was used instead of a paired dichronic mirror arrangement.

Regions of interest (ROI) were delineated and analyzed in comparison with corresponding sections of nanoparticle-treated and non-treated organs. The photocurrent intensities detected in the samples at different wavelengths (emission spectra in organs of treated animals) were plotted against the tissue autofluorescence spectra (negative control) and the spectra of nanoparticles seeded on control-tissue (positive control).

Additional files

Additional file 1: Raw results of characterization experiments. PS-NPs were incubated with inorganic or biological solutions. After 5-min, 4, 24 or 96 h incubation size distribution and zeta-potential measurements were carried out by a ZetasizerNano ZS90.

Additional file 2: Excitation of fluorophore labelled polystyrene nanoparticles. Nanoparticles were excited at 404, 457, 476 nm wavelengths to determine optimal excitation settings. Highest intensity of emitted light was reached when samples were excited at 457 nm.

Additional file 3: Intrinsic fluorescence of non-treated tissue sections. Spectral images showing autofluorescence of non-treated brain (A), placenta (B), kidney (C) and spleen (D) sections. Green curves represent particle fluorescence (positive controls); the red curve represents the autofluorescence. The autofluorescence spectra were used as negative controls for post hoc spectral identification of PS-NPs in the corresponding tissues. A', B', C', D': spectrum profiles of ROIs in the corresponding images. The spectrum of each ROI is marked with the same color as it is delineated in the microscopic image.

Additional file 4: Polystyrene NPs in brain vessels 5 min after systemic exposure. Fluorescence images of sections made from brain of PS-PEG (A) or PS-COOH (B) injected adult mice. Animals were sacrificed 5 min after intravenous injection. Sections were stained for Claudin V (red); cell nuclei are shown in blue. Scale bars: 50 µm.

Additional file 5: Embryonic tissues were free from nanoparticles 5 min after maternal NP-administration. Sections made from mouse embryonic (E 15) forebrain cortex (A, B, C) and liver (D, E) 5 min after the injection of carboxylated PS nanoparticles into the tail vein of the mother. Cell nuclei are stained with bisbenzimide (blue). Representative spectrum images (C, E) and spectrum profiles (C', E') showed no particles in the embryonic brain or liver tissues. Scale bars: 50 µm.

Additional file 6: PS-NP cleared from the brain and the placenta within the 4-day post-injection period. Spectral images of sections of mouse brain (A, B) and placenta (C, D) 4 days after injection of PS-COOH (A, C) or PS-PEG (B, D) nanoparticles into the tail vein of adult mice. A', B', C', D': spectrum profiles of ROIs in the corresponding images. The spectrum of each ROI is marked with the same color as it is delineated in the microscopic image. Green curves represent particle fluorescence (positive controls); red curve represents tissue autofluorescence (negative control). Scale bars: 50 µm.

Abbreviations

BSA: bovine serum albumin; DLS: dynamic light scattering; DMEM: Dulbecco's modified Eagle's cell culture medium; FBS: fetal bovine serum; MEM: minimum essential medium; NP: nanoparticles; PBS: phosphate buffered saline; PEG: polyethylene glycol; PS-NP: polystyrene nanoparticles; PS-COOH: carboxylated polystyrene nanoparticles; PS-PEG: PEGylated polystyrene nanoparticles; ROI: regions of interest; SDS-PAGE: sodium dodecyl sulfate polyacrylamide gel electrophoresis; TEM: transmission electron microscopy.

Authors' contributions

KK participated in the design of the study, optimized the spectral imaging and analysis for in vivo detection of nanoparticles, carried out the confocal and spectral microscopy studies and experiments on particle size, performed the statistical analysis, data interpretation and drafted as well as finalized the manuscript. KM performed the SDS-PAGE analyses on particle coronas and contributed to the histological work. ÁC was involved in the characterization of polystyrene nanoparticles. JP shared experiences in dynamic light scattering experiments, and was involved with particle size analysis. VP supervised the particle characterization experiments. EM initiated the study, supervised the in vivo experiments and data interpretations and was involved in the manuscript preparation and writing. All authors read and approved the final manuscript.

Author details

[1] School of PhD Studies, Semmelweis University, Üllői Street 26, Budapest 1085, Hungary. [2] Institute of Experimental Medicine, Hungarian Academy of Sciences, Szigony Street 43, Budapest 1083, Hungary. [3] Soft Flow Hungary Kft., Kedves u. 20, Pecs 7628, Hungary. [4] Catalan Institute of Nanoscience and Nanotechnology, Campus UAB, Bellaterra, 08193 Barcelona, Spain.

Acknowledgements

The authors wish to thank the Nikon Microscopy Center (NMC) at IEM, Nikon Austria GmbH and Auro-Science Consulting Ltd. for kindly providing microscopy support. We express our special thanks to László Barna, Head of NMC, for his professional help in fluorescence spectrum detection. We thank for the support in particle handling for Soft Flow Hungary Kft, and for the support in manuscript preparation to István Katona.

Competing interests

The authors declare that they have no competing interests.

Funding

This work was supported by FP7 projects QualityNano (INFRA-2010-262163, ICN-TAF-176) and by the Doctoral School of Semmelweis Univerity, Budapest to K.K; by the EU 7th framework programme, Marie Curie Actions, Network for Initial Training NanoTOES (PITN-GA-2010-264506) to K.M.; and by the Hungarian National Science Fund (OTKA; Grant No.: K106191) and Bio Surf (Nat. Office of Res. and Dev.; NKTH; Grant No.: TECH-08-A1-2008-0276) to EM. The funding bodies did not contribute to the design of the study, to collection, analysis, and interpretation of the data or to the writing of the manuscript.

References

1. Ostrowski A, Nordmeyer D, Boreham A, Holzhausen C, Mundhenk L, Graf C, Meinke MC, Vogt A, Hadam S, Lademann J, Ruhl E, Alexiev U, Gruber AD. Overview about the localization of nanoparticles in tissue and cellular context by different imaging techniques. Beilstein J Nanotechnol. 2015;6:263–80.
2. Leary J, Key J. Nanoparticles for multimodal in vivo imaging in nanomedicine. Int J Nanomedicine. 2014;9:711.
3. Liu Y, Hu Y, Huang L. Influence of polyethylene glycol density and surface lipid on pharmacokinetics and biodistribution of lipid-calcium-phosphate nanoparticles. Biomaterials. 2014;35:3027–34.

4. Choi HS, Liu W, Misra P, Tanaka E, Zimmer JP, Itty Ipe B, Bawendi MG, Frangioni JV. Renal clearance of quantum dots. Nat Biotechnol. 2007;25:1165–70.
5. Ye D, Dawson KA, Lynch I. A TEM protocol for quality assurance of in vitro cellular barrier models and its application to the assessment of nanoparticle transport mechanisms across barriers. Analyst. 2015;140:83–97.
6. Fagerland JA, Wall HG, Pandher K, LeRoy BE, Gagne GD. Ultrastructural analysis in preclinical safety evaluation. Toxicol Pathol. 2012;40:391–402.
7. Liba O, SoRelle E, Debasish S, de la Zerda A. Contrast-enhanced optical coherence tomography with picomolar sensitivity for functional in vivo imaging. Sci Rep. 2016;6:23337.
8. Cho M, Cho WS, Choi M, Kim SJ, Han BS, Kim SH, Kim HO, Sheen YY, Jeong J. The impact of size on tissue distribution and elimination by single intravenous injection of silica nanoparticles. Toxicol Lett. 2009;189:177–83.
9. Mahon E, Salvati A, Baldelli Bombelli F, Lynch I, Dawson KA. Designing the nanoparticle-biomolecule interface for "targeting and therapeutic delivery". J Control Release. 2012;161:164–74.
10. Da Silva CG, Rueda F, Löwik CW, Ossendorp F, Cruz LJ. Combinatorial prospects of nano-targeted chemoimmunotherapy. Biomaterials. 2016;83:308–20.
11. Yang Y, Yu C. Advances in silica based nanoparticles for targeted cancer therapy. Nanomedicine. 2015;12:1–16.
12. Bouccara S, Sitbon G, Fragola A, Loriette V, Lequeux N, Pons T. Enhancing fluorescence in vivo imaging using inorganic nanoprobes. Curr Opin Biotechnol. 2015;34:65–72.
13. Gambinossi F, Mylon SE, Ferri JK. Aggregation kinetics and colloidal stability of functionalized nanoparticles. Adv Colloid Interface Sci. 2014;222:332–49.
14. Murugan K, Choonara YE, Kumar P, Bijukumar D, du Toit LC, Pillay V. Parameters and characteristics governing cellular internalization and trans-barrier trafficking of nanostructures. Int J Nanomedicine. 2015;10:2191–206.
15. Firdessa R, Oelschlaeger TA, Moll H. Identification of multiple cellular uptake pathways of polystyrene nanoparticles and factors affecting the uptake: relevance for drug delivery systems. Eur J Cell Biol. 2014;93:323–37.
16. Al-Rawi M, Diabaté S, Weiss C. Uptake and intracellular localization of submicron and nano-sized SiO particles in HeLa cells. Arch Toxicol. 2011;85:813–26.
17. Naczynski DJ, Andelman T, Pal D, Chen S, Riman RE, Roth CM, Moghe PV. Albumin nanoshell encapsulation of near-infrared-excitable rare-Earth nanoparticles enhances biocompatibility and enables targeted cell imaging. Small. 2010;6:1631–40.
18. Hu X, Gao X. Silica–polymer dual layer-encapsulated quantum dots with remarkable stability. ACS Nano. 2010;4:6080–6.
19. Nel AE, Mädler L, Velegol D, Xia T, Hoek EMV, Somasundaran P, Klaessig F, Castranova V, Thompson M. Understanding biophysicochemical interactions at the nano-bio interface. Nat Mater. 2009;8:543–57.
20. Monopoli MP, Aberg C, Salvati A, Dawson KA. Biomolecular coronas provide the biological identity of nanosized materials. Nat Nanotechnol. 2012;7:779–86.
21. Casals E, Pfaller T, Duschl A, Oostingh GJ, Puntes V. Time evolution of the nanoparticle protein corona. ACS Nano. 2010;4:3623–32.
22. Casals E, Puntes VF. Inorganic nanoparticle biomolecular corona: formation, evolution and biological impact. Nanomedicine (Lond). 2012;7:1917–30.
23. Casals E, Pfaller T, Duschl A, Oostingh GJ, Puntes VF. Hardening of the nanoparticle-protein corona in metal (Au, Ag) and oxide (Fe3O4, CoO, and CeO2) nanoparticles. Small. 2011;7:3479–86.
24. Lundqvist M, Stigler J, Cedervall T, Berggård T, Flanagan MB, Lynch I, Elia G, Dawson K. The evolution of the protein corona around nanoparticles: a test study. ACS Nano. 2011;5:7503–9.
25. Tenzer S, Docter D, Kuharev J, Musyanovych A, Fetz V, Hecht R, Schlenk F, Fischer D, Kiouptsi K, Reinhardt C, Landfester K, Schild H, Maskos M, Knauer SK, Stauber RH. Rapid formation of plasma protein corona critically affects nanoparticle pathophysiology. Nat Nanotechnol. 2013;8:772–81.
26. Salvati A, Pitek AS, Monopoli MP, Prapainop K, Bombelli FB, Hristov DR, Kelly PM, Åberg C, Mahon E, Dawson KA. Transferrin-functionalized nanoparticles lose their targeting capabilities when a biomolecule corona adsorbs on the surface. Nat Nanotechnol. 2013;8:137–43.
27. Milani S, Bombelli FB, Pitek AS, Dawson KA, Rädler J. Reversible versus irreversible binding of transferrin to polystyrene nanoparticles: soft and hard corona. ACS Nano. 2012;6:2532–41.
28. Laurent S, Burtea C, Thirifays C, Rezaee F, Mahmoudi M. Significance of cell "observer" and protein source in nanobiosciences. J Colloid Interface Sci. 2013;392:431–45.
29. Izak-Nau E, Kenesei K, Murali K, Voetz M, Eiden S, Puntes VF, Duschl A, Madarász E. Interaction of differently functionalized fluorescent silica nanoparticles with neural stem- and tissue-type cells. Nanotoxicology. 2013;5390:1–11.
30. Sacchetti C, Motamedchaboki K, Magrini A, Palmieri G, Mattei M, Bernardini S, Rosato N, Bottini N, Bottini M. Surface polyethylene glycol conformation influences the protein corona of polyethylene glycol-modified single-walled carbon nanotubes: potential implications on biological performance. ACS Nano. 2013;7:1974–89.
31. Essa S, Rabanel JM, Hildgen P. Characterization of rhodamine loaded PEG-g-PLA nanoparticles (NPs): effect of poly(ethylene glycol) grafting density. Int J Pharm. 2011;411:178–87.
32. Peracchia MT, Harnisch S, Pinto-Alphandary H, Gulik A, Dedieu JC, Desmaële D, D'Angelo J, Müller RH, Couvreur P. Visualization of in vitro protein-rejecting properties of PEGylated stealth® polycyanoacrylate nanoparticles. Biomaterials. 1999;20:1269–75.
33. Murali K, Kenesei K, Li Y, Demeter K, Környei Z, Madarász EE. Uptake and bio-reactivity of polystyrene nanoparticles is affected by surface modifications, ageing and LPS adsorption: in vitro studies on neural tissue cells. Nanoscale. 2015;7:4199–210.
34. Peracchia M, Fattal E, Desmaële D, Besnard M, Noël J, Gomis J, Appel M, D'Angelo J, Couvreur P. Stealth® PEGylated polycyanoacrylate nanoparticles for intravenous administration and splenic targeting. J Control Release. 1999;60:121–8.
35. Chilukuri N, Sun W, Naik RS, Parikh K, Tang L, Doctor BP, Saxena A. Effect of polyethylene glycol modification on the circulatory stability and immunogenicity of recombinant human butyrylcholinesterase. Chem Biol Interact. 2008;175:255–60.
36. Gao X, Cui Y, Levenson RM, Chung LWK, Nie S. In vivo cancer targeting and imaging with semiconductor quantum dots. Nat Biotechnol. 2004;22:969–76.
37. Cohen JT, Carlson G, Charnley G, Coggon D, Delzell E, Graham JD, Greim H, Krewski D, Medinsky M, Monson R, Paustenbach D, Petersen B, Rappaport S, Rhomberg L, Ryan PB, Thompson K. A comprehensive evaluation of the potential health risks associated with occupational and environmental exposure to styrene. J Toxicol Environ Health B Crit Rev. 2002;5:1–265.
38. Code of federal regulations—food for human consumption. http://www.accessdata.fda.gov/scripts/cdrh/cfdocs/cfcfr/cfrsearch.cfm?fr=177.1640. Accessed 15 April 2015.
39. Janssen B, Debets J, Leenders P, Smits J. Chronic measurement of cardiac output in conscious mice. Am J Physiol Regul Integr Comp Physiol. 2002;282:R928–35.
40. Molino NM, Bilotkach K, Fraser DA, Ren D, Wang SW. Complement Activation and cell uptake responses toward polymer-functionalized protein nanocapsules. Biomacromolecules. 2012;13:974–81.
41. Moros M, Hernáez B, Garet E, Dias JT, Sáez B, Grazú V, González-Fernández A, Alonso C, de la Fuente JM. Monosaccharides versus PEG-functionalized NPs: influence in the cellular uptake. ACS Nano. 2012;6:1565–77.
42. Moghimi SM, Porter CJH, Muir IS, Illum L, Davis SS. Non-phagocytic uptake of intravenously injected microspheres in rat spleen: influence of particle size and hydrophilic coating. Biochem Biophys Res Commun. 1991;177:861–6.
43. Grafmueller S, Manser P, Diener L, Diener PA, Maeder-Althaus X, Maurizi L, Jochum W, Krug HF, Buerki-Thurnherr T, von Mandach U, Wick P. Bidirectional transfer study of polystyrene nanoparticles across the placental barrier in an ex vivo human placental perfusion model. Environ Health Perspect. 2015;123:1280–6.
44. Schlöndorff D, Banas B. The mesangial cell revisited: no cell is an island. J Am Soc Nephrol. 2009;20:1179–87.
45. Bronte V, Pittet MJ. The spleen in local and systemic regulation of immunity. Immunity. 2013;39:806–18.
46. SPHERO fluorescent particles. http://www.spherotech.com/fluorescent-particlescatalog2010–2011reva.pdf. Accessed 12 August 2014.

11

Observation of yttrium oxide nanoparticles in cabbage (*Brassica oleracea*) through dual energy K-edge subtraction imaging

Yunyun Chen[1], Carlos Sanchez[2], Yuan Yue[1], Mauricio de Almeida[3], Jorge M. González[3], Dilworth Y. Parkinson[4] and Hong Liang[1,2*]

Abstract

Background: The potential transfer of engineered nanoparticles (ENPs) from plants into the food chain has raised widespread concerns. In order to investigate the effects of ENPs on plants, young cabbage plants (*Brassica oleracea*) were exposed to a hydroponic system containing yttrium oxide (yttria) ENPs. The objective of this study was to reveal the impacts of NPs on plants by using K-edge subtraction imaging technique.

Results: Using synchrotron dual-energy X-ray micro-tomography with K-edge subtraction technique, we studied the uptake, accumulation, distribution and concentration mapping of yttria ENPs in cabbage plants. It was found that yttria ENPs were uptaken by the cabbage roots but did not effectively transferred and mobilized through the cabbage stem and leaves. This could be due to the accumulation of yttria ENPs blocked at primary-lateral-root junction. Instead, non-yttria minerals were found in the xylem vessels of roots and stem.

Conclusions: Synchrotron dual-energy X-ray micro-tomography is an effective method to observe yttria NPs inside the cabbage plants in both whole body and microscale level. Furthermore, the blockage of a plant's roots by nanoparticles is likely the first and potentially fatal environmental effect of such type of nanoparticles.

Keywords: Synchrotron X-ray micro-tomography, K-edge subtraction imaging, Yttria nanoparticles, Cabbage, Accumulation

Background

Engineered nanoparticles (ENPs) have attracted great interests in commercial applications due to their unique physical and chemical properties [1]. Increased usage of ENPs has raised concerns in the probability of nanoparticles exposure to environment and entry to food chain [2]. The potential health and environmental impact of ENPs need to be understood [3, 4].

Plants are essential components of ecosystems and they not only provide organic molecules for energy but they can also filter air and water, removing certain contaminants [5]. Definitively, plants play a very important role in uptake and transport of ENPs in the environment [6]. Once ENPs are uptaken by plants and translocated to the food chains, they could accumulate in organisms and even cause toxicity and bio magnification [7, 8]. Nanoparticles are known to interact with plants and some of those interaction have been studied to understand their potential health and environmental impact, including quantum dots [9], zinc oxide [10], cerium oxide [11], iron oxide [12], carbon nanotubes [13], among others [14, 15]. The uptake of various ENPs by different plants was summarized in Table 1. Nanoparticles are known to stimulate morphological and physiological changes in several edible plants [16]. Hawthorne et al. noted that the mass

*Correspondence: hliang@tamu.edu
[1] Materials Science and Engineering, Texas A&M University, College Station, TX 77843-3123, USA
Full list of author information is available at the end of the article

Table 1 The uptake of different ENPs by plants

ENPs	Plants	Uptake	Ref.
$NaYF_4$:Yb, Er	Pumpkin seedlings *(Cucurbita maxima)*	Root/stem/leaf	[8]
CdSe/ZnS QDs	A. Thaliana plant	Root	[9]
ZnO	Maize *(Zea mays L.)*	Root	[10]
CeO_2	Zucchini *(Cucurbita pepo L.)*	Root/stem/leaf/flower	[11]
Fe_3O_4	Pumpkin (*Cucurbita maxima*)	Root/stem/leaf	[12]
C_{70}	Rice *(Oryza sativa L.)*	Root/stem/leaf	[13]
AuNPs	Rice *(Oryza sativa)*	Root/shoot	[14]
	Radish *(Raphanus sativus)*	Root	
	Pumpkin (*Cucurbita maxima*)	Root	
	Ryegrass *(Lolium perenne L.)*	Root/shoot	
AgNPs	Soybean *(Glycine max)*	Root/shoot	[15]
	Wheat *(Triticum aestivum)*	Root/shoot	

of Zucchini's male flowers were reduced by exposed to CeO_2 NPs [11]. Quah et al. observed the browner roots and less healthy leaves of soybean treated by AgNPs, but less effects on wheat treated under same condition [15]. Qi et al. reported that the photosynthesis in tomato leaves could be improved by treated with TiO_2 NPs at appropriate concentration [17].

Yttrium oxide (Y_2O_3, yttria) ENPs have been broadly used in optics, electrics and biological applications due to their favorable thermal stability and mechanical and chemical durability [18–20]. One of the most common commercial applications is employed as phosphors imparting red color in TV picture tubes. The environmental effects of yttria ENPs have not been reported. Even though the effects of certain NPs have been studied on several plants [14], the uptake, translocation and bioaccumulation of yttria NPs in edible cabbage (Brassicaceae, *Brassica oleracea*) have not been addressed until this study. This plant species was chosen and tested as part of a closed hydroponic system designed to study nanoparticles movement and distribution in a substrate-plant-pest system as a model of a simple and controlled environment. The final test "substrate" used was plain distilled water (to avoid NPs to attach or react with other substrate elements), in which the tested NPs were mixed.

In order to observe the translocation and distribution of ENPs in plants, transmission electron microscopy (TEM) has been one of the most commonly used techniques to identify the localization at cellular scale in two-dimensions (2D), because it can be used to observe all kinds of ENPs [21, 22]. On the other hand, ENPs with special properties, such as upconversion NPs and quantum dots with a particular band gap can be studied with a confocal microscope with alternative excitation wavelengths to trace the ENPs [8, 23]. Several synchrotron radiation imaging techniques exploiting high energy X-ray have become widely used in plant science, which can measure both spatial and chemical information simultaneously, like micro X-ray fluorescence and computed tomography [24–26].

In this research, we use synchrotron X-ray microtomography (μ-XCT) with K-edge subtraction (KES) to investigate the interaction of yttria NPs with edible cabbage. By using the KES technique, the μ-XCT can not only detect the chemical and spatial information in 3D, but also analyze the concentration of target NPs. The uptake, accumulation, and distribution mapping of yttria NPs in both micro scale and relatively full view of cabbage roots and stem were investigated. We found that yttria NPs were absorbed and accumulated in the root but not readily transferred to the cabbage stem. Compared with yttria NPs, other minerals were observed along the xylem in both cabbage roots and stem. To the best of our knowledge, few reports have studied the impact of yttria NPs on cabbage plants. In addition, by using μ-XCT with KES technique, the distribution and concentration mapping of nanoparticles in full view of plant root have not been previously reported.

Results and discussion

Physical properties of yttria nanoparticles

The yttria NPs were characterized by using TEM and XRD (Fig. 1). The mean diameter of nanotubes is 31.3 ± 8.6 nm, and the mean length is 206.3 ± 77.3 nm. The average size of irregular nanoparticles is 64.9 ± 16.9 nm (Fig. 1a). The XRD pattern of as-synthesized NPs was finely indexed to a cubic phase of yttria (JCPDS card no. 83-0927), shown in Fig. 1b. The as-calcined yttira NPs did not have further surface modification, therefore, the NPs were not water-soluble.

Identifying nanoparticles in cabbage

The μ-XCT was carried out at Beamline 8.3.2 at the advanced light source, Lawrence Berkley National Laboratory. From scanning energies of 16.5 to 17.2 keV, below and above yttrium K-edge, the X-ray attenuation coefficient sharply increases by a factor of 5. Other elements decrease slightly in their attenuation coefficients over this energy range. The localization of yttria NPs can be identified by the subtraction between two reconstructed image datasets (17.2–16.5 keV), shown in Fig. 2. The slices collected above and below the K-edge were set with same brightness and contrast settings to fairly compare with each other. The grayscale values of reconstructed slices represent the absorption coefficient; therefore, the

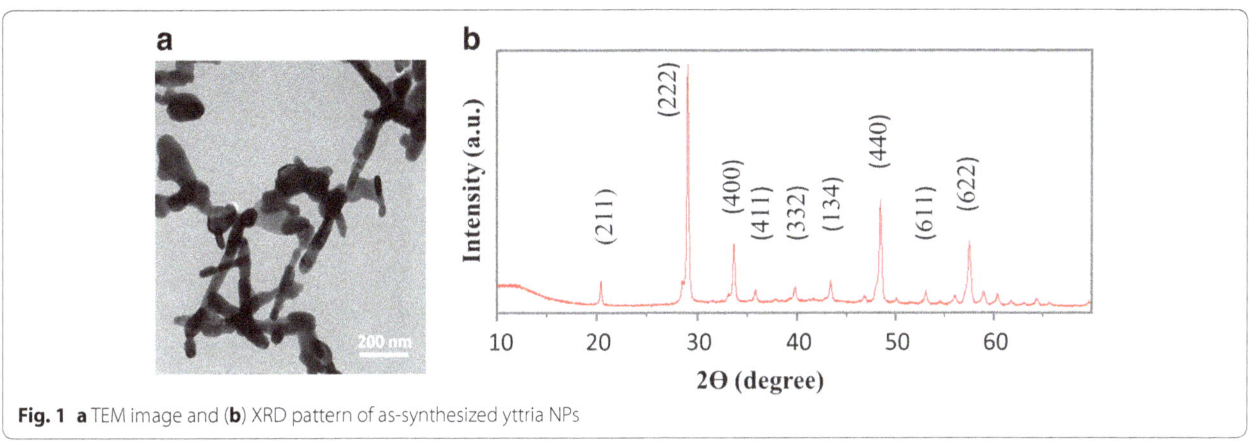

Fig. 1 **a** TEM image and (**b**) XRD pattern of as-synthesized yttria NPs

bright regions in subtracted slice denote the localization of yttria NPs (Fig. 2c arrowed). Other elements appear dark in subtracted slice marked with a red "▲" (Fig. 2f). These are inorganic elements which support the growth of cabbage. Some biological structures suffered radiation damage during scanning, resulting in a small amount of shrinkage. The bright regions circled in Fig. 2c were caused by such shrinkage, resulting in a registration mismatch between the images above and below the edge. To identify and map the distribution of yttria NPs, an image segmentation protocol was employed that could highlight regions with yttria without finding these regions corresponding to sample shrinkage. The detailed segmentation process is given in the "Method" section.

Three-dimensional distribution and quantification of nanoparticles in cabbage

By using K-edge subtracted image technique with Monochromatic X-ray tomography, the translocation and distribution of NPs in the cabbage root is clear (Fig. 3). Figure 3a and b were constructed by 17.2 keV and 16.5 keV reconstructed slice datasets, respectively. Their color maps were based on the transverse slice pixel values/absorption coefficients over the range from 0.2 to 17.8 cm^{-1}. An obvious difference between 17.2 and 16.5 keV visualization in absorption coefficient of yttria NPs was observed. The distribution of yttria NPs in root was segmented and colored in red (Fig. 3c). A large amount of NPs were found aggregated at left bottom of the root. Since yttria NPs were not water-soluble, the water that contained them was kept in constant movement with an air pump working 24/7. However, it seems that the dense roots formed a web-like structure that made the suspended NPs to accumulate and aggregate among the roots. Uptake of NPs by the root has been observed at primary and lateral root junction as well according to the transverse slice. Figure 2a is one transverse slice localized at the arrow in Fig. 3c (blue arrow) showing the junction between primary root and lateral root. We found that the yttria NPs were absorbed by the lateral roots, and particulates began to accumulate along the outer epidermis of primary roots with limited entrance into the vascular tissue (xylem and phloem) of the primary root. It might happen that endodermal cell walls were blocking the entrance of aggregated yttria NPs into vascular tissue [10]. This is shown in the upper section of the 3D visualization (Fig. 3c) where no yttria NPs were observed above the root system.

Besides the full view of the translocation in the cabbage root system, the distribution of yttria NPs at the micro-scale within a lateral root was detected and investigated (Fig. 4). Figure 4a shows the localization of the micro-scale lateral root visualization. The 3D visualization of micro-scale was built by the segmented transverse reconstructed slices, and the red regions were localized yttria NPs (Fig. 4b, c). It is clear that roots are able to uptake the yttria NPs in ground tissue (GT), which appear to accumulate in the root with limited entrance of yttria NPs into vascular tissue (VT) being transported through the xylem. Xylem vessels are small with diameters usually smaller than 1 μm in vegetables like cabbage plants to over 100 μm in vessels found in trunks of large trees [27]. Vessels allow nutrients contained in water to be distributed throughout the plant. For NPs, however, if they aggregate, the blockage is expected, that is what we have observed in this study. Long term studies might show that yttria NPs might provide more negative than positive effects on plant growth and development as found with other NPs (i.e., AuNPs, AgNPs) [16].

Using K-edge subtraction image technique with dual-energy X-ray scanning, the concentration of target NPs can be calculated. This method has been discussed elsewhere [28–30]. As attenuation coefficients of other

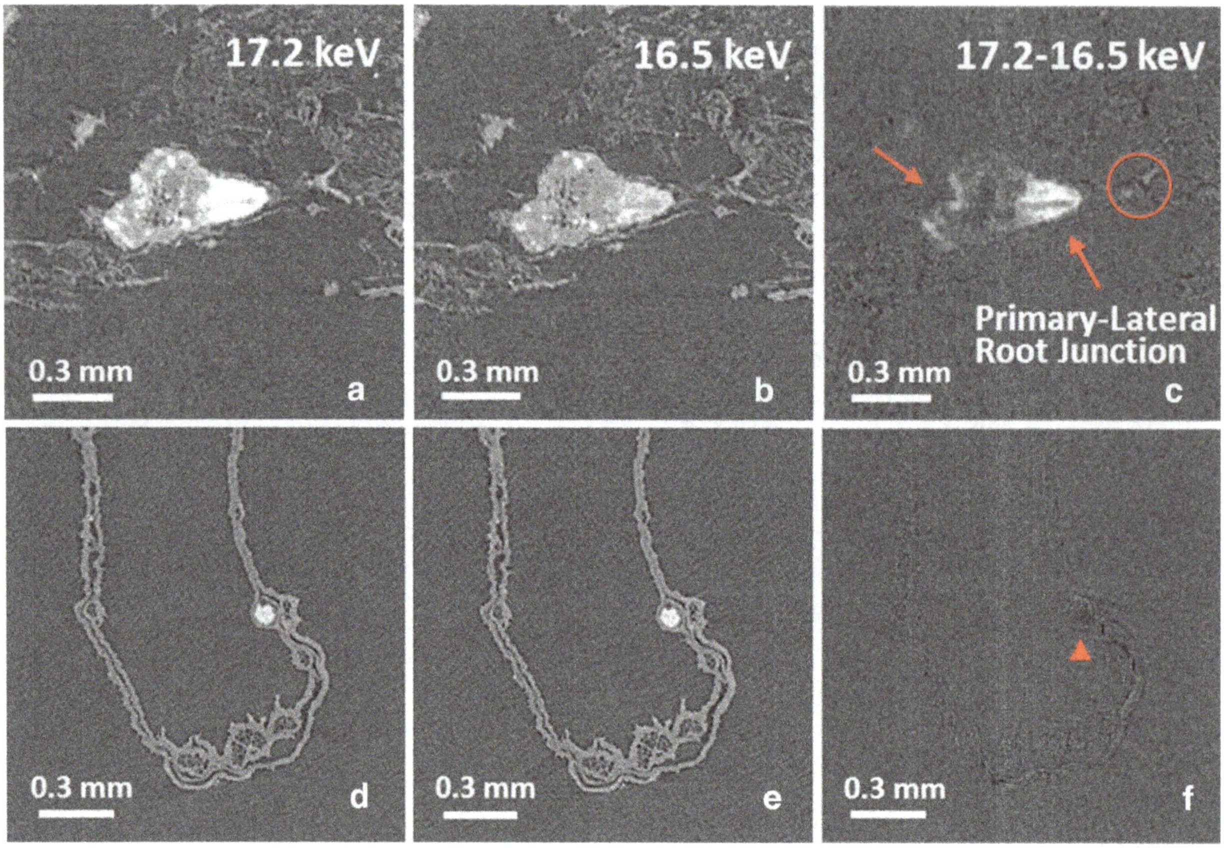

Fig. 2 Transverse reconstructed slices at the junction between primary root and lateral root scanned at (**a**) 17.2 keV, (**b**) 16.5 keV, and (**c**) subtracted slice. (**d–e**) The slices for leave section

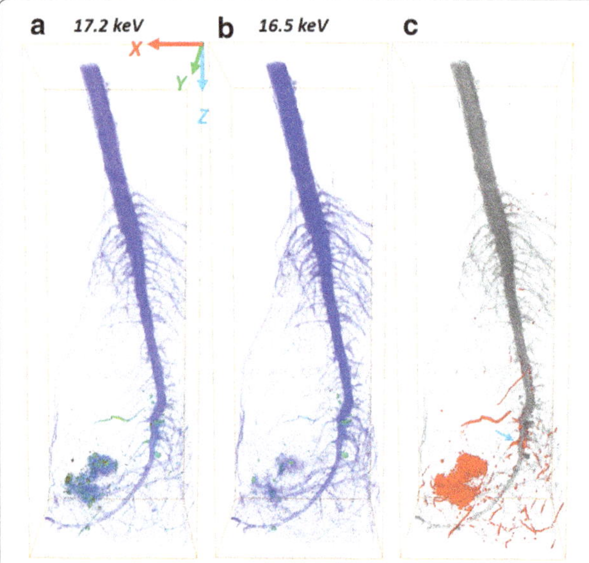

Fig. 3 The 3D visualization for a wide view of root built by (**a**) 17.2 keV transverse reconstructed slice datasets and (**b**) 16.5 keV datasets. (**c**) Distribution of yttria NPs (*red*) in root. The *grey* visualization was built by 17.2 keV; the *red* one was built by the subtracted datasets. The bounding box size is 6.77 × 5.10 × 19.40 mm

elements just have a slight decrease, the concentration (C_{NPs}) can be formulated in a simplified equation

$$C_{NPs} = \frac{\Delta\mu(x,y)}{\frac{\mu}{\rho}(17.2) - \frac{\mu}{\rho}(16.5)},$$

where Δμ is the difference in absorption coefficient obtained by subtraction between two energies, μ/ρ is the mass absorption coefficients. The value for Δμ is obtained from the voxel value of subtracted datasets, and the mass absorption coefficient is from Argonne National Labratory (Compute X-ray Absorption). The volume rendering enable the 3D visualization for the concentration map of yttria, shown in Fig. 5b. By using this formula, the calculated concentration is based on the voxel level. The minimum concentration was 44.12 mg/cc and the maximum was 551.47 mg/cc (to display the mapping colorful, the maximum set as 132.35 mg/cc). The grey visualization (setting 30 % transparent) of root shows the distribution and localization of yttria NPs. Using Avizo software with image segmentation and label-analysis, the total voxel volume of root is measured as 5.41604e + 07 voxels.

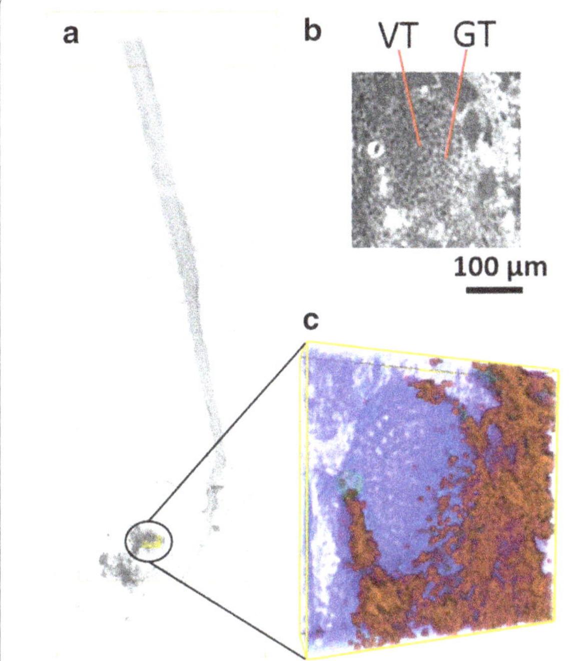

Fig. 4 The micro-scale of root segmentation localized at yttria NPs aggregated regions. **a** The selected *yellow* region in 17.2 keV visualization. **b** The top transverse slice of *yellow* frame region. **c** Magnified view of *yellow* frame. The vascular tissue (VT) and ground tissue (GT) are shown in (**b**). The *red* regions in (**c**) show the distribution of yttria NPs. The *yellow* frame size is 0.58 x 0.58 x 0.13 mm

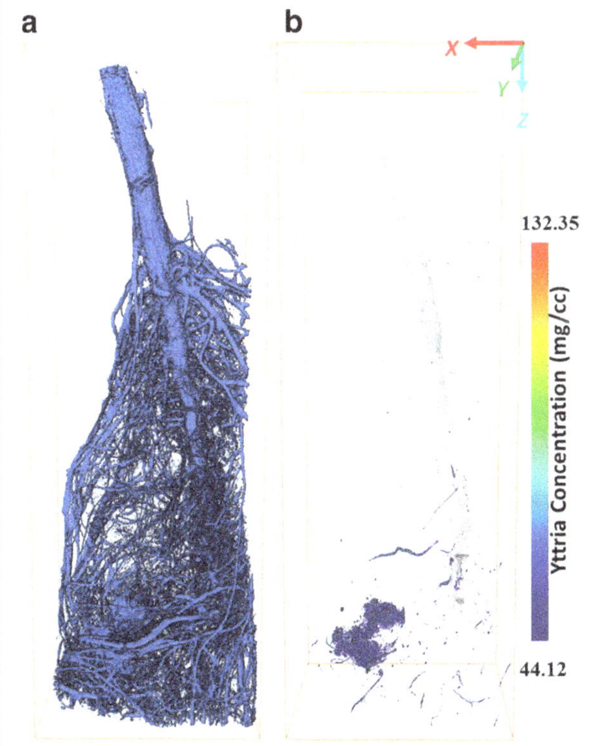

Fig. 5 (**a**) The full volume visualization of plant root. (**b**) Concentration map of yttria NPs in root (on the voxel level)

Figure 5a shows the full volume of the root section. As shown the concentration of nanoparticles at root was estimated in the range from 0.82 to 10.18 μg/L.

For the cabbage shoot, no yttria NPs were observed (Fig. 6), which means that no yttria NPs transported from roots to shoots. As we found no yttria NPs entering vascular tissues of primary root, the yttria NPs accumulated making it difficult to be transported by xylem from the root to the rest of the plant. Despite no clear evidence of yttria translocation, other elements were observed in the shoots. In general, the higher the atomic number (Z), the higher the absorption coefficient for a given X-ray energy (Fig. 2d–f). It is clearly to see some high-Z (compared with carbon) elements distributed in both roots and shoots. Crops require many mineral elements for their growth, such as calcium, magnesium, zinc, copper and iron [31, 32]. These high-Z elements could be the mineral elements absorbed by cabbage before the cabbage root exposed in the hydroponic system containing yttria NPs.

What are the possible uptake mechanisms based on the observation? As shown above, we observed that some yttria NPs were uptaken by the roots of cabbage plant (Figs. 2a–c and 4). The cell wall is considered as a tight and significant sieve which blocks the migration of NPs [33]. The typical pore sizes of a cell wall are in the range of 2–20 nm [2]. In our case, the yttria NPs sizes are larger than the pore sizes, therefore, the passage for NPs through the pores of cell walls should be difficult. On the contrary, the larger NPs were found to be taken up by roots or shoots that are in correlation with previous reports [12, 34, 35]. It is not that clear which route NPs can penetrate the cell wall for all these cases. Shen et al. reported that an endocytosis-like structure was observed in *Arabidopsis thaliana* leaf cells [36]. Therefore, the yttria NPs could penetrate the cell wall and be taken up by the roots. In addition, the dissolution rates of rare earth oxides are too low to be relevant [37]. Even though some yttria NPs penetrated into the ground tissues, the yttrium ions were not established. This could be the reason the yttria NPs only observed in ground tissue and blocked at primary-lateral-root junction.

Limitations in KES

Although a KES method can identify the localization of target NPs, if the concentration of root-to-shoot-transported yttria NPs was too low, the target NPs could not be detected. Furthermore, the KES method is based on the difference in attenuation coefficient of yttrium element over K-edge. This method is able to identify the yttrium-based NPs but it cannot distinguish the biotransformation of yttria.

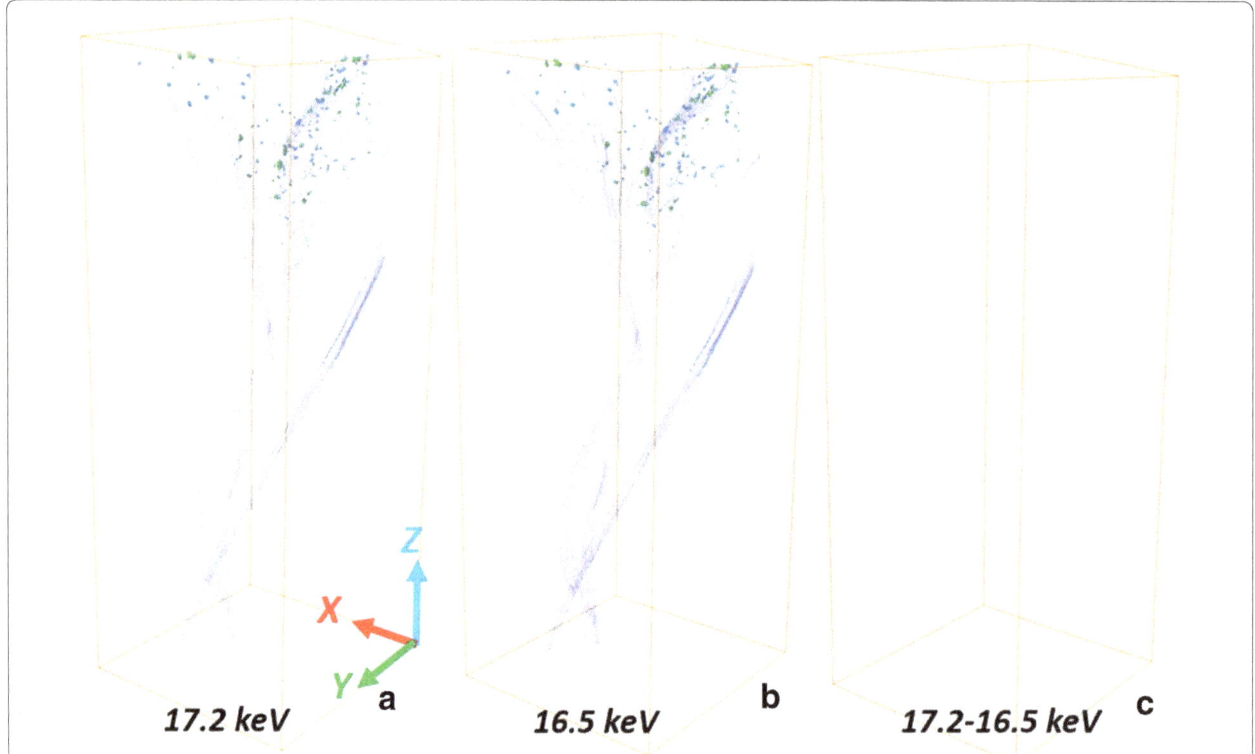

Fig. 6 The 3D visualization of shoot section for cabbage built by (**a**) 17.2 keV transverse reconstructed slice datasets and (**b**) 16.5 keV datasets. (**c**) Distribution of yttria NPs in shoot. The bounding box size is 7.65 × 7.29 × 19.40 mm

Conclusions

Synchrotron μ-XCT with KES image technique is a valid method to study the uptake, accumulation and spatial distribution mapping of yttria NPs in cabbage roots. Using the KES technique, the concentration mapping of yttria NPs was calculated and shown in 3D visualization. The yttria ENPs were uptaken by root but not found in the cabbage shoot. Instead, other non-yttria minerals were found in both cabbage root and shoot. The blockage of yttria NPs was mainly due to their accumulation at primary-lateral-root junctions.

Methods

Synthesis and characterization of nanoparticles

The Yttria nanoparticles were synthesized by using a hydrothermal method [38]. All chemicals were Sigma-Aldrich (USA). The 5.94 mmol Y_2O_3 and 0.02 mmol Al_2O_3 powders were dissolved in 250 mL HNO_3 solution (2.8 wt %) to attain a transparent solution at 60 °C, followed by adding 0.06 mmol $Er(NO_3)_3{\cdot}5H_2O$ and 0.06 mmol $Yb(NO_3)_3{\cdot}5H_2O$. By adding the 3 M KOH solution into the transparent solution, the solution pH value was adjusted to 10.5. When pH value was over 7, the white floccules were appeared. The obtained turbid solution was 900 mL. After stirred 10 min, the turbid solution was transferred to a 2 L general purpose pressure vessel and heated at 200 °C for 12 h without stirring. After cooling down to room temperature, the precipitate was attained by centrifuging at 5000 rpm for 15 min, continued by washing with DI water. The final Yttria-based NPs powders were acquired by drying the precipitate at 60 °C and heating the dried precipitate at 1000 °C for 3 h in the air.

A transmission electron microscopy (TEM, JEOL 1200 EX) was used to image the as-synthesized yttria NPs, using an accelerating voltage of 100 keV. The crystal structure of yttria NPs was measured by a Bruker-AXS D8 Advanced Bragg–Brentano X-ray powder diffractometer (XRD) operated at 40 mA and 40 kV with with Cu Kα radiation ($\lambda = 1.5418$ Å).

Cabbage culture and exposure to nanoparticles

Cabbage plants were reared in a hydroponic system as shown in Fig. 7. Seeds of cabbage were placed in 38 mm compressed (100 % peat) plugs and placed in a hydroponic mix containing water to which a 2-1-2 (NPK)

solution (118 mL per 20 gallons of water) was added every week. Once plants had four true leaves, they were extracted from the main culture system, cleaned and placed into two groups. One group was placed in a glass jar (1 pint) containing distilled water and yttria NPs (10 plants per jar), the other was placed in only distilled water (10 plants per jar; as control). The 0.120 g NPs were added to distilled water in a small Nalgene container, mixed with a mini vortexer, and then added to the distilled water up to 0.38 L in the final testing glass jar. All jars had an air pump in them which were running 24/7. The distilled water inside the glass jars containing NPs were kept in movement with the air pump working 24/7. NPs did not form conglomerates in the hydroponic testing system. The "substrate" used was plain distilled water (to avoid NPs to attach or react with other substrate elements), in which the tested NPs were mixed. Even though both groups showed clear sign of stress after 10 days, they were maintained in this system for a total of 22 days. About 30 % of the plants tested (with and without NPs) were wilted, the plants that were in better shape were collected, cleaned thoroughly with distilled water, dried and fixed with Kahle's, a fixing agent that provides sharp and clear preservation of nuclear structure of plant or animal tissues. Once received for imaging, plants were extracted from the container and let dried before placing them in the Synchrotron X-ray micro-tomography equipment.

Fig. 7 Hydroponic system for cabbage exposure to yttria NPs

Synchrotron X-ray micro-tomography

Synchrotron X-ray computed micro-tomography was conducted at Advanced Light Source beamline 8.3.2 facility, Lawrence Berkeley National Laboratory. Monochromatic X-ray at 16.5 and 17.2 keV were employed with calibration for transmission of yttria NPs at approximately 67 and 15 %, respectively. Radiographs were acquired by using a LuAG scintillator, 2× optical lens, and PCO_Edge scientific CMOS camera, yielding a pixel size of 0.00319 mm. The cabbage specimen was irradiated with 200 ms exposure time per frame and rotated over 180° with 512 projections. The datasets were reconstructed by (Fig. 8) using a Fourier method implemented in the commercial Octopus package and further processed using ImageJ. To investigate the translocation of NPs in cabbage, the three-dimensional (3D) visualization was built with Avizo software (FEI).

Image segmentation

The image segmentation was carried out by Avizo software to identify and display the distribution of NPs in 3D visualization. Figure 8 takes a root section image as an example to show the procedure and changes of segmentation. Figure 8a and b were the transverse slices scanning at 17.2 and 16.5 keV. Though the dual-energy slice datasets were scanned at the same anatomic location, the slight shrinkage and shift of the biological structure could take place during the hard X-ray radiation. Image registration was firstly employed to compensate for such shift and obtain the better quality of subtracted slices. Figure 8c and d are the subtracted slices obtained without and with image registration, respectively. The 17.2 keV reconstructed datasets were thresholded with the pixel value 3.8 corresponding to the Fig. 9 at marker "X", with count of 17.2 keV datasets more than that of 16.5 keV (Threshold A). The subtracted datasets were thresholded with three according to the Fig. 8e, as the pixel values less than 3 (light and deep blue labels) could be caused by the organic plant body or the noise (Threshold C). All pixel values above the threshold were labeled as 1, with candidate NPs; whereas non-labeled areas were set as 0. The shift due to the sample motion or shrink during scanning can be identified by the regions of increased darkness adjacent

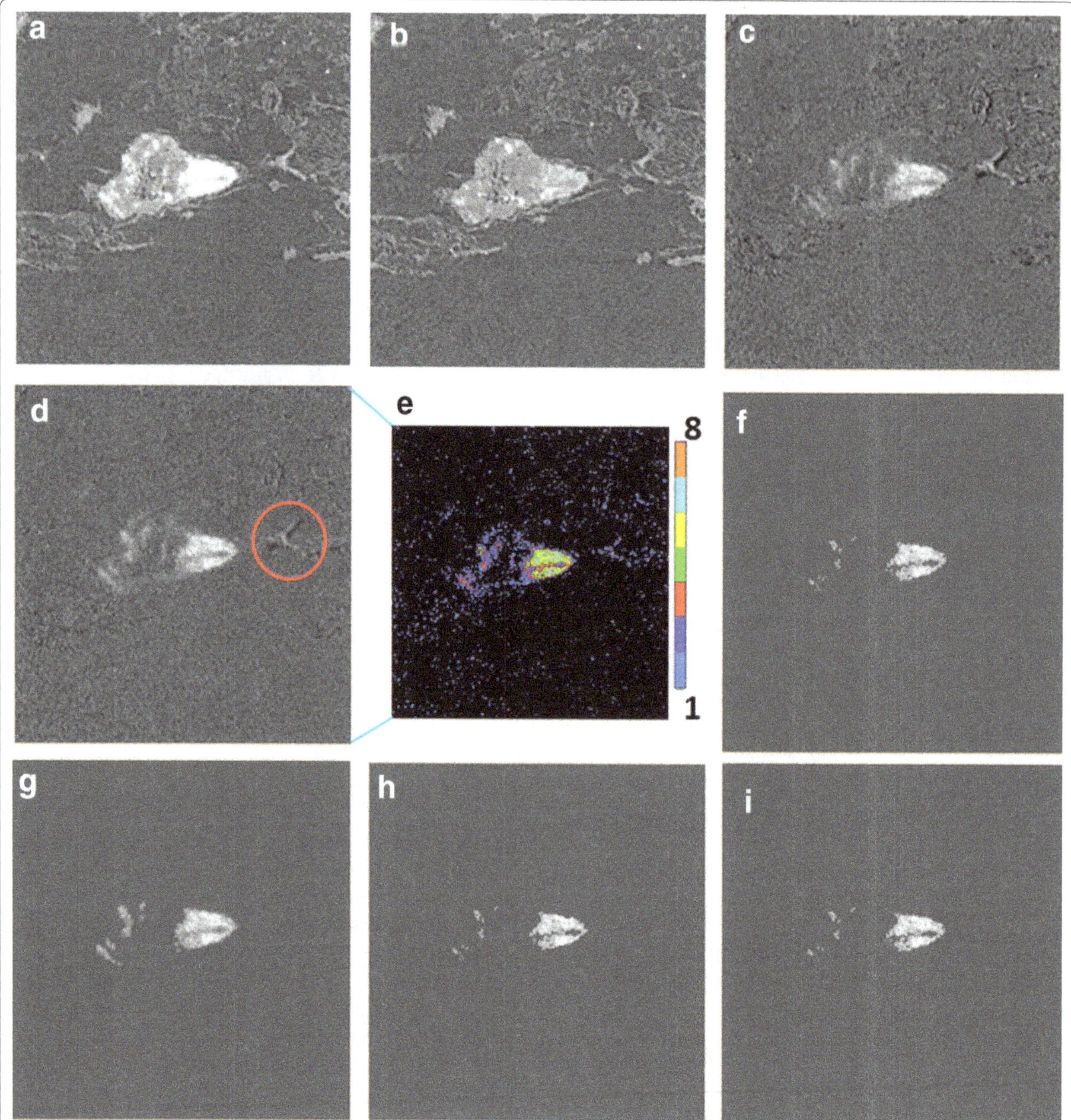

Fig. 8 The process of image segmentation. **a** Reconstructed image obtained 17.2 keV. **b** Reconstructed image at 16.5 keV. **c** The subtracted slice from (**a**) and (**b**). **d** Subtracted slice with an image registion fucntion. **e** Colors lable the intensity of pixel value of (**d**). **f–i** Image segmentation based on the threshold of (**a**, **b**). Detailed data of images (**f–i**) is listed in Table 2

to regions of increased brightness (Fig. 8d circled). The darkness regions (pixel value less than 0) will be selected and dilated in 3D with 26 adjacent voxels. The dilation regions were labeled as one (Threshold D). Figure 8f was derived by arithmetic with Threshold A, C and D as Table 2 shows. Figure 8g is the dilation of Fig. 8f with 26 adjacent voxels in 3D. Figure 8h was obtained by removing the pixel value over three in Fig. 8g (Table 2). The

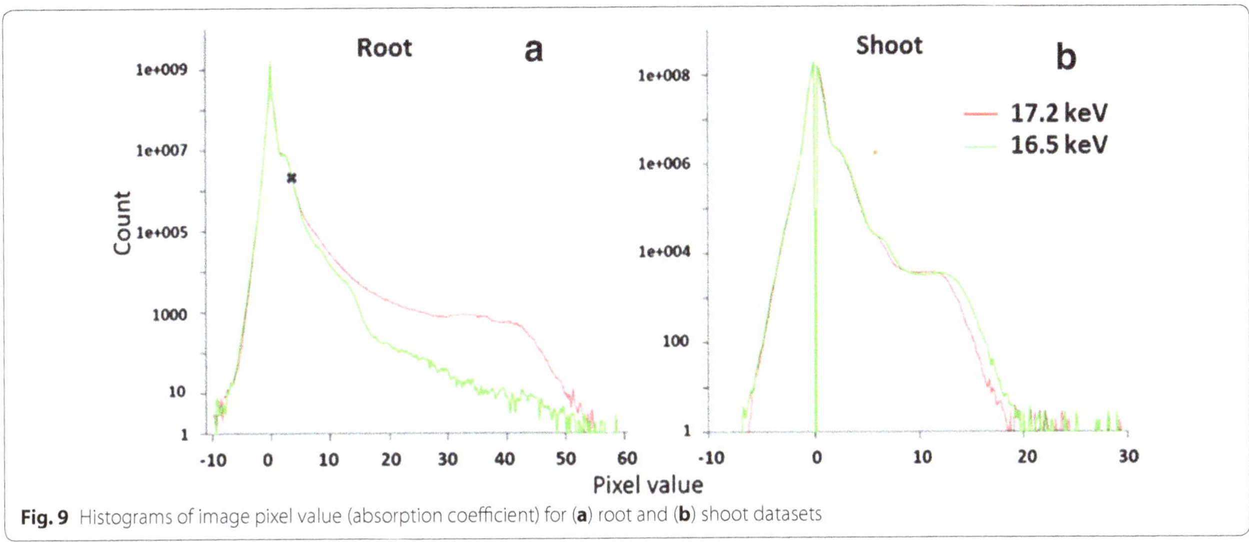

Fig. 9 Histograms of image pixel value (absorption coefficient) for (**a**) root and (**b**) shoot datasets

Table 2 The characterizations for the sub-Fig. 8

	Characterization
Figure 8a	17.2 keV; Threshold A: pixel value no less than 3.8; Threshold B: no less than 12
Figure 8b	16.5 keV; Threshold C: no less than 3; Threshold D: dilation for less than 0; Threshold E: between 3 and 4
Figure 8c	17.2–16.5 keV without registration
Figure 8d	17.2–16.5 keV with image registration
Figure 8e	Labeled color map due to pixel value
Figure 8f	Threshold A × C−D
Figure 8g	Dilation of Fig. 8f
Figure 8h	Figure 8g × Threshold C
Figure 8i	Figure 8h−(Threshold B × C)

final segment was derived via subtracting the non-yttria regions which were generated by the shifting of high-Z elements (Fig. 8i). The arithmetic was with Threshold B, E and Fig. 8h (Table 2).

Abbreviations

ENPs: engineered nanoparticles; TEM: transmission electron microscopy; Y_2O_3/Yttria: yttrium oxide; 2D: two-dimensional; 3D: three-dimensional; μ-XCT: X-ray micro-tomography; KES: K-edge subtraction; XRD: X-ray diffractometer; GT: ground tissue; VT: vascular tissue.

Authors' contributions

YYC, CS, YY, MdA and JMG conducted experiments. YYC and JMG analyzed data. DYP and HL designed experiments. YYC, JMG, DYP, and HL wrote the paper. All authors reviewed the manuscript. All authors read and approved the final manuscript.

Author details

[1] Materials Science and Engineering, Texas A&M University, College Station, TX 77843-3123, USA. [2] Mechanical Engineering, Texas A&M University, College Station, TX 77843-3123, USA. [3] Department of Plant Science, California State University, Fresno, CA 93740, USA. [4] Advanced Light Source, Lawrence Berkeley National Laboratory, Berkeley, CA 94720, USA.

Acknowledgements

YYC was partially sponsored by the ALS fellowship. JMG and MA were supported by the Provost's Assigned Time for Research (Summer 2015), California State University Fresno, Research, Scholarship and Creative proposal Awarded (2014–2015) and the CSUF Provost's undergraduate Research Grant (2014–2015). The Advanced Light Source is supported by the Director, Office of Science, Office of Basic Energy Sciences, of the U.S. Department of Energy under Contract No. DE-AC02-05CH11231.

Competing interests

The authors declare that they have no competing interests.

References

1. Kundu S, Wang K, Liang H. Size-controlled synthesis and self-assembly of silver nanoparticles within a minute using microwave irradiation. J Phys Chem C. 2009;113:134–41.
2. Rico CM, et al. Interaction of nanoparticles with edible plants and their possible implications in the food chain. J Agr Food Chem. 2011;59:3485–98.

3. Ma XM, et al. Interactions between engineered nanoparticles (ENPs) and plants: phytotoxicity, uptake and accumulation. Sci Total Environ. 2014;481:635.
4. Zhou Y, et al. Antibacterial activities of gold and silver nanoparticles against *Escherichia coli* and bacillus Calmette-Guerin. J Nanobiotechnol. 2012;10:19.
5. Wolverton BC, Mcdonald RC, Watkins EA. Foliage plants for removing indoor air-pollutants from energy-efficient homes. Econ Bot. 1984;38:224–8.
6. Ma XM, et al. Interactions between engineered nanoparticles (ENPs) and plants: phytotoxicity, uptake and accumulation. Sci Total Environ. 2010;408:3053–61.
7. Khan AG, et al. Role of plants, mycorrhizae and phytochelators in heavy metal contaminated land remediation. Chemosphere. 2000;41:197–207.
8. Nordmann J, et al. In vivo analysis of the size- and time-dependent uptake of NaYF4:Yb, Er upconversion nanocrystals by pumpkin seedlings. J Mater Chem B. 2015;3:144–50.
9. Navarro DA, Bisson MA, Aga DS. Investigating uptake of water-dispersible CdSe/ZnS quantum dot nanoparticles by *Arabidopsis thaliana* plants. J Hazard Mater. 2012;211:427–35.
10. Lv JT, et al. Accumulation, speciation and uptake pathway of ZnO nanoparticles in maize. Environ-Sci Nano. 2015;2:68–77.
11. Hawthorne J, et al. Particle-size dependent accumulation and trophic transfer of cerium oxide through a terrestrial food chain. Environ Sci Technol. 2014;48:13102–9.
12. Zhu H, et al. Uptake, translocation, and accumulation of manufactured iron oxide nanoparticles by pumpkin plants. J Environ Monitor. 2008;10:713–7.
13. Lin SJ, et al. Uptake, translocation, and transmission of carbon nanomaterials in rice plants. Small. 2009;5:1128–32.
14. Zhu ZJ, et al. Effect of surface charge on the uptake and distribution of gold nanoparticles in four plant species. Environ Sci Technol. 2012;46:12391–8.
15. Quah B, et al. Phytotoxicity, uptake, and accumulation of silver with different particle sizes and chemical forms. J Nanopart Res. 2015;17:1–13.
16. Siddiqui MH et al. Role of Nanoparticles in plants, in nanotechnology and plant sciences. Berlin: Springer; 2015. p. 19–35.
17. Qi M, Liu Y, Li T. Nano-TiO2 improve the photosynthesis of tomato leaves under mild heat stress. Biol Trace Elem Res. 2013;156:323–8.
18. Srinivasan R, et al. Structural and optical properties of europium doped yttrium oxide nanoparticles for phosphor applications. J Alloy Compd. 2010;496:472–7.
19. Li RB, et al. Surface interactions with compartmentalized cellular phosphates explain rare earth oxide nanoparticle hazard and provide opportunities for safer design. ACS Nano. 2014;8:1771–83.
20. Wu G, et al. Rolling up the sheet: constructing metal-organic lamellae and nanotubes from a [{Mn3(propanediolato)2}(dicyanamide)2]n Honeycomb Skeleton. J Am Chem Soc. 2013;135:18276–9.
21. Zhang DQ, et al. Uptake and accumulation of CuO nanoparticles and CdS/ZnS quantum dot nanoparticles by Schoenoplectus tabernaemontani in hydroponic mesocosms. Ecol Eng. 2014;70:114–23.
22. Palomo-Siguero M, et al. Accumulation and biotransformation of chitosan-modified selenium nanoparticles in exposed radish (Raphanus sativus). J Anal Atom Spectrom. 2015;30:1237–44.
23. Wang J, et al. Uptake, translocation, and transformation of quantum dots with cationic versus anionic coatings by populus deltoides x nigra cuttings. Environ Sci Technol. 2014;48:6754–62.
24. Lombi E, Scheckel KG, Kempson IM. In situ analysis of metal(loid)s in plants: state of the art and artefacts. Environ Exp Bot. 2011;72:3–17.
25. Tappero R, et al. Hyperaccumulator Alyssum murale relies on a different metal storage mechanism for cobalt than for nickel. New Phytol. 2007;175:641–54.
26. Zhu Y, et al. Synchrotron-based X-ray microscopic studies for bioeffects of nanomaterials. Nanomed-Nanotechnol. 2014;10:515–24.
27. Zach A, et al. Vessel diameter and xylem hydraulic conductivity increase with tree height in tropical rainforest trees in Sulawesi, Indonesia. Flora. 2010;205:506–12.
28. Dilmanian FA, et al. Single- and dual-energy CT with monochromatic synchrotron X-rays. Phys Med Biol. 1997;42:371–87.
29. Cooper DML, et al. Three dimensional mapping of strontium in bone by dual energy K-edge subtraction imaging. Phys Med Biol. 2012;57:5777–86.
30. Kruger RA, et al. Digital K-Edge subtraction radiography. Radiology. 1977;125:243–5.
31. White PJ. Greenwood, properties and management of cationic elements for crop growth. Soil Cond Plant Growth. 2013;19:160–94.
32. White PJ, Broadley MR. Biofortification of crops with seven mineral elements often lacking in human diets—iron, zinc, copper, calcium, magnesium, selenium and iodine. New Phytol. 2009;182:49–84.
33. Birbaum K, et al. No evidence for cerium dioxide nanoparticle translocation in maize plants. Environ Sci Technol. 2010;44:8718–23.
34. Corredor E, et al. Nanoparticle penetration and transport in living pumpkin plants: in situ subcellular identification. BMC Plant Biol. 2009;9:1–11.
35. Majumdar S, et al. Exposure of cerium oxide nanoparticles to kidney bean shows disturbance in the plant defense mechanisms. J Hazard Mater. 2014;278:279–87.
36. Shen C-X, et al. Induction of programmed cell death in *Arabidopsis* and rice by single-wall carbon nanotubes. Am J Bot. 2010;97:1602–9.
37. Takaya M, et al. Dissolution of functional materials and rare earth oxides into pseudo alveolar fluid. Ind Health. 2006;44:639–44.
38. He XL, Zhou Y, Liang H. Cun + -assisted synthesis of multi- and single-phase yttrium oxide nanosheets. J Mater Chem C. 2013;1:6829–34.

12

The agglomeration state of nanoparticles can influence the mechanism of their cellular internalisation

Blanka Halamoda-Kenzaoui, Mara Ceridono, Patricia Urbán, Alessia Bogni, Jessica Ponti, Sabrina Gioria and Agnieszka Kinsner-Ovaskainen*

Abstract

Background: Significant progress of nanotechnology, including in particular biomedical and pharmaceutical applications, has resulted in a high number of studies describing the biological effects of nanomaterials. Moreover, a determination of so-called "critical quality attributes", that is specific physicochemical properties of nanomaterials triggering the observed biological response, has been recognised as crucial for the evaluation and design of novel safe and efficacious therapeutics. In the context of in vitro studies, a thorough physicochemical characterisation of nanoparticles (NPs), also in the biological medium, is necessary to allow a correlation with a cellular response. Following this concept, we examined whether the main and frequently reported characteristics of NPs such as size and the agglomeration state can influence the level and the mechanism of NP cellular internalization.

Results: We employed fluorescently-labelled 30 and 80 nm silicon dioxide NPs, both in agglomerated and non-agglomerated form. Using flow cytometry, transmission electron microscopy, the inhibitors of endocytosis and gene silencing we determined the most probable routes of cellular uptake for each form of tested silica NPs. We observed differences in cellular uptake depending on the size and the agglomeration state of NPs. Caveolae-mediated endocytosis was implicated particularly in the internalisation of well dispersed silica NPs but with an increase of the agglomeration state of NPs a combination of endocytic pathways with a predominant role of macropinocytosis was noted.

Conclusions: We demonstrated that the agglomeration state of NPs is an important factor influencing the level of cell uptake and the mechanism of endocytosis of silica NPs.

Keywords: Silica nanoparticles, Cell uptake, Endocytosis route, Agglomeration/aggregation, In vitro

Background

A detailed understanding of the mechanisms of interaction between engineered nanoparticles (NPs) and biological systems is essential to properly assess the safety of newly developed nanotechnological and nanomedicinal products. Upon exposure, NPs may interact with the outer surface of the cellular membrane and subsequently enter the cells by different endocytic routes. Elucidation of the mechanism by which NPs are internalized into the cells can provide insights about the intracellular trafficking, fate and cytotoxic profile of NPs [1]. Targeting of specific cellular structures, the release of NPs by the cells or, on the contrary, their degradation in lysosomes are all key features that can significantly affect the NPs toxicity/safety, but also the efficacy of novel nanomedicines.

All these processes are highly dependent on the mechanism of NP internalisation [2, 3]. The involvement of various endocytic pathways in NP uptake has been investigated in several in vitro studies, and the obtained results show wide variability with respect to cell lines and nanoparticles used [4–6].

It is well recognised that an accurate characterisation of NPs in the biological medium and under the conditions

*Correspondence: agnieszka.kinsner-ovaskainen@ec.europa.eu
European Commission Joint Research Centre, Directorate for Health, Consumers and Reference Materials, Via E. Fermi 2749, TP 127, 21027 Ispra, VA, Italy

of an in vitro study is necessary to properly interpret the results and to enable correlation of the physicochemical parameters with the biological response [7]. Indeed, NP properties such as size, shape, surface chemistry and charge were shown to affect the level of cellular uptake and the mechanism of the endocytosis [8–11]. However up to date, little attention has been paid to the question of how this process is influenced by the agglomeration state of NPs. Though, the ability to agglomerate is one of the predominant features of a NP suspension. Changes to the pH and ionic strength or the presence of biomolecules, particularly proteins, can easily modify the NP surface properties, leading to the loss of colloidal stability and formation of agglomerates. That is why this study focused on understanding how the agglomeration state of NPs can influence the endocytic mechanism by which NPs can enter the cells. We have observed in our previous study [7] that silicon dioxide NPs showed a time-dependent tendency to agglomerate in the complete medium of CaCo-2 cells. Therefore, in this study we used the same amorphous fluorescent Rubipy-SiO_2 NPs of two different sizes, 30 and 80 nm that were either freshly dispersed in complete medium and added to the cells, or pre-incubated with the complete medium for 24 h in order to allow the agglomeration. Subsequently, for the agglomerated and non-agglomerated NPs of both sizes we compared the level of cellular uptake and the endocytosis route in Caco-2 cells. We used flow cytometry to quantify the cellular uptake, transmission electron microscopy (TEM) to visualise the NPs inside the cells and a high-throughput fluorescence microscopy technique to colocalise NPs with clathrin, caveolin-1 and SNX5 antibodies. We also performed blocking of different endocytic routes by chemical inhibitors and by gene silencing to assess the effect on NP internalisation.

Methods

Nanoparticle synthesis and characterization

The mono-dispersed fluorescent particles of silicon dioxide 30 and 80 nm labelled with Rubipy were synthetized as described previously [7]. Physicochemical characterization of the NPs including size distribution, zeta potential and fluorescence spectrum carried out in water, PBS and cell culture medium was already described [4]. Additionally, the size distribution of Rubipy-SiO_2 NPs was determined in complete CaCo-2 cell culture medium by centrifugal liquid sedimentation (CLS) in a sucrose density gradient using a CLS Disc Centrifuge model DC24000UHR (CPS Instruments Europe, Netherlands). The samples were prepared as a 1 mg/ml suspension of Rubipy-SiO_2 NPs in cell culture media (CCM) containing 10% of foetal bovine serum, and the measurements were performed either immediately or after 24 h incubation at 37 °C. In addition, the particle size in complete CaCo-2 cell culture medium was evaluated by TEM. Ultrathin Formvar-coated 200-mesh copper grids (Tedpella Inc.) were first functionalized by placing the carbon-coated side on a drop of 20 µl of Alcian blue (2% in water) deposited on a Parafilm. After 10 min incubation the grid was washed 5 times by the deposition on the drops of water placed on a Parafilm and the excess fluid was removed by blotting its edge on a strip of paper tissue, leaving a rest of humidity. Finally the grid was placed on a 20 µl drop of the corresponding sample, incubated for 10 min and the excess of fluid was removed again with a paper tissue. TEM (JEOL JEM 2100, Japan) at an accelerating voltage of 200 kV was used to visualize the nanoparticles.

Before the biological experiments, Rubipy-SiO_2 NPs were purified in centrifugal filters (Amicon Ultra, 10K, Milipore, Italy) at 3000 rpm for 10 min at room temperature (RT) and re-suspended in sterile PBS. The fluorescence spectrum of Rubipy-SiO_2 NPs before and after the centrifugation was compared and used for the calculation of the NP concentration. The fluorescence intensities of different concentrations of Rubipy-SiO_2 NPs in PBS and in complete CaCo-2 medium without phenol red were obtained using fluorescence spectrophotometer (Cary Eclipse, Varian, Australia Pty Ltd). The scan of the fluorescence emission between 500 and 700 nm was carried out at excitation 460 nm.

Cell lines and culture conditions

Colorectal adenocarcinoma CaCo-2 cells (ACC169, DSMZ, Braunschweig, Germany) were cultured under standard cell culture conditions in Dulbecco's modified Eagle's medium with high glucose (4.5 g/L), supplemented with 10% heat inactivated Foetal Bovine Serum (North American origin), 1% penicillin–streptomycin, 4 mM L-glutamine and 1% non-essential amino acids. All cell culture reagents were purchased from Thermo Fisher Scientific, Italy. The experiments were performed between passages 1–10 after defrosting of cells from liquid nitrogen storage.

Internalisation of endocytic markers

Cells were seeded in a 24-well plate (Costar, Italy) at a density of 6×10^4 cells/well, grown until 80–90% confluence, washed, and after serum starvation for 2 h, pre-incubated for 30 min with endocytosis inhibitors: chlorpromazine 50 µM, dynasore 80 µM, methyl-beta-cyclodextrin 5 mM, nystatin 40 µg/ml, genistein 200 µM, or EIPA 75 µM (all from Sigma-Aldrich, Italy). The following endocytic markers were then added to the cells for 30 min: Transferrin Alexa Fluor 488 conjugate 50 µg/ml (clathrin-mediated endocytosis), Bodipy-FL C5-Lactosylceramide (LaCer) complexed to bovine serum albumin (BSA) 1 μg/ml (caveolae-mediated endocytosis), and Alexa Fluor 488-labeled Dextran 10 kDa, 200 μg/ml (macropinocytosis) (all from Thermo Fisher Scientific, Italy).

After exposure the cells were washed twice in cold washing buffer (Hepes 10 mM, NaN_3 5 mM) and the remaining membrane bound label was removed according to previously published methods [12, 13]. Briefly, 1 min incubation in ice-cold acidic wash (0.2 M acetic acid, 0.2 M NaCl, pH 2.0), followed by two washes in cold washing buffer was employed to remove surface-bound Transferrin. For the LaCer uptake the "back exchange" procedure was applied, consisting on 1 h washing with 5% fatty acid-free BSA in washing buffer at 4 °C changing for fresh BSA every 10 min. After the washing procedure, the cells were detached by trypsinisation and analysed immediately by flow cytometry.

Analysis of uptake of Rubipy-SiO_2 NPs by flow cytometry

Cells were seeded in a 24-well plate (Costar) at a density of 6×10^4 cells/well, grown until 80–90% confluence and exposed to 200 μg/ml of 30 and 80 nm Rubipy-SiO_2 NPs freshly dispersed in complete medium or pre-incubated in complete medium for 24 h at 37 °C. Depending on the purpose of the assay the exposure was done at 37 °C or at 4 °C. For the study of endocytic pathways before the exposure to Rubipy-SiO_2 NPs, the cells were pre-incubated for 30 min with endocytosis inhibitors, as described above. Following the exposure to Rubipy-SiO_2 NPs the cells were washed 3 times in PBS, trypsinised and blocked with a complete cell culture medium, washed again in PBS and analysed immediately by flow cytometry.

Evaluation of cell associated fluorescence, forward scattering (FSC) and side scattering (SSC) were carried out using CyFlow space flow cytometer (Partec, Munster, Germany) and the data were analysed using FCS Express 4 software (De Novo, Los Angeles, CA). Laser excitation was 488 nm and emission bandpass wavelength was 590/50 nm for Rubipy-SiO_2 NPs related fluorescence. A minimum of 15,000 cells per sample were analysed; cells debris, nanoparticles and doublets were excluded from the analysis by gating on the FSC versus SSC log graph and on the FL-2 area versus FL-2 width graph, respectively. The median cell associated fluorescence after the subtraction of cell autofluorescence was averaged between three independent experiments (2 replicas each). The results were then normalised according to the reference values of fluorescence intensity of 30 and 80 nm Rubipy-SiO_2 NPs to allow the comparison of cell uptake of both sizes of NPs. The efficiency of inhibitors is calculated as a percentage of the uptake by the control cells without any inhibitor.

Analysis of uptake of Rubipy-SiO_2 NPs by transmission electron microscopy

Following 3 and 6 h of exposure to 200 μg/ml of 30 and 80 nm Rubipy-SiO_2 NPs freshly dispersed in complete medium or pre-incubated in complete medium for 24 h at 37 °C, the cells were washed 3 times in PBS, trypsinised, blocked with a complete cell culture medium and washed again 3 times in PBS by centrifugation (200×*g*, 5 min). The supernatant was discarded and the cells were fixed using a Karnovsky 2% v/v solution (glutaraldehyde and paraformaldehyde in 0.05 M cacodylate, pH 7.3, Sigma-Aldrich, Italy) over night. Cells were then washed 3 times with 0.05 M cacodylate, pH 7.3 and post-fixed in osmium tetroxide solution in 0.1 M cacodylate, pH 7.3 (both from Sigma-Aldrich, Italy) for 1 h. After 3 washes in cacodylate 0.05 M of 10 min each, cells were dehydrated in a graded series of ethanol solution in MilliQ water (30; 50; 75; 95% for 15 min each, and 100% for 30 min), incubated in absolute propylene oxide (Sigma-Aldrich, Italy) for 20 min (2 changes of 10 min each) and embedded in a solution of 1:1 epoxy resin (Sigma-Aldrich, Italy) and propylene oxide for 90 min. This mixture was renewed with pure epoxy resin (Sigma-Aldrich, Italy) over night at room temperature and later polymerized at 60 °C for 48 h. Ultrathin sections (60–90 nm) were obtained using Leica UCT ultramicrotome (Leica, Italy) and stained with uranyl acetate for 25 min and lead citrate for 20 min (both from Sigma-Aldrich, Italy), washed and dried. Ultrathin sections were imaged using JEOL-JEM 2100 TEM (JEOL, Milan, Italy) at 120 kV.

Colocalisation with endocytic proteins

Cells were seeded in a 96-well plate (black, glass bottom, Greiner Bio-one) at a density of 1×10^4 cells/well and grown until 80–90% confluence was reached. Afterwards the cells were exposed to 200 μg/ml of 30 or 80 nm Rubipy-SiO_2 NPs freshly dispersed in complete medium or pre-incubated in complete medium for 24 h at 37 °C. After 3 h exposure the cells were washed 3 times in PBS, fixed with 3.7% paraformaldehyde in PBS, permeabilised with cold methanol at −20 °C and blocked for 30 min at RT with 3% BSA in PBS. Overnight incubation at 4 °C with primary antibodies: either anti-clathrin heavy chain (Abcam, rabbit polyclonal, 1:2000), anti-caveolin-1 (Abcam, rabbit polyclonal, 1:800) or anti-SNX5 (Abcam, goat polyclonal, 1:90) was followed by 45 min incubation at 37 °C with secondary antibodies: anti-rabbit IgG (Invitrogen, goat, Cy® 5 conjugate, 1:300), or anti-goat IgG (Invitrogen, rabbit, Alexa Fluor® 647, 1:300) and DAPI (Molecular Probes, 2,5 μg/ml final concentration). Cells were imaged using IN Cell Analyzer 2200 (GE Healthcare). During acquisition, a minimum of 20 fields per well were imaged using a 60× objective. The representative images were selected, DAPI, Cy3 and Cy5 channels were merged and the qualitative analysis of colocalisation between images from channel Cy3 (Rubipy-SiO_2 NPs) and from channel Cy5 (secondary antibodies) was performed using ImageJ software.

Transfection of cells with si-RNA

The cells were seeded in a 24-well plate (Costar) at a density of 5×10^4 cells/well and cultured for 24 h to be 30–40% confluent on the day of transfection. The si-RNA procedure was performed according to the manufacturer's protocol using Silencer® Select validated si-RNA and Lipofectamine™ RNAiMAX at the concentrations selected from the optimization assay: 5 nM si-RNA (CAV1 silencing, Ambion, s2446), or 8 nM si-RNA (PAK1 silencing, Ambion, s10019) and 1 µl Lipofectamine/well. In parallel, a negative control si-RNA was used at the same concentration. Both si-RNA and Lipofectamine were diluted in Opti-MEM® I Reduced Serum Medium, combined, mixed gently and incubated for 15 min at RT before being added drop-wise to the wells. All reagents were purchased from Thermo Fisher Scientific, Italy. After 24 h incubation the medium in the wells was exchanged with the standard cell culture medium containing 10% of serum and no antibiotics. 48 h after transfection the silencing efficiency was tested using TaqMan® gene expression assay and the protein expression was evaluated by western blot method. The cell uptake of Rubipy-SiO_2 NPs was quantified by flow cytometry after 3 and 6 h exposure in cells transfected with silencing-RNA in comparison to the cells transfected with negative control si-RNA.

Real-time PCR amplification

Following transfection cells were washed with cold PBS and lysed in 350 µl RLT lysis buffer (Qiagen, Germantown, MD, USA). Samples were stored at −80 °C until RNA extraction was carried out. RNA was extracted from cells and purified using the RNeasy Plus kit (Qiagen, Germantown, MD, USA). The RNA quantification was done by a ND-1000 UV–Vis Spectrophotometer (NanoDrop Technologies), and the RNA integrity was assessed with the Agilent 2100 Bioanalyzer (Agilent), according to the manufacturer's instructions. All RNA samples used in this study had a 260/280 ratio above 1.9 and a RNA Integrity Number (RIN) above 9.0.

1 µg of total RNA was reverse transcribed using the high-capacity cDNA Reverse Transcription Kit (Thermo Fisher Scientific, Italy), following the manufacturer's protocol. Real-Time PCR was performed with a total of 10 ng of cDNA for each reaction, using TaqMan® gene expression assay for CAV1 (Hs00971716_m1, Applied Biosystems), for PAK1 (Hs00945621_m1, Applied Biosystems) for GAPDH as a negative control (Hs02758991_g1, Applied Biosystems) and TaqMan® Universal PCR Master Mix (Applied Biosystems). The reaction was performed on Applied Biosystems 7900HT Fast Real-Time PCR System, with standard mode thermal cycling conditions, according to the manufacturer's instructions. Analysis of real-time PCR data to quantify gene expression level was done by comparative Ct methods [14].

Western blot

Following transfection (48 or 72 h) or exposure to Rubipy-SiO_2 NPs (6 h), cells were rinsed twice with cold PBS and incubated with ice-cold lysis buffer (20 mM Tris–HCl, 100 mM NaCl and 1 mM EDTA, 0.5% Triton X-100) containing protease inhibitors cocktail (all from Sigma-Aldrich, Italy). After 10 min incubation on ice to ensure complete lysis, cell lysates were scrapped and centrifuged at 18,000×*g* for 15 min at 4 °C. The supernatant containing the cytoplasmatic protein fraction was transferred to a new tube. Protein concentration was measured by Bicinchoninic acid assay (BCA kit, Sigma-Aldrich, Italy). Equal amount of protein extracts (20 µg) were loaded onto a 10% SDS–polyacrylamide gel electrophoresis (SDS-PAGE) (Mini-PROTEAN® BIORAD). Separated proteins were transferred to a methanol-activated Hybond-P membrane (Amersham Biosciences, USA) (Mini Trans-Blot BIORAD®). The PVDF membrane was probed with a primary rabbit polyclonal antibody against clathrin heavy chain (Abcam, 1:1000), anti-caveolin-1 (Abcam, 1:800), anti-PAK1 (Prestige Antibodies, Sigma-Aldrich, 1:250), anti-SNX5 (Abcam, 1:1000) or anti-GAPDH (Millipore Cat MAB374, Italy, 1:7500) as loading control. The membrane was then incubated with the secondary anti-rabbit (Sanzta-Cruz, 1:5000) or anti-mouse (Zymax antibodies, 1:3000) antibodies IgG-horseradish peroxidase-conjugated and detected by enhanced chemiluminescence (ECL, Amersham Biosciences, USA).

Fluorescence microscopy

CaCo-2 cells were seeded at a density of 10^5 cells/well in 4-chamber slides (Falcon), grown for 24 h and left untreated or incubated with chlorpromazine 50 µM, dynasore 80 µM, methyl-beta-cyclodextrin 5 mM, nystatin 40 µg/ml, genistein 200 µM, or EIPA 75 µM for 1 h at 37 °C. To investigate the energy dependence of NP uptake, CaCo-2 cells were exposed to 200 µg/ml of 30 and 80 nm-sized fluorescent Rubipy-SiO_2 NPs for 3 h at 37 or 4 °C in complete cell culture medium.

Following exposure, cells were washed 3 times in PBS, fixed with 4% (v/v) paraformaldehyde in PBS and permeabilised with 0.1% (v/v) Triton X-100 in PBS (Sigma-Aldrich, Italy) before staining with AlexaFluor 488-conjugated Phalloidin (Thermo Fisher Scientific,

Italy), diluted 1:100 for 40 min at RT. The nuclei were counterstained with the Hoechst 33342 dye (Dako, Italy). After staining, the cells were washed in PBS and mounted for microscopy. Images were acquired with an Axiovert 200 M inverted microscope equipped with a ApoTome slide module and Axiovision 4.8 software (Carl Zeiss; Jena, Germany), using a 40×/1.0 objective lens.

Evaluation of cell metabolic activity (MTT assay)

Cells were grown in 96-well cell culture plates (Costar) until 75% confluent, exposed to Rubipy-SiO_2 NPs for 48 h or to chemical inhibitors for 3.5 h and then washed in PBS. Cell viability was evaluated using MTT [3-(4,5-dimethylthiazol-2-yl)-2,5-diphenyl-2H tetrazolium bromide] (Sigma-Aldrich, Italy) added to the cells in fresh complete culture medium at a 250 µg/ml final concentration. After 2 h of incubation at 37 °C the supernatant was removed, the precipitated formazan crystals were dissolved in 0.1 M HCl in propan-2-ol and the absorbance was quantified at 540 nm in a multiwell plate reader (FluoStar, Omega, BMG Labtech, Offenburg, Germany). In parallel, to evaluate the possibility of interference of NPs with the assay, the PBS washing containing the silica NPs residues from each well was transferred to empty wells, incubated with MTT reagent in the conditions of the experiment and after 2 h the absorbance at 540 nm was read in a multiwell plate reader.

Results

Characterization of the size distribution and agglomeration state of Rubipy-SiO_2 NPs

Amorphous, negatively charged fluorescent Rubipy-SiO_2 NPs of 30 and 80 nm were synthetized and characterized in water, PBS and cell culture medium as described previously [7]. The size distribution of Rubipy-SiO_2 NPs in the complete CaCo-2 medium was measured by CLS immediately after preparing the NP suspension and after 24 h incubation at 37 °C (Fig. 1a; Table 1). In case of freshly prepared NP suspensions we observed a narrow size distribution of 80 nm NPs and a slightly larger peak of 30 nm NPs, indicating the initiation of the agglomeration already at this point. After 24 h incubation in the complete medium the size distribution has become much larger, and the average size of the particles similar for both types of Rubipy-SiO_2 NPs. Moreover, visual inspection of both suspensions indicated agglomeration, and precipitation was visible to the naked eye.

The CLS results were confirmed by TEM images (Fig. 1b) showing in fresh suspensions well dispersed 80 nm NPs and the 30 nm NPs initiating to agglomerate, and in suspensions incubated for 24 h the agglomerates of both sizes of Rubipy-SiO_2 NPs together with the clumps of the proteins.

Therefore, taking into account the difference in agglomeration state and in order to facilitate the comprehension of the manuscript we will refer to the NPs incubated for 24 h in the complete medium as "agglomerated", and to the NPs freshly suspended in the complete medium as "non-agglomerated", even if NP agglomeration had already started in freshly prepared suspensions, particularly in those of 30 nm.

Fluorescence characteristics of Rubipy-SiO_2 NPs were described in our previous study [7]. To ensure that the agglomeration process is not interfering with fluorescence measurement we carried out a scan of the fluorescence emission of both types of Rubipy-SiO_2 NPs freshly dispersed in the complete Caco-2 medium and after 24 h incubation at 37 °C. No major changes in the fluorescence signal were noted with the increase of agglomeration state (Additional file 1: Figure S1), however the background fluorescence of the cellular medium was interfering slightly with the measurements. Since the intensity of fluorescence emission of 80 nm NPs was higher than the one of 30 nm NPs at the same mass concentration, we performed a calibration curve of the fluorescence intensity versus the mass concentration for each size of Rubipy-SiO_2 NPs (Additional file 1: Figure S2) and we calculated the reference values that enabled the comparison of the cellular uptake of both sizes of Rubipy-SiO_2 NPs (Additional file 1: Table S1).

Cell uptake of agglomerated and non-agglomerated of Rubipy-SiO_2 NPs

Rubipy-SiO_2 NPs did not induce any toxicity to CaCo-2 cells for up to 48 h, as assessed by MTT assay (Additional file 1: Figure S3). For the quantification of the cellular uptake the cells were exposed to Rubipy-SiO_2 NPs pre-incubated in the complete medium for 24 h and, in parallel, to Rubipy-SiO_2 NPs freshly added to the complete medium, at the same concentration (200 µg/ml) and for the same exposure time (3 h). Measurements of the cell-associated fluorescence by flow cytometry, normalized to the fluorescence of NPs per mass unit, indicated a much higher cell uptake of the agglomerated form, particularly of 80 nm Rubipy-SiO_2 NPs, than of the non-agglomerated form of NPs (Fig. 2).

TEM images were obtained after 3 and 6 h exposure to Rubipy-SiO_2 NPs and showed an increased presence of NPs inside the cells after 6 h comparing to 3 h exposure (not shown). Both the agglomerated and non-agglomerated forms of 30 nm and the agglomerated 80 nm Rubipy-SiO_2 NPs were observed as clusters, either interacting with the cellular membrane or in big endosomal vacuoles inside the cells (Fig. 3), whereas non-agglomerated 80 nm NPs were observed as small groups or single particles, thus showing a difference in the agglomeration

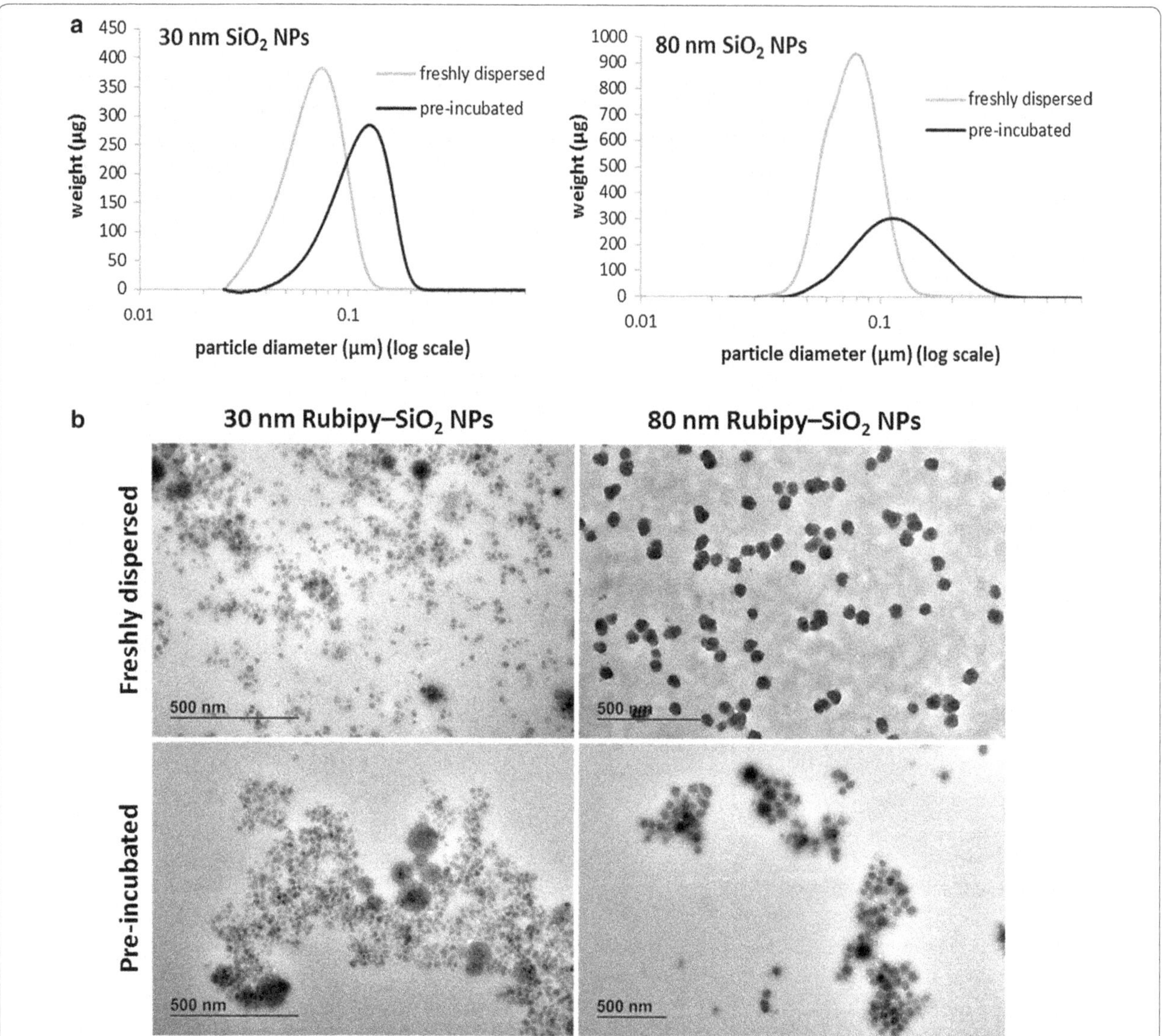

Fig. 1 Size distribution of Rubipy-SiO_2 NPs in complete cell culture medium. Rubipy-SiO_2 NPs were suspended in CaCo-2 complete cell culture medium (10% of serum) at concentration of 1 mg/ml and the size distribution was evaluated either immediately or after 24 h pre-incubation at 37 °C by CLS (**a**) and by TEM (**b**)

Table 1 Average size and polydispersity of Rubipy-SiO_2 NPs in water and complete cell culture medium measured by CLS (by weight), reported in nm

	H_2O	Complete CaCo-2 medium (immediately)	Complete CaCo-2 medium (24 h incubation)
30 nm Rubipy-SiO_2 NPs	26.1 ± 11	84.5 ± 64	129 ± 92
80 nm Rubipy-SiO_2 NPs	68.8 ± 22	75 ± 49	116 ± 114

state also at that level. The presence of many large clusters of 80 nm agglomerated Rubipy-SiO_2 NPs inside the cells was particularly striking. Beside the large vacuoles, NPs were also observed in early and late endosomes (sometimes interestingly in flask-shaped endosomes) and lysosomes. In many areas and under all conditions it was possible to see characteristic ruffles of the plasma membrane involved in the macropinocytotic process (Fig. 3) or smaller invaginations of the membrane associated with clathrin- or caveolae-mediated endocytosis.

The evaluation of the kinetics of cellular uptake of Rubipy-SiO_2 NPs was performed after 1, 3 and 5 h exposure at 37 and at 4 °C using flow cytometry (Additional file 1: Figure S4A). At 37 °C the interaction of the agglomerated 80 nm NPs with the cells was very intense up to 3 h and then reached a plateau, suggesting saturation, whereas the cell uptake of 30 nm Rubipy-SiO_2 NPs was

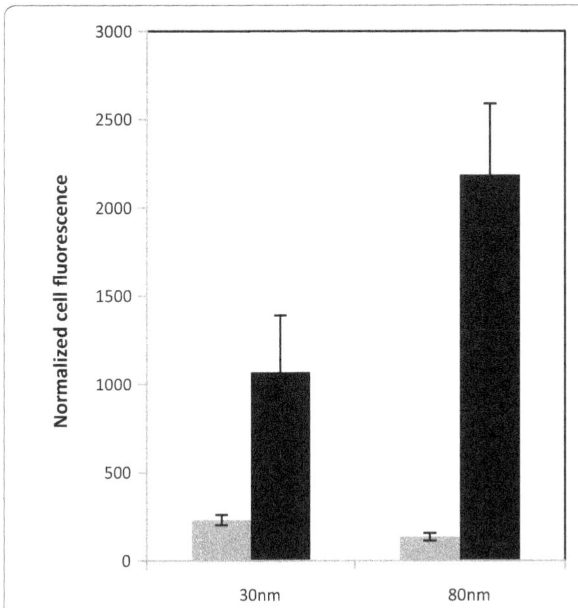

Fig. 2 Effect of agglomeration state of Rubipy-SiO_2 NPs on the level of cell uptake. The cellular uptake of Rubipy-SiO_2 NPs was quantified by flow cytometry after 3 h exposure of CaCo-2 cells to 200 µg/ml of 30 and 80 nm Rubipy-SiO_2 NPs either added to complete medium immediately before exposure (*grey bars*) or pre-incubated in complete medium for 24 h at 37 °C (*black bars*)

increasing proportionally to the time of exposure. Experiments carried out at 4 °C showed a strong inhibition of the cell uptake of Rubipy-SiO_2 NPs indicating that the mechanism of the internalization was energy-dependent (Additional file 1: Figure S4 A, B). Still, a slight cell-associated fluorescence was detected in these conditions, though not increasing in time, suggesting that NPs were interacting with the cell membrane also at 4 °C.

Effect of chemical inhibitors on cell uptake of Rubipy-SiO_2 NPs

The use of chemical inhibition of the endocytic pathways, however widespread, is frequently criticised for lack of specificity, toxicity and multiple side effects [15, 16]. Consequently, a careful evaluation of the inhibitors' effects on different endocytic pathways and for each tested cell line must be undertaken before actual experiments. Here, we used the classical markers of the endocytic pathways to assess the efficacy and the specificity of the employed inhibitors (Additional file 1: Figure S5): transferrin for clathrin-mediated endocytosis (CME), Bodipy-Lactosylceramide complexed to BSA (LaCer) for caveolae-mediated pathway and Dextran 10 kDa for macropinocytosis. We also evaluated the effect of the inhibitors on cellular morphology and on cell metabolism (Additional file 1: Figures S6, S7) to guarantee the absence of toxic effects on the cells in the range of the concentrations used in the study.

Fluorescence of the cells exposed to Rubipy-SiO_2 NPs after the treatment with the endocytosis inhibitors was compared to the fluorescence of the cells not pre-treated with the inhibitors. The inhibitors of CME did not decrease the uptake of Rubipy-SiO_2 NPs, except a slight but significant effect of chlorpromazine on the uptake of agglomerated NPs (Fig. 4). This inhibitor was shown to act strongly and specifically on the CME (Additional file 1: Figure S5). Methyl-β-cyclodextrin (MβCD), which is inducing cholesterol depletion in the lipid rafts, blocked the cell uptake of Rubipy-SiO_2 NPs almost completely and was the most potent among the tested inhibitors. However, it is not a specific inhibitor, acting on both caveolae-mediated endocytosis and macropinocytosis pathway (Additional file 1: Figure S5). Nystatin, a specific inhibitor of caveolae-mediated pathway was more efficient for non-agglomerated form of NPs (both 30 and 80 nm, ~40–50% remaining cell uptake) than for their agglomerated form (~60% remaining cell uptake). The opposite effect was observed after the treatment with genistein, which reduced the uptake of agglomerated 30 nm Rubipy-SiO_2 NPs by almost 80%, while of non-agglomerated form of these NPs only by around 20%. However, in our experiments with endocytosis markers we observed the lack of specificity of this inhibitor, since it was acting on all three tested pathways (Additional file 1: Figure S5). Similarly, also ethyl-isopropyl amiloride (EIPA), frequently used to inhibit macropinocytosis, was shown to be non-specific. In our study, EIPA demonstrated a higher efficacy in the reduction of cell uptake of 30 nm NPs than 80 nm NPs, no matter what their agglomeration state was (Fig. 4).

Colocalisation study and protein expression

In the next step we investigated if Rubipy-SiO_2 NPs internalized by the cells were colocalising with the proteins involved in different endocytic pathways: clathrin (for CME), caveolin-1 (CAV1, for caveolae-mediated endocytosis) and sorting nexin 5 (SNX5) involved in macropinocytosis. After 3 h exposure of Caco-2 cells to Rubipy-SiO_2 NPs we observed that, independently of their size and agglomeration state, NPs colocalised to a great extent with CAV1 (Fig. 5), even if the presence of 80 nm non-agglomerated NPs was rather modest. We have seen also some colocalisation with clathrin, mainly for 30 nm NPs, but at much lower degree than with CAV1 (Fig. 6). All forms of Rubipy-SiO_2 NPs except non-agglomerated 80 nm NPs colocalised well also with SNX5 confirming the implication of macropinocytosis in the internalization process of agglomerated NPs (Fig. 7). However, the expression of proteins in exposed cells was not in complete alignment with these results (Additional file 1:

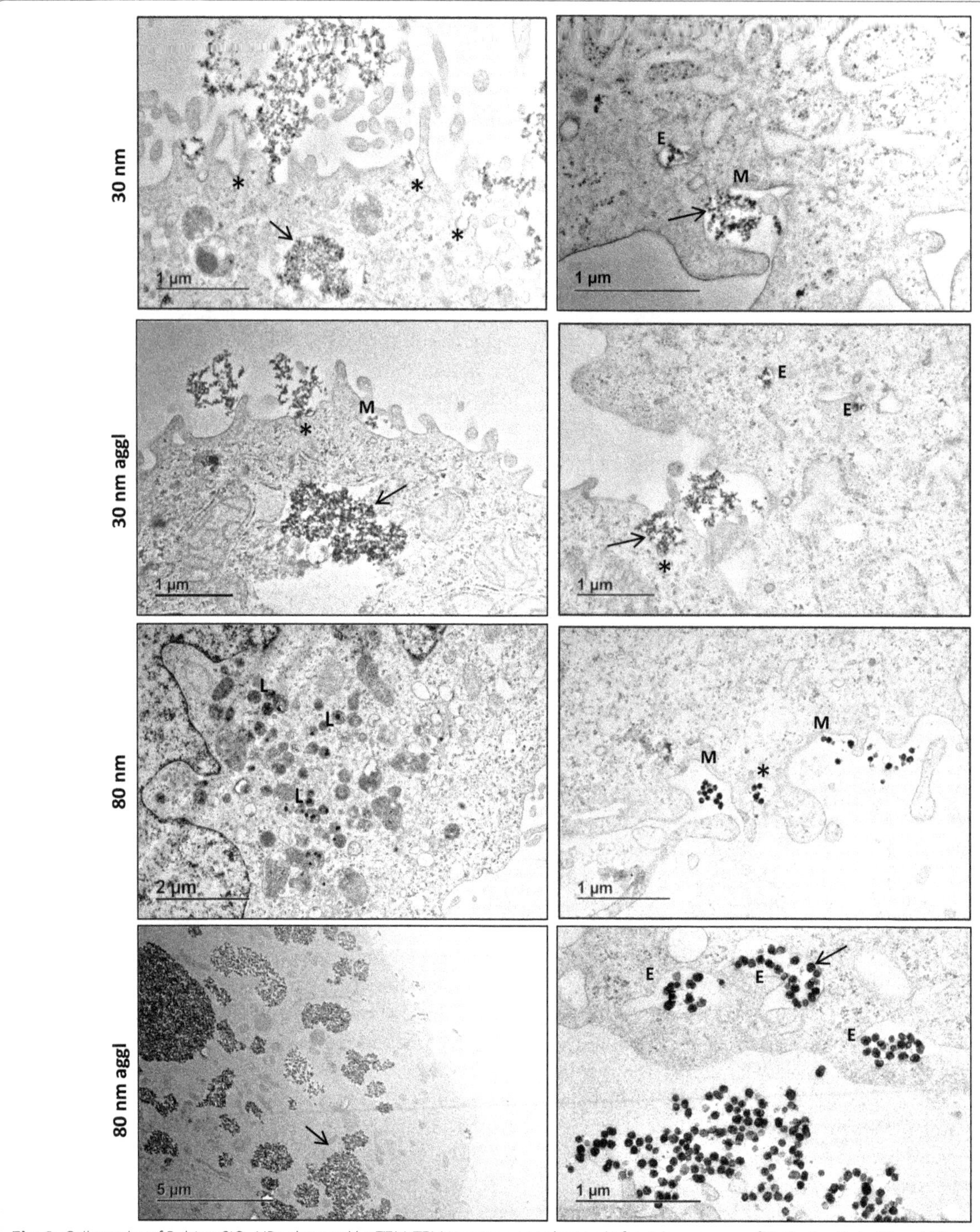

Fig. 3 Cell uptake of Rubipy-SiO_2 NPs observed by TEM. TEM images were obtained after 6 h exposure of CaCo-2 cells to 30 and 80 nm Rubipy-SiO_2 NPs, either freshly added to cell medium or pre-incubated in the complete medium for 24 h (agglomerates); on the images we can observe clusters of NPs (*black arrows*), endosomes (*E*) and lysosomes (*L*) containing NPs, membrane ruffling typical for macropinocytosis (M), membrane invaginations (*Asterisk*). Note the flask-shaped endosomes on the *right-bottom image* (80 nm aggl)

Figure S8), showing only very slight increase of CAV1 in cells exposed to 80 nm Rubipy-SiO_2 NPs and of SNX5 in all treated cells. The expression of clathrin-HC was slightly decreased after the exposure to 80 nm agglomerated NPs, whereas the expression of PAK1 was at the same level in all tested conditions.

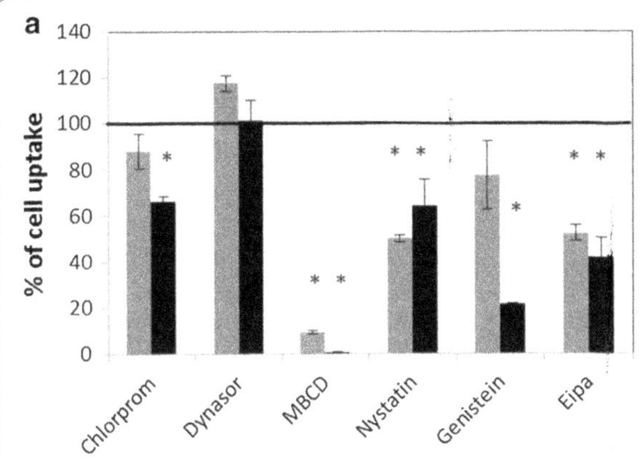

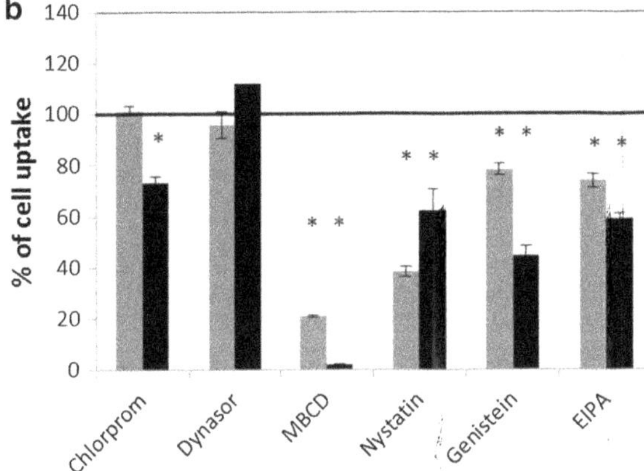

Fig. 4 Inhibition of cell uptake of Rubipy-SiO_2 NPs by endocytosis inhibitors. CaCo-2 cells were treated with the endocytosis inhibitors for 30 min, then exposed to 30 nm (**a**) and 80 nm (**b**) Rubipy-SiO_2 NPs either freshly added to cell medium (*grey bars*) or pre-incubated in the complete medium for 24 h (*black bars*). The cellular uptake of NPs was quantified after 3 h exposure by flow cytometry and calculated as a percentage of the cell uptake in absence of the inhibitors, evaluated in parallel. Statistical significance was assessed by a Student's *t* test. *$p < 0.05$, treated cells compared to untreated cells

Gene silencing

The results of the studies using chemical inhibitors and of the colocalisation assay suggested that mainly two pathways are implicated in the cellular uptake of Rubipy-SiO_2 NPs: caveolae-mediated endocytosis and macropinocytosis. In order to focus on these pathways using a more specific method we employed silencing-RNA (si-RNA) targeting sequences of CAV1 and of p21-activated kinase (PAK1) implicated in macropinocytosis.

The cells were first transfected with si-RNA targeting CAV1 (si-cav1) and with a si-RNA nonsense sequence used as a negative control (si-ctrl). 48 h after transfection RNA levels of the targeted gene were measured by the real-time PCR and the proteins expression was assessed by Western Blot. Despite a high rate of gene knock-down (>80%) (Fig. 8a) the depletion of CAV1 at the protein level was not very successful (Fig. 8b) since the protein was still present in the transfected cells, probably due to efficient recycling. Consequently caveolae-mediated internalization of LaCer was only slightly reduced compared to non-silenced cells (Fig. 8c). We evaluated the uptake of Rubipy-SiO_2 NPs by the cells transfected with si-cav1 in comparison to the cells transfected with si-ctrl. However, we could observe only a slight reduction in the level of cell-associated Rubipy-SiO_2 NPs after 3 h exposure and after 6 h exposure (Fig. 8d, e), suggesting that the partial depletion of CAV1 was not sufficient to effectively inhibit the endocytosis of NPs.

The same procedure was used to evaluate the effect of PAK1 gene silencing. Here, 48 h after transfection the cells were demonstrating ~70% decrease in gene expression (Fig. 9a), and the depletion of PAK1 protein was not complete, but the upper band in the western blot was strongly reduced (Fig. 9b). Yet, this reduction was not sufficient to block the basal uptake of Dextran, which is usually used as a marker of the macropinocytosis (Fig. 9c). Indeed, depending on the cell line, incubation time and choice of analytical tool the internalisation of dextran was shown to be associated not only with macropinocytic process but also with CME, phagocytosis and other mechanisms [17, 18]. The uptake of the agglomerated 80 nm Rubipy-SiO_2 NPs was significantly reduced in PAK1-silenced cells after 3 h exposure already, and after 6 h exposure the reduction of uptake was observed also for non-agglomerated 80 nm NPs and both forms of 30 nm NPs (Fig. 9d).

Discussion

Silica NPs are finding multiple applications in nanomedicine and industry, as they are inexpensive, biocompatible, stable and easy to functionalize [19, 20]. In the area of drug delivery [21], but also food contact materials [22], the intestinal tract could be the first physiological barrier exposed to new nano-formulations. Even if perfectly stable in pristine conditions silica NPs can easily loose the colloidal stability in a new environment or in the presence of biomolecules. Here, we investigated how the agglomeration of silica NPs can impact on the interaction of NPs with the living cells. Our first observation was that the agglomeration of Rubipy-SiO_2 NPs led to a major enhancement of their uptake by CaCo-2 cells, and that both agglomeration and cellular internalization were more pronounced for 80 nm than for 30 nm NPs. Similar findings, related to the increased cell uptake of agglomerated NPs, were already reported and explained by the modified kinetics of the deposition of the agglomerates on the cell layer compared to the monodispersed particles [23, 24].

However, in our study the agglomerates of silica NPs demonstrated not only modified kinetics but also a

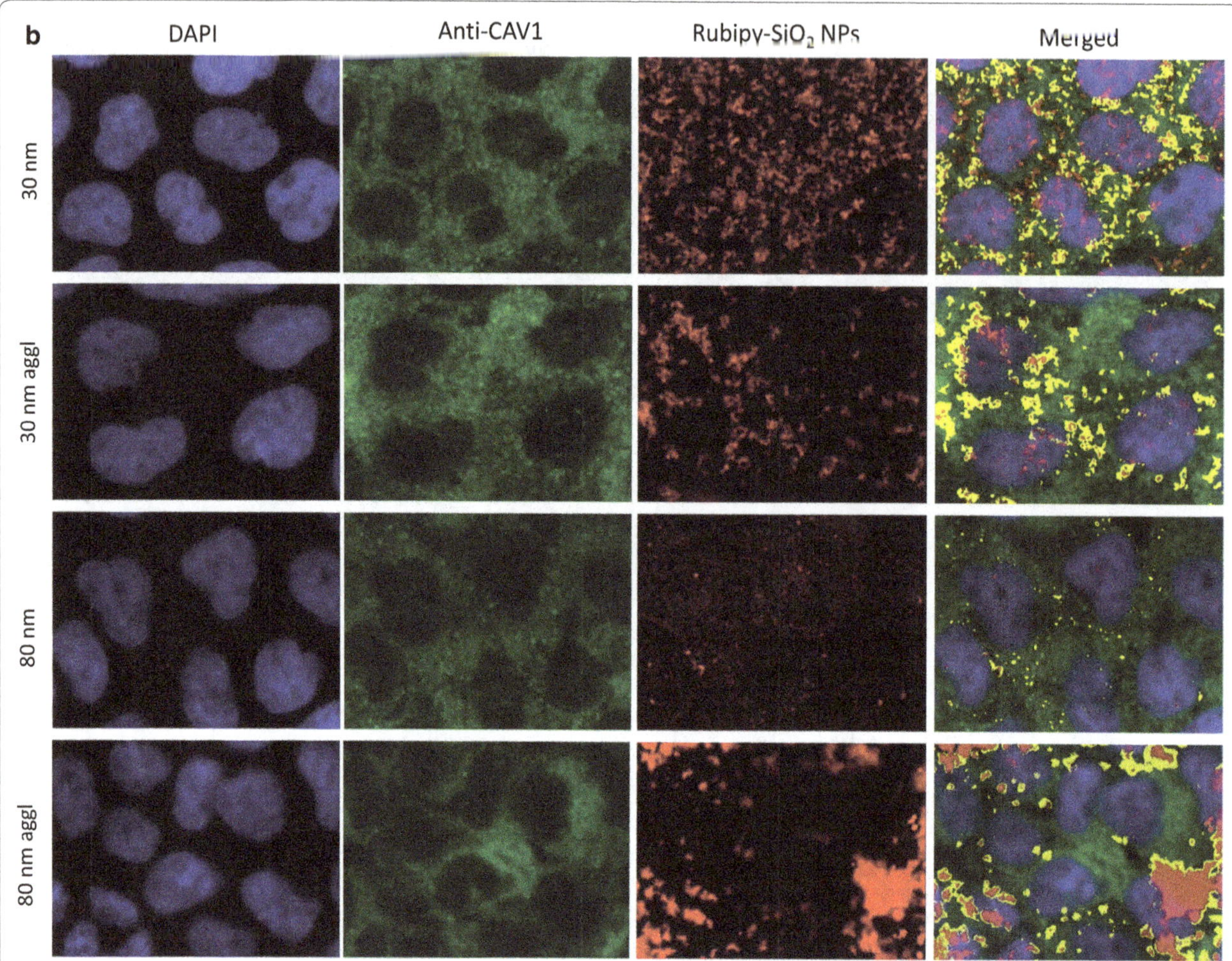

Fig. 5 Colocalisation of Rubipy-SiO_2 NPs with the CAV1 protein. After 3 h exposure to Rubipy-SiO_2 NPs the cells were fixed and incubated with anti-CAV1. The images of different channels were merged and the colocalisation of signal from channel Cy3 (Rubipy-SiO_2 NPs, shown in *red*) and of signal from channel Cy5 (secondary antibodies, shown in *green*) obtained by ImageJ is shown in *yellow*

slightly modified mechanism of cellular uptake. Endocytosis is a complex process that can occur through several distinct but interdependent pathways, of which many are not well described yet [25, 26]. Here, we limited our research to the best studied pathways such as CME, caveolae-mediated endocytosis and macropinocytosis, but other clathrin- and caveolae-independent mechanisms could be involved as well in the internalisation process.

The clathrin-mediated pathway, the most extensively described pathway, is a receptor-mediated endocytosis. The involved proteins recruit cargo (approximately 100–200 nm) together with the receptor into developing clathrin-coated pits that are cut from the membrane invaginations via dynamin scission and form clathrin-coated vesicles. The individual vesicles can then fuse and form early endosomes, late endosomes and end up in lysosomes or can be recycled to the plasma membrane surface. CME is widely used for the specific uptake of certain substances required by the cell such as nutrients, antigens and growth factors including transferrin or LDL [27]. An upper size limit reported for particles entering via this pathway is 200 nm [8, 11]. Silica NPs of 200 nm and polymeric NPs of 100–200 nm were reported to internalize predominantly through CME [11, 28], and the positive charge on the surface of quantum dots [29], dendrimers [30] and polymer NPs [10, 31] was shown to increase the probability of internalization via CME rather than the use of other endocytic pathway. In our study, the inhibitors of CME were not very successful to inhibit the uptake of Rubipy-SiO_2 NPs by CaCo-2 cells; however the effect of chlorpromazine and the colocalisation with clathrin suggests a minor role of this pathway in the internalization of 30 nm Rubipy-SiO_2 NPs.

The caveolae-mediated pathway is characterized by the presence of small flask-shaped invaginations of the plasma membrane, responsible for uptake and transport

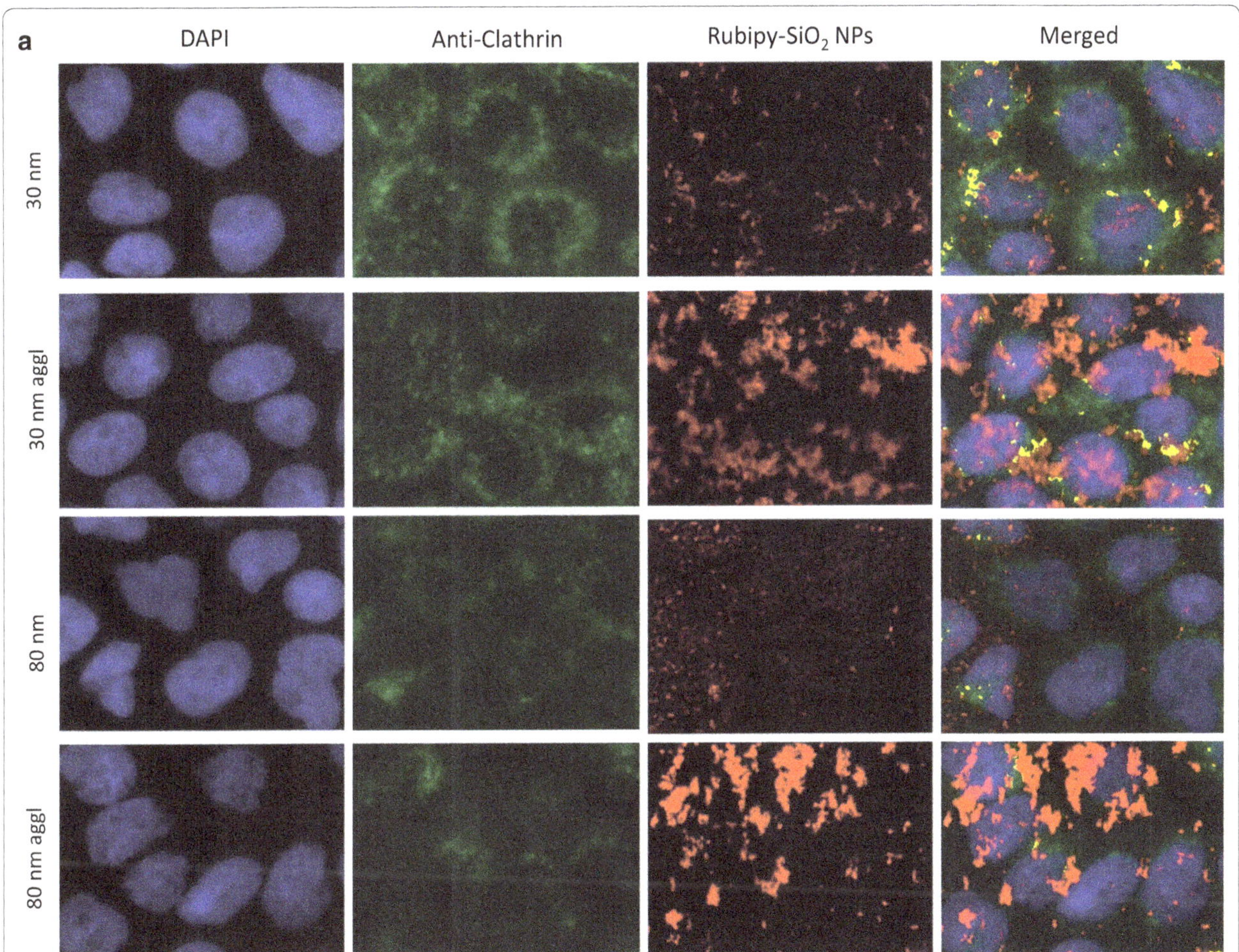

Fig. 6 Colocalisation of Rubipy-SiO_2 NPs with clathrin. After 3 h exposure to Rubipy-SiO_2 NPs the cells were fixed and incubated with anti-clathrin. The images of different channels were merged and the colocalisation of signal from channel Cy3 (Rubipy-SiO_2 NPs, shown in *red*) and of signal from channel Cy5 (secondary antibodies, shown in *green*) obtained by ImageJ is shown in *yellow*

of smaller (20–100 nm) molecules. Caveolae are lipid raft enriched and contain cholesterol-binding caveolins (mainly caveolin-1), essential for caveolae formation and function [26, 32, 33]. Once pinched off from the plasma membrane the caveolae vesicles transport and fuse with pH neutral caveosomes and are then transported to multivascular bodies, the Golgi apparatus or endoplasmic reticulum but not necessarily to acidic lysosomes [34]. Its potential to bypass the lysosomal degradation has been recently explored in nanomedicine as route for intracellular delivery of proteins and genes [35–37].

A negative surface charge of NPs has been found to trigger the cellular internalization predominantly via caveolae [10, 35, 38]. In agreement with these reports, also our study showed that the caveolae-mediated pathway was strongly implicated in the cellular internalization of the negatively-charged [7] Rubipy-SiO_2 NPs. Colocalisation with CAV1 after 3 h exposure was obvious for all forms of NPs, and the CAV1 gene silencing, even if not very efficient at the protein level, induced a slight reduction of the cellular uptake of Rubipy-SiO_2 NPs. Both immunofluorescence and gene silencing experiments indicated an abundant intracellular presence of CAV1 protein, which did not increase significantly upon exposure to Rubipy-SiO_2 NPs. Interestingly, nystatin, which was a specific inhibitor of the caveolae-mediated pathway, decreased the uptake of non-agglomerated NPs more than agglomerated NPs, particularly those of 80 nm that according to characterization data displayed good monodispersity. MβCD was by far the most efficient to inhibit the uptake of all Rubipy-SiO_2 NPs, however its mode of action based on the cholesterol depletion from the lipid rafts was not specific only to the caveolae-mediated pathway but also to other internalization mechanisms, mainly macropinocytosis (Additional file 1: Figure

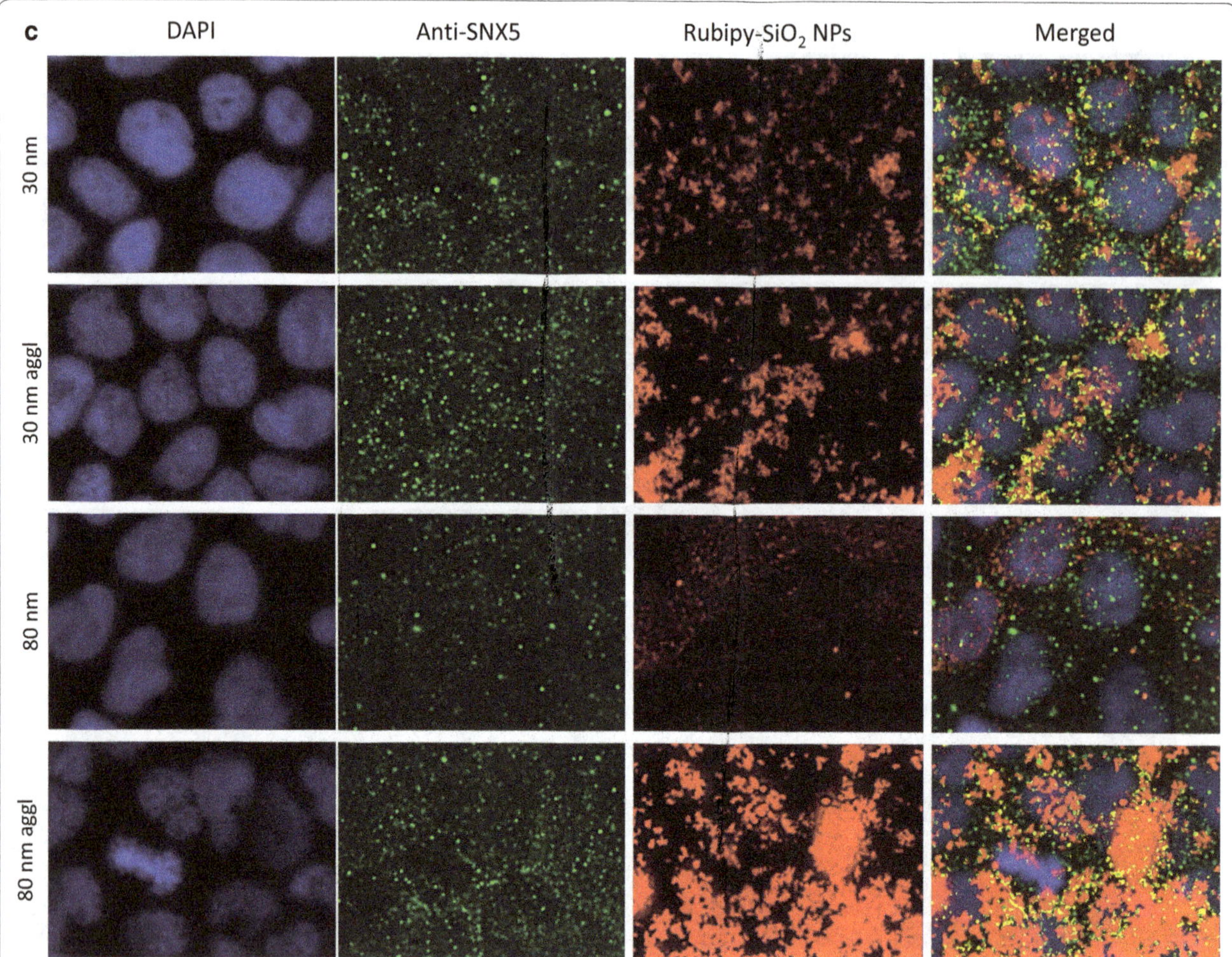

Fig. 7 Colocalisation of Rubipy-SiO_2 NPs with the SNX5 protein. After 3 h exposure to Rubipy-SiO_2 NPs the cells were fixed and incubated with anti-SNX5 antibodies. The images of different channels were merged and the colocalisation of signal from channel Cy3 (Rubipy-SiO_2 NPs, shown in *red*) and of signal from channel Cy5 (secondary antibodies, shown in *green*) obtained by ImageJ is shown in *yellow*

S5), a non-selective, clathrin and caveolae-independent endocytic mechanism.

Macropinocytosis is characterized by changes in the actin cytoskeleton and the membrane ruffles forming invaginations that enclose a region of extracellular fluid with suspended molecules [39, 40]. The internalized vesicles or macropinosomes have a size up to 5 μm, they undergo acidification and fuse with lysosomes. The non-specific inhibitor of micropinocytosis, EIPA, and other non-specific inhibitors in our study: MβCD and genistein, were more potent to block the internalization of agglomerated form than non-agglomerated form of NPs, suggesting the involvement of more than one pathway in the cell uptake of agglomerated NPs. The combination of endocytic pathways including the macropinocytosis was also suggested by TEM images showing characteristic macropinocytic ruffles of the plasma membrane accompanying the entrance of NPs but also small invaginations typical for the clathrin- or caveolae-mediated endocytosis. Study of the colocalisation of Rubipy-SiO_2 NPs with SNX5 protein, involved in the formation of macropinosomes [39, 41], confirmed the involvement of macropinocytosis in the internalization of all forms of Rubipy-SiO_2 NPs except 80 nm non-agglomerated NPs. Finally, we performed silencing of the PAK1 gene, an actin regulator shown to be required for both stimulated and basal fluid phase uptake [42]. After 3 h exposure to Rubipy-SiO_2 NPs we observed a reduction in uptake of 80 nm agglomerated NPs by PAK1-silenced cells, and after 6 h the reduction of internalization of both forms of 30 nm of Rubipy-SiO_2 NPs and of 80 nm non-agglomerated Rubipy-SiO_2 NPs. This result was in agreement with the

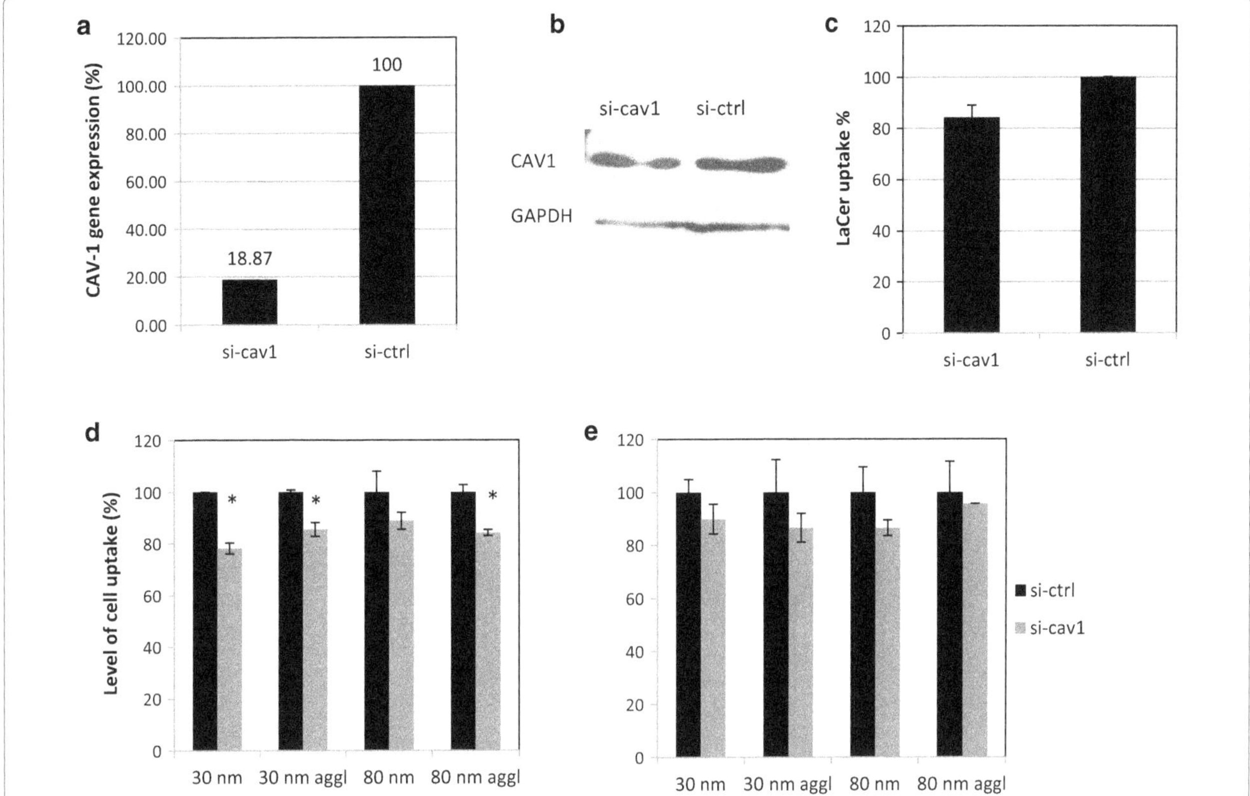

Fig. 8 Inhibition of cell uptake of Rubipy-SiO_2 NPs in CAV1-silenced cells. CaCo-2 cells were transfected with si-ctrl or si-cav1 and after 48 h the remaining RNA expression was evaluated with real-time PCR (**a**), protein expression was assessed with western blot (**b**), the cell uptake of LaCer was measured after 30 min incubation (**c**) and the cell uptake of Rubipy-SiO_2 NPs was measured after 3 h (**d**) and 6 h (**e**) exposure by flow cytometry. Statistical significance was assessed by a Student's *t* test; $^*p < 0.05$, si-cav1 silenced cells compared to si-ctrl treated cells

colocalisation study. As the process of internalization of agglomerated 80 nm Rubipy-SiO_2 NPs was much faster in the control cells than in silenced cells and reached saturation after 3 h exposure (Additional file 1: Figure S4A), no difference in uptake between the silenced and control cells was visible after 6 h exposure. This fact points out the importance of selecting appropriate time-points in the study of the endocytosis process.

Macropinocytosis contributes to the internalization of bacteria, viruses, necrotic cells and larger particles, often in conjunction with other entry mechanisms [13, 43, 44]. However, smaller silica NPs of 20–100 nm were also shown to be internalized mainly by the macropinocytosis [45–47]. However, their size and state of agglomeration in complete medium was unfortunately not studied and it cannot be excluded that, similarly to Rubipy-SiO_2 NPs, the initial small monodispersed NPs became large agglomerates in the presence of serum proteins.

Conclusions

Taking into account the results of physicochemical characterisation of NP dispersion in the cell culture medium we demonstrated that the level of cell uptake and the mechanism of endocytosis of silica NPs were strongly dependent on their agglomeration state. Both caveolae-mediated endocytosis and macropinocytosis were implicated in the process of NP internalisation, whereas the clathrin-mediated pathway was involved to a minor extent. Well dispersed 80 nm Rubipy-SiO_2 NPs were internalized mainly by the caveolae-mediated endocytosis, whereas 30 nm Rubipy-SiO_2 NPs entered the cells via a combination of different endocytic pathways. Interestingly, with the increase of NP agglomeration we observed a highly enhanced cellular uptake and slightly modified mechanism of endocytosis with a predominant role of macropinocytosis. This finding highlights the importance of a careful evaluation of NP dispersion in relevant

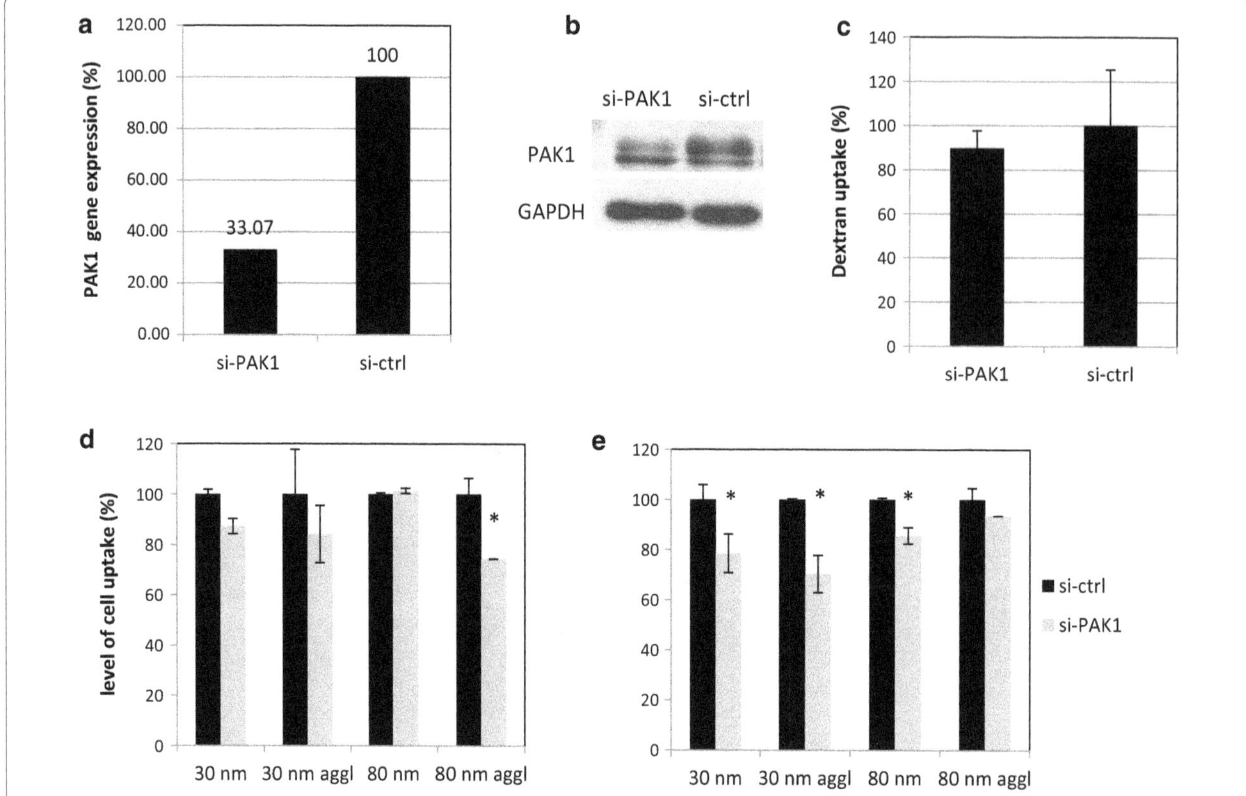

Fig. 9 Effect of PAK1 depletion on cell uptake of Rubipy-SiO_2 NPs. CaCo-2 cells were transfected with si-ctrl or si-PAK1 and after 48 h the remaining RNA expression was evaluated with real-time PCR (**a**), protein expression was assessed with western blot (**b**), the cell uptake of Dextran 10 kDA was measured after 30 min incubation (**c**) and the cell uptake of Rubipy-SiO_2 NPs was measured after 3 h (**d**) and 6 h (**e**) exposure by flow cytometry. Statistical significance was assessed by a Student's *t* test; *$p < 0.05$, si-PAK1 silenced cells compared to si-ctrl treated cells

experimental conditions since a modified environment can easily induce NP agglomeration and consequently influence a biological response.

Abbreviations

NPs: nanoparticles; DLS: dynamic light scattering; CLS: centrifugal liquid sedimentation; CCM: cell culture medium; TEM: transmission electron microscopy; CME: clathrin-mediated endocytosis; CAV1: caveolin-1; SNX5: sorting nexin 5; PAK1: p21-activated kinase; BSA: Bovine serum albumin.

Authors' contributions

BHK contributed to the design of the research, performed the experimental work and data analysis, wrote the manuscript, MC was involved in the conception of the study, performed the experimental work, contributed to the description of the methods and results and to the revision of the manuscript, PU, AB, JP and SG performed the experimental work and contributed to the description of the corresponding methods, and AKO contributed to the design of the research and was involved in discussion of data, manuscript writing and revisions. All authors read and approved the final manuscript.

Acknowledgements

The authors would like to acknowledge Dr. Douglas Gilliland who kindly provided Rubipy-SiO_2 NPs used in this work and Dr. Kirsten Rasmussen for reading the manuscript and providing critical comments.

Competing interests

The authors declare that they have no competing interests.

Funding

The work was funded by the institutional budget of the European Commission`s Joint Research Centre. Blanka Halamoda's work was supported by a Post-doctoral grant from the European Commission's Joint Research Centre (Contract No. 2012-IPR-I-30-000 00645).

References

1. Elkin SR, Lakoduk AM, Schmid SL. Endocytic pathways and endosomal trafficking: a primer. Wien Med Wochenschr. 2016;166:196–204.
2. Mustata RC, Grigorescu A, Petrescu SM. Encapsulated cargo internalized by fusogenic liposomes partially overlaps the endoplasmic reticulum. J Cell Mol Med. 2009;13:3110–21.
3. Yameen B, Choi W, Vilos C, Swami A, Shi J, Farokhzad OC. Insight into nanoparticle cellular uptake and intracellular targeting. J Control Release. 2014;190:485–99.
4. Zhao F, Zhao Y, Liu Y, Chang X, Chen C, Zhao Y. Cellular uptake, intracellular trafficking, and cytotoxicity of nanomaterials. Small. 2011;7:1322–37.
5. Dos Santos T, Varela J, Lynch I, Salvati A, Dawson K. Effects of transport inhibitors on the cellular uptake of carboxylated polystyrene nanoparticles in different cell lines. PLoS ONE. 2011;6:e24438.
6. Cleal K, He L, Watson PD, Jones AT. Endocytosis, intracellular traffic and fate of cell penetrating peptide based conjugates and nanoparticles. Curr Pharm Des. 2013;19:2878–94.
7. Halamoda-Kenzaoui B, Ceridono M, Colpo P, Valsesia A, Urbán P, Ojea-Jiménez I, et al. Dispersion behaviour of silica nanoparticles in biological media and its influence on cellular uptake. PLoS ONE. 2015;10:e0141593.
8. Rejman J, Oberle V, Zuhorn IS, Hoekstra D. Size-dependent internalization of particles via the pathways of clathrin- and caveolae-mediated endocytosis. Biochem J. 2004;377:159–69.
9. Andar AU, Hood RR, Vreeland WN, Devoe DL, Swaan PW. Microfluidic preparation of liposomes to determine particle size influence on cellular uptake mechanisms. Pharm Res. 2014;31:401–13.
10. Bannunah AM, Vllasaliu D, Lord J, Stolnik S. Mechanisms of nanoparticle internalization and transport across an intestinal epithelial cell model: effect of size and surface charge. Mol Pharm. 2014;11:4363–73.
11. Herd H, Daum N, Jones AT, Huwer H, Ghandehari H, Lehr CM. Nanoparticle geometry and surface orientation influence mode of cellular uptake. ACS Nano. 2013;7:1961–73.
12. Marks DL, Singh RD, Choudhury A, Wheatley CL, Pagano RE. Use of fluorescent sphingolipid analogs to study lipid transport along the endocytic pathway. Methods. 2005;36:186–95.
13. Al Soraj M, He L, Peynshaert K, Cousaert J, Vercauteren D, Braeckmans K, et al. siRNA and pharmacological inhibition of endocytic pathways to characterize the differential role of macropinocytosis and the actin cytoskeleton on cellular uptake of dextran and cationic cell penetrating peptides octaarginine (R8) and HIV-Tat. J Control Release. 2012;161:132–41.
14. Schmittgen TD, Livak KJ. Analyzing real-time PCR data by the comparative C(T) method. Nat Protoc. 2008;3:1101–8.
15. Vercauteren D, Vandenbroucke RE, Jones AT, Rejman J, Demeester J, De Smedt SC, et al. The use of inhibitors to study endocytic pathways of gene carriers: optimization and pitfalls. Mol Ther. 2010;18:561–9 **(Nature Publishing Group)**.
16. Ivanov AI. Pharmacological inhibition of endocytic pathways: is it specific enough to be useful? Methods Mol Biol. 2008;440:15–33.
17. Pustylnikov S, Sagar D, Jain P, Khan ZK. Targeting the C-type lectins-mediated host-pathogen interactions with dextran. J Pharm Pharm Sci. 2014;17:371–92.
18. Gold S, Monaghan P, Mertens P, Jackson T, Mayor S, Pagano R, et al. A clathrin independent macropinocytosis-like entry mechanism used by bluetongue virus-1 during infection of BHK cells. PLoS ONE. 2010;5:e11360.
19. Bitar A, Ahmad NM, Fessi H, Elaissari A. Silica-based nanoparticles for biomedical applications. Drug Discov Today. 2012;17:1147–54.
20. Slowing II, Vivero-Escoto JL, Wu C-W, Lin VS-Y. Mesoporous silica nanoparticles as controlled release drug delivery and gene transfection carriers. Adv Drug Deliv Rev. 2008;60:1278–88.
21. Paris JL, de la Torre P, Manzano M, Cabañas MV, Flores AI, Vallet-Regí M. Decidua-derived mesenchymal stem cells as carriers of mesoporous silica nanoparticles. In vitro and in vivo evaluation on mammary tumors. Acta Biomater. 2016;33:275–82.
22. Peters R, Kramer E, Oomen AG, Herrera Rivera ZE, Oegema G, Tromp PC, et al. Presence of nano-sized silica during in vitro digestion of foods containing silica as a food additive. ACS Nano. 2012;6:2441–51.
23. Hinderliter PM, Minard KR, Orr G, Chrisler WB, Thrall BD, Pounds JG, et al. ISDD: a computational model of particle sedimentation, diffusion and target cell dosimetry for in vitro toxicity studies. Part Fibre Toxicol. 2010;7:36.
24. Cho EC, Zhang Q, Xia Y. The effect of sedimentation and diffusion on cellular uptake of gold nanoparticles. Nat Nanotechnol. 2011;6:385–91.
25. Doherty GJ, McMahon HT. Mechanisms of endocytosis. Annu Rev Biochem. 2009;78:857–902.
26. Chaudhary N, Gomez GA, Howes MT, Lo HP, McMahon K-A, Rae JA, et al. Endocytic crosstalk: cavins, caveolins, and caveolae regulate clathrin-independent endocytosis. PLoS Biol. 2014;12:e1001832.
27. Kirchhausen T. Clathrin. Annu Rev Biochem. 2000;69:699–727.
28. Reix N, Parat A, Seyfritz E, Van der Werf R, Epure V, Ebel N, et al. In vitro uptake evaluation in Caco-2 cells and in vivo results in diabetic rats of insulin-loaded PLGA nanoparticles. Int J Pharm. 2012;437:213–20.
29. Chakraborty A, Jana NR. Clathrin to lipid raft-endocytosis via controlled surface chemistry and efficient perinuclear targeting of nanoparticle. J Phys Chem Lett. 2015;6:3688–97.
30. Vidal F, Vásquez P, Díaz C, Nova D, Alderete J, Guzmán L. Mechanism of PAMAM dendrimers internalization in hippocampal neurons. Mol Pharm. 2016;13:3395–403.
31. Harush-Frenkel O, Debotton N, Benita S, Altschuler Y. Targeting of nanoparticles to the clathrin-mediated endocytic pathway. Biochem Biophys Res Commun. 2007;353:26–32.
32. Pelkmans L, Bürli T, Zerial M, Helenius A. Caveolin-stabilized membrane domains as multifunctional transport and sorting devices in endocytic membrane traffic. Cell. 2004;118:767–80.
33. Parton RG. Caveolae meet endosomes: a stable relationship? Dev Cell. 2004;7:458–60.
34. Pfeffer SR. Caveolae on the move. Nat Cell Biol. 2001;3:E108–10.
35. Sahay G, Alakhova DKA. Endocytosis of nanomedicines. J Control Release. 2010;3:182–95.
36. Sundaramoorthy P, Ramasamy T, Mishra SK, Jeong K-Y, Yong CS, Kim JO, et al. Engineering of caveolae-specific self-micellizing anticancer lipid nanoparticles to enhance the chemotherapeutic efficacy of oxaliplatin in colorectal cancer cells. Acta Biomater. 2016;42:220–31.
37. Li H, Quan J, Zhang M, Yung BC, Cheng X, Liu Y, et al. Lipid-albumin nanoparticles (LAN) for therapeutic delivery of antisense oligonucleotide against HIF-1α. Mol Pharm. 2016;13:2555–62.
38. Voigt J, Christensen J, Shastri VP. Differential uptake of nanoparticles by endothelial cells through polyelectrolytes with affinity for caveolae. Proc Natl Acad Sci USA. 2014;111:2942–7.
39. Kerr MC, Teasdale RD. Defining macropinocytosis. Traffic. 2009;10:364–71.
40. Falcone S, Cocucci E, Podini P, Kirchhausen T, Clementi E, Meldolesi J. Macropinocytosis: regulated coordination of endocytic and exocytic membrane traffic events. J Cell Sci. 2006;119:4758–69.
41. Lim JP, Wang JTH, Kerr MC, Teasdale RD, Gleeson PA. A role for SNX5 in the regulation of macropinocytosis. BMC Cell Biol. 2008;9:58.
42. Dharmawardhane S, Schürmann A, Sells MA, Chernoff J, Schmid SL, Bokoch GM. Regulation of macropinocytosis by p21-activated kinase-1. Mol Biol Cell. 2000;11:3341–52.
43. Gratton SEA, Ropp PA, Pohlhaus PD, Luft JC, Madden VJ, Napier ME, et al. The effect of particle design on cellular internalization pathways. Proc Natl Acad Sci USA. 2008;105:11613–8.
44. Khalil IA, Kogure K, Futaki S, Harashima H. High density of octaarginine stimulates macropinocytosis leading to efficient intracellular trafficking for gene expression. J Biol Chem. 2006;281:3544–51.
45. Vranic S, Boggetto N, Contremoulins V, Mornet S, Reinhardt N, Marano F, et al. Deciphering the mechanisms of cellular uptake of engineered nanoparticles by accurate evaluation of internalization using imaging flow cytometry. Part Fibre Toxicol. 2013;10:2.
46. Nowak JS, Mehn D, Nativo P, García CP, Gioria S, Ojea-Jiménez I, et al. Silica nanoparticle uptake induces survival mechanism in A549 cells by the activation of autophagy but not apoptosis. Toxicol Lett. 2014;224:84–92.
47. Zielinski J, Möller AM, Frenz M, Mevissen M. Evaluation of endocytosis of silica particles used in biodegradable implants in the brain. Nanomedicine. 2016;12:1603–13.

13

Labeling mesenchymal cells with DMSA-coated gold and iron oxide nanoparticles: assessment of biocompatibility and potential applications

Luisa H. A. Silva[1], Jaqueline R. da Silva[1], Guilherme A. Ferreira[2], Renata C. Silva[3], Emilia C. D. Lima[2], Ricardo B. Azevedo[1] and Daniela M. Oliveira[1*]

Abstract

Background: Nanoparticles' unique features have been highly explored in cellular therapies. However, nanoparticles can be cytotoxic. The cytotoxicity can be overcome by coating the nanoparticles with an appropriated surface modification. Nanoparticle coating influences biocompatibility between nanoparticles and cells and may affect some cell properties. Here, we evaluated the biocompatibility of gold and maghemite nanoparticles functionalized with 2,3-dimercaptosuccinic acid (DMSA), Au-DMSA and γ-Fe_2O_3-DMSA respectively, with human mesenchymal stem cells. Also, we tested these nanoparticles as tracers for mesenchymal stem cells in vivo tracking by computed tomography and as agents for mesenchymal stem cells magnetic targeting.

Results: Significant cell death was not observed in MTT, Trypan Blue and light microscopy analyses. However, ultrastructural alterations as swollen and degenerated mitochondria, high amounts of myelin figures and structures similar to apoptotic bodies were detected in some mesenchymal stem cells. Au-DMSA and γ-Fe_2O_3-DMSA labeling did not affect mesenchymal stem cells adipogenesis and osteogenesis differentiation, proliferation rates or lymphocyte suppression capability. The uptake measurements indicated that both inorganic nanoparticles were well uptaken by mesenchymal stem cells. However, Au-DMSA could not be detected in microtomograph after being incorporated by mesenchymal stem cells. γ-Fe_2O_3-DMSA labeled cells were magnetically responsive in vitro and after infused in vivo in an experimental model of lung silicosis.

Conclusion: In terms of biocompatibility, the use of γ-Fe_2O_3-DMSA and Au-DMSA as tracers for mesenchymal stem cells was assured. However, Au-DMSA shown to be not suitable for visualization and tracking of these cells in vivo by standard computed microtomography. Otherwise, γ-Fe_2O_3-DMSA shows to be a promising agent for mesenchymal stem cells magnetic targeting.

Keywords: Mesenchymal stem cells, Iron oxide nanoparticle, Gold nanoparticles, Biocompatibility, Computed microtomography, Magnetic targeting, DMSA-nanoparticles

Background

Due to the progression of nanotechnology, there are new materials at the nanometer scale that have been introduced to the Medicine [1]. One promising medical application of nanomaterials is the use of inorganic nanoparticles within mesenchymal stem cells (MSCs)-based therapies: nanomaterials have facilitated not only the investigation of stem cells' biology but also the development of new approaches for their expansion, differentiation and transplantation. Some of these nanomaterials

*Correspondence: dmoliveira@unb.br
[1] IB-Departamento de Genética e Morfologia, Universidade de Brasília-UNB, Campus Universitário Darcy Ribeiro-Asa Norte, Brasília, DF CEP 70910-970, Brazil
Full list of author information is available at the end of the article

possess chemical, optical or magnetic properties which can be used for visualization and tracking of MSCs [2–6].

Among the nanoparticles, iron oxide nanoparticles (IONPs) have great prominence because of their superparamagnetism, a property highly valued for biomedical applications [7]. Due to the strong signal that IONPs generate in magnetic resonance imaging, it is possible to visualize cells in microscopic levels and get specific information about their distribution in vivo [3, 7–9]. In addition, IONPs' magnetic properties can be explored for magnetically assisted cell delivery and retention in target organs [10–12], which is one of major current challenges in cell therapy. Lastly, IONPs superparamagnetism has also been applied in hyperthermia therapies in tumors, taking advantage of MSCs tropism to tumor cells [9].

Recently, another class of inorganic nanoparticles, gold nanoparticles (Au-NPs), has been explored in cell based therapies. It is known that nanoscale gold strongly absorbs and scatters visible light—a phenomenon that is based on the occurrence of surface plasmons [13]. Therefore, Au-NPs have been used for stem cells marking, which can be detected in vivo by fluorescence or photothermal imaging [6, 14–16]. Moreover, as Au-NPs scatter X-rays efficiently, labeled MSCs can also be detected by computed tomography, a technique that provides greater spatial resolution compared to magnetic resonance imaging [5, 6, 17, 18].

Despite their potential in clinical practice, the impacts of IONPs and Au-NPs on MSCs are not entirely clear. First, both materials can be toxic to cells. It is known that transition metals such as iron, when retained in excess in cell cytoplasm in a non-complexed form, act as catalysts for oxidation reactions of biomolecules, then increasing the rate of free radicals generation [19, 20]. On the other hand, even though macroscopic gold appears chemically inert, at the nanometric scale it may induce oxidative stress and it can bind permanently to nuclear and mitochondrial DNA [21–26]. There are also some studies showing unexpected physiological alterations in nanoparticle labeled MSCs, such as changes in differentiation ability [25–29] and in growth rates [26, 30]. These observations reinforce the importance of conducting previous biocompatibility tests of these nanoparticles before in vivo procedures [31, 32].

Coating nanoparticles with a proper surface modification is a strategy largely used to decrease potential toxic effect on cells. Biomedical applications of nanoparticles require surface modifications of nanoparticles in order to make them non-toxic, biocompatible, non-agregable and stable [33]. Surface functionalization of inorganic particles with 2, 3-dimercaptosuccinic acid (DMSA) is considered to be a promising strategy to increase biocompatibility [34, 35].

Here, we aimed to investigate the biocompatibility and potential use of two inorganic nanoparticles coated with DMSA, iron oxide nanoparticles coated with DMSA (γ-Fe_2O_3-DMSA) and gold nanoparticles coated with DMSA (Au-DMSA) used to label human MSC. Further, these nanoparticles were tested on two practical applications: Au-DMSA nanoparticles were tested as tracers for MSC in vivo tracking by computed tomography; and γ-Fe_2O_3-DMSA as agents for magnetic targeting of MSC.

Methods

Nanoparticles synthesis and characterization

Maghemite nanoparticles were prepared via oxidation of precursor magnetite nanoparticles, as described in the literature. γ-Fe_2O_3 nanoparticles were synthesized by mixing ferric and ferrous chloride aqueous solutions (2:1 molar ratio) with concentrated ammonia aqueous solution followed by vigorous stirring. The black magnetite precipitate was washed several times with water and collected by a magnet. The oxidation of magnetite to maghemite was carried out by refluxing the nanoparticles in 0.5 mol/L hydrochloric acid (HCl) solution under oxygen flux at 96 °C, yielding a brownish colloidal suspension. The brown precipitate was extensively washed using the 1 mol/L HCl solution and decanted by a magnet. The sample was redispersed in water and dialyzed against demineralized water to produce an aqueous acidic magnetic dispersion. The meso-2,3-dimercaptosuccinic acid (DMSA) coated maghemite nanoparticles (γ-Fe_2O_3-DMSA) was prepared according the early described protocol [36]. Five millilitre of DMSA (Acros Chemicals) stock solution (0.3 mol/L) were added to the 25 mL of the dispersion of maghemite nanoparticles in a molar ratio DMSA/Fe of 11 %. The dispersion was shacked for 12 h at room-temperature. Then, the dispersion was dialyzed for 12 h against demineralized water to eliminate the free DMSA out from the bulk solution. The pH was adjusted to the range of 7.0–7.2 and the suspension containing the maghemite nanoparticles functionalized with DMSA was purified against large aggregates by centrifugation at 5000 rpm for 10 min.

The total iron and gold content in the suspensions were determined by Atomic absorption spectrophotometry in a commercial Perkin-Elmer 5000 system (Perkin-Elmer, Norwalk, USA). The Fe^{2+}/Fe^{3+} ratio was determined by the 1–10 phenanthroline colorimetric method. X-ray powder diffraction (XRD) data were collected by a XRD-6000 diffractometer (Shimadzu, Kyoto, Japan). The average diameter of the nanocrystalline domain (*d*) was estimated using the Scherrer's equation [37]. Electronic micrographs of maghemite nanoparticles were obtained with a JEOL JEM 2100 Transmission Electron Microscopy (TEM). Hydrodynamic diameter and zeta potential measurements were performed using the Malvern Zetasizer Nano-ZS (Malvern Instruments Ltd., Worcestershire, UK).

The synthesis of the DMSA coated gold nanoparticles (Au-DMSA) was carried out by following the method purposed by Gao et al. [38]. Shortly, 5 mL of an aqueous DMSA solution 1.8×10^{-3} mol/L was added to 25 mL of an aqueous $HAuCl_4$ solution 6×10^{-4} mol/L at the boiling point, and the system was maintained under stirring by 15 min. After cooling to room temperature the colloidal suspension was against demineralized water and stored in the dark.

The total gold content in the metallic suspensions were determined by Atomic absorption spectrophotometry in a commercial Perkin-Elmer 5000 system (Perkin-Elmer, Norwalk, USA). Au-DMSA nanoparticle morphology was examined using a JEOL JEM 2100 TEM. Au-DMSA particle size was measured by dynamic light scattering (DLS) and zeta potential measurements were performed using Malvern Zetasizer Nano-ZS.

Cells

Dental pulp tissues were obtained from the permanent teeth of patients (17–43 years of age) under approval of the Ethical Committee of Health Sciences Faculty of the University of Brasília (Brazil) (Project number 023/08), as previously described [39]. All pulp tissues were washed with a-MEM, digested with 3 mg/mL collagenase type I (Gibco) in supplemented medium for 60 min at 37 °C. After enzymatic digestion, cell suspension was washed three times by centrifugation (10 min at 750*g*) in culture medium and placed into 6-well plates. The human MSC obtained were cultured in Low-Glucose Dulbecco's Modified Eagle Medium (DMEM-LG) (GIBCO®, Invitrogen,Carlsbad, CA) supplemented with 1 % L-glutamine (GIBCO®, Invitrogen, Carlsbad, CA), 1 % antibiotic–antimycotic (10,000 UI/mL penicillin, 10,000 mg/mL streptomycin and 25 μg/mL Amphotericin B) (GIBCO®, Invitrogen, Carlsbad, CA) and 10 % fetal bovine serum. MSC were grown under standard cell culture conditions (37 °C, 5 % CO_2) and have been maintained to their confluence below 80 %. Only passage 3–4 cells were used in this study.

Assessment of cell viability and morphology

The MSC were exposed to γ-Fe_2O_3-DMSA (15, 30, 60 and 80 μg iron/mL) and to Au-DMSA (52, 90 and 130 μg gold/mL) in growth medium for 02, 06 or 24 h. After each exposition time, labeled MSC viability was evaluated by MTT (3-[4,5-dimethylthiazol-2-yl]-2,5-diphenyltetrazolium bromide) assay [40], by Trypan blue exclusion assay [41] and by cell morphology analysis. For MTT assay, MSC were exposed to γ-Fe_2O_3-DMSA and Au-DMSA during 02, 06 or 24 h. In Trypan blue exclusion assay, MSC were incubated with nanoparticles only during 24 h. For evaluation of cell morphology, labeled MSC were incubated for 24 h with DMEM-LG with serum (negative control group), with DMEM-LG without serum (positive control group), with γ-Fe_2O_3-DMSA nanoparticles (80 μg iron/mL) or with Au-DMSA nanoparticles (90 μg gold/mL). After incubation time, the cells were stained with Instant Prov Kit (Newprov) according to the manufacturer's recommendations and observed analyzed using an inverted microscope (Axiovert 100, Zeiss).

Nanoparticle uptake

After 24 h of incubation, the cells were washed, ressuspended in PBS and counted using an automatic counter (Scepter ™, Millipore). Prussian blue staining was used to quantify the amount of uptaken γ-Fe_2O_3-DMSA, as described by Boutry et al. [42], with some modifications: 100 μl of 5 N HCl were added to the samples, incubating them at 80 ℃ for 4 h to lyse cells. After that, the samples were transferred to a 96-well polystyrene plate and 100 mL of 5 % potassium ferrocyanide were added. The absorbance of samples at 630 nm was measured and data were compared with a standard curve whose function relates the Prussian blue $OD_{630\ nm}$ with iron concentration in the sample. The results are expressed as iron per cell.

Inductively coupled plasma optical emission spectrometry (ICP-OES) technique (Spectro Arcos, Ametek) was used to measure the amount of intracellular gold after exposure to the Au-DMSA (90 μg gold/mL) for 24 h.

Transmission electron microscopy analysis

After incubation of MSC with Au-DMSA (90 μg gold/mL) and γ-Fe_2O_3-DMSA (80 μg iron/mL), as described previously, they were fixed in modified Karnovsky's fixative (2 % paraformaldehyde, 2.5 % glutaraldehyde in 0.1 M sodium cacodylate buffer, pH 7.2) for 2 h at room temperature. Samples were postfixed in solution containing 1 % osmium tetroxide, 0.8 % potassium ferricyanide, and 5 mM calcium chloride and contrasted in bloc with 0,5 % uranyl acetate. Samples were then dehydrated in acetone and embedded in Spurr. Semi-thin sections (3 μm) were stained with toluidine blue and examined under a light microscope to localize cells with visible nucleus. Ultrathin sections (70 nm) were examined using a Tecnai Spirit G2 TEM (FEI, USA).

MSC differentiation

To verify if both nanoparticles interfere with the MSC ability to differentiate, the cells were cultured in medium enriched with inducing agents, after their exposure to γ-Fe_2O_3-DMSA (80 μg/mL) and Au-DMSA (90 μg/

mL) for 24 h. For the experiments of differentiation into osteoblasts, the cell inducers used were dexamethasone (5×10^{-6} M); ascorbic acid (2.8×10^{-4} M); and β-glycerol phosphate (10^{-2} M). Subsequently, cytochemical analyses with specific labeling with dyes were performed. The MSCs differentiated into osteoblasts were fixed in 50 % ethanol for 15 min at 4 °C and stained with a solution of Alizarin Red S 1 %. Quantitative analysis of osteogenic differentiation was performed by quantification of Alizarin Red S adhered to calcified tissues, following the protocol adopted by Gregory et al. [43], and measurement of alkaline phosphatase (ALP) activity, by the colorimetric method of para- nitrophenol [44], using the kit SIGMAFAST *p*-Nitrophenyl phosphate tablets (Sigma) according to the manufacturer's recommendations. The corresponding values of enzyme activity in milliunits (mIU) per milliliter were divided by the total protein content (in μg) of the monolayer, estimated by the Lowry method [45].

To induce differentiation into adipocytes, the inducers used were dexamethasone (5×10^{-6} M); 0.3-isobutylmethylxanthine (4.5×10^{-4} M); insulin (5 μg/mL); and indomethacin (3×10^{-4} M). After 24 days, the cells differentiated into adipocytes were fixed with formaldehyde solution for 15 min, and stained with a solution of "Oil Red O" at 0.3 % for 20 min for cytochemical analysis. An indirect measurement of adipogenesis was performed by quantification of Oil Red O in MSC monolayers adding 100 % isopropanol to extract the dye and measure its absorbance at 510 nm.

MSC growth curve

MSC were cultivated in 75 cm^2 culture flasks until they reached 80 % confluence. The growth curve was performed by incubating the cells with DMEM-LG (control), with γ-Fe_2O_3-DMSA (80 μg/mL), and with Au-DMSA (90 μg/mL). The cells were washed, disassociated from the flasks and seeded in 12-well cell culture plates at a starting concentration of 10,000 cells per well. Then, they were collected after six different times of incubation (2, 4, 6, 8, and 10 days), stained with Trypan blue dye and counted. The results are expressed as percentage of live cells.

Suppression of lymphocyte proliferation by MSC

Human mononuclear cells were isolated by Ficoll-Paque gradient. Briefly, blood were collected, stored in tubes containing anticoagulants, diluted in phosphate buffered saline (PBS) and transferred to a centrifuge tube, over a Ficoll-Paque layer ($\rho = 1.077$ g/mL). The tubes were centrifuged at 2000 rpm for 25 min, and the mononuclear cell layer obtained was then transferred to a new centrifuge tube. The mononuclear cells were counted using a Neubauer chamber (Gibco), resuspended at a final concentration of 10^7 cells/mL and labeled with Carboxyfluorescein succinimidyl ester (CFSE), according to manufacturer recommendations (CellTrace™ CFSE dye, Life Technologies). Lymphocytes then received allogeneic stimulus in the presence or absence of MSC for 5 days.

Murine model of silicosis

All experimental protocols with animals in this study were approved by the animal experimentation ethics committee of University of Brasilia (certificate # 99769/2012). C57BL/6 mice, 8 weeks old, were randomly divided into: control group, instilled intratracheally with 50 μL of sterile saline; and silicosis group, instilled intratracheally with a silica particle suspension (20 mg/50 μL of saline). In this model, the pathophysiological characteristics of silicosis are observed 15 days after installation of crystalline silica [46].

Computed tomography analysis

The application of Micro-CT to tracking Au-DMSA labeled MSCs in vivo was tested using a 1076 Skyscan microtomography device (Skyscan, Aartselaar, Belgium). The mice were analyzed daily in the equipment until the 7th day after the inoculation of Au-DMSA labeled MSCs or saline. Images were acquired using voltage 50 kV, current 180 mA, 0.5 mm aluminum filter and isotropic voxel size of 18 μm. For the two-dimensional image reconstruction, we used NRecon software (V 1.6.9, 64 bit version with GPU acceleration, Skyscan, Kontich, Belgium). For three-dimensional reconstructions, we used CTVox software (V 1.5.0, 64 bit version, Skyscan, Kontich, Belgium) and CTVol software (2.2 V, 64 bit version, Skyscan, Kontich, Belgium). Analyses of reconstructions were performed using the software CTAnalyzer (V 1.5.0, 64 bit version, Skyscan, Kontich, Belgium). The most appropriate parameters of smoothing, ring artifacts correction and beam-hardening correction were used. All acquisition and reconstruction parameters were the same for all mice.

Magnetic targeting of γ-Fe_2O_3-DMSA labeled MSC

γ-Fe_2O_3-DMSA nanoparticles were tested as potential agents for MSC magnetic targeting to injured lungs. The cells were incubated with 80 μg/mL of γ-Fe_2O_3-DMSA for 24 h and inoculated into silicotic mice. Neodymium circular magnets (20 mm in diameter and 2 mm height) were held in the thoracic region of some animals for up to 24 h. 48 h after inoculation of MSC, the animals were euthanized and their lungs were collected for iron quantification following the protocol described by Boutry et al. [42], and histological analysis. Slides containing lung sections were also stained with Prussian blue.

Statistical analysis

The normality of the data was analyzed by the Kolmogorov–Smirnov test. Then, the parametric test ANOVA, followed by post hoc Tukey's test was performed for statistical comparison of data of the following experiments: Differences were considered statistically significant with $p < 0.05$.

Results

Fe-DMSA and Au-DMSA characterization

According to Atomic absorption spectrophotometry data, the γ-Fe_2O_3-DMSA solution used in this study has 4.09 mg of iron per mL. In addition, γ-Fe_2O_3-DMSA nanoparticles present zeta potential of −43 ± 0.66 mV, irregular shape from square to sphere in TEM micrographs (Fig. 1a) and nanocrystalline diameter in a range of 5–18 nm, as determined by Scherrer's equation with XRD data (data not showed).

On the other hand, Au-DMSA solution presented low gold concentration—0.07 mg/mL—hindering XRD analysis; thus, Au-DMSA particle size was determined by Dynamic Light Scattering: 26.4 ± 0.96 nm. Similarly to γ-Fe_2O_3-DMSA nanoparticles, Au-DMSA also has negative zeta potential—40.8 ± 3.70 mV—and has spherical morphology in TEM micrographs (Fig. 1b).

Cytotoxicity assays

According to the MTT assay, MSC exposed to γ-Fe_2O_3-DMSA nanoparticles (15, 30, 60 and 80 µg iron/mL) remained viable, with no difference between experimental groups and their respective control groups at any incubation time (Fig. 2a). Differently, there was a 20–25 % reduction in the cell viability when they were exposed to Au-DMSA for 24 h, compared to the control cells (Fig. 2b). However, no difference was observed after 48 and 72 h of exposure (Fig. 2c).

The data of Trypan Blue dye test also demonstrate that γ-Fe_2O_3-DMSA (60 and 80 µg/mL) is not cytotoxic to MSCs, approximately 98 % of cells remained alive after 24 h of incubation. Interestingly, this test also indicated that at least 97.6 % of the cells exposed to Au-DMSA (52 and 90 µg/mL) also survived (Fig. 2d).

The morphological analysis of MSC under light microscope (Fig. 3) showed, as expected,that the negative control group MSC (Fig. 3a) presented a spindle form and a large nucleus. Otherwise, cells in the positive control group (Fig. 3b) showed pyknotic nuclei, a sign of apoptosis, after being cultured in serum-depleted media for 24 h. MSCs exposed to 80 µg/mL Fe_2O_3-DMSA (Fig. 3c) and to 90 µg/mL Au-DMSA (Fig. 3d) for 24 h were similar to the negative control MSCs, without cell shrinkage or pyknosis.

Nanoparticle uptake

The amount of γ-Fe_2O_3-DMSA nanoparticles (80 µg/mL) uptaken by MSCs after 24 h, measured by Prussian Blue colorimetric quantification indicated that each cell contain approximately 17 pg of iron, while control cells only 5 pg. ICP-OES measurement indicated that each MSC retained approximately 4 pg of gold after exposure to Au-DMSA (90 µg/mL) for 24 h.

In the Fig. 3, TEM analysis confirms the uptake of nanoparticles by MSC (Fig. 4a–c). Fe_2O_3-DMSA (Fig. 4b) and Au-DMSA (Fig. 4c) are free in the cell cytoplasm, inside vesicles, or in different cellular compartments, particularly into mitochondria.

Compared to unlabeled MSCs, Au-DMSA labeled cells presented a similar ultrastructure, however, few differences were observed: concentric electron dense myelin figures (Fig. 4d) and electron-lucent vesicles (Fig. 4e). Likewise, γ-Fe_2O_3-DMSA labeled cells presented swollen and degenerated mitochondria, full of iron in their ridges (Fig. 4f) and myelin figures (Fig. 4g). Lastly, γ-Fe_2O_3-DMSA labeled MSC had higher amounts of nanoparticles into cytoplasmic vesicles, different to Au-DMSA labeled cells (Fig. 4g).

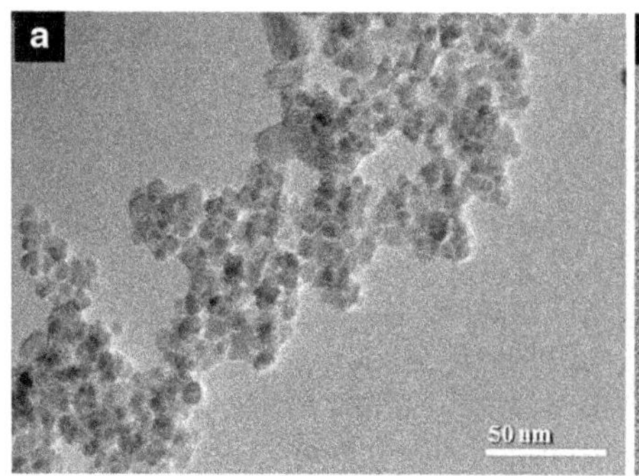

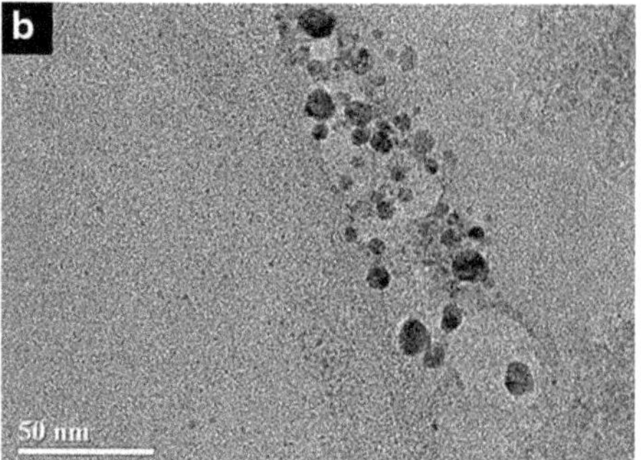

Fig. 1 Transmission electron microscopy analysis of γ-Fe_2O_3-DMSA and Au-DMSA. **a** TEM micrographs of Fe-DMSA nanoparticles, *bars* 50 nm. **b** TEM micrographs of Au-DMSA nanoparticles, *bars* 50 nm

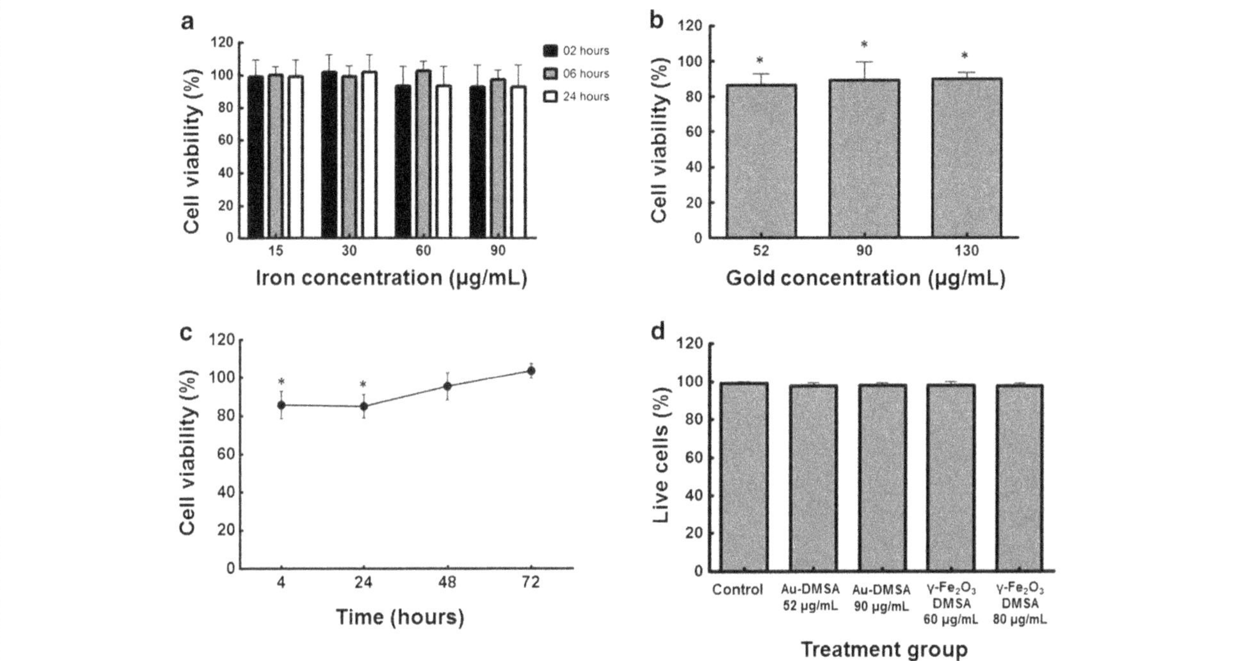

Fig. 2 Cell viability assessment by MTT and Trypan-blue staining. **a** MTT assay of γ-Fe_2O_3-DMSA labeled MSC. The data express average percentage and standard deviation of MSC that have remained viable after exposure to γ-Fe_2O_3-DMSA in four different concentrations at three different exposure times. **b** MTT assay of Au-DMSA labeled MSC. The average percentage and standard deviation of viable MSC are represented after 24 h of exposure to three different concentrations of Au-DMSA. (*) Significant reduction between the cells of the three experimental groups compared to the control group ($p < 0.01$). **c** MTT assay of Au-DMSA labeled MSC 4, 24, 48 and 72 h after incubation with the nanoparticles (90 μg/mL) during 24 h. There were significant differences between control cells and labeled cells only 4 and 24 h after exposure ($p < 0.05$). **d** Cell viability test by trypan-blue staining. The data express the mean percentage and standard deviation of MSC that remained alive after 24 h of exposure to 52 and 90 μg/mL of Au-DMSA and 60 and 80 μg/mL of γ-Fe_2O_3-DMSA

Impact of DMSA-nanoparticles in MSC physiology

MSC differentiation

After 24 days of treatment with osteogenic medium, we observed the formation of calcified nodules in MSC monolayers of all experimental groups (Fig. 5a–c). However, it was possible to see that there were fewer calcified nodes in the cells with Au-DMSA, while there were more in MSC monolayers with γ-Fe_2O_3-DMSA. This difference was confirmed after the measurement of the amount of Alizarin Red (ARS) absorbed by mineralized nodules (Fig. 5d).

In order to corroborate this data, we performed the measurement of alkaline phosphatase (ALP) activity in labeled and unlabeled MSC, using *p*-nitrophenylphosphate as substrate. Contrary to what was seen in ARS measurement test, there was no statistically significant difference between control cells and both experimental groups (Fig. 5e).

Intracellular lipid vacuoles were observed in MSCs after 24 days of incubation with adipogenic medium, showing that differentiation occurred in control and experimental groups (Fig. 6a–c). No difference in the amount of Oil Red O dye from differentiated MSC monolayers was observed in all experimental groups (Fig. 6d).

MSC proliferation and lymphocyte suppression

The results of total cell number by trypan blue staining, in order to assess the proliferative potential of labeled MSC are expressed in Fig. 7a, b. According to the individualized data analysis, there were no significant differences among the number of cells labeled with γ-Fe_2O_3-DMSA and control cells in any times (Fig. 7a). Au-DMSA labeled MSC showed cell number increase after 2 days of incubation (Fig. 7b).

To verify whether nanoparticle uptake influence the characteristic capability of MSC to cause unspecific lymphocyte suppression in vitro a lymphocyte proliferation test was performed. Lymphocytes were labeled with CFSE and their proliferation rates were analyzed by flow cytometry (Fig. 7c). Their proliferation leads to an intracellular reduction of the fluorescent tracer, decreasing its intensity, as shown in the Fig. 7c (red line). Because of the activation (and subsequent division) of these mononuclear cells in the control group, there are several populations with different amounts of marker, therefore,

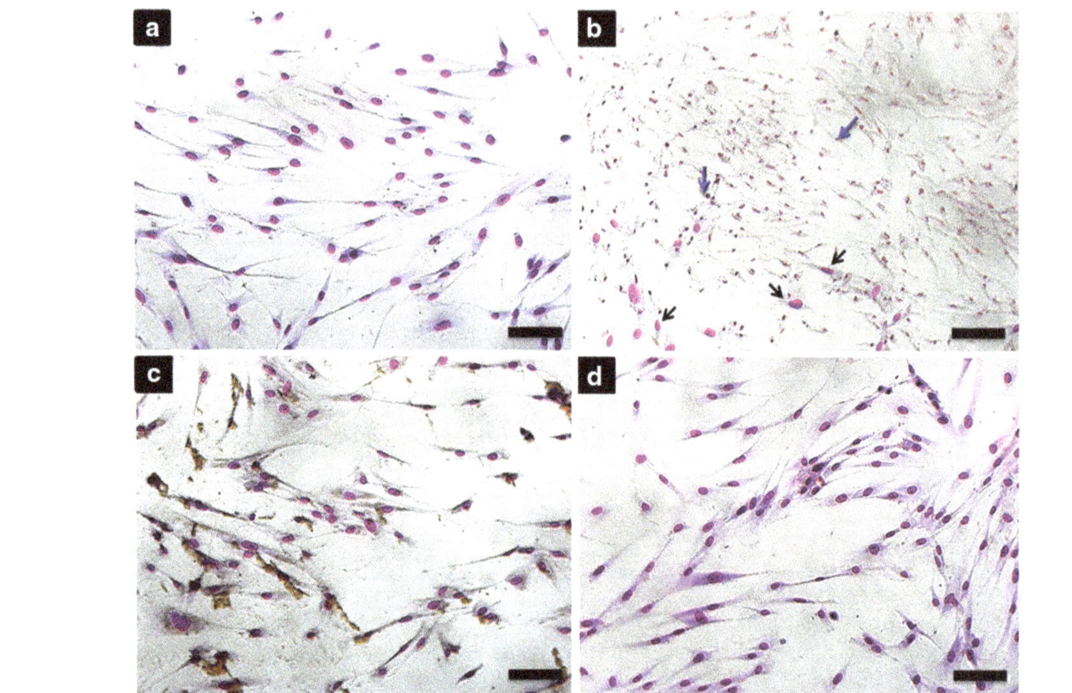

Fig. 3 Analysis of MSC morphology by Instant Prov staining kit. **a** Negative control group. **b** Positive control group, with pyknotic nuclei (*blue arrows*) and normal nuclei (*black arrows*). **c** MSC incubated with Au-DMSA (90 μg/mL) and **d** with γ-Fe_2O_3-DMSA (80 μg/mL) for 24 h. *Bars* 50 micrometers (μm)

the red line span multiple fluorescence intensity values. Based on this, the representative curve of the negative control group that does not contain activated lymphocytes (black line), remained narrow. Finally, it was found in this experiment that lymphocytes, either after the co-cultivation with unmarked MSC (blue line), either with γ-Fe_2O_3-DMSA labeled MSC (green line) or Au-DMSA marked cells (yellow line), not proliferated.

Thus, all these results suggest that exposure to γ-Fe_2O_3-DMSA and Au-DMSA nanoparticles, at the tested concentrations, does not cause toxic effects to MSC and do not change their physiology. Therefore, both inorganic nanoparticles are biocompatible with MSCs.

Computed tomography analysis

Firstly, a test was performed in which tubes containing precipitates of unlabeled MSC or Au-DMSA labeled cells (90 μg/mL, 24 h) were scanned in a micro-CT equipment to assess whether nanoparticles generate adequate contrast (Fig. 8a–c). The parameters of acquisition and image reconstruction were adjusted similarly to those used in analysis with mice. A tube containing water served as control, and the signal generated by the liquid was then considered as 0 (zero) Hounsfield (HU). The values in Hounsfield scale for each sample were: 284.70 HU in control MSC and 352.79 HU in Au-DMSA labeled MSC.

Although the Au-DMSA has generated a visible contrast in images of cell precipitates, labeled MSCs were not detected by the device after its inoculation in mice in any analyzed times (Fig. 8d–g).

MSC magnetic targeting

In order to verify if γ-Fe_2O_3-DMSA labeled MSC (80 μg/mL, 24 h) became magnetically responsive, a test was previously performed in vitro (Fig. 9a–d). Labeled cells were maintained in culture in the presence of an external circular magnet or of a similar size plastic (control). It can be seen that the cells migrated toward the edge of the circular magnet (Fig. 9b, d), unlike the control group cells (Fig. 9a, c).

Next, we test the potential of in vivo magnetic targeting of γ-Fe_2O_3-DMSA labeled MSCs in an experimental mice model of lung silicosis. Histological analysis of experimental silicosis mice inoculated with γ-Fe_2O_3-DMSA labeled MSC and with magnets fixed in their thoracic region presented higher iron content in their lung tissue than animals that did not hold external magnets (Fig. 9e). The light microscopy images also corroborate this data (Fig. 9f–h): animals treated with marked cells and holding magnets had more blue spots, corresponding to γ-Fe_2O_3-DMSA, compared to animals without magnets.

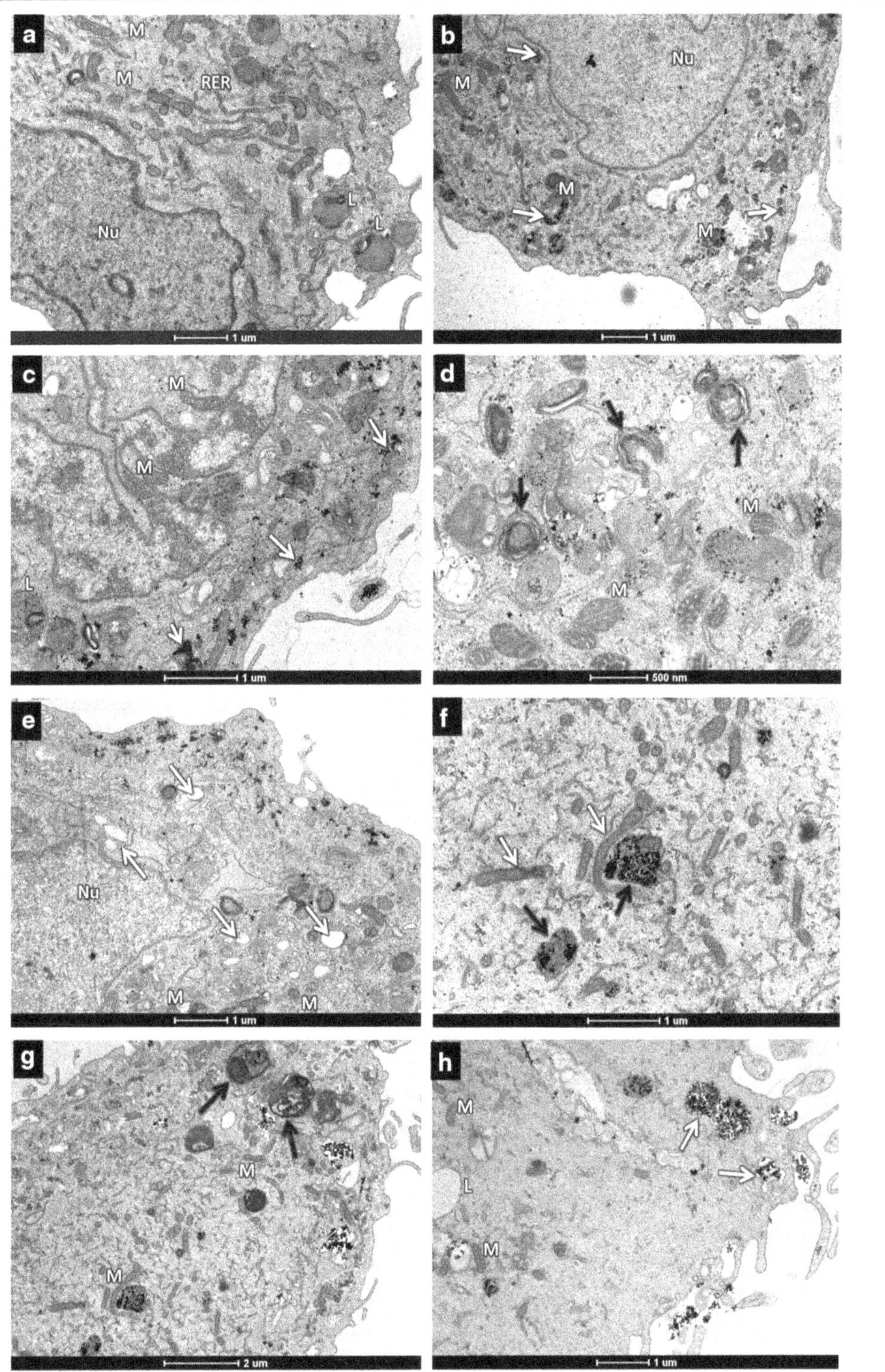

Fig. 4 Transmission electron microscopy micrographs of labeled MSC. **a** Unlabeled MSC (control). **b** γ-Fe_2O_3-DMSA labeled MSC (80 μg/mL); the white arrows point some of the uptaken nanoparticles. **c** MSC labeled with Au-DMSA (90 μg/mL); the *white arrowheads* point some of the uptaken nanoparticles. **d** Cells exposed to Au-DMSA for 24 h presented myelin figures (*black arrows*) and **e** more electron-lucent structures (*white arrows*) compared to unlabeled cells. **f** In turn, after 24 h of exposure toγ-Fe_2O_3-DMSA, MSC presented signs of mitochondrial toxicity: mitochondria full of nanoparticles are swollen and degenerated (*black arrows*), as compared to organelles without them (*white arrows*). **g** These cells also presented some myelin figures (*black arrows*). **h** Lastly, γ-Fe_2O_3-DMSA nanoparticles were stored in vesicles (*white arrows*). *Nu* nucleus, *M* mitochondria, *RER* rough endoplasmic reticulum, *L* lipid

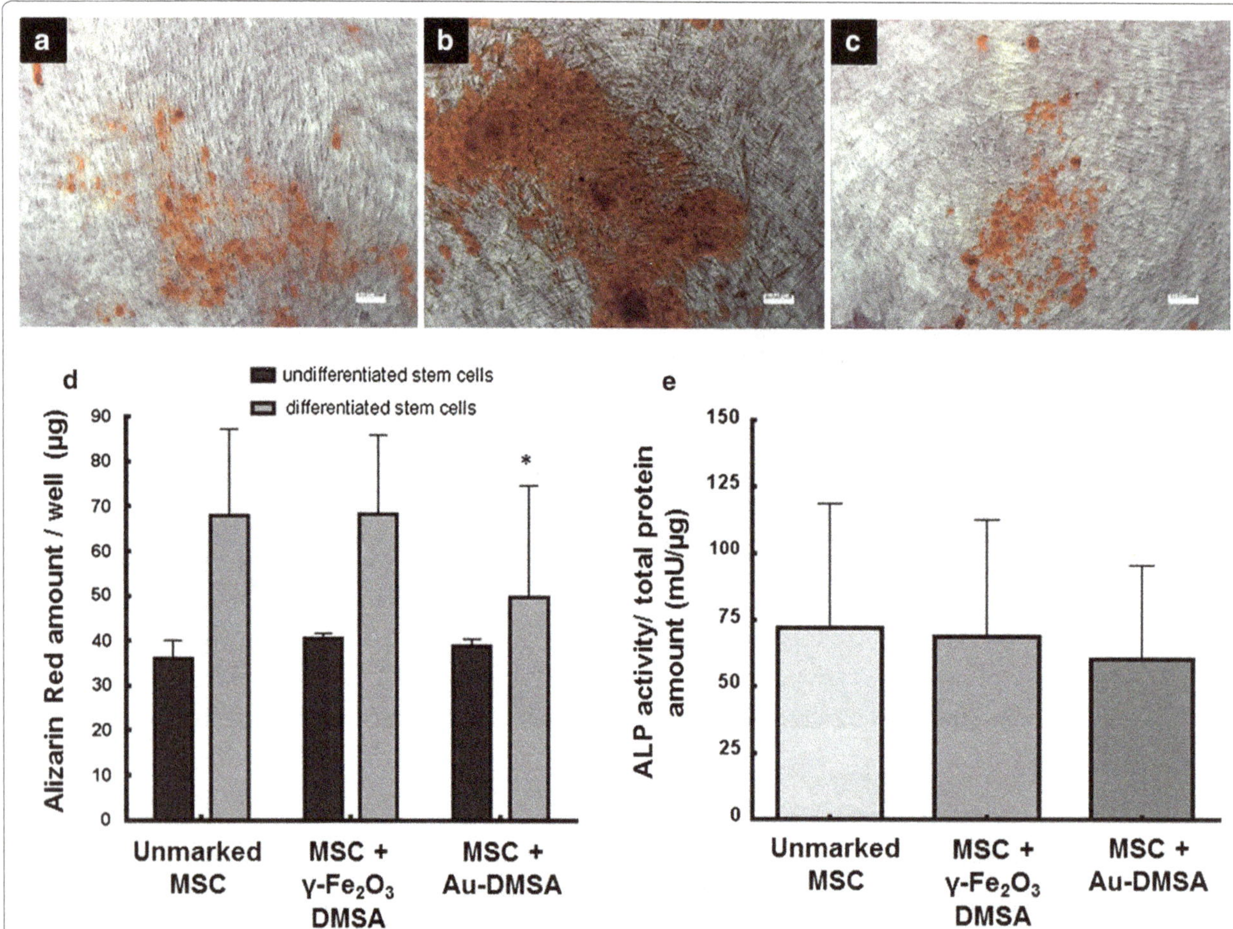

Fig. 5 MSC osteogenic differentiation assay. **a–c** Cytochemical analysis of differentiated MSC monolayers in light microscopy, with Alizarin Red, in order to evidence the formation of mineralized nodules. These nodules were present in control cells (**a**), in γ-Fe_2O_3-DMSA labeled cells, (**b**) and in Au-DMSA labeled cells (**c**). **d** Quantification of alizarin red incorporated in monolayers of differentiated and non-differentiated MSC; (*) Significant reduction in the group "MSC + Au-DMSA" compared to the groups "Unmarked MSC" and "MSC + γ-Fe_2O_3-DMSA" ($p < 0.05$). **e** Measurement of alkaline phosphatase (ALP) activity: data are represented as the average ratio of ALP activity and total protein content, with the respective standard deviations. There was no significant difference between control and experimental groups ($p > 0.05$). *Bars* 100 μm

Discussion

This work had two main purposes: (1) to evaluate biocompatibility between MSC and DMSA-coated inorganic nanoparticles, verifying cytotoxic effects or physiological alterations; and (2) to test Au-DMSA and γ-Fe_2O_3-DMSA as agents for MSC in vivo tracking and for MSC magnetic targeting, respectively. Firstly, our results demonstrated the absence of toxic effects on MSC and suggested no significant changes in physiological parameters such as cell differentiation, proliferation and immunomodulation, at the concentrations tested. In addition, Au-DMSA nanoparticles had a poor performance as MSC tracers when analyzed on a microtomograph; while γ-Fe_2O_3-DMSA showed to be good agent for magnetic targeting.

The results of MTT and trypan blue analysis demonstrated the low toxicity of γ-Fe_2O_3-DMSA on MSC, corroborating with other studies [34, 35, 47, 48]. Although iron oxide catalyze free radicals production, chemical surface modifications make them safer materials for biological applications [34, 49, 50]. For example, Auffan et al. [34] suggested that is difficult to remove DMSA coating from the nanostructure (different than dextran or albumin), preventing direct contact of the cells with iron, protecting them from possible toxic effects. Thus, if the nanoparticle coating is easily degraded, the core can then react with cellular biomolecules [49].

Moreover, Chen et al. [47] demonstrated that IONPs oxidative activity depends on the acidity of intracellular microenvironment in which they are located. In lysosomes (low pH), IONPs produce more hydroxyl radicals (OH^-), increasing cell damage induced by H_2O_2. In neutral environments, IONPs break H_2O_2 into H_2O and

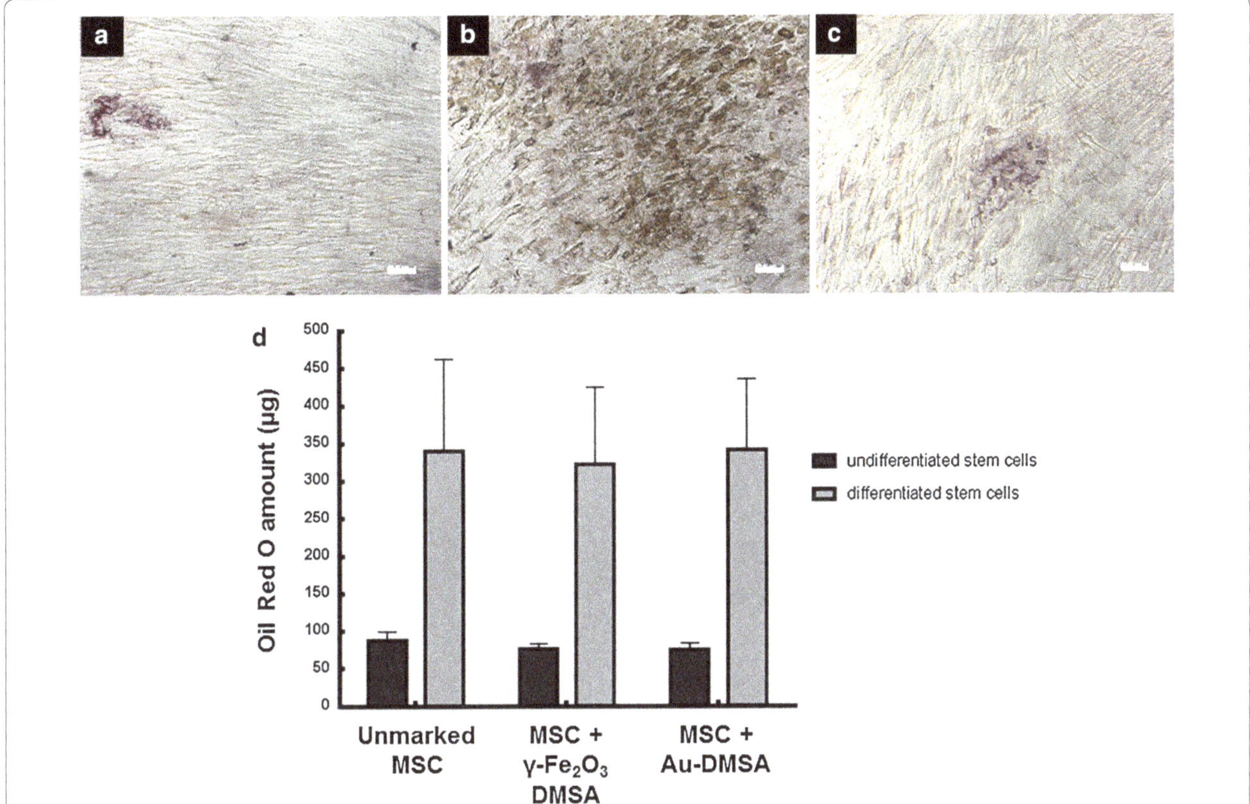

Fig. 6 MSC adipogenic differentiation assay. **a–c** Oil Red O cytochemical analysis of differentiated MSC monolayers in light microscopy, with in order to evidence the formation of intracellular lipid vesicles. These vesicles were seen in control cells (**a**), in γ-Fe_2O_3-DMSA labeled cells, (**b**) and in Au-DMSA labeled cells (**c**). **d** Quantification of Oil Red O incorporated in monolayers of differentiated and non-differentiated MSC, with no significant difference between control and experimental groups ($p > 0.05$). *Bars* 50 μm

O_2 [47]. In our work, TEM images showed few γ-Fe_2O_3-DMSA in structures similar to lysosomes; most was located in MSC mitochondria, which are neutral pH organelles. Considering that, γ-Fe_2O_3-DMSA could not exert toxic effects to receptor MSC.

Otherwise, MTT tests showed significant differences between unlabeled MSC and Au-DMSA labeled MSC, suggesting harmful effects in their mitochondria. In spite of this, our data followed the same pattern found by Fan et al. in 2009, which studied biocompatibility between gold nanoparticles (Au-NP) and MSC, using slightly lower gold concentrations (71,1 μg/mL) [51]. This similarity suggests that MSC have a natural sensitivity to Au-NP, that is, the toxic effects observed are not related to Au-DMSA features, but to MSC cellular mechanisms that make them more or less vulnerable to label [50]. In addition, Fig. 1c shows that MSC viability increases 48 and 72 h after exposure to Au-DMSA, which illustrates *cellular recovery* described by Mironava et al. [24, 26]: damage caused by Au-NP marking are not permanent because, after nanoparticle exposure, gold cytoplasmic levels diminish and cells can completely recover their structures and/or altered functions. Interestingly, despite deleterious effects on mitochondrial metabolism (Fig. 2b), trypan blue tests (Fig. 2d) and cell morphology analysis (Fig. 3) suggested that the Au-DMSA did not cause MSC death 24 h after exposure. So, we had to verify if these mitochondrial damages led to changes in important physiological parameters of cells; what was accomplished in the following experiments.

Some reports suggest that nanoparticles actively interact with plasma membrane receptors, modulating signal transduction pathways, and inducing proliferation, immunomodulation, apoptosis or differentiation [52]. These harmful effects caused by altered cellular communication pathways cannot be detected only with viability tests, such as MTT and Trypan Blue. Therefore, MSC differentiation, MSC proliferation and lymphocyte suppression tests were also performed. It is important to note that our study was the first that investigated nanoparticles effects on MSC immunomodulatory capacity. At the concentrations tested, both γ-Fe_2O_3-DMSA and Au-DMSA did not change this intrinsic property of cells, essential for the success of cellular therapies.

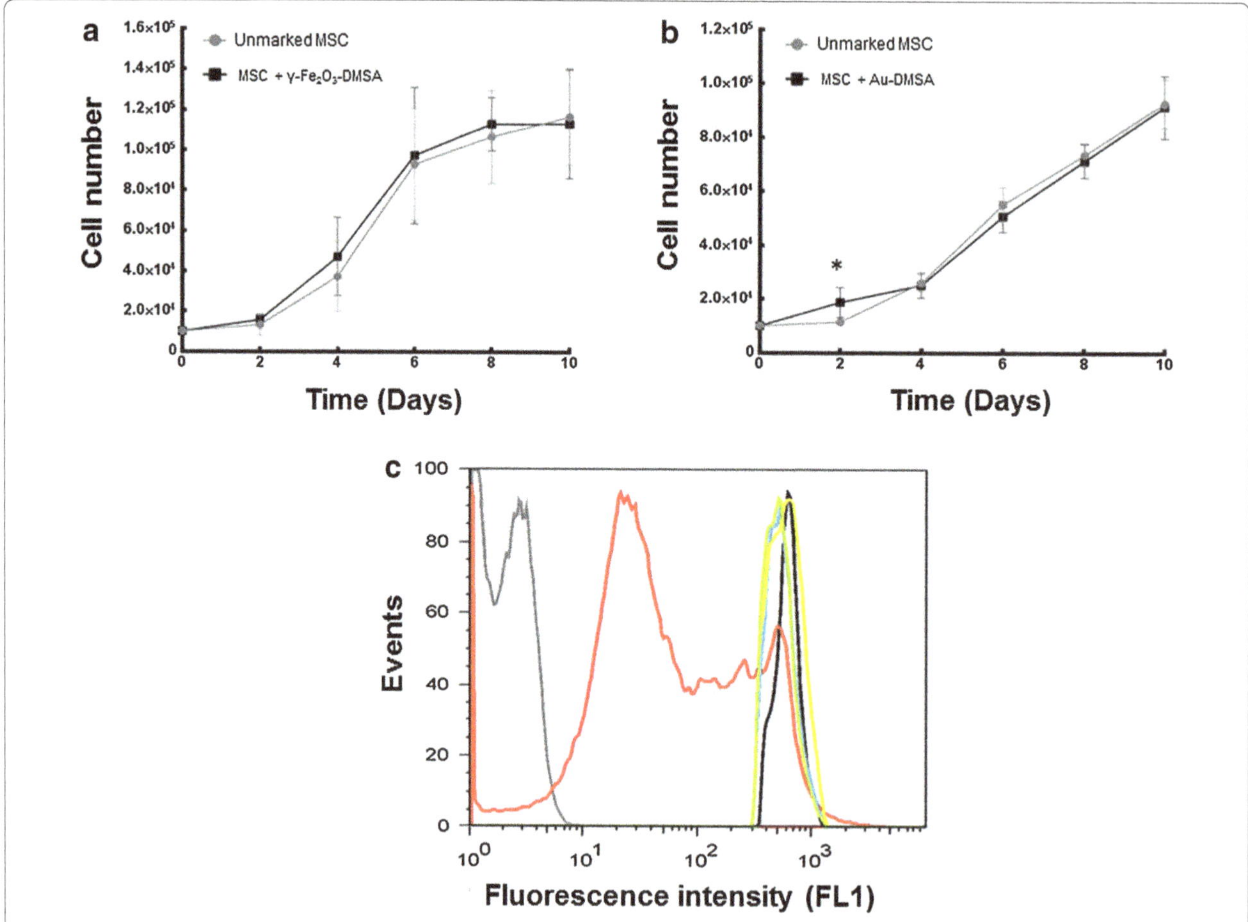

Fig. 7 **a**, **b** MSC proliferation curves. **a** Cells were incubated for 24 h with DMEM-LG (*filled circle*), or with DMEM-LG with diluted γ-Fe_2O_3-DMSA (*filled square*) (80 μg/mL), then were plated and counted after different times. There was no significant difference between the experimental groups ($p > 0.05$) in any count times. **b** Cells were incubated for 24 h with DMEM-LG (*filled circle*), or with DMEM-LG with diluted Au-DMSA (*filled square*) (90 μg/mL), then were plated and counted after different times. (*) Significant increase in Au-DMSA labeled MSC amount compared to the control group only on the second day ($p < 0.05$). **c** Analysis by flow cytometry of CFSE-marked lymphocytes, co-cultured with labeled and unlabeled MSC. The spectra shown are representative of assays performed in triplicate. *Gray Line* Lymphocytes not marked with CFSE; *Black line* not activated marked lymphocytes; *Red Line* activated marked lymphocytes; *Blue line* activated lymphocytes co-cultured with MSC; *Yellow line* activated lymphocytes co-cultured with γ-Fe_2O_3-DMSA labeled MSC; *Green line* activated lymphocytes co-cultured with Au-DMSA labeled MSC

In differentiation tests, MSC incubated with γ-Fe_2O_3-DMSA showed no changes in adipogenesis and osteogenesis capacity, confirming previously published work [27]. Unlikely, Au-DMSA reduced MSC osteogenesis, corroborating Fan et al. data [51]. Although many studies in the literature describe Au-NP stimulus on osteogenic differentiation and mineralization [29, 53, 54], Fan et al. verified a reduction in ALP activity and in calcium deposition, similar to our findings (Fig. 4). This disagreement may be caused by the Au-NP concentration used to label MSC: while 1.97×10^{-4} μg/mL induced differentiation in Zhang et al. report [54], 71.1 μg/mL inhibited osteogenesis in Fan et al. study [51]. The higher Au-NP amount caused cytotoxic effects by oxidative stress, which represses both osteogenic and adipogenic differentiation [51].

In proliferation assays, labeled MSC proliferation rates presented no differences compared to respective control groups, except a significant increase in Au-DMSA labeled MSC amount, on the 2nd day. This result disagrees with data presented by Mironava et al. [26]: MSC were incubated for 72 h with Au-NP (45 nm; 13, 20 and 26 μg/mL) and proliferation rates were lower than controls in all samples. The most notable difference between our study and Mironava et al. study was the incubation times—MSC were exposed to Au-NP during 24 and 72 h, respectively. Importantly, Mironava et al. aimed to observe long-term effects (3–6 days) and *cellular*

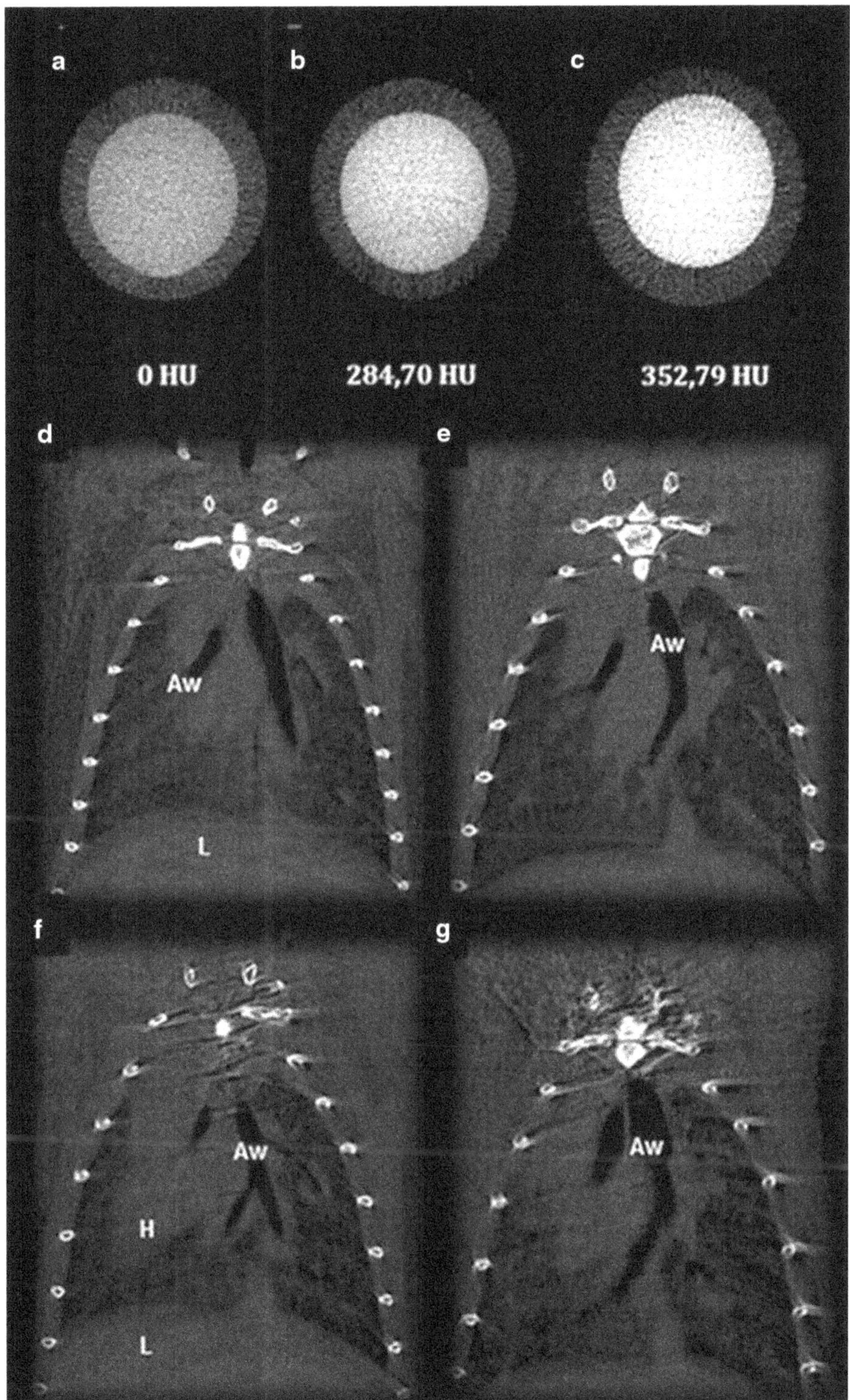

Fig. 8 Au-DMSA potential as a tracer for MSC tracking in computed microtomography. **a–c** Analysis of MSC precipitates in Sky-Scan 1640 microtomograph in which cross-sections of samples in Eppendorf tubes are represented: **a** water, **b** unlabeled MSC, **c** MSC labeled with Au-DMSA. **d–g** Sky-Scan 1640 images of longitudinal sections of mice. The animals were immediately analyzed after intranasal instillation with unlabeled cells (**d**) or with Au-DMSA labeled cells (**e**). Five days later, animals instilled with unlabeled cells (**f**) and Au-DMSA labeled cells (**g**) were analyzed again. *Aw* airways; *H* heart; *L* liver

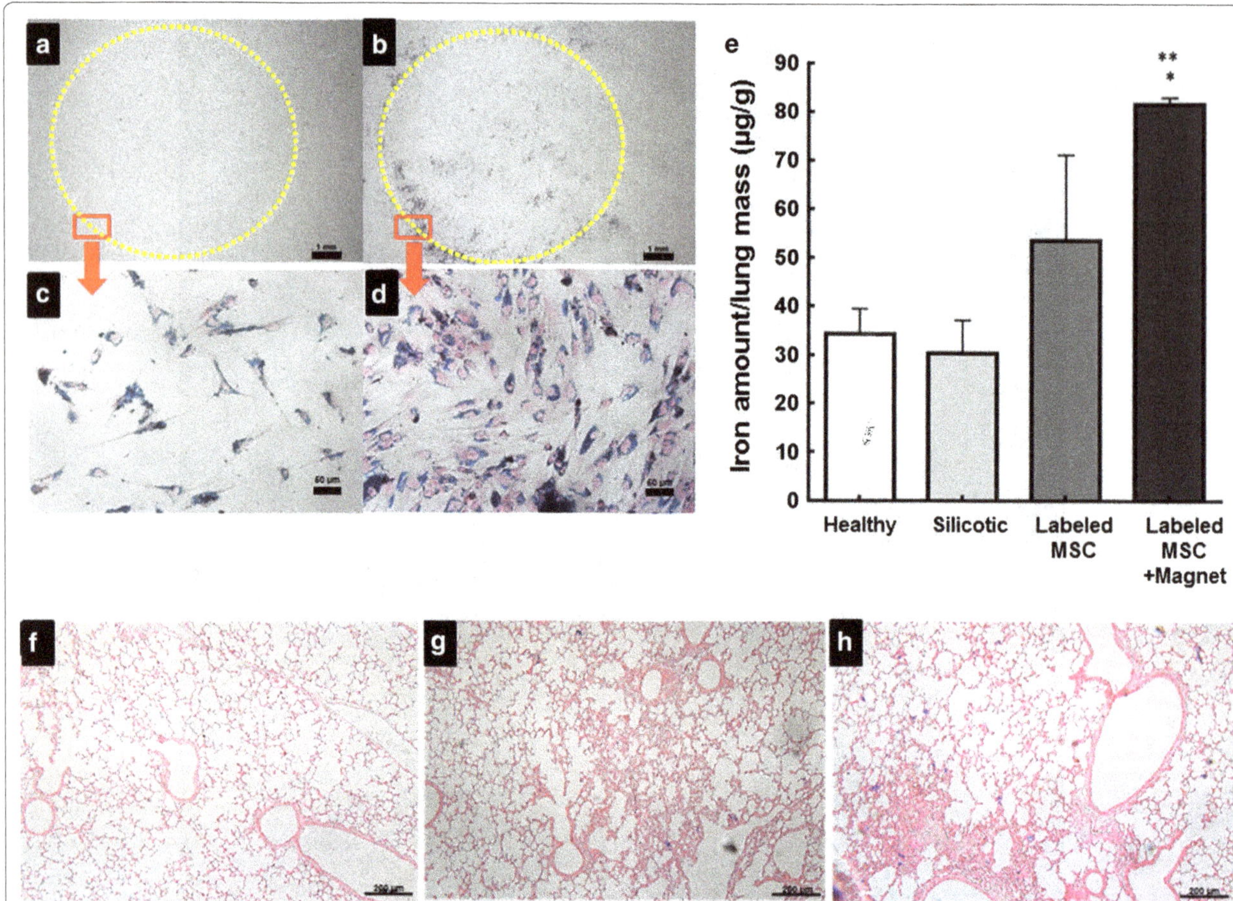

Fig. 9 γ-Fe_2O_3-DMSA as a potential agent for MSC magnetic targeting. **a–d** Magnetic responsiveness test in vitro. **a, b** γ-Fe_2O_3-DMSA labeled cells were seeded in culture plates with a fixed plastic piece (**a**) or a circular magnet (**b**). The region in *orange* corresponds to the site where the materials were fixed. **c, d** The culture plates were examined by light microscopy in order to demonstrate the difference in amount of cells present in regions *highlighted in red*. **e** Iron measurement by colorimetric dosage of Prussian blue. Data refer to the mean ± SD from the mass of iron present in lungs, divided by lung's weight. **f–h** Histological analysis of MSC retention in silicotic mice lungs. The slides were stained with Prussian Blue technique and contrasted with neutral red, evidencing iron from the γ-Fe_2O_3-DMSA in blue. **f** Healthy animal treated with saline **g** Animals treated with γ-Fe_2O_3-DMSA labeled MSC without external magnets **h** Animal treated with labeled MSC and with external magnets

recovery on different cell lines exposed to Au-NP, highlighting the cytotoxic effects over long incubation times. On the other hand, to date, there are no reports of stimulation of MSC proliferation by Au-NP. Together, all these data indicate that Au-DMSA nanoparticles are innocuous labels for MSC.

Some reports have described that Au-NPs enable cell tracking in vivo by using computed tomography [5, 17]; however in our study we could not detect the signals generated by the Au-DMSA in a microtomograph. Some facts may explain this disagreement: (1) It is possible Au-DMSA provided contrast to MSC, but not enough to distinguish them from the connective tissue associated with bronchi or the fibrous own tissue; (2) In our study, we injected an amount of labeled MSC lower than that used in other studies, for example, Menk et al. [17] inoculated 10^7 cells , while we used between 0.5 and 1×10^6 cells, an amount we considered safe to prevent lung occlusions; (3) The different tomography equipments used in our study and Menk et al. [17] study may also be a cause for the discrepancy observed. Menk et al. used an equipment connected to a source of synchrotron radiation, resulting in images with higher contrast and better visibility of details such as cells labeled with Au-NPs [55]. In addition, Au-NPs-labeled MSC are usually inoculated systemically in tracking studies [5, 17]. However, in this work, we decided to check the feasibility of inoculating MSC via intranasal route which is a more direct route for respiratory disease treatments. On the other hand, our in vitro and in vivo assays demonstrated γ-Fe_2O_3-DMSA potential for magnetic targeting of MSC. Qualitative (histological sections analysis) and quantitative assays (Prussian blue colorimetric dosage) showed a higher amount of iron in mice lungs which had magnets fixed in their bodies, suggesting a greater MSC retention. This was the first study that explored magnetic targeting of cells to the lung and our results raise the possibility of using this technique to enhance the effects of cellular therapy in lung diseases such as silicosis. Although it is known that pulmonary capillaries retains much of the systemically infused cells, there is evidence that this retention is not permanent [56], which may explain why beneficial effects of these cells in models of lung diseases do not last.

The findings of this study showed that iron and gold nanoparticles functionalized with DMSA, in the tested concentrations, were effectively uptaked by MSC, did not exert toxic effects, and did not induced changes in MSCs function (differentiation capacity, proliferation and inhibition of T lymphocytes). In addition, our results suggest the use of γ-Fe_2O_3-DMSA as agents for magnetic targeting of MSCs.

Abbreviations

ALP: alkaline phosphatase; Au-DMSA: meso-2,3-dimercaptosuccinic acid coated gold nanoparticles; CFSE: carboxyfluorescein succinimidyl ester; DLS: dynamic light scattering; DMEM-LG: low-glucose Dulbecco's modified Eagle medium; DMSA: meso-2,3-dimercaptosuccinic acid; ICP-OES: inductively coupled plasma optical emission spectrometry; IONPs: iron oxide nanoparticles; MSCs: mesenchymal stem cells; MTT: 3-[4,5-dimethylthiazol-2-yl]-2,5-diphenyltetrazolium bromide; TEM: transmission electron microscopy; XRD: X-ray powder diffraction; γ-Fe_2O_3-DMSA: meso-2,3-dimercaptosuccinic acid coated maghemite nanoparticles.

Authors' contributions

Experiments were designed by LHAS, JRS, EMCO, RBA and DMO; and conducted by LHAS, GAF, and RCS. Data was analyzed by LHAS, RCS, JRS, EMCO and DMO. Manuscript was prepared by LHAS and edited by JRS, EMCO, RBA and DMO. All authors read and approved the final manuscript.

Author details

[1] IB-Departamento de Genética e Morfologia, Universidade de Brasília-UNB, Campus Universitário Darcy Ribeiro-Asa Norte, Brasília, DF CEP 70910-970, Brazil. [2] Instituto de Química, Universidade Federal de Goias, Goiânia, GO, Brazil. [3] Instituto Nacional de Metrologia, Rio de Janeiro, RJ, Brazil.

Acknowledgements

Not applicable.

Competing interests

The authors declare that they have no competing interests.

Funding

Conselho Nacional de Pesquisa (Brazil) funded this study. The funders had no role in study design, data collection and analysis, decision to publish, or preparation of the manuscript.

of the Ethical Committee of Health Sciences Faculty of the University of Brasília (Brazil) (Project number CAAE 0020.0.012.000-08).

References

1. Nikalje AP. Nanotechnology and its Applications in Medicine. Med chem. 2015;5:081–089. doi:10.4172/2161-0444.1000247.
2. Ferreira L, Karp JM, Nobre L, Langer R. New opportunities: the use of nanotechnologies to manipulate and track stem cells. Cell Stem Cell. 2008;3:136–46.
3. Cromer Berman SM, Walczak P, Bulte JWM. Tracking stem cells using magnetic nanoparticles. Wiley Interdiscip Rev Nanomed Nanobiotechnol. 2011;3:343–55.
4. Guenoun J. Cationic Gd-DTPA liposomes for highly efficient labeling of mesenchymal stem cells and cell tracking with MRI. Cell Transpl. 2011;21(1):191–205.
5. Astolfo A, Schültke E, Menk RH, Kirch RD, Juurlink BHJ, Hall C, Harsan LA, Stebel M, Barbetta D, Tromba G, Arfelli F. In vivo visualization of gold-loaded cells in mice using x-ray computed tomography. Nanomed Nanotechnol Biol Med. 2013;9:284–92.
6. Bao C, Conde J, Polo E, Del Pino P. A promising road with challenges: where are gold nanoparticles in translational research? Nanomedicine. 2014;9(15):2353–70.
7. Colombo M, Carregal-Romero S, Casula MF, Gutiérrez L, Morales MP, Böhm IB, Heverhagen JT, Prosperi D, Parak WJ. Biological applications of magnetic nanoparticles. Chem Soc Rev. 2012;41:4306–34.
8. Jasmin, Torres ALM, Nunes HMP, Passipieri JA, Jelicks LA, Gasparetto EL, Spray DC, de Campos Carvalho AC, Mendez-Otero R. Optimized labeling of bone marrow mesenchymal cells with superparamagnetic iron oxide nanoparticles and in vivo visualization by magnetic resonance imaging. J Nanobiotechnol. 2011;9:4.
9. Ruan J, Ji J, Song H, Qian Q, Wang K, Wang C, Cui D. Fluorescent magnetic nanoparticle-labeled mesenchymal stem cells for targeted imaging and hyperthermia therapy of in vivo gastric cancer. Nanoscale Res Lett. 2012;7:1–12.
10. Yanai A, Häfeli UO, Metcalfe AL, Soema P, Addo L, Gregory-Evans CY, Po K, Shan X, Moritz OL, Gregory-Evans K. Focused magnetic stem cell targeting to the retina using superparamagnetic iron oxide nanoparticles. Cell Transpl. 2012;21:1137–48.
11. Vaněček V, Zablotskii V, Forostyak S, Růžička J, Herynek V, Babič M, Jendelová P, Kubinová S, Dejneka A, Syková E. Highly efficient magnetic targeting of mesenchymal stem cells in spinal cord injury. Int J Nanomed. 2012;7:3719–30.
12. Vandergriff AC, Hensley TM, Henry ET, Shen D, Anthony S, Zhang J, Cheng K. Magnetic targeting of cardiosphere-derived stem cells with ferumoxytol nanoparticles for treating rats with myocardial infarction. Biomaterials. 2014;35:8528–39.
13. Murphy CJ, Gole AM, Stone JW, Sisco PN, Alkilany AM, Goldsmith EC, Baxter SC. Gold nanoparticles in biology: beyond toxicity to cellular imaging. Acc Chem Res. 2008;41:1721–30.
14. Jokerst JV, Thangaraj M, Kempen PJ, Sinclair R, Gambhir SS. Photoacoustic imaging of mesenchymal stem cells in living mice via silica-coated gold nanorods. ACS Nano. 2012;6:5920–30.
15. Nam SY, Ricles LM, Suggs LJ, Emelianov SY. In vivo ultrasound and photoacoustic monitoring of mesenchymal stem cells labeled with gold nanotracers. PLoS One. 2012;7:e37267.
16. Cheong S-K, Jones BL, Siddiqi AK, Liu F, Manohar N, Cho SH. X-ray fluorescence computed tomography (XFCT) imaging of gold nanoparticle-loaded objects using 110 kVp x-rays. Phys Med Biol. 2010;55:647–62.
17. Menk RH, Schültke E, Hall C, Arfelli F, Astolfo A, Rigon L, Round A, Ataelmannan K, MacDonald SR, Juurlink BHJ. Gold nanoparticle labeling of cells is a sensitive method to investigate cell distribution and migration in animal models of human disease. Nanomed Nanotechnol Biol Med. 2011;7:647–54.
18. Mizutani R, Suzuki Y. X-ray microtomography in biology. Micron. 2012;43:104–15.

19. Halliwell B, Gutteridge JM. Oxygen toxicity, oxygen radicals, transition metals and disease. Biochem J. 1984;219:1–14.
20. Stohs SJ, Bagchi D. Oxidative mechanisms in the toxicity of metal ions. Free Rad Biol Med. 1995;18:321–36.
21. Pernodet N, Fang X, Sun Y, Bakhtina A, Ramakrishnan A, Sokolov J, Ulman A, Rafailovich M. Adverse effects of citrate/gold nanoparticles on human dermal fibroblasts. Small. 2006;2:766–73.
22. Pan Y, Neuss S, Leifert A, Fischler M, Wen F, Simon U, Schmid G, Brandau W, Jahnen-Dechent W. Size-dependent cytotoxicity of gold nanoparticles. Small. 2007;3:1941–9.
23. Brown CL, Whitehouse MW, Tiekink ERT, Bushell GR. Colloidal metallic gold is not bio-inert. Inflammopharmacology. 2008;16:133–7.
24. Mironava T, Hadjiargyrou M, Simon M, Jurukovski V, Rafailovich MH. Gold nanoparticles cellular toxicity and recovery: effect of size, concentration and exposure time. Nanotoxicology. 2010;4:120–37.
25. Fan J-H, Li W-T, Hung W-I, Chen C-P, Yeh J-M. Cytotoxicity and differentiation effects of gold nanoparticles to human bone marrow mesenchymal stem cells. Biomed Eng Appl Basis Commun. 2011;23:141–52.
26. Mironava T, Hadjiargyrou M, Simon M, Rafailovich MH. Gold nanoparticles cellular toxicity and recovery: adipose derived stromal cells. Nanotoxicology. 2014;8:189–201.
27. Kostura L, Kraitchman DL, Mackay AM, Pittenger MF, Bulte JWM. Feridex labeling of mesenchymal stem cells inhibits chondrogenesis but not adipogenesis or osteogenesis. NMR Biomed. 2004;17:513–7.
28. Chen Y-C, Hsiao J-K, Liu H-M, Lai I-Y, Yao M, Hsu S-C, Ko B-S, Chen Y-C, Yang C-S, Huang D-M. The inhibitory effect of superparamagnetic iron oxide nanoparticle (Ferucarbotran) on osteogenic differentiation and its signaling mechanism in human mesenchymal stem cells. Toxicol Appl Pharmacol. 2010;245:272–9.
29. Yi C, Liu D, Fong C-C, Zhang J, Yang M. Gold nanoparticles promote osteogenic differentiation of mesenchymal stem cells through p38 MAPK pathway. ACS Nano. 2010;4:6439–48.
30. Huang D-M, Hsiao J-K, Chen Y-C, Chien L-Y, Yao M, Chen Y-K, Ko B-S, Hsu S-C, Tai L-A, Cheng H-Y, Wang S-W, Yang C-S, Chen Y-C. The promotion of human mesenchymal stem cell proliferation by superparamagnetic iron oxide nanoparticles. Biomaterials. 2009;30:3645–51.
31. Marquis BJ, Love SA, Braun KL, Haynes CL. Analytical methods to assess nanoparticle toxicity. Analyst. 2009;134:425–39.
32. Soenen SJH, De Cuyper M. How to assess cytotoxicity of (iron oxide-based) nanoparticles: a technical note using cationic magnetoliposomes. Contrast Media Mol Imaging. 2011;6:153–64.
33. Barrow M, Taylor A, Murray P, Rosseinsky MJ, Adams DJ. Design considerations for the synthesis of polymer coated iron oxide nanoparticles for stem cell labelling and tracking using MRI. Chem Soc Rev. 2015;44:6733–48.
34. Auffan M, Decome L, Rose J, Orsiere T, De Meo M, Briois V, Chaneac C, Olivi L, Berge-lefranc J, Botta A, Wiesner MR, Bottero J. In vitro interactions between DMSA-coated maghemite nanoparticles and human fibroblasts: a physicochemical and cyto-genotoxical study. Environ Sci Technol. 2006;40:4367–73.
35. Wang Y, Wang L, Che Y, Li Z, Kong D. Preparation and evaluation of magnetic nanoparticles for cell labeling. J Nanosci Nanotechnol. 2011;11:3749–56.
36. Soler MAG, Lima ECD, Nunes ES, Silva FLR, Oliveira AC, Azevedo RB, Morais PC. Spectroscopic study of maghemite nanoparticles surface-grafted with DMSA. J Phys Chem A. 2011;115:1003–8.
37. Cullity BD, Stock SR. Elements of X-ray diffraction. 3rd ed. Prentice Hall; 2001. Chapter 1.
38. Gao J, Huang X, Liu H, Zan F, Ren J. Colloidal stability of gold nanoparticles modified with thiol compounds: bioconjugation and application in cancer cell imaging. Langmuir. 2012;28:4464–71.
39. Pereira LO, Longo JPF, Azevedo RB. Laser irradiation did not increase the proliferation or the differentiation of stem cells from normal and inflamed dental pulp. Arch Oral Biol. 2012;57:1079–85.
40. Borenfreund E, Babich H, Martin-Alguacil N. Comparisons of two in vitro cytotoxicity assays-The neutral red (NR) and tetrazolium MTT tests. Toxicol In Vitro. 1988;2:1–6.
41. Reich-Slotky R, Colovai AI, Semidei-Pomales M, Patel N, Cairo M, Jhang J, Schwartz J. Determining post-thaw CD34+ cell dose of cryopreserved haematopoietic progenitor cells demonstrates high recovery and confirms their integrity. Vox Sang. 2008;94:351–7.
42. Boutry S, Forge D, Burtea C, Mahieu I, Murariu O, Laurent S, Vander Elst L, Muller RN. How to quantify iron in an aqueous or biological matrix: a technical note. Contrast Media Mol Imaging. 2009;4:299–304.
43. Gregory CA, Gunn WG, Peister A, Prockop DJ. An Alizarin red-based assay of mineralization by adherent cells in culture: comparison with cetylpyridinium chloride extraction. Anal Biochem. 2004;329:77–84.
44. Jaiswal N, Haynesworth SE, Caplan AI, Bruder SP. Osteogenic differentiation of purified, culture-expanded human mesenchymal stem cells in vitro. J Cell Biochem. 1997;64:295–312.
45. Lowry OH, Rosebrough NJ, Farr AL, Randall RJ. Protein measurement with the Folin phenol reagent. J Biol Chem. 1951;193:265–75.
46. Faffe DS, Silva GH, Kurtz PMP, Negri EM, Capelozzi VL, Rocco PRM, Zin WA. Lung tissue mechanics and extracellular matrix composition in a murine model of silicosis. J Appl Physiol. 2001;90:1400–6.
47. Chen Z, Yin J, Zhou Y, Zhang Y, Song L, Song M, Hu S, Gu N, Al CET. Dual enzyme-like activities of iron oxide nanoparticles and their implication for diminishing cytotoxicity. ACS Nano. 2012;6(5):4001–12.
48. Sun J-H, Zhang Y-L, Qian S-P, Yu X-B, Xie H-Y, Zhou L, Zheng S-S. Assessment of biological characteristics of mesenchymal stem cells labeled with superparamagnetic iron oxide particles in vitro. Mol Med Rep. 2012;5:317–20.
49. Auffan M, Rose J, Wiesner MR, Bottero J-Y. Chemical stability of metallic nanoparticles: a parameter controlling their potential cellular toxicity in vitro. Environ Pollut. 2009;157:1127–33.
50. Li J, Chang X, Chen X, Gu Z, Zhao F, Chai Z, Zhao Y. Toxicity of inorganic nanomaterials in biomedical imaging. Biotechnol Adv. 2014;32(4):727–43.
51. Fan J, Hung W, Li W, Yeh J. Biocompatibility study of gold nanoparticles to human cells. In: 13th international conference. Berlin: Springer; 2009.
52. Rauch J, Kolch W, Laurent S, Mahmoudi M. Big signals from small particles: regulation of cell signaling pathways by nanoparticles. Chem Rev. 2013;113:3391–406.
53. Ricles LM, Nam SY, Sokolov K, Emelianov SY, Suggs LJ. Function of mesenchymal stem cells following loading of gold nanotracers. Int J Nanomed. 2011;6:407–16.
54. Zhang D, Liu D, Zhang J, Fong C, Yang M. Gold nanoparticles stimulate differentiation and mineralization of primary osteoblasts through the ERK/MAPK signaling pathway. Mater Sci Eng C. 2014;42:70–7.
55. Yu Y, Ning R, Cai W, Liu J, Conover D. Performance investigation of a hospital-grade X-ray tube-based differential phase-contrast cone beam CT system. Proc SPIE. 2012;. doi:10.1117/12.911400.
56. Wang H, Cao F, De A, Cao Y, Contag C, Gambhir SS, Wu JC, Chen X. Trafficking mesenchymal stem cell engraftment and differentiation in tumor-bearing mice by bioluminescence imaging. Stem Cells. 2009;27:1548–58.

14

Vertical transport and plant uptake of nanoparticles in a soil mesocosm experiment

Alexander Gogos[1,2], Janine Moll[1], Florian Klingenfuss[1], Marcel van der Heijden[1], Fahmida Irin[3], Micah J. Green[4], Renato Zenobi[2] and Thomas D. Bucheli[1*]

Abstract

Background: Agricultural soils represent a potential sink for increasing amounts of different nanomaterials that nowadays inevitably enter the environment. Knowledge on the relation between their actual exposure concentrations and biological effects on crops and symbiotic organisms is therefore of high importance. In this part of a joint companion study, we describe the vertical translocation as well as plant uptake of three different titanium dioxide (nano-)particles (TiO_2 NPs) and multi-walled carbon nanotubes (MWCNTs) within a pot experiment with homogenously spiked natural agricultural soil and two plant species (red clover and wheat).

Results: TiO_2 NPs exhibited limited mobility from soil to leachates and did not induce significant titanium uptake into both plant species, although average concentrations were doubled from 4 to 8 mg/kg Ti at the highest exposures. While the mobility of MWCNTs in soil was limited as well, microwave-induced heating suggested MWCNT-plant uptake independent of the exposure concentration.

Conclusions: Quantification of actual exposure concentrations with a series of analytical methods confirmed nominal ones in soil mesocosms with red clover and wheat and pointed to low mobility and limited plant uptake of titanium dioxide nanoparticles and carbon nanotubes.

Keywords: Nanomaterials, Black carbon, Soil leachate, Multi-angle light scattering, Microwave induced heating, Wheat, Red clover

Background

It is scientifically ascertained that, due to their increased production and use, nanomaterials (NMs) will inevitably enter the environment [1], including soils. The currently most produced NMs are titanium dioxide nanoparticles (TiO_2 NPs) [2]. They are used in diverse applications such as paints, UV-protection, photovoltaics and photocatalysis [3], but also as a food additive [4]. Carbon nanotubes (CNTs) are closing the gap in the last years, with 10-fold increased production volumes since 2006 [5]. Due to their extraordinary mechanical and electrical properties, CNTs are mostly used as building blocks in light-weight composite materials as well as electronics.

*Correspondence: thomas.bucheli@agroscope.admin.ch
[1] Agroscope, Institute for Sustainability Sciences ISS, 8046 Zurich, Switzerland
Full list of author information is available at the end of the article

These particles can enter soils via different pathways [1, 6]. Application of biosolids to landfills and irrigation with surface waters is most likely for TiO_2 NPs, while CNTs may enter soils via landfills and atmospheric deposition [7]. These types of release are unintentional, however, also applications in plant protection and fertilization have been foreseen [8, 9], which may lead to severely increased fluxes of these NP into soils. Apart from the positive effects and functions that are envisioned for agricultural applications of TiO_2 NPs and CNTs [8, 9], such as protection of active ingredients and increased plant growth, respectively, also negative effects on microorganisms and plants have been reported [10–12].

The enduring uncertainty regarding the environmental safety of NMs highlights the need for a thorough risk assessment of these materials, which includes the study of their effects on organisms and the ecosystem as well as their fate. However, the analysis of NMs such as TiO_2 and

CNTs in complex systems such as real soils is challenging in many ways. For both, elemental analysis alone is not sufficient to trace the particles due to high elemental background concentrations of Ti and carbon.

Therefore, most studies until now used simplified laboratory systems as well as specifically labeled particles for eased detection to investigate both NM transport through porous media as well as plant uptake, often without confirmation of actual exposure concentrations. For example, TiO_2 NP transport was investigated in sand columns under well controlled conditions [13, 14]. Fang et al. [15] studied TiO_2 NP transport through soil columns at very high concentrations (40 g/kg). However, vertical translocation of both TiO_2 NPs and CNTs has neither been investigated yet in large pot experiments or field studies, nor in the presence of plants. Plant uptake was shown for TiO_2 NP in hydroponic exposure systems at high concentrations [16, 17]. In contrast, in a more realistic exposure setting using natural soil amended with TiO_2 NPs, Du et al. [12] found no uptake of Ti into wheat. Also, CNTs were shown to be taken up into plants [18–20] from hydroponic systems. However, until now, no data is available for CNT uptake from natural soils, in which CNT transport and subsequent availability to plants could be different due to their high interactions with the soil matrix [21–23].

Here, we investigated the vertical distribution and leaching behavior of three different TiO_2 (nano-)particles [P25, E171 and a non-nanomaterial TiO_2 (NNM TiO_2)] and the vertical distribution of a multi-walled CNT (MWCNT) within two elaborate pot exposure studies with red clover (*Trifolium pratense*) [24] and spring wheat (*Triticum spp.*) [25] in natural soil, and quantified their fractions in aboveground parts of the plants. We used recently developed methods such as microwave induced heating (MIH) [26] and asymmetric flow field-flow fractionation coupled to multi-angle light scattering (aF4-MALS) [27] to detect and quantify unlabeled MWCNTs in plant and soil samples, respectively. We additionally imaged root cross sections of exposed plants using (scanning) transmission electron microscopy. All data from this study were gathered to accompany two corresponding effect studies with actual, rather than nominal exposure concentrations. These studies examined the functionality of an agricultural ecosystem in presence of the NMs with regard to nitrogen fixation by the red clover-rhizobium symbiosis, as well as root colonization by arbuscular mycorrhizal fungi of both red clover [24] and wheat [25].

Methods

Chemicals and nanoparticles

Food grade E171 TiO_2 particles were obtained from Sachtleben Chemie GmbH (Duisburg, Germany). All other chemicals and TiO_2 nanoparticles were purchased from Sigma-Aldrich (Buchs, Switzerland). Uncoated titanium containing NPs were selected to represent different primary particle size ranges; average primary particle sizes were determined by TEM image analysis and were 29 ± 9 (P25, n = 92), 92 ± 31 (E171, n = 52) and 145 ± 46 nm (NNM TiO_2, n = 49), see also Additional file 1: Figure S1. Anatase was the dominating crystal structure in all of the used particles. However, P25 also contains 20 % rutile, according to the manufacturer.

Multi-walled carbon nanotubes were purchased from Cheap Tubes Inc. (Brattleboro, VT). They were declared to have a length of 10–30 µm, and outer diameter of 20–30 nm, a purity of >95 % and an elemental carbon content of >98 %. The MWCNTs were used as received without further purification. Further characterization of the MWCNTs used was carried out and described in [27, 28]. All parameters were confirmed to be within the specified ranges with the exception of CNT length. The latter could only be determined in suspension, where it may have been altered due to sonication necessary for dispersing the particles.

Soil

A natural soil was collected from an agricultural field at the facility of Agroscope, Zurich (N47° 25′ 39.564″ E8° 31′ 20.04″). The soil was classified as brown earth with a sandy loamy to loamy fine fraction. The top layer (5 cm) of the soil was removed and approximately 0.9 m^3 of the underlying 15 cm topsoil were sampled. The soil was then sieved <5 mm, homogenized by shoveling it three times from one soil pile to another, and stored in a dry place until it was used in both red clover and wheat experiments.

Spiking of the soil with NPs

Particle concentrations were selected to represent potential agricultural exposure scenarios as well as analytically accessible and potentially toxicologically effective concentrations. In a potential agricultural exposure scenario, fluxes from pesticide or fertilizer formulations may range from several micrograms to grams of NMs per kilogram of soil, depending on the formulation [8]. Thus, low doses (1, 10 mg/kg) were included as well as high doses.

For the spiking process, the soil was firstly blended with quartz sand (50 % v/v) to facilitate the recovery of below-ground plant organs after harvest. The properties of the soil-sand mixture are listed in Table 1. First, 300 g of the sand-soil mixture were each mixed with (i) 0.03 g (wheat experiment only), 0.3 g (red clover experiment only), 3 and 30 g of TiO_2 NPs (both experiments), and (ii) 90 mg and 88 g MWCNT powder (clover experiment only), each in a 500 mL glass bottle which was rotated in

Table 1 Properties of the soil-quartz mixture (50:50 v/v) administered to the pots

	Value	StDev
Org. C %	0.55	0.03
CEC mmol+/kg	6	
$CaCO_3$ %	2.6	
pH	7.7	
Max. WHC g H_2O/g dry soil	0.308	
Sand %	86.1	0
Silt %	6.3	0
Clay %	6.7	0.5

a powder mixer (Turbula® T 2 F, Willy A. Bachofen AG, Basel, Switzerland) for 30 min. For P25 and MWCNTs, the highest particle amounts resulted in a volume too big for the glass bottles. Therefore, these were split in two and four aliquots, respectively, and each aliquot mixed with 300 g sand-soil mixture.

Into a cement mixer, 30 kg (including the pre-mixture) of a fresh sand-soil mixture (50 % v/v) were added, to yield final nominal NP concentrations of 1, 10, 100 or 1000 mg/kg, respectively, for TiO_2 NPs, and 3 or 2933 mg/kg for MWCNTs. The mixing chamber was covered with a plastic sheet to avoid dust formation and run for 6 h. The soil was not dried before mixing to avoid changes to the microbial community structure, also investigated in Moll et al. [25]. Actual exposure concentrations were verified by X-ray fluorescence spectroscopy (XRF, for TiO_2) and chemo-thermal oxidation at 375 °C [28] [CTO-375, for MWCNTs/Black Carbon (BC)] analysis as described below.

General experimental design

A detailed description of the general setup, design and execution of the underlying exposure experiments is given in [24, 25]. In brief, for each plant type seven pot replicates were generated for each NP treatment, consisting of seven plants per pot for red clover and three for wheat. Non-plant controls were not performed because these two studies were primarily designed to observe possible biological effects of the NP treatments. Each pot was filled with a drainage layer of sand (0.5 L, 520 g) and 3.3 kg soil (corresponding to 2.9 L). Each pot was kept at 50–60 % (wheat) and 60–70 % (red clover) of the total water holding capacity (WHC, Table 1) during the entire experiment. Plants were grown over a period of 14 weeks (red clover) and 12 weeks (wheat) in a greenhouse with a 16 h light period (light intensity of 300 W m^{-2}) and a 25/16 °C light/dark temperature regime. Wheat plants were fertilized weekly starting after week 3. Red clover plants were fertilized after 6 and 9 weeks, respectively. The composition of the nutrient solutions is given in the Additional file 1.

Sampling of soil cores

Soil cores were sampled at the day of harvest from each pot using a conventional soil driller with a 2 cm diameter. Two cores were taken per pot and each divided into three depths (0–5, 5–10 and 10–15 cm). For each depth, both subsamples were joined into one and stored in plastic bags at 4 °C until further processing.

Titanium analysis in soils with XRF

The soil samples from the cores were dried at 60 °C until a constant weight resulted, and ground to a fine powder using a Retsch ZM400 Ball Mill (Retsch GmbH, Haan, Germany) with a tungsten carbide bead at a frequency of 25/s for 5 min. Four grams of ground soil were homogenously mixed with 0.9 g of wax and pressed to a 32 mm tablet at 15 tons. Tablets were analyzed using an energy-dispersive XRF spectrometer (XEPOS, SPECTRO Analytical Instruments GmbH, Kleve, Germany). For correction of matrix effects, standard additions of the respective material to the soil were performed. For quality assurance we also analyzed a certified lake sediment reference sample (LKSD1, CANMET Mining and Mineral Sciences Laboratories, Ontario, Canada) with recoveries for Ti of >95 %.

Titanium analysis in leachates with ICP-OES

A week before harvest, each pot was watered with 520 mL tap water, leading to approx. 110 % WHC. Consequently, 45 mL of leachate were collected through a valve at the bottom of the pots. The leachate was analyzed on the same day without any further treatment using inductively-coupled plasma optical emission spectrometry (ICP-OES) (ARCOS, SPECTRO Analytical Instruments GmbH). For quality control, an external Ti containing standard solution (ICAL, Bernd Kraft GmbH, Duisburg, Germany) was analyzed. The instrumental limit of quantification for Ti was determined at 22 µg/L.

MWCNT analysis of soil with CTO-375

The CTO-375 procedure used in this study is described in detail in Sobek and Bucheli [28] as well as specifically for this work in the Additional file 1. This method quantifies total soil BC, which also encompasses MWCNT-carbon. We analyzed the soil samples taken from the cores, as well as the bulk spiked soil before the experiment. For the latter, six random grab samples of approx. 10 g were taken from the spiked pile.

MWCNT analysis of soil with aF4-MALS

The method for MWCNT detection using aF4-MALS is described in detail by Gogos et al. [27]. Briefly, 120 mg of dry and ground soil from the cores were extracted with 10 mL of a 2 % sodium deoxycholate/0.05 % sodium azide solution, sonicated three times for 10 min using a high power sonication bath (720 W, Bandelin, Switzerland) and centrifuged at 17,500 g for 10 min. The supernatant was then used as a working suspension. This procedure was performed for each replicate of each soil depth. Afterwards, the replicates of each depth were joined to form a collective sample and analyzed using aF4-MALS, which generates a shape factor ρ from the radius of gyration and the hydrodynamic radius for each time point in the aF4 fractogram. The difference in ρ (Δρ) compared to native soil is then used to detect the MWCNTs [27]. The method detection limit (MDL) of the present study is presented and further discussed in the "Results and discussion" section.

Titanium analysis of plants with ICP-OES

Due to their high importance for agricultural scenarios, from both plants, the parts used as food or feed were analyzed, i.e. the whole aboveground red clover, and the wheat grains. Dried plant samples were ground to a fine powder using a Retsch ZM200 centrifugal mill (Retsch GmbH). Subsamples (100 mg) were digested in a mixture of 0.2 mL hydrofluoric acid, 1.5 mL nitric acid and 0.2 mL hydrogen peroxide using a microwave (Ultraclave, MLS, Germany). The sample volume was subsequently adjusted to 50 mL. Digested samples were analyzed using ICP-OES (CIROS, SPECTRO Analytical Instruments GmbH). For quality assurance we also analyzed an industrial sludge reference sample (standard reference material SRM 2782, NIST, Gaithersburg, US) with recoveries for Ti of >85 %.

MWCNT analysis of plants with MIH

Dry plant material was ground to a fine powder as described before. The amount of MWCNT uptake was then quantified by MIH, which is described in detail by Irin et al. [26]. MWCNTs have a high microwave absorption capacity, which results in a rapid rise in temperature within a very short microwave exposure time. Original method development included the generation of a calibration curve using the thermal response as a function of known CNTs spiked into Alfalfa (*Medicago sativa)* root samples.

Utilizing the data from Irin et al. [26], a new calibration curve was generated, where the slope of the curve depends on the respective nanomaterial and the intercept on the sample type. To this end, first, the initial slope was corrected using a factor based on the ratio of the source nanomaterials (MWCNTs of this study) microwave sensitivity and the one of the Irin et al. study. The sensitivity was determined by exposing ~1 mg of MWCNT powder to 30 W microwave power (2.45 GHz frequency) and recording the final temperature rise immediately (within 1 s) with a temperature rise (ΔT) of 346 °C. Second, the intercept was corrected based on the control plant microwave response. Additional file 1: Figure S2 shows the renormalized calibration curve for MWCNTs at 50 W (6 s). The plant samples from the controls and the two MWCNT treatments were then tested at 50 W over 6 s and the quantity of MWCNT uptake were calculated using this new calibration curve. The limit of detection (LOD) as well as the limit of quantification (LOQ) where calculated based on the temperature rise from five measurements of control plant samples (blank signal) according to Keith et al. [29] (3 and 10 σ above the blank signal, respectively).

Transmission electron microscopy of root cross sections

Fresh root samples were washed with tap water and prefixed in 2.5 % glutaraldehyde in phosphate buffered saline directly on the day of harvest and stored at 4 °C until processing. Ultrathin cross Sects. (70 nm thickness) were obtained by cutting root samples embedded in epoxyresin using an ultramicrotome (Ultracut E, Leica, Wetzlar, Germany). The detailed sample preparation steps are provided in the Additional file 1. Ultrathin sections were imaged using a TEM (Tecnai G2 Spirit, FEI, Hillsboro, USA), coupled to an energy-dispersive X-ray (EDX) spectroscope (X-Max, 80 mm^2, Oxford Instruments, Abingdon, UK) as well as a STEM (HD-2700-Cs, Hitachi, Japan) coupled to an EDX system as well (EDAX, NJ).

Statistics

In the case of normal distributed residuals and homogenous data, an analysis of variance (ANOVA) was applied. If these model assumptions were not fulfilled, a Mann–Whitney test was conducted. All statistical analyses were done with the software R (version 3.01, the R Foundation for Statistical Computing) integrated in RStudio (version 0.97.551, RStudio, Boston, MA).

Results and discussion

Vertical soil distribution and leaching of Ti

Only the highest exposure concentration (1000 mg/kg) was analytically accessible using XRF, i.e., standard deviations among the replicates were in the order of the added Ti amount in samples spiked with <1000 mg/kg TiO_2. Actual dry weight exposure concentrations of Ti were almost always slightly higher at the time of harvest than the initial nominal ones predicted from native and added Ti amounts, probably due to the residual water content in soils at the time of spiking (Fig. 1b, c, e, f). However, the differences were minimal (2.5–7.6 %) and overall not statistically significant (except for Fig. 1c, P25 1000 mg/kg, 5–10 cm), indicating that the employed spiking procedure was rather reliable. The control soils in the wheat experiment were systematically—though not significantly—lower in Ti content and showed higher standard deviations compared to the controls in the red clover experiment. This unexpected result may be explained by the fact that the two experiments were conducted independently using different subsets of the native soil and also highlights the necessity to verify actual exposure concentrations.

No statistically significant difference could be found between the different soil layers in any of the treatments (Fig. 1). Still, some trends could be observed; the distribution profiles of Ti in the control and in the P25

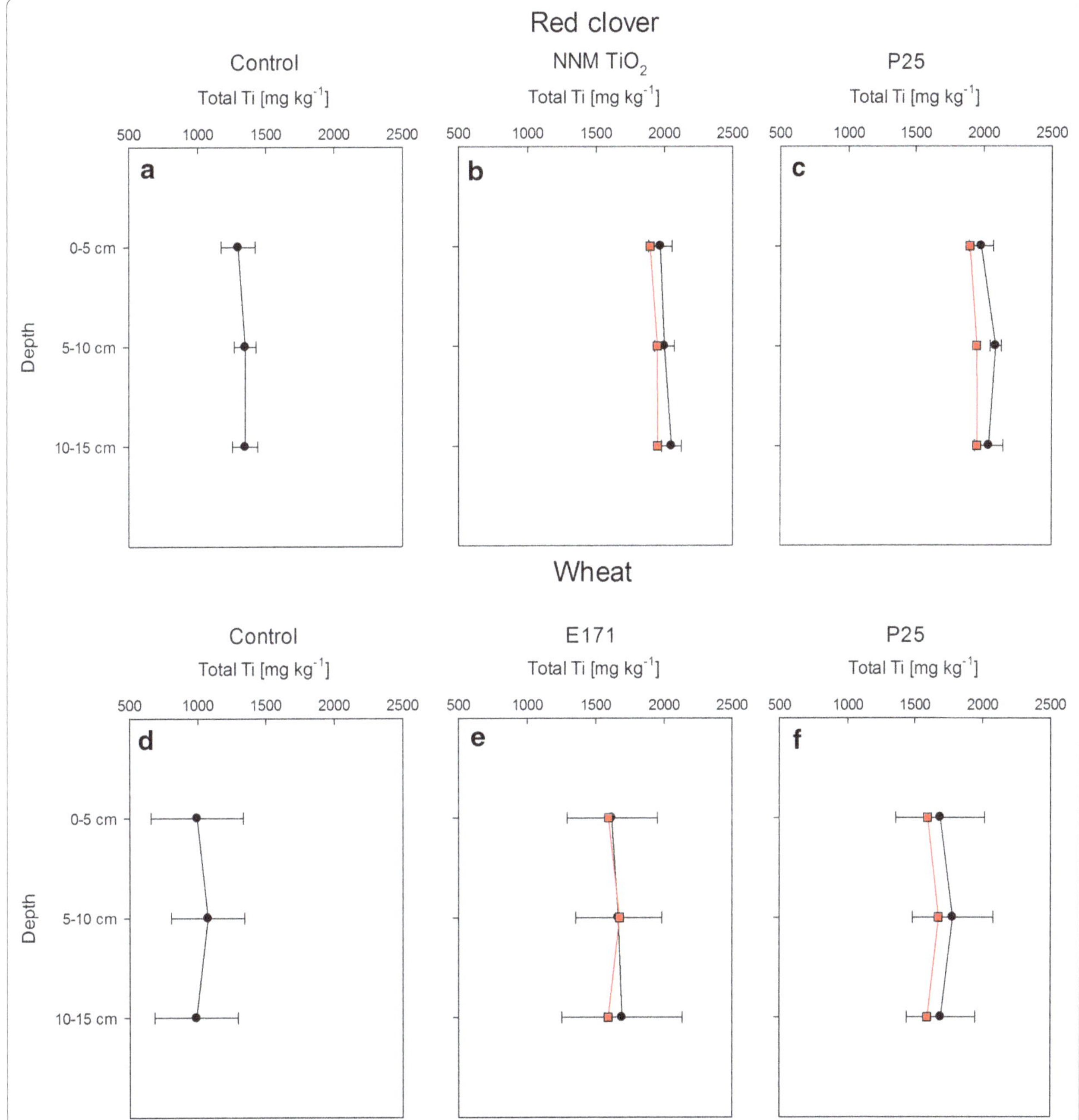

Fig. 1 Vertical distributions of elemental Ti as determined by XRF analysis for three depths and for two different exposure experiments: **a–c** Red clover controls and red clover exposed to 1000 mg/kg of NNM TiO_2 and P25 and **d–f** Wheat controls and wheat exposed to 1000 mg/kg of E171 and P25. *Error bars* show the standard deviation of seven replicates. *Red squares* show the predicted concentrations based on the control values and the nominal amount of Ti that was added as TiO_2 NPs

(80 % anatase, 20 % rutile) treatments were similar, with a tendency to slightly higher concentrations in the middle layer in both red clover and wheat pots. In contrast, the distribution profiles of the two pure anatase particles (NNM and E171) both tended towards elevated concentrations in the lowest part.

In addition, Ti concentrations in leachates of these two treatments were significantly elevated compared to the controls (Fig. 2, $p < 0.05$), thus it can be assumed that the elevated Ti originated from eluting TiO_2 NPs. However, the leached Ti amount—even in the treatments showing significantly higher concentrations—was very low and constituted not more than 10^{-4} % of the initial spiked Ti amount. In a dedicated transport study by Fang et al. [15], a soil with comparable properties (sandy loam, denoted as "JS soil") showed a medium to high permeability for TiO_2 NPs, attributed to the soil's high sand content. A breakthrough of Ti in this soil started to occur after 1 pore volume. In our case, 520 mL of water was added to the pots (equivalent to 30 mm of precipitation) to collect the leachate, which correspond to 0.4 pore volumes only (1.24 L pore volume at full WHC). Thus, the added water amount was too low to initiate quantitative elution and would therefore explain the relatively low Ti concentration in the leachate after collection.

The observed difference in mobility (both in terms of Ti profiles and leachate content) may partly be explained by differences in the isoelectric point (IEP) of the TiO_2 particles: while the more mobile NNM TiO_2 and E171 exhibited a very low IEP of 2.2 (see Additional file 1: Figure S3), the one of P25 was 5.1, being much closer to the soil pH (7.7, see Table 1) and indicating a lesser colloidal stability [30]. TiO_2 NPs with low IEPs may thus have a higher tendency to reach the groundwater and should thus be avoided in applications where this might be of relevance, e.g., when used as a component of a plant protection product [8, 9].

Vertical soil distribution of BC/MWCNTs

Figure 3 shows the BC distribution as well as the shape factor difference ($\Delta\rho$) for the different soil depths of the 2933 mg/kg MWCNT amended red clover pots. As with Ti, only the highest MWCNT concentration was analytically accessible. The total background BC in the control soil was 0.50 ± 0.06 mg/g ($n = 4$). The specific recovery of the employed MWCNT in the soil over the

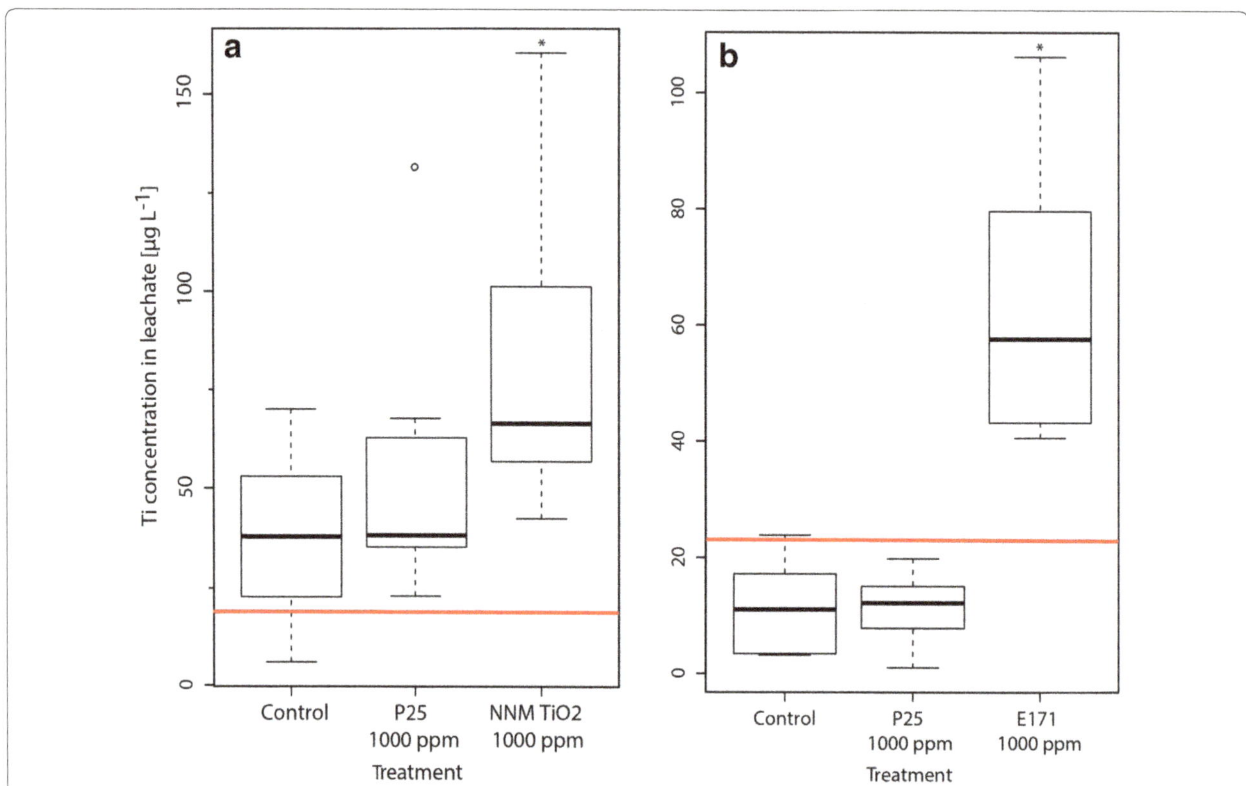

Fig. 2 *Boxplots* (*solid line* = median) showing the Ti content of the leachates in the clover (**a**, each treatment n = 7) and wheat (**b**, each treatment n = 6) experiment. The LOQ is indicated with a *solid red line*. Significant difference ($p < 0.05$) of a treatment compared to the respective controls is indicated with an *asterisk*. The *lower* and *upper* borders of the *boxes* represent the 25th and 75th percentile, respectively. *Whiskers* represent maximum and minimum values, *circles* indicate outliers

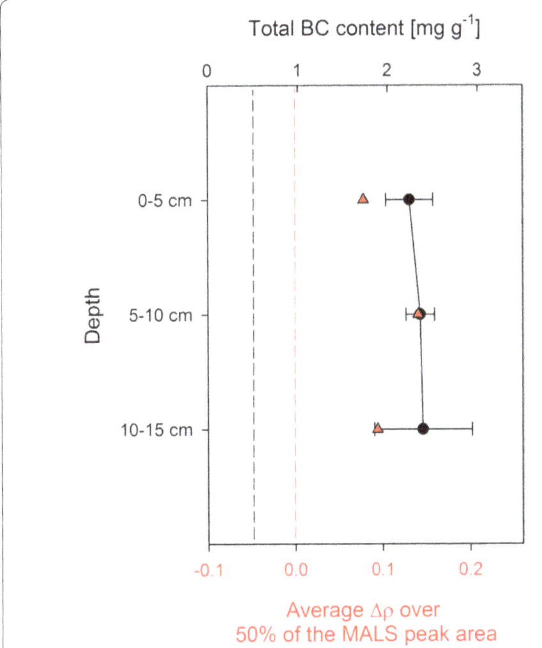

Fig. 3 Vertical distribution of BC content in the *red clover pots*, determined by CTO-375 (*black circles*), and of Δρ values, indicative of the presence of MWCNT, determined by aF4-MALS (*red triangles*). The *dashed black line* shows the native BC content of the soil, while the *dashed red line* shows the MWCNT-free soil baseline in aF4-MALS (Δρ = 0). *Error bars* show the standard deviation of five replicates. Δρ values were determined once from pooled extracts of the five replicates

CTO-375 method was 85 ± 13 % (n = 18, determined by standard addition). Therefore, the expected total BC concentration in the 2933 mg/kg MWCNT amended pots after CTO-375 can be calculated as follows: (2933 × 0.85) + 500 = 2993 mg/kg. However, the average BC content in the spiked soil before filling into the pots was lower than expected, with 2400 ± 100 mg/kg (n = 6), corresponding to 80 % of the expected BC concentration. Eventually, losses during the large scale mixing procedure could have contributed to these lower values. The variability of 4 % however suggests that the employed spiking procedure still resulted in a rather homogenous MWCNT distribution before the experiment. After the experiment, the average BC content quantified over all soil depths was 2330 ± 280 mg/kg, corresponding to 78 ± 12 % (n = 15) of the total expected BC concentration, with no significant difference between the layers. The average value was comparable to the BC content quantified before the experiment. However, precision, expressed by relative standard deviations, increased from 4 % (original spiked soil) to 12 % (aged soil). This increase in variability of the BC content may be associated with partial transport and/or aging (i.e. physiochemical modification of the particles, influencing their survival in CTO-375) of MWCNTs during the experiment.

To orthogonally observe the MWCNT behavior between the different layers with a second method, we also measured the cores with aF4-MALS [27]. With the soil of the present study, the MDL was at a Δρ of 0.099, corresponding to a CTO-determined MWCNT content of approx. 2 mg/g (Fig. 3), which is slightly lower than with the soil used in Gogos et al. (4 mg/g) [27]. The soil layers showed Δρ values of 0.078, 0.141 and 0.094 in descending order (Fig. 3). Thus, only the value of the middle layer was above the MDL. In combination with the results from CTO-375 and the increase in variability compared to the initial spike, this suggests a limited transport of the MWCNTs in the experiment. Such a low mobility would be in accordance to a dedicated soil transport study by Kasel et al. [22]. Using 14-C labeled functionalized MWCNTs, they found no detectable breakthrough in a comparable soil (loamy sand, denoted as "KAL" soil) even at water contents close to saturation (96 %).

Plant uptake of Ti

With 4.1 mg/kg, the determined Ti concentration in the red clover control plant material (Fig. 4a) was in the range of literature values for a plant species of the same family (*M. sativa*, a legume which also forms a symbiosis with rhizobia) and total soil Ti [31]. After treatment with TiO_2 (nano-)particles, the average shoot Ti content of the red clover plants increased to 8 mg/kg at the highest exposure concentration of both NNM TiO_2 and P25 (Fig. 4a). For NNM TiO_2, the average Ti content was rising with the exposure concentration, whereas for P25 no such trend could be observed. However, variability within the treatments was relatively high, and no statistical difference between the different treatments was observed. Therefore, the Ti-content in the red clover plants was not dependent on a NNM or NM exposure.

To elucidate whether the nevertheless elevated Ti contents within the red clover shoots was related to the uptake of actual TiO_2 (nano-)particles, we investigated cross sections of these roots with TEM and EDX elemental analysis. In red clover roots treated with NNM TiO_2, Ti containing particles with a similar morphology to the employed particles (Additional file 1: Figure S1A) were observed at the root surface (Fig. 5a, A1) but never inside the root cells. Some of these particles also contained Si (Fig. 5 A1, Particle 2) pointing to a possible natural origin of the particles. However, the absence of NNM TiO_2 particles within the investigated thin sections does not necessarily disprove particle uptake, as it is not possible to representatively sample a whole plant root in this way.

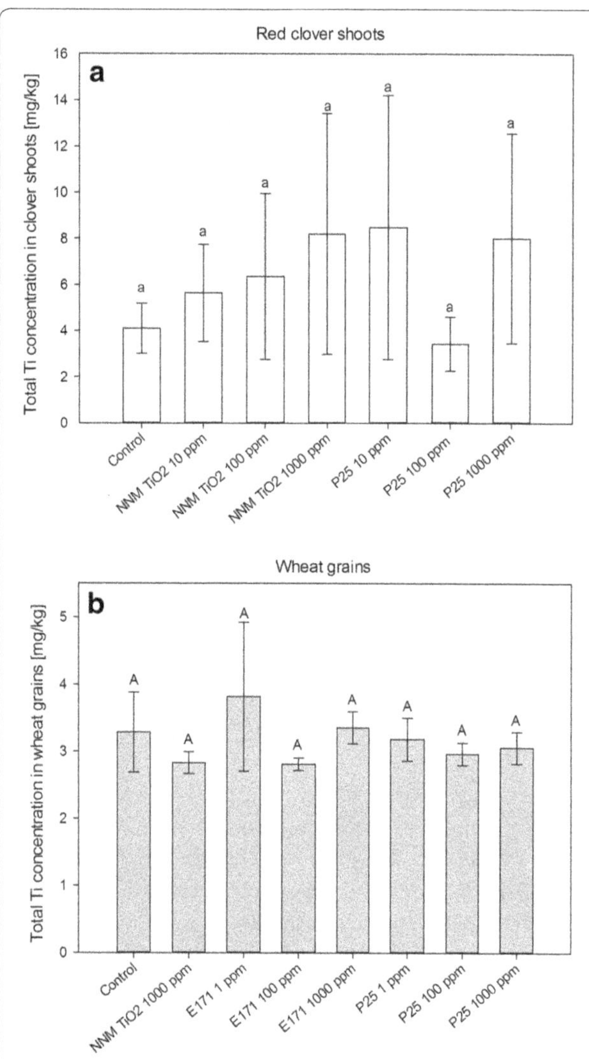

Fig. 4 Total Ti concentration in red clover shoots (**a**) and wheat grains (**b**) for the different soil exposures. *Error bars* indicate one standard deviation ($n = 4$). *Different letters* above the *bars* indicate significant statistical difference ($p < 0.05$)

In red clover roots treated with P25, only very few Ti containing nano-sized particles were found inside plant cells. The particle B1 in Fig. 5 shows a clear Ti EDX peak and is morphologically similar to the employed P25 particles (Additional file 1: elongated hexagon/Figure S1C). In addition, the oxygen peak in particle B1 is more distinct than in the other particles/objects, suggesting that the particle may consist of titanium-oxide/dioxide.

With an average of 3.3 mg/kg, the Ti content in the control wheat grains was slightly lower compared to red clover. In this case however, after treatment with TiO_2 NPs, the average Ti content in the grains remained approx. constant (Fig. 4b). Thus, both for red clover shoots and wheat grains, no significant difference in Ti uptake between the different treatments and the controls could be found.

While no data is available for red clover plants, Larue et al. [32] and Servin et al. [16] demonstrated that nano-TiO_2 can be taken up into wheat and cucumber, respectively, under extreme conditions (direct hydroponic exposure, high concentrations). Larue et al. [32] reported contents of up to 109 mg/kg Ti inside wheat roots, whereas Ti content in wheat leaves was below their LOD. To date, quantitative uptake data for aboveground plant material grown in natural TiO_2 NP spiked soil however is available only from one study performed with wheat plants [12]. Therein, the Ti content of wheat grains was in the same range as in our study, with no significant uptake, confirming our observations. However, only one exposure concentration was employed (approx. 100 mg/kg TiO_2 NPs), so no comparison can be made with regard to concentration dependent trends.

Altogether, our results suggest that Ti (-NP) uptake to red clover plants from real soils is insignificant. The biological data [24, 25] may represent another indirect piece of evidence, as for all endpoints (root and shoot biomass, number of flowers, nitrogen fixation and arbuscular mycorrhizal colonization), no significant effect of the treatments were observed for both plants.

Plant uptake of MWCNTs

Figure 6 shows the temperature rise (ΔT, °C) of dry red clover shoot material from the two MWCNT treatments. The LOD of the MIH method [26] was calculated to be at $\Delta T = 76$ °C (corresponding to a 16 μg/g MWCNT content) and the LOQ at $\Delta T = 117$ °C (corresponding to a 55 μg/g MWCNT content).

A large fraction of the values was located in the region between LOD and LOQ, and can thus be considered as MWCNT detections (60 % of the values in case of the 3 mg/kg treatment and 43 % in case of the 2933 mg/kg treatment). The values above the LOQ represent MWCNT contents of 68 (3 mg/kg treatment, $n = 1$) and 99 μg/g (2933 mg/kg treatment, $n = 1$).

Taking into account the average dry weight of the red clover plants (14.3 g for the 3 mg/kg treatment and 15.3 g for the 2933 mg/kg treatment, see also Moll et al. [24]), the two cases with values above the LOD would correspond to a total amount of MWCNTs of 0.97 and 1.5 mg taken up into the plants per pot in the two treatments, respectively. This means that 9.8 % of the initial MWCNT amount in the soil would have been translocated to the shoots in the 3 mg/kg treatment. Conversely, in the 2933 mg/kg treatment, only 0.015 % of the initial amount would have been translocated. It is interesting to note that the MWCNT uptake was independent from the applied MWCNT concentration. In addition,

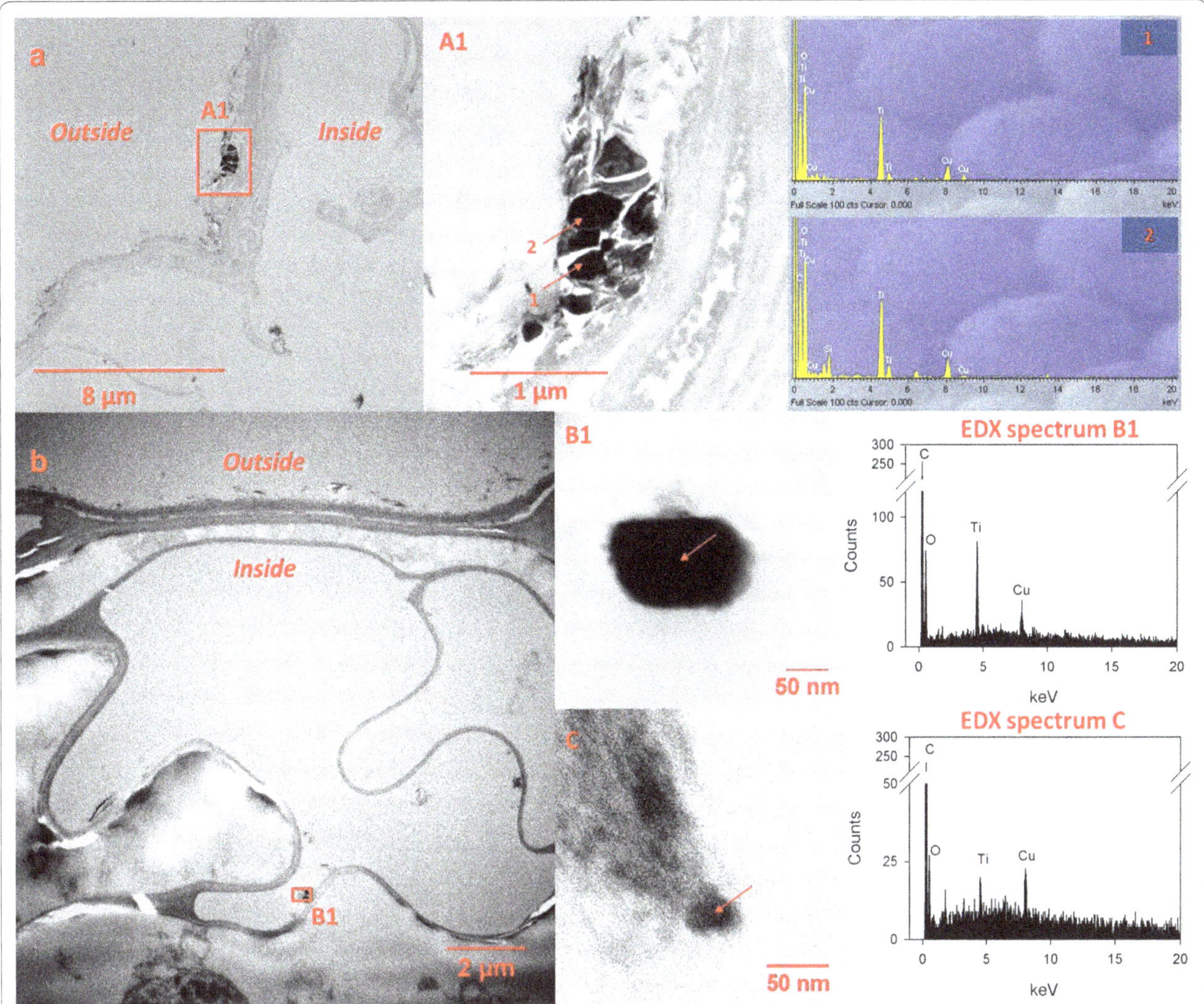

Fig. 5 Electron microscopy micrographs of **a** an ultrathin-section of a root treated with 1000 mg/kg NNM TiO_2 (imaged with TEM) together with a magnification (**A1**, outside of the root) and corresponding EDX spectra of selected spots (Spectrum 1 and 2) and **b** an ultrathin-section of a root treated with 1000 mg/kg P25 TiO_2 (imaged with STEM) together with a magnification (**B1**, inside of the root) and corresponding EDX spectrum of the selected particle. **c** Represents a particle at a location different from **b**, but also inside a cell. EDX spectra were collected from the center of the particles. The copper (Cu) peak that is present in all spectra originates from the grid material

we observed that within the MWCNT treatments, a significant reduction of flowering occurred (see Moll et al. [24]), which was not concentration dependent as well.

Uptake of CNTs into a plant cell is likely to be limited to the fraction dispersed in water. MWCNTs however are highly hydrophobic and prone to homo- as well as hetero-agglomeration with soil constituents. This in turn may result in a very small fraction of MWCNTs that remains well dispersed in the soil pore water. In addition, the plant surface may act as a filter that becomes clogged over time. However, further experiments are needed to explain this intriguing result.

We tried to orthogonally confirm the observed MWCNT uptake by using TEM imaging on cross sections of the plant roots. Khodakovskaya et al. [18] and Tripathi et al. [19] provided such optical evidence for CNT uptake from hydroponic solutions. However, in our case, the sole use of TEM was not conclusive. Additional file 1: Figure S4A shows a MWCNT-like particle that was observed within a plant root cell of the MWCNT treatment. This particle showed structural and dimensional similarity to the native MWCNTs administered to the pots (Additional file 1: Figure S4B). Still, this observation remained the only one

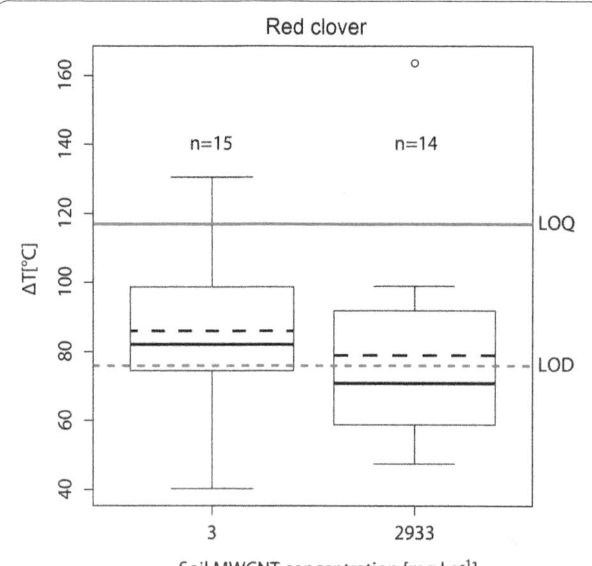

Fig. 6 *Boxplots* (mean = *dashed*, median = *solid*) of the temperature increase at 50 W, 6 s for the red clover plant samples of the two MWCNT treatments (3 and 2933 mg/kg). The LOD (at $\Delta T = 76$ °C, corresponding to 16 μg/g) is indicated with the *red dashed line* and the LOQ (at $\Delta T = 117$ °C, corresponding to 55 μg/g) is indicated with the *solid red line*. Both LOD and LOQ have been determined 3 and 10 σ above the blank signal (control plants), respectively. All seven replicates of the treatments have been measured at least twice. The total number of measurements is indicated above the respective *boxplot*. The *lower* and *upper* borders of the *boxes* represent the 25th and 75th percentile, respectively. *Whiskers* represent maximum and minimum values

within a number of cross sections that were manually inspected.

We then made additional attempts to screen the samples for the presence of MWCNTs with confocal Raman spectroscopy (Additional file 1: Figure S5). However, this approach requires that the sample is free (or almost free) of carbon allotropes (native carbon or contaminations), such as soot and amorphous carbon. In principle, Raman spectroscopy has enough sensitivity to detect single MWCNTs, but we observed that the spectra of MWCNTs and other carbon allotropes as well as cell wall material (i.e. lignin [33], which is present in clover roots [34]) had a large overlap which made the screening difficult.

While the exact amount of MWCNTs taken up could not be fully quantified and optical confirmation is still not entirely affirmed, based on the specificity of the MIH method, it is still suggested that MWCNTs were taken up and translocated to the aboveground part of the plant in some cases. Studies that reported plant uptake or cellular localization of CNTs until now were performed in hydroponic cultures, where the particles were freely available for interactions with the root [18, 19, 35, 36]. Uptake from soil would thus constitute a novelty; however, due to the lack of an orthogonal confirmation of the observed uptake, this result should be interpreted with care.

Conclusions

In this part of a combined effect and exposure study we placed emphasis on a rigorous confirmation of actual NP exposure concentrations. To achieve this goal we applied an array of analytical techniques to the soil and plant samples, of which some are novel and used for the first time in this kind of effect studies. In particular, this includes the combination of CTO-375 and aF4-MALS that showed that MWCNTs exhibited a rather limited mobility in the soil, as well as MIH that showed a concentration independent uptake of MWCNTs into some plants. In addition, the battery of analytical techniques confirmed the relatively constant exposure situation in both TiO_2 NP and MWCNT treatments over several months, with only subtle changes in concentrations, which could however be explained qualitatively with underlying NP/soil properties, distribution processes and experimental conditions.

Additional file

Additional file 1. Figure S1. Bright field TEM micrographs and size information of TiO_2 particles. **Figure S2.** MIH calibration curve. **Figure S3.** Dependence of the ζ-potential [mV] of TiO_2 particles on pH. **Figure S4.** Transmission electron microscopy micrographs of a potential CNT structure. **Figure S5.** Raman spectra of the employed MWCNT powder and plant samples. **Text.** Composition of fertilizers, MWCNT analysis of soil with CTO-375. Detailed sample preparation steps of root cross sections for analysis using transmission electron microscopy.

Authors' contributions

AG carried out physical–chemical analyses (CTO-375, aF4-MALS, XRF, ICP-OES) as well as the transmission electron microscopy and wrote the manuscript. JM designed and conducted the red clover exposure experiment. FK performed the leachate measurements and conducted the wheat exposure experiment. MvdH participated in the design of the exposure study. FI conducted MIH measurements, and FI and MJG analyzed the data. MJG and RZ edited parts of the manuscript. TDB conceived of the study, and participated in its design and coordination and helped to write the manuscript. All authors read and approved the final manuscript.

Author details

[1] Agroscope, Institute for Sustainability Sciences ISS, 8046 Zurich, Switzerland. [2] Department of Chemistry and Applied Biosciences, ETH Zurich, 8093 Zurich, Switzerland. [3] Department of Chemical Engineering, Texas Tech University, Lubbock, TX, USA. [4] Artie McFerrin Department of Chemical Engineering, Texas A&M University, College Station, TX, USA.

Acknowledgements

This work is part of the project "Effects of NANOparticles on beneficial soil MIcrobes and CROPS (NANOMICROPS)", within the Swiss National Research Programme NRP 64 "Opportunities and Risks of Nanomaterials". We thank the Swiss National Science Foundation (SNF) for financial support. Andres Kaech, Ursula Lüthi and the team at the center for microscopy and image analysis (ZMB), University of Zurich are gratefully acknowledged for TEM support. We also thank Jacek Szczerbiński for his support in the confocal Raman microscopy, Franziska Blum for help with the BC analysis as well as Ralf Kaegi and Brian Sinnet for their help with the plant Ti determination and the possibility to carry out the XRF analyses in their lab.

Competing interests

The authors declare that they have no competing interests.

References

1. Nowack B, Ranville JF, Diamond S, Gallego-Urrea JA, Metcalfe C, Rose J, et al. Potential scenarios for nanomaterial release and subsequent alteration in the environment. Environ Toxicol Chem. 2012;31(1):50–9.
2. Piccinno F, Gottschalk F, Seeger S, Nowack B. Industrial production quantities and uses of ten engineered nanomaterials in Europe and the world. J Nanopart Res. 2012;14(9):1–11.
3. Chen X, Mao SS. Titanium dioxide nanomaterials: synthesis, properties, modifications, and applications. Chem Rev. 2007;107(7):2891–959.
4. Weir A, Westerhoff P, Fabricius L, Hristovski K, von Goetz N. Titanium dioxide nanoparticles in food and personal care products. Environ Sci Technol. 2012;46(4):2242–50.
5. De Volder MFL, Tawfick SH, Baughman RH, Hart AJ. Carbon nanotubes: present and future commercial applications. Science. 2013;339(6119):535–9.
6. Petersen EJ, Zhang L, Mattison NT, O'Carroll DM, Whelton AJ, Uddin N, et al. Potential release pathways, environmental fate, and ecological risks of carbon nanotubes. Environ Sci Technol. 2011;45(23):9837–56.
7. Gottschalk F, Sonderer T, Scholz RW, Nowack B. Possibilities and limitations of modeling environmental exposure to engineered nanomaterials by probabilistic material flow analysis. Environ Toxicol Chem. 2010;29(5):1036–48.
8. Gogos A, Knauer K, Bucheli TD. Nanomaterials in plant protection and fertilization: current state, foreseen applications, and research priorities. J Agric Food Chem. 2012;60(39):9781–92.
9. Kah M, Beulke S, Tiede K, Hofmann T. Nanopesticides: state of knowledge, environmental fate, and exposure modeling. Crit Rev Environ Sci Technol. 2012;43(16):1823–67.
10. Tong Z, Bischoff M, Nies LF, Myer P, Applegate B, Turco RF. Response of soil microorganisms to as-produced and functionalized single-wall carbon nanotubes (SWNTs). Environ Sci Technol. 2012;46(24):13471–9.
11. Ge YG, Schimel JP, Holden PA. Evidence for negative effects of TiO2 and ZnO nanoparticles on soil bacterial communities. Environ Sci Technol. 2011;45(4):1659–64.
12. Du WC, Sun YY, Ji R, Zhu JG, Wu JC, Guo HY. TiO2 and ZnO nanoparticles negatively affect wheat growth and soil enzyme activities in agricultural soil. J Environ Monit. 2011;13(4):822–8.
13. Chen G, Liu X, Su C. Transport and retention of TiO2 rutile nanoparticles in saturated porous media under low-ionic-strength conditions: measurements and mechanisms. Langmuir. 2011;27(9):5393–402.
14. Solovitch N, Labille J, Rose J, Chaurand P, Borschneck D, Wiesner MR, et al. Concurrent aggregation and deposition of TiO2 nanoparticles in a sandy porous media. Environ Sci Technol. 2010;44(13):4897–902.
15. Fang J, Shan XQ, Wen B, Lin JM, Owens G. Stability of titania nanoparticles in soil suspensions and transport in saturated homogeneous soil columns. Environ Pollut. 2009;157(4):1101–9.
16. Servin AD, Castillo-Michel H, Hernandez-Viezcas JA, Diaz BC, Peralta-Videa JR, Gardea-Torresdey JL. Synchrotron micro-XRF and micro-XANES confirmation of the uptake and translocation of TiO2 nanoparticles in cucumber (*Cucumis sativus*) plants. Environ Sci Technol. 2012;46(14):7637–43.
17. Larue C, Veronesi G, Flank A-M, Surble S, Herlin-Boime N, Carriere M. Comparative uptake and impact of TiO2 nanoparticles in wheat and rapeseed. J Toxicol Environ Health Part A. 2012;75(13–15):722–34.
18. Khodakovskaya M, Dervishi E, Mahmood M, Xu Y, Li ZR, Watanabe F, et al. Carbon nanotubes are able to penetrate plant seed coat and dramatically affect seed germination and plant growth. ACS Nano. 2009;3(10):3221–7.
19. Tripathi S, Sonkar SK, Sarkar S. Growth stimulation of gram (*Cicer arietinum*) plant by water soluble carbon nanotubes. Nanoscale. 2011;3(3):1176–81.
20. Liu QL, Chen B, Wang QL, Shi XL, Xiao ZY, Lin JX, et al. Carbon nanotubes as molecular transporters for walled plant cells. Nano Lett. 2009;9(3):1007–10.
21. Jaisi DP, Elimelech M. Single-walled carbon nanotubes exhibit limited transport in soil columns. Environ Sci Technol. 2009;43:9161–6.
22. Kasel D, Bradford SA, Šimůnek J, Pütz T, Vereecken H, Klumpp E. Limited transport of functionalized multi-walled carbon nanotubes in two natural soils. Environ Pollut. 2013;180:152–8.
23. Tian Y, Gao B, Wang Y, Morales VL, Carpena RM, Huang Q, et al. Deposition and transport of functionalized carbon nanotubes in water-saturated sand columns. J Hazard Mater. 2012;213–214:265–72.
24. Moll J, Gogos A, Bucheli TD, Widmer F, van der Heijden MGA. Effect of nanoparticles on red clover and its symbiotic micro-organisms. J Nanobiotechnol. 2016;14:36.
25. Moll J, Klingenfuss F, Widmer F, Gogos A, Bucheli TD, et al. Assessing the effects of titanium dioxide nanoparticles on soil microbial communities and wheat growth. in preparation.
26. Irin F, Shrestha B, Canas JE, Saed MA, Green MJ. Detection of carbon nanotubes in biological samples through microwave-induced heating. Carbon. 2012;50(12):4441–9.
27. Gogos A, Kaegi R, Zenobi R, Bucheli TD. Capabilities of asymmetric flow field-flow fractionation coupled to multi-angle light scattering to detect carbon nanotubes in soot and soil. Environ Sci Nano. 2014;1(6):584–94.
28. Sobek A, Bucheli TD. Testing the resistance of single- and multi-walled carbon nanotubes to chemothermal oxidation used to isolate soots from environmental samples. Environ Pollut. 2009;157(4):1065–71.
29. Keith LH, Crummett W, Deegan J, Libby RA, Taylor JK, Wentler G. Principles of environmental analysis. Anal Chem. 1983;55(14):2210–8.
30. Dunphy Guzman KA, Finnegan MP, Banfield JF. Influence of surface potential on aggregation and transport of titania nanoparticles. Environ Sci Technol. 2006;40(24):7688–93.
31. Dumon JC, Ernst WHO. Titanium in Plants. J Plant Physiol. 1988;133(2):203–9.
32. Larue C, Laurette J, Herlin-Boime N, Khodja H, Fayard B, Flank A-M, et al. Accumulation, translocation and impact of TiO2 nanoparticles in wheat (Triticum aestivum spp.): influence of diameter and crystal phase. Sci Total Environ. 2012;431:197–208.
33. Gierlinger N, Schwanninger M. The potential of Raman microscopy and Raman imaging in plant research. Spectrosc-Int J. 2007;21(2):69–89.
34. de Neergaard A, Hauggaard-Nielsen H, Jensen LS, Magid J. Decomposition of white clover (*Trifolium repens*) and ryegrass (*Lolium perenne*) components: C and N dynamics simulated with the DAISY soil organic matter submodel. Eur J Agron. 2002;16(1):43–55.
35. Lin SJ, Reppert J, Hu Q, Hudson JS, Reid ML, Ratnikova TA, et al. Uptake, translocation, and transmission of carbon nanomaterials in rice plants. Small. 2009;5(10):1128–32.
36. Serag MF, Kaji N, Gaillard C, Okamoto Y, Terasaka K, Jabasini M, et al. Trafficking and subcellular localization of multiwalled carbon nanotubes in plant cells. ACS Nano. 2011;5(1):493–9.

15

Antioxidant capacities of the selenium nanoparticles stabilized by chitosan

Xiaona Zhai[1], Chunyue Zhang[1], Guanghua Zhao[1], Serge Stoll[2], Fazheng Ren[1] and Xiaojing Leng[1*]

Abstract

Backgrounds: Selenium (Se) as one of the essential trace elements for human plays an important role in the oxidation reduction system. But the high toxicity of Se limits its application. In this case, the element Se with zero oxidation state (Se^0) has captured our attention because of its low toxicity and excellent bioavailability. However, Se^0 is very unstable and easily changes into the inactive form. By now many efforts have been done to protect its stability. And this work was conducted to explore the antioxidant capacities of the stable Se^0 nanoparticles (SeNPs) stabilized using chitosan (CS) with different molecular weights (Mws) (CS-SeNPs).

Results: The different Mws CS-SeNPs could form uniform sphere particles with a size of about 103 nm after 30 days. The antioxidant tests of the DPPH, ABTS, and lipid peroxide models showed that these CS-SeNPs could scavenge free radicals at different levels. And the 1 month old SeNPs held the higher ABTS scavenging ability that the value could reach up to 87.45 ± 7.63% and 89.44 ± 5.03% of CS(l)-SeNPs and CS(h)-SeNPs, respectively. In the cell test using BABLC-3T3 or Caco-2, the production of the intracellular reactive oxygen species (ROS) could be inhibited in a Se concentration-dependent manner. The topical or oral administration of CS-SeNPs, particularly the Se nanoparticles stabilized with low molecular weight CS, CS(l)-SeNPs, and treated with a 30-day storage process, could efficiently protect glutathione peroxidase (GPx) activity and prevent the lipofusin formation induced by UV-radiation or D-galactose in mice, respectively. Such effects were more evident in viscera than in skin. The acute toxicity of CS(l)-SeNPs was tenfold lower than that of H_2SeO_3.

Conclusions: Our work could demonstrate the CS-SeNPs hold a lower toxicity and a 30-day storage process could enhance the antioxidant capacities. All CS-SeNPs could penetrate the tissues and perform their antioxidant effects, especially the CS(l)-SeNPs in mice models. What's more, the antioxidant capacities of CS-SeNPs were more evident in viscera than in skin.

Keywords: Chitosan, Selenium nanoparticles, ROS, Lipofuscin, UV radiation, D-Galactose

Background

Selenium (Se) is involved in the antioxidant defense systems of the liver and plays an important role in protecting against oxidative stress. Many studies demonstrated that Se supplementation can increase the level of enzymes such as GPx etc., prevent the accumulation of free radical species, and reduce the cellular damage [1–4]. However, the narrow margin between the effective and toxic doses limited the application of this substance [5]. The Se^0 has thus gained more attention because of its low toxicity and excellent bioavailability compared with Se(IV) and Se(VI), since both having a strong ability to capture free radicals [6, 7]. Nevertheless, poor water solubility and the ability to easily transform into a grey analogue that is thermodynamically stable but biologically inert, makes Se^0 difficult to be used in food and medicine fields [8, 9].

The water solubility of an insoluble substance can be greatly improved by reducing the size and increased the specific surface with convenient nanotechnology. In the past decades, nanotechnology has been used to prepare antioxidant products using minerals including silver [10], gold [11], cerium oxide [12], and platinum [13] etc., based

*Correspondence: lengxiaojingcau@163.com
[1] Beijing Advanced Innovation Center for Food Nutrition and Human Health, Beijing Laboratory for Food Quality and Safety, Beijing Dairy Industry Innovation Team, College of Food Science & Nutritional Engineering, China Agricultural University, Beijing 100083, China
Full list of author information is available at the end of the article

upon their red-ox abilities. Selenium was also considered owing its multiple valence states (+6, +4, +2, 0, −1, −2) and more complex antioxidant activities [14]. In a quest to use Se^0, many efforts have been made to design such nano-vehicles using polysaccharides, proteins, and/or lipids etc. as stabilizers [15–17]. The obtained Se nanoparticles are reported as novel compounds with excellent antioxidant properties and lower toxicity compared with other selenospecies [18]. It should be noted that in these reports the data about the effects of the stabilizers on the antioxidant functionalities of the nanosystem are still incomplete, especially on the relationships between the microstructure features and bio-activities of the whole system in vitro and in vivo.

Chitosan (CS), the N-deacetylated form of natural chitin found widely in the exoskeleton of crustaceans, insects, and fungi, has been often used as the Se^0-stabilizer not only because of its low toxicity and bioavailability, but it can also withstand pepsin and pancreatin to a great extent [19, 20]. This naturally helps to enhance the stability of the Se^0 system in the digestive enzyme environment. In our previous work, we compared the physicochemical properties of the Se^0 spherical nanoparticles with a size at about 103 nm prepared through the reduction of seleninic acid with ascorbic acid in the presence of chitosan with different molecular weights [21]. We found that, although SeNPs could be stabilized using both the chitosan with low [CS(l)-SeNPs] or high molecular weight [CS(h)-SeNPs] in 30 days, the microstructure of the former seemed more compact than the latter. This divergence caused the Se release of the former more slowly than the latter in the simulated gastric, intestinal, and sweat environment. This raises a question as to whether such difference in the microstructure of SeNPs between CS(l)-SeNPs and CS(h)-SeNPs affects the bioactivities of these nanoparticles in vitro and in vivo.

As side-products of the normal metabolism, the accumulation of random molecular damage due to ROS promoted by oxidative stress is widely believed to cause cellular aging [22]. Lipofusin (LF) as the hallmark of aging is a membrane-bound cellular waste by oxidation that can be neither degraded nor ejected from the cell but can only be diluted through cell division and subsequent growth which is often found in skin and viscera [23–25]. In spite of LF formation involving complex intracellular reactions, it can be retarded by various antioxidant systems including enzymatic (e.g., GPx, SOD, etc.) and nonenzymatic antioxidant systems (e.g., vitamins E and C etc.) [26]. Many works pointed out that the level of the GPx activity could represent the state of Se uptake [3]. In addition, some reports indicated that a low status of Se was related with LF accumulation, and topical and oral Se administration of L-selenomethionine or sodium selenite could prevent LF formation induced by UV irradiation [27–29]. Therefore, the detection of GPx activity and LF levels can be used to study the antioxidant activities of CS-SeNPs.

In this work, CS-SeNPs were manufactured using chitosan with different molecular weights and with different storage times according to our previous work [21]. The inhibition of the intracellular ROS by CS-SeNPs was examined in the BABLC-3T3 and Caco-2 cell lines, designed as skin or viscera cell models, respectively. The former cell has been scientifically validated for the skin phototoxicity test [30], and the latter can represent drug intestinal absorption. The effects of CS-SeNPs on LF in skin and viscera were investigated using mice models treated with UV-radiation and D-galactose, respectively. The concerned acute toxicity of the nanoparticles was also verified.

Methods

Reagents

The seleninic acid (H_2SeO_3), 2,2′-Azino-bis(3-ethylbenzothiazoline-6-sulfonic acid (ABTS), 2,2-Diphenyl-1-picrylhydrazyl (DPPH), D-(+)-galactose, reduced L-glutathione (buffered aqueous solution, ≥10 units/mg protein, recombinant, expressed in *E. coli*), 2,3-Diaminonaphthalene (DAN), 2′,7′-Dichlorofluorescein diacetate (DCFH-DA), 2,4,6-Tris(dimethylaminomethyl) phenol(DMP-30), and glutaraldehyde were purchased from Sigma Aldrich, Inc. (St. Louis, MO, USA). Dulbecco's Modified Eagle Medium (DMEM), fetal bovine serum, Penicillin-Streptomycin Solution (100×), GlutaMAXTM-1 (100×), MEM Nonessential Amino Acid Solution (NEAA, 100×), dimethyl sulfoxide (DMSO), potassium phosphate (PBS, pH 7.4), Trypsin–EDTA, formalin, hematoxylin, and eosin were purchased from Solarbio Science & Technology Co., Ltd. (Beijing, China). The chitosan with a molecular weight of less than 3 kDa (CS3) and 200 kDa (CS200) (Poly-β-(1,4)-D-glucosamine, DD > 85%) were purchased from Jinan Haidebei Co., Ltd. (Shandong, China). The other regents included acetic acid, ascorbic acid, $HClO_4$, HNO_3, HCl, H_2SO_4, EDTA, ethanol, methanol, acetone, cyclohexane, potassium persulfate ($K_2S_2O_8$), $Na_2HPO_4 \cdot 12H_2O$, $NaH_2PO_4 \cdot 2H_2O$, egg lecithin, $FeSO_4$, trichloroacetic acid (TCA), 2-Thiobarbituric acid (TBA), NaCl, hydroxylamine hydrochloride, cresol red, quinine sulphate, ammonium hydroxide, stearic acid, white petrolatum, propylene glycol, triethanolamine, and edetate disodium dehydrate were of analytical grade. The edible oil, wax, and rosin were from the local market.

Preparation and characterization of CS-SeNPs

CS-SeNPs were manufactured according to the method described in our previous work [21]. These nanoparticles stabilized with CS3 and CS200 were denoted as CS(l)-SeNPs and CS(h)-SeNPs, respectively. The numbers 0 and 30 in CS(l)-SeNPs-0 day, CS(l)-SeNPs-30 days, CS(h)-SeNPs-0 day, and CS(h)-SeNPs-30 days were used to distinguish the nanoparticles manufactured immediately and those followed by 30-days storage, respectively. The Se concentration of all of the CS-SeNPs stock was adjusted to 0.1 mol/L.

The morphology of these nanoparticles was observed by means of scanning transmission electron microscopy (STEM). The sample solution was dropped on a carbon-coated copper grid for 5 min and the excess solution was removed and dried in the air for 30 min. The observations were performed using a Hitachi S-5500 STEM (Hitachi High Technologies America, Inc. IL, USA) with an operation voltage of 30 kV. The images were acquired using a Gatan high-angle annular bright field scintillating detector. The hydrodynamic size and zeta potential of the nanoparticles were measured using a Delsa–Nano Particle Analyzer (A53878, Beckman Coulter, Inc., CA, USA).

Assay for antioxidant activities of CS-SeNPs in vitro

The antioxidant abilities of the CS-SeNPs samples were presented as the radicals scavenging activity (RSC%) in DPPH, ABTS, or lipid peroxide. The value of RSC% was calculated using the following formula:

$$\mathrm{RSC}\,(\%) = \frac{A_0 - A_1}{A_0} \times 100\% \tag{1}$$

where A_0 is the absorbance of the control and A_1 is the absorbance of the mixed solution of the antioxidant and free radical agent.

The RSC% in DPPH was determined according to the method described in the work of Xu [31]. A 0.2 mL dose of the nanoparticle sample was mixed vigorously with 3.8 mL of DPPH radical ethanol solution (final DPPH concentration: 0.1 mmol/L), and then kept at room temperature in the dark for 30 min. The absorbance was measured at 517 nm with a UV spectrophotometer (UVmini-1240; Shimadzu, Japan).

The RSC% in ABTS was determined according to the work of Re [32]. The stock was prepared by mixing 0.5 mL of 14 mmol/L ABTS and 0.5 mL of 4.9 mmol/L $K_2S_2O_8$, and then keeping them in the dark at room temperature for at least 12 h in a 1.5 mL tube. The absorbance of the ABTS solution was adjusted by PBS buffer (pH 7.4, 150 mmol/L) to 0.70 ± 0.02 at 734 nm. The measurement was performed at 734 nm exactly 4 min after mixing 900 μL of the diluted ABTS solution with 100 μL of the nanoparticle sample.

A modified TBA-reactive species assay was used to measure the formed lipid peroxide with egg yolk lecithin homogenates as a lipid-rich media [33]. The occurrence of malondialdehyde (MDA), a secondary end product of the oxidation of polyunsaturated fatty acids, was used as an index of lipid peroxidation. The MDA reacted with TBA to yield a pinkish-red chromogen with an absorbance maximum at 532 nm. One gram of egg lecithin was sonicated in 50 mL PBS buffer (pH 7.4, 150 mmol/L) at 4 °C for 30 min. After mixing 0.5 mL of this solution with 0.1 mL of the nanoparticle sample, the total volume was made up to 1 mL with distilled water. The obtained mixture was added into 0.05 mL of $FeSO_4$ (70 mmol/L) and then incubated at 37 °C for 30 min. We added 0.5 mL of TCA (10%, w/w) into the above incubated solution, followed by 0.5 mL of TBA (1%, w/w). The final mixture was vortexed and heated in a boiling water bath for 60 min. After cooling, the solution was centrifuged at $3000\times g$ for 10 min. The upper organic layer was collected and measured at 532 nm.

Cell lines and culture

Two types of cell lines, purchased from China Infrastructure of Cell Line Recourses (Beijing, China), were used in this work. One was the mouse embryonic fibroblast BABLC 3T3 cells cultured in DMEM media supplemented with 10% (v/v) bovine calf serum and 1% (v/v) GlutaMAX, and the other was Caco-2 cells cultured in DMEM containing 10% (v/v) bovine calf serum and 1% (v/v) NEAA. Both cell lines were incubated at 37 °C in a humidified incubator with 5% CO_2.

Cell viability assay

The MTT assay was used to determine the cytotoxicity of the CS-NPs [18] and a MTT [3-(4, 5-dimethylthiazol-2-yl)-2, 5-diphenyltetrazolium bromide] cell viability/cytotoxicity assay kit (Beyotime Biotechnology, Jiangsu, China) was used to determine cell viability. Healthy cells can reduce the MTT to a purple formazan dye. Both cells were seeded in a 96-well microplate with 5×10^3 cell/well and 0.1 mL growth medium/well for 24 h, respectively. After that, each cell line was treated by incubating with CS(l)-SeNPs, CS(h)-SeNPs, and H_2SeO_3, respectively. The Se concentrations varied between 50 and 500 μmol/L. The incubation was performed for another 24 h. The control groups were left untreated. The absorbance was measured at 570 nm with a Thermo Fisher Scientific Varioskan® Flash Multimode Reader (Thermo Scientific, USA); the viability was determined based on the manufacturer's instructions.

Measurement of the intracellular ROS generation

The intracellular ROS accumulation was evaluated using a DCF fluorescence assay [34]. The BABLC-3T3 and Caco-2 cells were seeded in a 96-well microplate with 9×10^4 cell/well and 0.1 mL of growth medium/well for 24 h, respectively. After that, the growth medium was removed and the wells were washed with the PBS buffer (pH 7.4, 10 mmol/L). The cells were then incubated with CS(l)-SeNPs, CS(h)-SeNPs, and H_2SeO_3, respectively. The Se concentrations varied between 50 and 500 μmol/L. The control groups were treated without the above Se samples. The incubation was performed for another 24 h. At the end of the incubation, the cells were rinsed three times with a cold PBS buffer (4 °C) in order to remove the excess nanoparticles around the cells. Finally, these cells were incubated with DCFH-DA at a final concentration of 20 μm at 37 °C for 60 min. The level of the intracellular ROS was examined by detecting the fluorescence intensity conducted with a Thermo Scientific Varioskan® Flash Multimode Reader (with the excitation and emission wavelength set at 488 and 525 nm, respectively).

Animals and treatments

The Kunming (KM) mice (Strain code: 202, initial weight: 20 g to 25 g) were purchased from Vital River Laboratories Co., Ltd. (Beijing, China). These mice were allowed free access to food and water. All animal procedures were conducted in accordance with the Animal Care and Use Guidelines of the China Council on Animal Care (Regulations on the Administration of Laboratory Animals, 2013 Revision published by the State Council on July 18, 2013). The protocol complied with the guidelines of China Agriculture University for the care and use of laboratory animals.

Acute toxicity

A total of 120 KM mice were randomly divided into 12 groups, with equal numbers of female and male in each group. The CS(l)-SeNPs and H_2SeO_3 were administered by single intragastric administration with increasing doses (1.43-fold), and the mortalities were recorded within 14 days. The values of LD50 and 95% confidence were calculated by Trimmed Spearman-Kaber's Method [35].

Transdermal tests of CS-SeNPs

The transdermal tests were conducted using a vertical Franz diffusion cell system (TP-6, Tianguang Photoelectric Instrument Co., Tianjin, China) equipped with 6 identical diffusion cells. Each cell contained a donor compartment and a receptor compartment filled with 17 mL normal saline. These two compartments were connected through a circular channel with a cross-sectional area of 3.4 cm^2 (Fig. 1). A piece of mouse dorsal skin, free of subcutaneous fat, tissues, blood vessels, and epidermal hairs, was mounted on the channel as a diffusion membrane with the stratum corneum facing the donor compartment. The sample solutions were added in the donor compartment for 6 h, and the substance through the skin was collected with the normal saline stirred at a rate of 600 rpm at 37 °C.

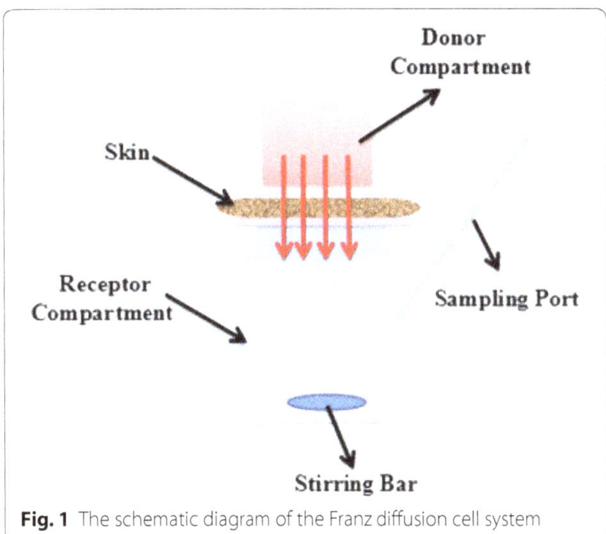

Fig. 1 The schematic diagram of the Franz diffusion cell system

The Se concentration in the donor compartment was kept at 2 mM, and the Se through the skin was collected and determined by means of hydride-generation atomic fluorescence spectrometry (AFS-230E, Beijing Haiguang Instrument Co., Beijing, China) as the following procedure noted in literature [36]: the collected solutions were filtered with 0.45 μm Millipore filter and then heated with 5 mL $HClO_4/HNO_3$ (1/3, V/V) mixture and 3 mL HCl to eliminate the organic impurities. After cooling, 5 mL deionized water, 1 mL EDTA (1%), 1 mL hydroxylamine hydrochloride (10%), and 0.2 mL cresol red (0.02%) was added successively into the filtrates. The pH was adjusted to 1.5 with HCl or NH_4OH. The solutions were incubated at 60 °C for 30 min after adding 1 mL DAN (0.1%), and then 5 mL cyclohexane was added into the cooled solutions by shaking. After standing for 30 min, the supernatant was collected and then measured by AFS with excitation and emission wavelengths at 376 and 520 nm, respectively.

Bioactivity of CS-SeNPs in the UV-induced skin damage model

A total of 48 male mice were randomly divided into 6 groups with 8 mice in each group. More details of the procedure were noted in Table 1. The dorsal skin of the mice was denuded with a wax/rosin mixture (1:1, w/w) every 10 days [37]. The drug vehicle was prepared using a standard low-Sun Protection Factor (SPF) cosmetic base formula [38]. The samples were stirred to smooth pastes with the vehicle. The paste was used 30 min before the UV treatment. The irradiation was made using a UV lamp (TL12rs 40 W UVB lamp, Philips, Poland) at a dose of 1.0 kJ/m^2 and lasted for 15 days [39]. Then the mice were sacrificed and the dorsal skins were carefully removed

Table 1 Group, drug dose and UV treatment parameters for the topical tests of the CS-SeNPs

Group	Se sample	Dose [mg/kg body weight]	UV radiation
UV-induced skin damage			
1	–	/	–
2	Drug vehicle	/	+
3	CS(l)-SeNPs (30 days)	1	+
4	CS(l)-SeNPs (30 days)	10	+
5	CS(h)-SeNPs (30 days)	1	+
6	H_2SeO_3	1	+

Table 2 Group, drug dose, and D-galactose parameters for the topical tests of CS-SeNPs

Group	Se sample	Dose [mg/kg body weight]	D-Galactose
D-Galactose induced aging			
1	–	/	–
2	Drug vehicle	/	+
3	CS(l)-SeNPs (30 days)	1	+
4	CS(l)-SeNPs (30 days)	10	+
5	CS(h)-SeNPs (30 days)	1	+
6	H_2SeO_3	1	+

and collected to determine LF content and GPx activity. The pathological study of skins was also performed.

Bioactivity of CS-SeNPs in the D-galactose induced mouse aging model

A total of 48 male mice were randomly divided into 6 groups with 8 mice in each group. More details of the procedure were noted in Table 2. Along with the oral supplementation of the tested samples, a dose of 200 mg/kg D-galactose (drug/body weight) per day was intraperitoneally injected for 4 weeks. The normal saline was used as the blank. Then the mice were sacrificed and the livers and kidneys were immediately collected to determine the LF content and GPx activity.

LF and GPx assessment

LF content was determined by a modified fluorescence method described in the work of Harvey et al. [40]. A saline solution containing of 10% (w/w) skin or viscera was freshly homogenized in an ice-water bath. After mixing 2 mL of this homogenate with a 4 mL of the $CHCl_3$/MeOH (2:1, v/v) extraction agent, the solution was sonicated for 30 min and then centrifuged at 5000 rpm for 1 min. The lower chloroform phase in the tube was carefully collected with a syringe for the following measurement. The LF content was determined using the following relationship:

$$\text{Lipofuscin content}\ (\text{mg/g tissue}) = \frac{I_{\text{sample}} - I_{\text{control}}}{I_{\text{standard}}} \times C_{\text{standard}}\ (0.1\ \text{mg/mL}) \times \frac{V_{\text{extract}}\ (4\ \text{mL})}{W_{\text{tissue}}\ (\text{g})} \quad (2)$$

where I_{sample} is LF, $I_{control}$ is the $CHCl_3$/MeOH extraction agent, and $I_{standard}$ is the calibration against a quinine sulfate solution (1 μg/mL, 0.1 mol/L H_2SO_4). The wavelengths of the excitation and emission were 365 and 435 nm, respectively.

The GPx activity, expressed as NU/mg protein, was determined using a Total Glutathione Peroxidase Assay Kit according to the manufacturer's protocol (Beyotime Biotechnology, Jiangsu, China). The protein concentrations were determined by means of Bradford dye-binding assay using bovine serum albumin as the standard [41].

Histological measurements and ultrathin sections for SEM

The histological tests of dorsal skin from the mice used for the UV-radiation test were performed in accordance with standard laboratory procedures. The biopsy skin samples (2 cm × 3 cm) were cut into small pieces, fixed in 10% formalin, and then embedded in paraffin. The samples were sliced into 2-μm-thick sections and then stained with hematoxylin and eosin staining. The observations were performed using an optical microscope controlled with TSView software in version 7.0 (Chong Qing Optical and Electrical Instrument Co., Ltd. Chongqing, China).

KM mice were deprived of food for over 24 h and were orally administered the CS-SeNPs solution and the CS-SeNPs lotion at a dosage of 25 mg Se/kg mice on the skin. After 6 h of exposure, biopsy samples from the small intestines and dorsal skin were immediately obtained for SEM observation. The ultrathin sections were made as following [42]. The small intestines and dorsal skin were quickly sliced into small pieces (1 mm × 1 mm), and then washed and fixed with 2.5% glutaraldehyde in PBS buffer (pH 7.4, 10 mmol/L). The fixed samples were dehydrated with graded ethanol solutions (70, 80, 90, and 100%, v/v, ethanol/water) and 50% acetone (v/v, acetone/ethanol), and then dehydrated twice with pure acetone. Each dehydration process lasted for 15 min. These samples were embedded in graded QUETOL 651 resin solutions (1/3, 1/1, 3/1, v/v, resin/acetone) and pure resin (with DMP-30) overnight. After standing for 24 h at 60 °C, the samples were cut into ultrathin pieces of about 70-nm thickness with a Leica EMUC6 ultramicrotome and then placed on a carbon-coated copper grid. Digital images were acquired with a Zeiss Merlin Scanning Electric Microscope (Germany) and elementary analysis was conducted with a Horiba INCA 450 energy dispersive x-ray analysis spectroscopy.

Statistical analysis

All experiments were conducted in triplicate and expressed as mean ± SD. Statistical analysis was performed using Origin 8.5 and SPSS 16.0. The comparison was performed with χ^2 or one-way ANOVA, followed by Dunnett's multiple comparison tests. Statistically significant differences between groups were defined as $p < 0.05$.

Results and discussion

Characterization of chitosan stabilized selenium nanoparticles

The preparation of the CS-SeNPs was performed according to our previous work [21]. The CS(l)-SeNPs (Fig. 2a) formed uniform small sphere particles with a size of about 50 nm, while the CS(h)-SeNPs (Fig. 2b) formed loose and irregular aggregates with an average size of more than 350 nm caused by the bridging effect of the long macromolecular chains during the initial stage of their formation. In 30 days, via a "bottom-up" growth or "top-down" shrinkage process, respectively, both the size of the CS(l)-SeNPs and CS(h)-SeNPs tended to be about 103 nm. As depicted in the previous work, the cores of CS(h)-SeNPs were more scattered than those of CS(l)-SeNPs. The zeta potential of CS(l)-SeNPs decreased from 49.5 ± 0.9 to 33.5 ± 1.0 mv and that of CS(h)-SeNPs changed from 65.9 ± 0.1 to 44.8 ± 0.6 mv. The comparison of the zeta potential values indicated that the loose microstructure of CS(h)-SeNPs was caused by its relatively high intermolecular electrostatic repulsion and bridging effects. It was believed that such a loose microstructure of CS(h)-SeNPs led to its relatively higher Se release rate compared with that of CS(l)-SeNPs [21].

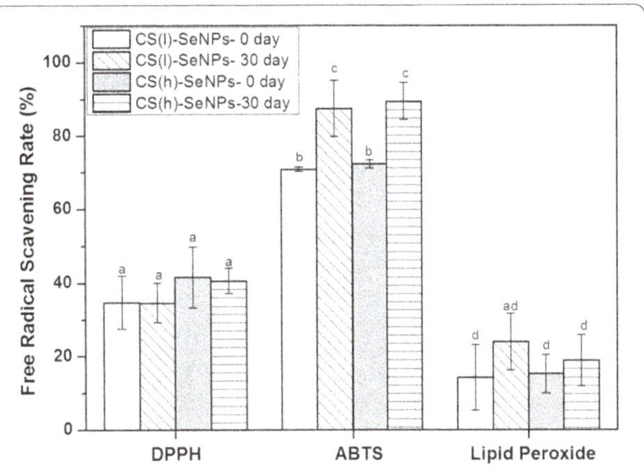

Fig. 3 Comparison of the antioxdiant capacities of CS-SeNPs in DPPH, ABTS and lipid peroxidation systems. The *different letter* markers denote the significant mean difference at $p < 0.05$

Antioxidant capacities of CS-SeNPs in vitro

The antioxidant capacities in vitro of the CS-SeNPs were investigated using the assays reported in literature, most of them can be classified into two types [44]: assays based on electron transfer (ET-based) such as DPPH and ABTS, and assays based on hydrogen atom transfer (HAT-based) reactions such as lipid peroxidation, depending upon the chemical reactions involved. Among them, DPPH and lipid peroxidation were carried out in hydrophobic media, while ABTS was in hydrophilic media. In order to understand the multifaceted aspects of the CS-SeNPs,

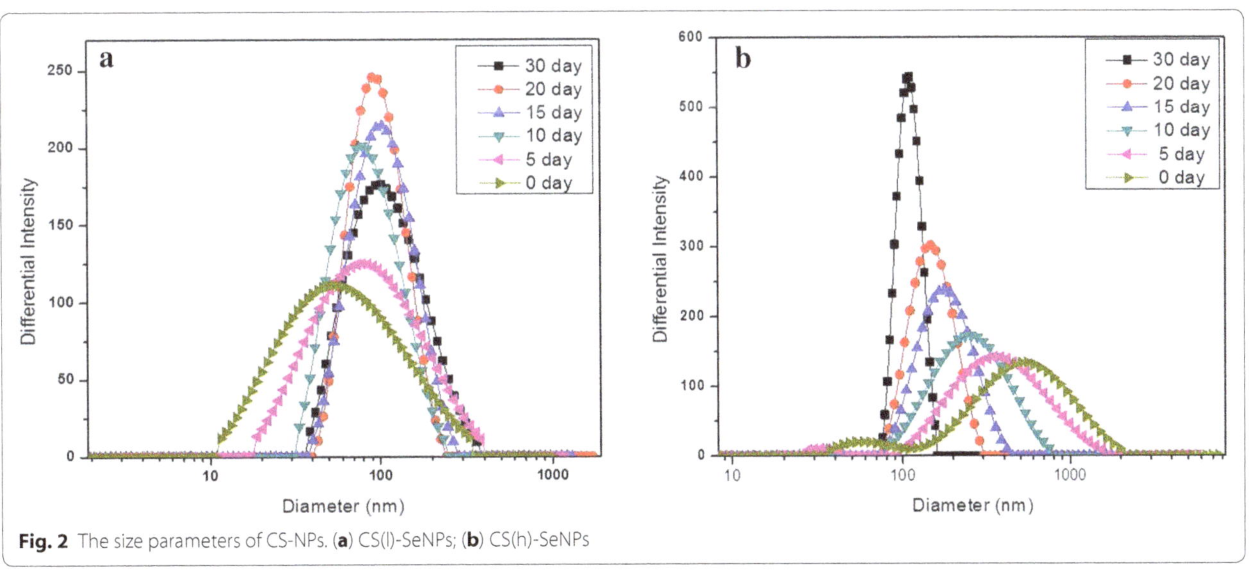

Fig. 2 The size parameters of CS-NPs. (**a**) CS(l)-SeNPs; (**b**) CS(h)-SeNPs

the above assays were all used and results were compared in Fig. 3.

It was observed that the values of RSC% in ABTS were higher than those in DPPH and lipid peroxidation. This feature was due to the effect of the high water solubility of the nanoparticles, which led to the separation of the Se nanoparticle-rich water phase from the free radical-rich lipid phase, and thus reduced the ability of Se^0 to capture the free radicals. The somewhat higher values of RSC% in DPPH than in lipid peroxidation indicated that the nanoparticles were more likely to ET-based reaction rather than HAT-based reaction. This behavior was different from that of some organic antioxidants such as rutin, which could react quickly with lipid peroxyl radicals but not nitrogen radicals [42]. The discussion about the difference between Se and organic antioxidants was not discussed because it was beyond the scope of this work.

Figure 3 also revealed the storage effect on the antioxidant capacities of CS-SeNPs. It was observed that the RSC% of the nanoparticles was enhanced by approximately 25% after a treatment of 30 days storage in ABTS assay. Such enhancement was normally caused by the protection of the stabilized CS shell on the antioxidant activity of Se during storage. This effect was not significant in DPPH and lipid peroxidation ($p < 0.05$), which was probably due to the low level of RSC% concealing the difference between these assays. Anyway, the use of the

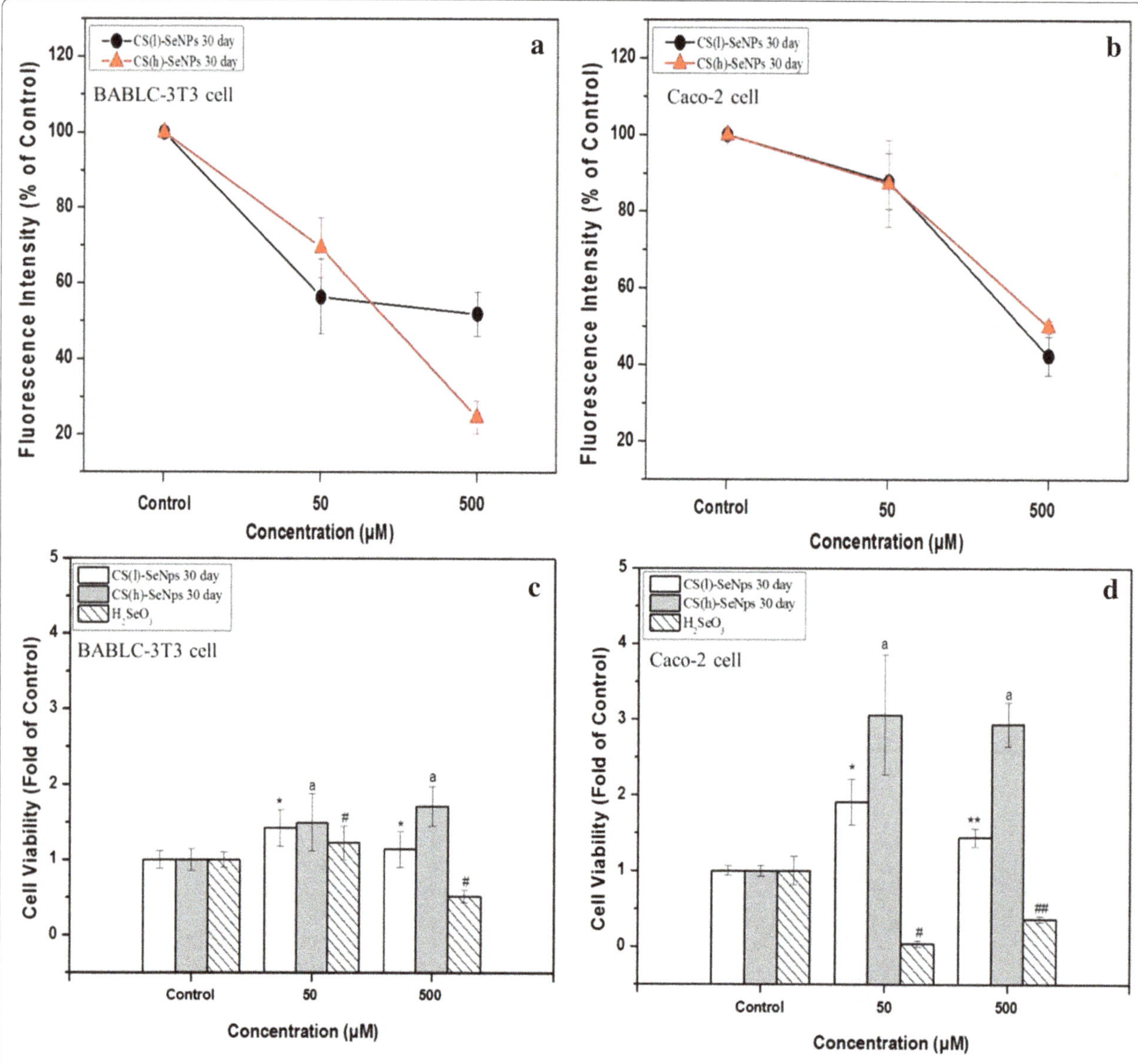

Fig. 4 Comparison of the CS-SeNPs effects on ROS accumulations and cell viabilities in cell lines. **a**, **c** BABLC-3T3 cell, **b**, **d** Caco-2 cell lines. The *different markers* denote the significant mean difference at $p < 0.05$

stabilized nanoparticles was the best choice for the following experiments. No effect of CS molecular weight was observed in all tests in vitro. Although the CS molecular weight could affect the Se release rate via the modification of the nano-carrier microstructure, the minor difference of the released Se quantities was not serious enough to disturb the antioxidant capacities in the present experimental conditions.

ROS inhibition effects and cytotoxicity of CS-SeNPs in vitro

An appropriate antioxidant capacity of Se nanoparticles can be used to inhibit ROS accumulation in cell, while an excessive one will result in cytotoxicity. Since the skin and digestive tract were targeted in this work, the BABLC 3T3 (a) and Caco-2 (b) cell lines were used as skin [30] and intestinal cell [45] models to test ROS inhibition effects and cytotoxicity of CS-SeNPs (Fig. 4).

Both CS(l)-SeNPs and CS(h)-SeNPs could inhibit ROS accumulation in BABLC-3T3 cell lines in a dose-dependent manner, but a larger rate of divergence existed, and an abnormal strong inhibition effect was observed with a high dose (500 μM) of CS(h)-SeNPs. While in Caco-2 cell lines, the inhibitory effects of both nanoparticles increased at almost the same rate as the drug dose.

The Se nanoparticles were cytotoxic. Zheng et al. have studied the properties of a grey Se stabilized with polyethylene glycol, PEG-SeNPs, in HepG2 cell lines [43]. They found that the PEG and Se had a synergetic effect on cell apoptosis via the induction of mitochondrial dysfunction. However, compared with H_2SeO_3, it was obviously observed that CS nano-systems could effectively reduce the selenium cytotoxicity in BABLC-3T3 (Fig. 4c) and Caco-2 (Fig. 4d) cell lines, respectively. The values of the cell viability of CS(h)-SeNPs were generally higher than those of CS(l)-SeNPs. Nevertheless a dose-dependent manner could be observed for CS(l)-SeNPs, but not for CS(h)-SeNPs. The properties of the former needs to be further studied, where the relationship between the physicochemical properties of the nanoparticles and biochemical properties of the cells should be considered. In this work, the CS(l)-SeNPs were preferentially used in the subsequent tests.

Penetration tests of CS-SeNPs

The bio-activities of nanoparticles are related to their penetration ability. This ability can be affected by the nanoparticle surface coatings and also the biochemical characteristics of target organelle or tissue [46]. Since the skin contact and intestinal intake were concerned in this work, the dorsal skin and intestinal tissues of mice were used as models to test the transdermal capacities of the CS-SeNPs, respectively.

Figure 5 exhibited SEM observation of Se nanoparticles in the dorsal skin and intestinal tissues of mice (a) and compared the transdermal kinetic data of CS(l)-SeNPs, CS(h)-SeNPs, and H_2SeO_3 (b) in the skin. The kinetic analysis of the intestinal system was not performed, because the Se quantity before the penetration process could be affected by the portion of Se released in intestine tract [21], and the state of Se accumulation on the surface of intestine wall should be also considered. As shown in Fig. 5a, it was observed that Se nanoparticles could easily penetrate cell membrane and stay near the rough endoplasmic reticulum and mitochondria. In contrast, recent reports have shown the interactions of Se nanoparticles with mitochondria [47] and lysosome [48]. Obviously, it appears that the distribution of nanoparticles in cell was quite broad.

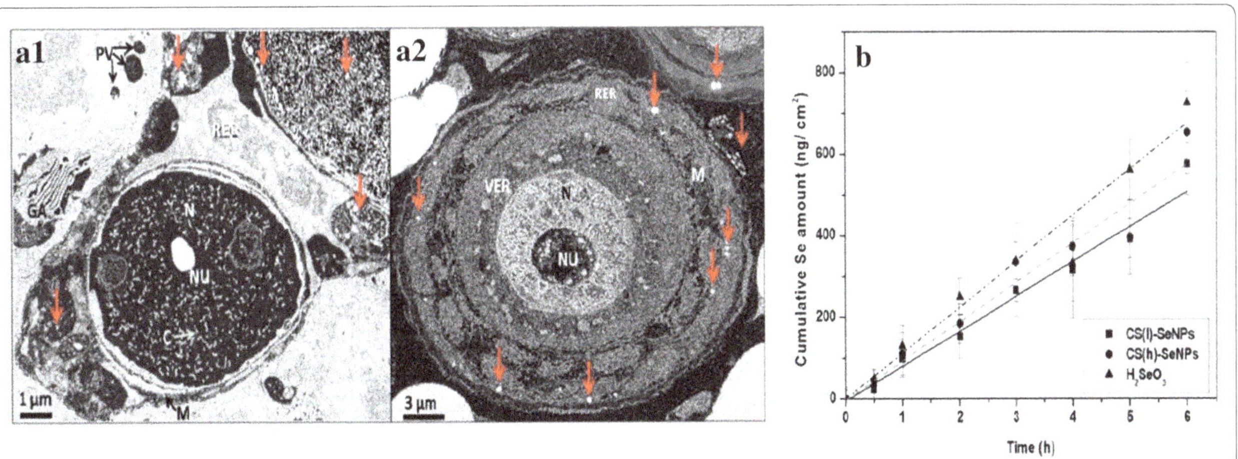

Fig. 5 SEM observation and transdermal tests of CS-SeNPs in mice tissue. **a** Small intestine (**a1**) and skin (**a2**). *N* nucleus, *NU* nucleolus, *C* chromatin, *RER* rough endoplasmic reticulum, *M* mitochondria, *VER* vacuolization of the endoplasmic reticulum, *GA* Golgi apparatus, and *PV* pinocytic vesicle. The Se particles in the cell are marked with *red arrows*. **b** Comparison of the transdermal kinetics of the CS-SeNP and H_2SeO_3

Passive diffusion was denoted as the principle penetration mode of SeO_3^{2-} [49] while endocytosis often happened in nanosystems [14]. Nevertheless, the transdermal Se amount of CS(l)-SeNPs, CS(h)-SeNPs, and H_2SeO_3 was very close and approximately showed a linear relationship with time. The rate values were at about 85.7 ± 5.5, 95.7 ± 3.5 and 112.9 ± 0.8 ng/cm^{2}h, respectively (Fig. 5b), indicating that Se delivery efficiency of the CS-SeNPs was considerable to that of selenite ion diffusion.

Antioxidant capacities of CS-SeNPs in vivo

The investigations of the CS-SeNPs antioxidant capacities in vivo were conducted using the KM mouse skin (Fig. 6) or viscera (Fig. 7) treated with UV-radiation or D-galactose, respectively.

UV-radiation system

Figure 6 exhibited the optical micrographs of the mouse skin (a, b, c, and d: skin surfaces; e, f, g, and h: skin cross-sections) and compared the GPx (i) activity and LF level

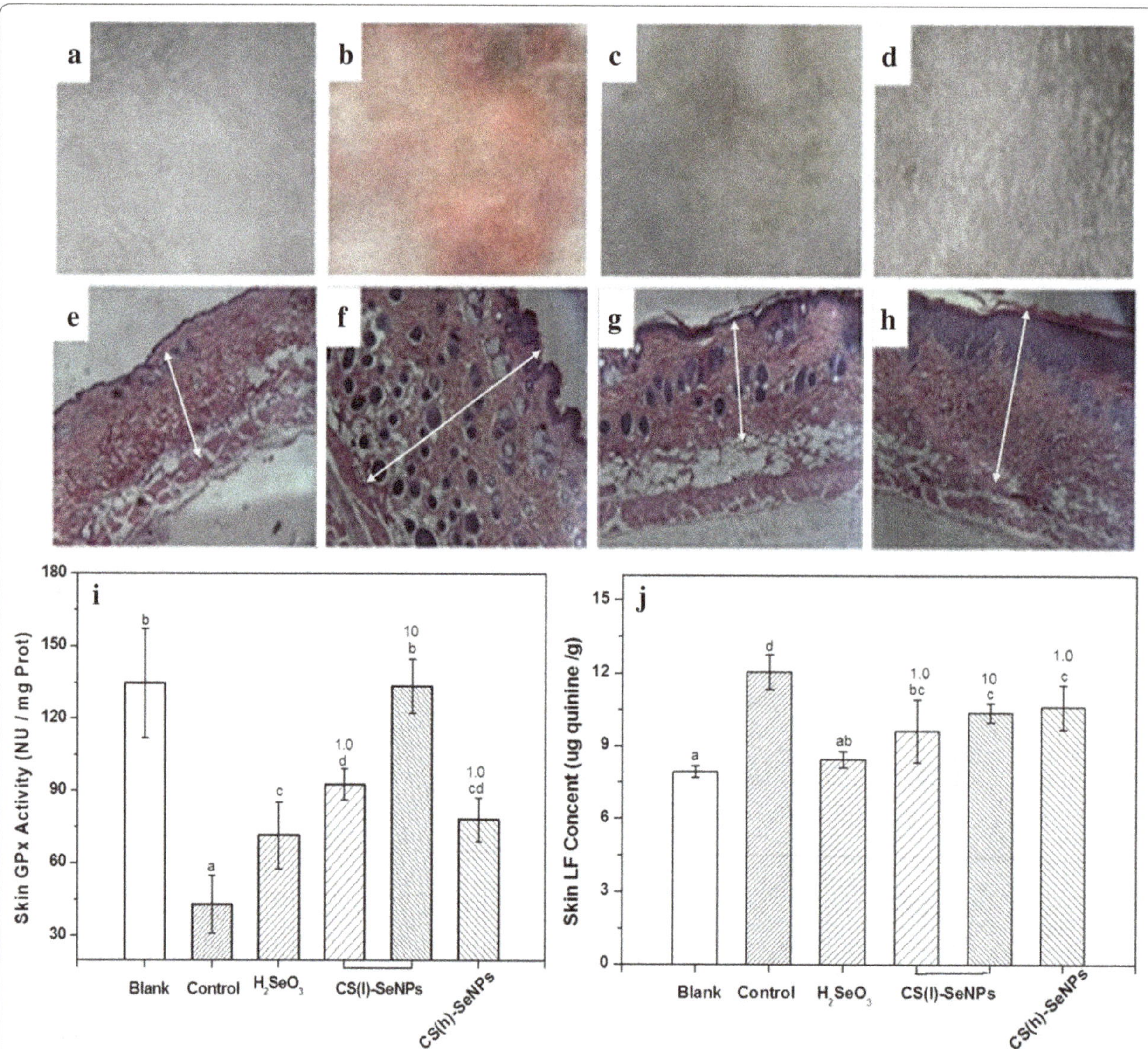

Fig. 6 Optical micrographs of the mice skin and comparison of the GPx (i) and LF levels (j) after UV-radiation. **a**–**d** skin surface; **e**–**h** skin cross section. **a**, **e** unirradiated group; **b**, **f** UV-irradiated and treated with blank lotion; **c**, **g** UV-irradiated and treated with the CS-SeNPs; and **d**, **h** UV-irradiated and treated with H_2SeO_3. H&E stain, 400× magnification, the thickness of the granular layers is marked with the *double-headed arrows*. The *different letter* markers denote the significant mean difference at $p < 0.05$

(j) after UV-radiation. The surface (Fig. 6a) and cross-section (Fig. 6e) of the unirradiated skin were used as the blank, which had relatively high GPx activity and low LF. After a 15-day UV-radiation treatment, a palpable pathological pigmentation could be observed on the surface (Fig. 6b), and a number of dark granules appeared in the cross section (Fig. 6f). Meanwhile, the level of GPx activity was sharply reduced and LF increased. This irradiated group, without any antioxidant treatment, was used as the control. Under the same dose of UV radiation, the level of the pathological pigmentation and the thickness of the granular layer were greatly reduced for the skin treated with CS-SeNPs (Fig. 6c, g) or H_2SeO_3 (Fig. 6d, h), respectively. For these two Se samples, the former was better to protect GPx activity than the latter in respect to the same Se dose (1 mg/kg, drug/body weight). As for the two types of CS-SeNPs, CS(l)-SeNPs was better than CS(h)-SeNPs, and the concerned effect could be improved in a dose-dependent manner. The LF level could be reduced by CS-SeNPs. However, no significant dose-dependence was observed in the present dose range.

D-galactose system

Figure 7 compares the effects of CS-SeNPs on GPx activity and LF level in KM mouse livers and kidneys treated with D-galactose, respectively. In both viscera models, D-galactose increased the LF level and reduced GPx

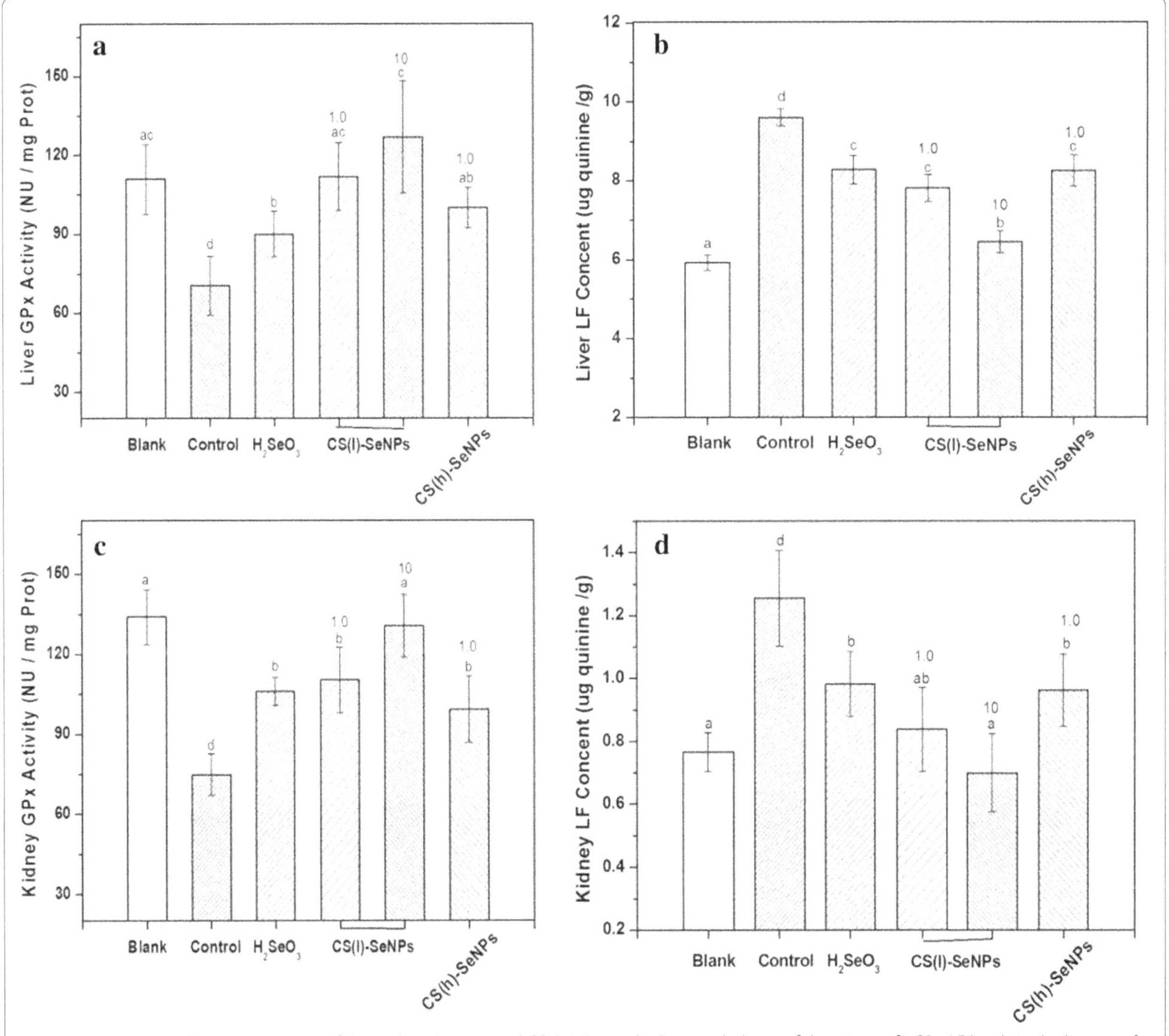

Fig. 7 Comparison of the bio-activities of the orally administered CS-SeNPs on the livers or kidneys of the mice. **a**, **b** GPx, LF levels in the livers; **c**, **d** GPx, LF levels in the kidneys. The *different letter* markers denote the significant mean difference at $p < 0.05$

Table 3 Acute lethal effect of Se samples by oral administration

H_2SeO_3		CS(l)-SeNPs	
Selenium dose (mg drug/body)	Mouse mortality (%)	Selenium dose (mg drug/body)	Mouse mortality (%)
7.1	0	74.1	0
10.7	10	111.1	10
16.0	40	166.7	10
24.0	60	250.0	50
36.0	90	375.0	90

activities. Well, the presence of the Se substance could weaken the effects of D-galactose. Similar results were reported elsewhere [50]. Among these Se substances, CS(l)-SeNPs was better than both H_2SeO_3 and CS(h)-SeNPs to protect the GPx activities (Fig. 7a, c) and reduce LF accumulation (Fig. 7b, d) with the same Se dose (1 mg/kg, drug/body weight). The effects of the CS(l)-SeNPs could also be improved in a dose-dependent manner.

The dosages of the drugs were limited by their toxicity. The toxic doses of H_2SeO_3 and CS(l)-SeNPs were compared in Table 3. According to literature [51], the moderately toxic dose and highly toxic doses were categorized as 50–500 and 5–50 mg/kg (drug/body weight), respectively. The LD50 of H_2SeO_3 was 22.0 mg/kg with 95% confidence from 15.9 to 30.4 and was highly toxic while the CS(l)-SeNPs, with LD50 of 258.2 mg/kg with 95% confidence between 193.9 and 343.9, belonged to a moderately toxic substance. A similar result was reported by Wang et al. in which the LD50 of Se nanoparticle was at the level of 113.0 mg/kg with 95% confidence being 89.9–141.9 [6].

Conclusions

In conclusion, the antioxidant abilities of the Se nanoparticles stabilized with different CS, i.e. CS(l)-SeNPs and CS(h)-SeNPs, could be enhanced by a 30-day storage process. The transdermal Se delivery efficiency of these CS-SeNPs was equivalent to that of selenite. The good abilities to penetrate cell or tissue have made these nanoparticles to be able to effectively inhibit ROS accumulation, reduce Se cytotoxicity, protect GPx activity and prevent LF accumulation, in vitro or in vivo. The UV-radiation or D-galactose tests indicated that the antioxidant capacities of CS-SeNPs were more evident in viscera than in skin of mice. However, regarding the aspect of dose effect control, CS(l)-SeNPs was found more efficient than CS(h)-SeNPs. From a more prospective point of view, we believe that further studies will be needed to explore the metabolic fate and long-term fate, stability and potential transformation of the chitosan selenium nanoparticles in vivo.

Abbreviations

Se: selenium; Se^0: the element Se with zero oxidation state; SeNPs: Se^0 nanoparticles; CS: chitosan; Mws: molecular weigthts; CS-SeNPs: selenium nanoparticles stabilized with chitosan; CS(l)-SeNPs: selenium nanoparticles with the low molecular weight chitosan; CS(h)-SeNPs: selenium nanoparticles with the high molecular weight chitosan; ROS: reactive oxygen species; GPx: glutathione peroxidase; LF: lipofusin; H_2SeO_3: the seleninic acid; ABTS: 2, 2′-Azino-bis (3-ethylbenzothiazoline-6-sulfonic acid; DPPH: 2, 2-diphenyl-1-picrylhydrazyl; DAN: 2, 3-diaminonaphthalene; DCFH-DA: 2′,7′-dichlorofluorescein diacetate; DMP-30: 2,4,6-Tris(dimethylaminomethyl) phenol; DMEM: Dulbecco's modified Eagle medium; NEAA: MEM nonessential amino acid solution; DMSO: dimethyl sulfoxide; PBS: potassium phosphate; CS3: chitosan with a molecular weight of less than 3 kDa; CS200: chitosan with a molecular weight of less than 200 kDa; $K_2S_2O_8$: potassium persulfate; TCA: trichloroacetic acid; TBA: 2-thiobarbituric acid; STEM: scanning transmission electron microscopy; RSC%: the radicals scavenging activity; MDA: malondialdehyde; MTT: 3-(4, 5-dimethylthiazol-2-yl)-2, 5-diphenyltetrazolium bromide; AFS: hydride-generation atomic fluorescence spectrometry; ET-based: assays based on electron transfer; HAT-based: assays based on hydrogen atom transfer.

Authors' contributions

XJL, GHZ, SS, and FZR coordinated the experiments, and provided important advice for each designed the study. CYZ performed the nanoparticles preparing and transdermal tests, XNZ performed the in vitro antioxidant activity assays and the cell experiments, and they completed the animal texts together. All authors contributed to the interpretation of data, the preparation of the paper and approval of the final version. All authors read and approved the final manuscript.

Author details

[1] Beijing Advanced Innovation Center for Food Nutrition and Human Health, Beijing Laboratory for Food Quality and Safety, Beijing Dairy Industry Innovation Team, College of Food Science & Nutritional Engineering, China Agricultural University, Beijing 100083, China. [2] Group of Environmental Physical Chemistry, F.-A. Forel Institute, University of Geneva, Geneva, Switzerland.

Acknowledgements

We would like to acknowledge Professors Yan and Wei (Beijing National Center for Microscopy, Department of Chemistry, Tsinghua University) for their technical advice. We would also like to thank Professor Jia at the Center for Biological Imaging (CBI), Institute of Biophysics, Chinese Academy of Science for her help in making ultrathin section samples.

Competing interests

The authors declare they have no competing financial interest.

Funding

The study was supported by the National Key Research and Development Program (No. 2016YFD0400804) and Beijing Dairy Industry Innovation Team.

References

1. Tapiero H, Townsend DM, Tew KD. The antioxidant role of selenium and seleno-compounds. Biomed Pharmacother. 2003;57:134–44.
2. Liu W, Zheng WJ, Zhang YB, Cao WQ, Chen TF. Selenium nanoparticles as a carrier of 5-fluorouracil to achieve anticancer synergism. ACS Nano. 2012;6:6578–91.
3. McKenzie RC. Selenium, ultraviolet radiation and the skin. Clin Exp Dermatol. 2000;25:631–6.
4. Qin S, et al. Effects of selenium-chitosan on blood selenium concentration, antioxidation status, and cellular and humoral immunity in mice. Biol Trace Elem Res. 2015;165:145–52.
5. Wang WF, et al. Dietary selenium requirement and its toxicity in juvenile abalone *Haliotis discus hannai* Ino. Aquaculture. 2012;330:42–6.
6. Wang H, Zhang J, Yu H. Elemental selenium at nano size possesses lower toxicity without compromising the fundamental effect on selenoenzymes: comparison with selenomethionine in mice. Free Radic Biol Med. 2007;42:1524–33.
7. Torres SK, et al. Biosynthesis of selenium nanoparticles by *Pantoea agglomerans* and their antioxidant activity. J Nanopart Res. 2012;14:1–9.
8. Zhang JS, Wang HL, Yan XX, Zhang LD. Comparison of short-term toxicity between Nano-Se and selenite in mice. Life Sci. 2005;10:1099–109.
9. Chen HY, Shin DW, Nam JG, Kwon KW, Yoo JB. Selenium nanowires and nanotubes synthesized via a facile template-free solution method. Mater Res Bull. 2010;45:699–704.
10. Rai M, Yadav A, Gade A. Silver nanoparticles as a new generation of antimicrobials. Biotechnol Adv. 2009;27:76–83.
11. Daniel MC, Astruc D. Gold nanoparticles: assembly, supramolecular chemistry, quantum-size-related properties, and applications toward biology, catalysis, and nanotechnology. Chem Rev. 2004;104:293–346.
12. Schubert D, Dargusch R, Raitano J, Chan SW. Cerium and yttrium oxide nanoparticles are neuroprotective. Biochem Biophys Res Commun. 2006;342:86–91.
13. Kim J, et al. Effects of a potent antioxidant, platinum nanoparticle, on the lifespan of *Caenorhabditis elegans*. Mech Aging Dev. 2008;129:322–31.
14. Peters RJB, et al. Nanomaterials for products and application in agriculture, feed and food. Trends Food Sci Technol. 2016;54:155–64.
15. Kaur G, Iqbal M, Bakshi MS. Biomineralization of fine selenium crystalline rods and amorphous spheres. J Phys Chem C. 2009;113:13670–6.
16. Zhang YF, Wang JG, Zhang LN. Creation of highly stable selenium nanoparticles capped with hyperbranched polysaccharide in water. Langmuir. 2010;26:17617–23.
17. Wu SS, Sun K, Wang X, Wang DX, Wan XC, Zhang JS. Protonation of epigallocatechin-3-gallate (EGCG) results in massive aggregation and reduced oral bioavailability of EGCG-dispersed selenium nanoparticles. J Arg Food Chem. 2013;61:7268–75.
18. Estevez H, Garcia-Lidon JC, Luque-Garcia JL, Cmara C. Effects of chitosan-stabilized selenium nanoparticles on cell proliferation, apoptosis and cell cycle pattern in HepG2 cells: comparison with other selenospecies. Colloid Surf B. 2014;122:184–93.
19. Anal AK, Stevens WF, Remuñán-López C. Ionotropic cross-linked chitosan microspheres for controlled release of ampicillin. Int J Pharm. 2006;312:166–73.
20. Roncal T, Oviedo A, Armentia IL, Fernández DL, Villarán MC. High yield production of monomer-free chitosan oligosaccharides by pepsin catalyzed hydrolysis of a high deacetylation degree chitosan. Carbohyd Res. 2007;342:2750–6.
21. Zhang CY, Zhai XN, Zhao GH, Ren FZ, Leng XJ. Synthesis, characterization, and controlled release of selenium nanoparticles stabilized by chitosan of different molecular weights. Carbohyd Polym. 2015;134:158–66.
22. Blagosklonny MV. Aging: Ros or tor. Cell Cycle. 2008;7:3344–54.
23. Brunk UT, Terman A. Lipofuscin: mechanisms of age-related accumulation and influence on cell function. Free Radic Biol Med. 2002;33:611–9.
24. Family F, Mazzitello KI, Arizmendi CM, Grossniklaus HE. Dynamic scaling of lipofuscin deposition in aging cells. J Stat Phys. 2011;144:332–43.
25. Höhn A, König J, Grune T. Protein oxidation in aging and the removal of oxidized proteins. J Proteom. 2013;92:132–59.
26. Terman A, Brunk ULFT. Lipofuscin: mechanisms of formation and increase with age. Apmis. 1998;106:265–76.
27. Thorling EB, Overvad K, Bjerring P. Oral selenium inhibits skin reactions to UV light in hairless mice. Acta Pathologica Microbiologica Scandinavica Series A: Pathol. 1983;91:81–4.
28. Overvad K, Thorling EB, Bjerring P, Rbbesen P. Selenium inhibits UV-light-induced skin carcinogenesis in hairless mice. Cancer Lett. 1985;27:163–70.
29. Burke KE, Combs-Jr GF, Gross EG, Bhuyan KC, Abu-Lideh H. The effects of topical and oral L-selenomethionine on pigmentation and skin cancer induced by ultraviolet irradiation. Nutr Cancer. 1992;2:123–37.
30. Lynch AM, Wilcox P. Review of the performance of the 3T3 NRU in vitro phototoxicity assay in the pharmaceutical industry. Exp Toxicol Pathol. 2011;63:209–14.
31. Xu BJ, Chang SKC. A comparative study on phenolic profiles and antioxidant activities of legumes as affected by extraction solvents. J Food Sci. 2007;72:S159–66.
32. Re R, Pellegrini N, Proteggente A, Pannala A, Yang M, Rice-Evans C. Antioxidant activity applying an improved ABTS radical cation decolorization assay. Free Radic Biol Med. 1999;26:1231–7.
33. Tsuda T, et al. Antioxidative activity of the anthocyanin pigments cyanidin 3-O-.β-D-glucoside and cyaniding. J Agric Food Chem. 1994;42:2407–10.
34. Wolfe KL, Hai LR. Cellular antioxidant activity (CAA) assay for assessing antioxidants, foods, and dietary supplements. J Agric Food Chem. 2007;55:8896–907.
35. Hamilton MA, Russo RC, Thurston RV. Trimmed Spearman-Karber method for estimating median lethal concentrations in toxicity bioassays. Environ Sci Technol. 1977;11:714–9.
36. Olson OE, Palmer IS, Cary EE. Modification of the official fluorometric method for selenium in plants. J Am Heart Assoc. 1975;58:117–21.
37. Müller-Röver S, et al. A comprehensive guide for the accurate classification of murine hair follicles in distinct hair cycle stages. J Invest Dermatol. 2001;117:3–15.
38. Zhao T. Hygienic standard for cosmetics health. Beijing; 2007.
39. Pence BC, Delver E, Dunn DM. Effects of dietary selenium on UVB-induced skin carcinogenesis and epidermal antioxidant status. J Invest Dermatol. 1994;102:759–61.
40. Harvey H, Ju SJ, Son S, Feinberg L, Shaw C, Peterson W. The biochemical estimation of age in Euphausiids: laboratory calibration and field comparisons. Deep-Sea Res Part II. 2010;57(7):663–71.
41. Bradford MM. A rapid and sensitive method for the quantitation of microgram quantities of protein utilizing the principle of protein-dye binding. Anal Biochem. 1976;72:248–54.
42. Mei L, et al. Bioconjugated nanoparticles for attachment and penetration into pathogenic bacteria. Biomaterials. 2013;34:10328–37.
43. Zheng SY, et al. PEG-nanolized ultrasmall selenium nanoparticles overcome drug resistance in hepatocellular carcinoma HepG2 cells through induction of mitochondria dysfunction. Int J Nanomed. 2012;7:3939–49.
44. Huang DJ, Ou BX, Prior R. The chemistry behind antioxidant capacity assays. J Agr Food Chem. 2005;53:1841–56.
45. Sambuy Y, De Angelis I, Ranaldi G, Scarino ML, Stammati A, Zucco F. The Caco-2 cell line as a model of the intestinal barrier: influence of cell and culture-related factors on Caco-2 cell functional characteristics. Cell Biol Toxicol. 2005;21:1–26.
46. Lewinski N, Colvin V, Drezek R. Cytotoxicity of nanoparticles. Small. 2008;4:26–49.
47. Chen TF, Wong YS, Zheng WJ, Bai Y, Huang L. Selenium nanoparticles fabricated in *Undaria pinnatifida* polysaccharide solutions induce mitochondria-mediated apoptosis in A375 human melanoma cells. Colloid Surf B. 2008;67:26–31.
48. Feng YX, et al. Differential effects of amino acid surface decoration on the anticancer efficacy of selenium nanoparticles. Dalton Trans. 2014;3:1854–61.
49. Daniels LA. Elenium metabolism and bioavailability. Biol Trace Elem Res. 1996;54:185–99.
50. Kennedy GL, Ferenz RL, Burgess BA. Estimation of acute oral toxicity in rates by determination of the approximate lethal dose rather than the LD50. J Appl Toxicol. 1986;6:145–8.
51. Zhou X, et al. Enhancement of endogenous defenses against ROS by supra-nutritional level of selenium is more safe and effective than antioxidant supplementation in reducing hypertensive target organ damage. Med Hypotheses. 2007;68:952–95.

16

Study on the mechanism of antibacterial action of magnesium oxide nanoparticles against foodborne pathogens

Yiping He[1*], Shakuntala Ingudam[2], Sue Reed[1], Andrew Gehring[1], Terence P. Strobaugh Jr.[1] and Peter Irwin[1]

Abstract

Background: Magnesium oxide nanoparticles (MgO nanoparticles, with average size of 20 nm) have considerable potential as antimicrobial agents in food safety applications due to their structure, surface properties, and stability. The aim of this work was to investigate the antibacterial effects and mechanism of action of MgO nanoparticles against several important foodborne pathogens.

Results: Resazurin (a redox sensitive dye) microplate assay was used for measuring growth inhibition of bacteria treated with MgO nanoparticles. The minimal inhibitory concentrations of MgO nanoparticles to 10^4 colony-forming unit/ml (CFU/ml) of *Campylobacter jejuni*, *Escherichia coli* O157:H7, and *Salmonella* Enteritidis were determined to be 0.5, 1 and 1 mg/ml, respectively. To completely inactivate 10^{8-9} CFU/ml bacterial cells in 4 h, a minimal concentration of 2 mg/ml MgO nanoparticles was required for *C. jejuni* whereas *E. coli* O157:H7 and *Salmonella* Enteritidis required at least 8 mg/ml nanoparticles. Scanning electron microscopy examination revealed clear morphological changes and membrane structural damage in the cells treated with MgO nanoparticles. A quantitative real-time PCR combined with ethidium monoazide pretreatment confirmed cell membrane permeability was increased after exposure to the nanoparticles. In a cell free assay, a low level (1.1 μM) of H_2O_2 was detected in the nanoparticle suspensions. Consistently, MgO nanoparticles greatly induced the gene expression of KatA, a sole catalase in *C. jejuni* for breaking down H_2O_2 to H_2O and O_2.

Conclusions: MgO nanoparticles have strong antibacterial activity against three important foodborne pathogens. The interaction of nanoparticles with bacterial cells causes cell membrane leakage, induces oxidative stress, and ultimately leads to cell death.

Keywords: MgO nanoparticles, Foodborne pathogens, Antimicrobial mechanism, H_2O_2, Oxidative stress

Background

Campylobacter jejuni, *Escherichia coli* O157:H7, and *Salmonella* are the most common foodborne pathogens responsible for millions of cases of illnesses and hundreds of deaths each year in the United States [1]. *C. jejuni* is a spiral shaped and oxygen-sensitive microaerophile, whereas *E. coli* O157:H7 and *Salmonella* are rod shaped and facultative anaerobic bacteria [2]. The main natural reservoirs for these bacteria are intestinal tracts of birds (e.g. chicken) and other animals (e.g. cattle) [3–5]. Transmission of these pathogens from animal feces or the environment to food can occur during harvesting, processing, distribution, and preparation of food. Pathogen contamination has been frequently found in various food products including meat, fresh produce, dairy products, and ready-to-eat foods. The sporadic prevalence of microbial pathogens in food and increased incidence of antibiotic resistant strains have posed serious concerns to public health [6, 7]. Hence, there is a need to develop alternative strategies for effective control of microbial pathogens in food and the environment.

*Correspondence: yiping.he@ars.usda.gov
[1] Molecular Characterization of Foodborne Pathogens Research Unit, Eastern Regional Research Center, Agricultural Research Service, United States Department of Agriculture, 600 East Mermaid Lane, Wyndmoor, PA 19038, USA
Full list of author information is available at the end of the article

Metal oxide nanoparticles such as ZnO, MgO, CuO, CaO, Ag_2O, and TiO_2 are a new class of antimicrobial agents that have been increasingly studied for their antibacterial properties and potential applications in food, the environment, and healthcare [8, 9]. As nanoscale (<100 nm) inorganic materials, metal oxide nanoparticles have distinct features including broad spectrum antibacterial activity, large surface area of interaction with cells, low possibility for bacteria to develop resistance, high stability even under harsh conditions, and tunable sizes, shapes, surface properties, and chemical compositions, which lead to great potential for developing nanomaterials as effective antimicrobial agents. Among these, nanostructured MgO is particularly interesting due to its strong antibacterial activity, but high thermal stability and low cost.

The mechanism of metal oxide nanoparticle action on bacteria is complicated and not fully understood. It has been reported that the antibacterial activity of MgO nanoparticles is attributed to the production of reactive oxygen species (ROS) which induce lipid peroxidation in bacteria [10]. In contrast, non-ROS mediated bacterial toxicity was also found in MgO nanoparticles, suggesting oxidative stress might not be the primary mechanism of cell death [11]. Furthermore, the antibacterial effect not only depends upon the sizes, shapes, chemical composition, and surface properties (e.g. hydrophobicity) of the nanoparticles, but also varies with bacterial species [12, 13]. Several studies have shown that smaller particles have greater antibacterial activity due to higher reactive surface area [9]. However, aggregations of very small nanoparticles (~5 nm) could reduce the efficiency of interaction with bacteria. It also has been reported that MgO and CuO nanoparticles had substantially higher antibacterial activities on Gram-positive (G+) than Gram-negative (G−) bacteria, presumably due to the differences in cell membrane structure between these organisms. Our previous study showed that ZnO nanoparticles displayed extremely strong activity against *C. jejuni* compared to *E. coli* O157:H7 or *Salmonella* likely due to the different tolerances of these organisms to oxidative stress induced by nanoparticles [13].

Cytotoxicity is a major concern in the development of antimicrobial agents. MgO has been used as a mineral supplement for magnesium, an essential nutrient for the human body. As a medicine, MgO is used for the relief of cardiovascular disease and stomach problems. At low concentrations (0.3 mg/ml), MgO nanoparticles were reported to not be toxic to human cells [14]. However, toxic effects are greatly dependent on the physical and chemical properties of nanoparticles as well as the types of cells tested [15, 16]. Hence, extensive evaluation of nanoparticles on different biological systems is needed to determine the toxicity of nanoparticles. Understanding of the mechanism of nanoparticle action on bacteria could provide useful guidelines for rational design and assembly of effective antibacterial derivatives. Advanced strategies, such as packaging multiple antimicrobial agents into the same nanoparticle, coating nanoparticles with biodegradable materials, and engineering target-specific nanoparticles for delivery to infection site, have emerged to improve antimicrobial activities and reduce undesirable side effects of nanoparticles.

The aim of this research was to study the antimicrobial activity and mechanism of action of MgO nanoparticles on three major foodborne pathogens. Through scanning electron microscopy (SEM) examination of cell morphology and membrane structure, ethidium monoazide combined with quantitative real-time PCR (EMA-qPCR) measurement of membrane permeability, transcriptional analysis of oxidative stress defense genes, and quantification of H_2O_2 produced in nanoparticle suspensions, we suggested the most conceivable mechanism of action of MgO nanoparticles on bacteria.

Methods

MgO nanoparticles

MgO nanoparticles with an average size of 20 nm were purchased from Nanostructured & Amorphous Materials, Inc. (Houston, TX, USA). ZnO nanoparticles (average size of 30 nm) were from Inframat Advanced Materials LLC (Manchester, CT, USA). A stock suspension (8 mg/ml) was freshly prepared by resuspending 80 mg of the nanoparticles into 10 ml ddH_2O for H_2O_2 production assay or Mueller Hinton broth (MH broth; Becton–Dickinson Co., Sparks, MD, USA) for cell culture experiments. All of the nanoparticle suspensions were homogenized by vigorous vortexing prior to use in the following experiments.

Bacterial culture conditions

Campylobacter jejuni 81–176 was statically grown at 42 °C in MH broth in a microaerophilic workstation (Don Whitley Scientific, Ltd., Shipley, UK) maintaining an atmosphere of 5 % O_2, 10 % CO_2, 85 % N_2, and 82 % relative humidity. *Salmonella enterica* serovar Enteritidis ATCC 13076 and *E. coli* O157:H7 EDL 933 were aerobically grown at 37 °C in MH broth with shaking at 190 revolution per minute (rpm) (Innova 42, New Brunswick, Enfield, CT, USA).

Minimum inhibitory concentration of MgO nanoparticles

The viability of bacterial cells when exposed to varying concentrations of MgO nanoparticles was analyzed in a 96-well plate using the Resazurin Cell Viability Assay Kit (Bio Trend Chemicals LLC, Destin, FL, USA).

Resazurin indicates cell viability by changing from a blue/non-fluorescent state to a pink/highly fluorescent state upon chemical reduction resulting from aerobic respiration due to cell growth. Overnight cultures of *C. jejuni*, *E. coli* O157:H7, and *S.* Enteritidis were diluted to approx. 10^4 CFU/ml. MgO nanoparticles were diluted 1:2 in MH broth from a starting concentration of 8 mg/ml in successive columns of a microtiter plate to an ending concentration of 0.03 mg/ml. To each well containing 100 µl of MgO nanoparticle suspension and 100 µl of the diluted bacteria, 20 µl of Resazurin dye was added and mixed thoroughly. Each microorganism was tested in a different plate with eight replicates of each concentration of MgO nanoparticles. Three controls without nanoparticles (100 µl each of MH broth without cells, and live and heat-killed cells at 10^4 CFU/ml) were also included in each plate. The plates were then incubated aerobically at 37 °C for *E. coli* and *Salmonella* or under microaerophilic conditions overnight at 42 °C for *Campylobacter*. After incubation, the plates were subjected to fluorescence measurement at an excitation wavelength of 530 nm and emission wavelength of 590 nm using a Tecan Safire2™ microplate reader (Männedorf, Switzerland) as well as visual inspection for color change. The lowest concentration of nanoparticle suspension that inhibited cell growth (dye did not convert to red) was defined as the minimum inhibitory concentration (MIC).

Antimicrobial effects of MgO nanoparticles

The antimicrobial effects of MgO nanoparticles were studied by exposing 10^{8-9} CFU/ml of *C. jejuni*, *E. coli* O157:H7, and *S.* Enteritidis to 0, 0.5, 1, 2, 4 and 8 mg/ml nanoparticles. The samples were incubated aerobically at 37 °C for *E. coli* and *Salmonella* or microaerobically at 42 °C for *Campylobacter* for a total of 24 h. At specific time intervals (0, 0.5, 1, 2, 4, 6 and 8 h), 1 ml of each sample was collected to determine colony-forming units (CFU) on MH agar by the 6 × 6 drop plate method [17]. The average number of CFU/ml ($\log_{10}$) and standard deviation from 6 replicates were used to plot each data point.

Morphological analysis by scanning electron microscope

Bacterial cultures of *E. coli* O157:H7, *S.* Enteritidis, and *C. jejuni* were treated with 0, 1 and 2 mg/ml MgO nanoparticles for 8 h. Aliquots of 1 ml samples were centrifuged for 2 min at 4000 rpm and the cell pellets were resuspended in 0.1 ml MH broth. Subsequently, 20 µl of each concentrated sample was deposited and spread onto a glass coverslip pre-washed with acetone and ethanol. After drying the slips for 15 min at 37 °C, the bacterial cells were subjected to fixation, dehydration, and critical point drying for SEM analysis as described previously [13].

EMA-qPCR assessment of cell membrane integrity

EMA-qPCR analysis of *C. jejuni*, *E. coli* O157:H7, and *S.* Enteritidis cells treated and untreated with MgO nanoparticles was carried out as described previously [18]. Briefly, late log phase cultures of *Campylobacter* were treated with 0, 1 and 2 mg/ml of MgO nanoparticles for 8 h at 42 °C in microaerophilic conditions. *E. coli* and *Salmonella* were treated with 0, 2 and 4 mg/ml of nanoparticles for 8 h aerobically at 37 °C. After incubation, 1 ml of each sample was treated with 20 µg/ml EMA in the dark for 5 min and subsequently exposed to a 600 W halogen light for 1 min on ice. Cells were immediately centrifuged at 8000 rpm for 5 min and washed with phosphate-buffered saline. Genomic DNA was extracted from the samples using the DNeasy Blood and Tissue kit (Qiagen, Valencia, CA, USA) and qPCR analyzed using a 7500 Real-Time PCR system (Applied Biosystems, Carlsbad, CA, USA). In the qPCR assay, organism-specific gene targets (*hipO*, *rfbE*, and *invA*), primers and TaqMan probes, and DNA standard curves were chosen for the detection of *C. jejuni*, *E. coli*, and *Salmonella*, respectively [18, 19]. The threshold cycle (Ct) values obtained from the EMA-qPCR assay were converted to DNA copy numbers based on the linear regression equations of the DNA standard curves. The copy number of the DNA target is equivalent to the genome copy of the bacterial cells and provides a good estimate for the number of cells.

Gene expression analysis of nanoparticle treated and untreated cells

Total cellular RNA was extracted from 100 ml of late-log phase (13 h microaerobic incubation) *C. jejuni* culture treated with 0 and 1 mg/ml MgO nanoparticles for 30 min by using TRI-Reagent® (Molecular Research Center, Inc. Cincinnati, OH, USA). DNase I treatment and reverse transcription of the RNA samples were carried out as previously described [20]. The expression of stress response genes was quantified by a real-time PCR assay using SYBR green master mix (Applied Biosystems, Foster City, CA, USA) and the primer sets described in our previous report [13]. Housekeeping genes *tsf*, *gyrA*, and *16S rRNA* were included as references for data normalization. The difference of gene expression between nanoparticle treated and untreated cells was calculated using $2^{-\Delta\Delta Ct}$ formula: $\Delta\Delta Ct = \Delta Ct$ (treated sample) $-\Delta Ct$ (untreated sample), $\Delta Ct = Ct$ (target gene) $-Ct$ (*16S rRNA*), and Ct is the threshold cycle value of the amplified target or reference gene [21].

Quantification of H_2O_2 production in MgO and ZnO nanoparticle suspensions

The assay of H_2O_2 released from nanoparticle suspensions was performed in a 96-well plate using the Red

Hydrogen Peroxide Assay Kit (Enzo Life Sciences, Inc. Farmingdale, NY, USA). The assay utilizes horseradish peroxidase (HRP) to catalyze the conversion of a red peroxidase substrate into resorufin (a highly colored and fluorescent compound) during the reduction of H_2O_2 to O_2 and H_2O. Scanning of resorufin absorbance was used for the detection of H_2O_2 production. Briefly, a 5 ml reagent mix was prepared by adding 50 µl of red peroxidase substrate (100×) and 200 µl of HRP (20 units/ml) into 4.75 ml assay buffer. To each well of a microtiter plate, 100 µl each of the freshly prepared reagent mixture and nanoparticle suspensions (0.5, 1, 2 and 4 mg/ml) were added. After 1-h incubation at room temperature, the plate was centrifuged at 4000 rpm for 5 min to pellet the nanoparticles. Subsequently, 100 µl supernatant from each well was transferred into an ultraviolet (UV) transparent plate and scanned for absorbance between 475 and 600 nm wavelengths using a Tecan Safire2™ microplate reader. To estimate the H_2O_2 concentrations in the nanoparticle suspensions, a standard curve was prepared using a set of 1:3 serial dilutions of H_2O_2 in distilled water (10, 3.33, 1.11, 0.37, 0.12, 0.04 and 0.01 µM) and scanned in the same assay. Distilled water was used as a blank for scanning. Each of the samples including the standards were analyzed in duplicate.

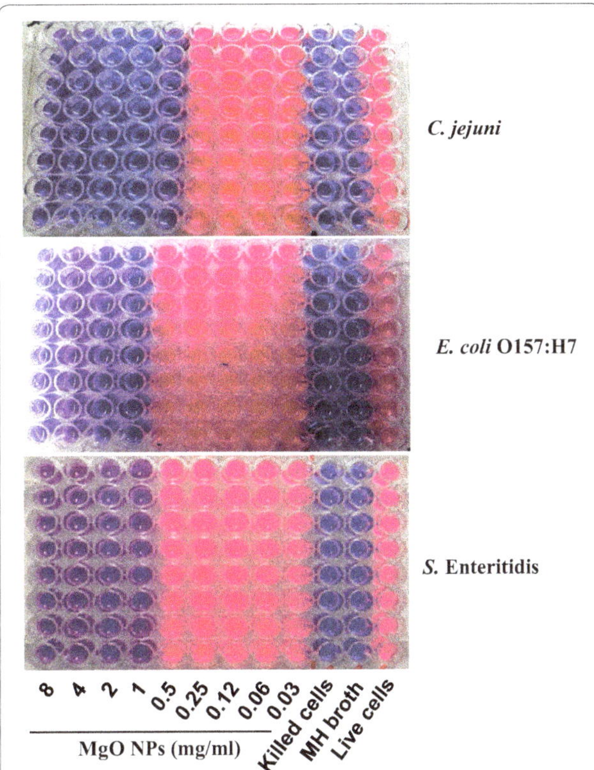

Fig. 1 MIC of MgO nanoparticless for *C. jejuni*, *E. coli* O157:H7, and *S.* Enteritidis. Freshly grown and diluted bacterial cultures (approx. 10^4 CFU/ml) were treated with various concentrations of MgO nanoparticles. The visible color change from *nonfluorescent blue* (resazurin) to *fluorescent red* (resorufin) shows reductive activities of the cells, indicating no inhibition on cell growth

Results

Minimum inhibitory concentration of MgO nanoparticles

The viability of *C. jejuni*, *E. coli* O157:H7, and *S.* Enteritidis when exposed to varying concentrations of MgO nanoparticles was determined using an oxidation–reduction assay. Resazurin, a redox sensitive dye, was used to indicate cell viability. Metabolically active cells reduce the non-fluorescent blue resazurin to fluorescent red resorufin. Non-living cells do not reduce the resazurin, and thus indicate cell death. This visible change in color and fluorescence indicates the cells are viable. Freshly grown and diluted bacterial cultures in MH broth (approx. 10^4 CFU/ml) were exposed to MgO nanoparticles at concentrations ranging from 0.03 to 8 mg/ml. *C. jejuni* cells were rendered non-viable after treatment with 0.5 mg/ml of MgO nanoparticles. *E. coli* O157:H7 and *S.* Enteritidis were inactivated with 1 mg/ml of MgO nanoparticles (Fig. 1). This indicates that *C. jejuni* is more susceptible to the antimicrobial effects of MgO nanoparticles than *E. coli* O157:H7 or *S.* Enteritidis.

Lethal effect of MgO nanoparticles

The observed antimicrobial effect was further investigated by exposing 10^8 CFU/ml *C. jejuni* and 10^9 CFU/ml *E. coli* O157:H7 and *S.* Enteritidis to 0, 0.5, 1, 2, 4 and 8 mg/ml MgO nanoparticles over a set time trial (Fig. 2). Live cells were measured by the colony forming units on MH agar. At a concentration of 2 mg/ml MgO nanoparticles, *C. jejuni* was reduced 6 orders of magnitude after 2 h and completely killed after 4 h. At 4 mg/ml, *C. jejuni* was completely killed within 1 h. On the contrary, 8 mg/ml MgO nanoparticles were required to kill all *E. coli* O157:H7 and *S. Enteritidis* cells in 4 h and 4 mg/ml in 6 h. In addition, *E. coli* O157:H7 could also be killed by 2 mg/ml in 8 h, whereas *S.* Enteritidis was only reduced 5 logs after the same exposure. This demonstrates again that MgO nanoparticles are effective at killing *C. jejuni* at low concentrations in short periods of time. They are also advantageous at killing *E. coli* O157:H7 and *S.* Enteritidis within 4 h.

MgO nanoparticles alter bacterial cell morphology and membrane structure

To explore antimicrobial mechanism of the nanoparticles, scanning electron microscopy was used to examine the morphological and membrane structure changes of *C. jejuni*, *E. coli* O157:H7, and *S.* Enteritidis induced by MgO nanoparticles. Bacterial cells in late-log growth

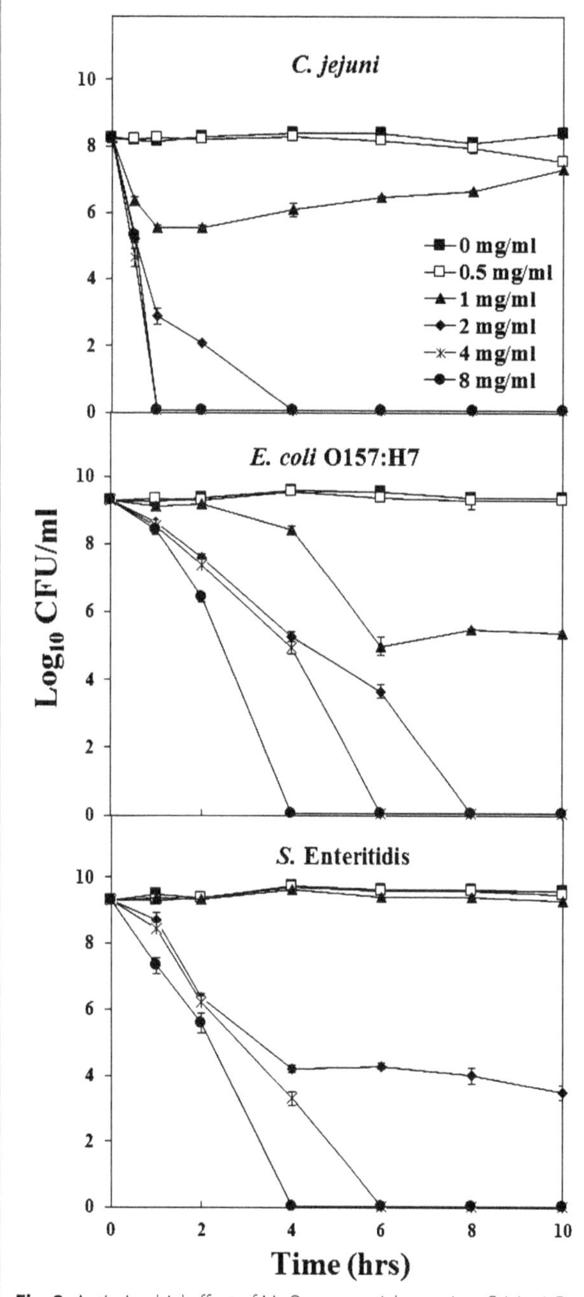

Fig. 2 Antimicrobial effect of MgO nanoparticles against *C. jejuni*, *E. coli* O157:H7, and *S.* Enteritidis. Various concentrations of nanoparticles were applied to approx. 10^8 CFU/ml of *C. jejuni* and 10^9 CFU/ml of *E. coli* O157:H7 or *S.* Enteritidis. At different times after the treatment, viable cell counts were measured by culturing bacterial colonies on MH agar plates. Each of the CFU/ml value represents the mean of six replicates

were treated with sub-lethal doses of MgO nanoparticles (1 and 2 mg/ml) for 4 h and collected for SEM study. Both treated and untreated cells were incubated under the same conditions and analyzed by SEM in parallel in order to observe the differences between the control and cells exposed to nanoparticles. SEM images in Fig. 3 show all of the untreated cells have intact and smooth surfaces. As expected, *C. jejuni* cells are spiral-shaped, whereas *E. coli* O157:H7 and *S.* Enteritidis are rod-shaped. After incubation with a sub-lethal concentration of nanoparticles, *C. jejuni* cells underwent significant morphological changes from spiral to coccoid form, but *E. coli* O157:H7 and *S.* Enteritidis remained rod shaped. Noticeably, all of the treated cells displayed some deep craters on their membrane surface, indicating a degree of membrane structure damage. These cells appear to be shorter and more compact, suggesting there could be some leakage of the cellular contents caused by the treatment. No cell lysis was noticed after the treatment of sub-lethal concentrations of nanoparticles.

MgO nanoparticles increase *C. jejuni* membrane permeability

To further investigate the effect of MgO nanoparticles on cell membrane integrity and permeability, ethidium monoazide-qPCR assay was employed. Ethidium monoazide (EMA) is a DNA-binding intercalator that enters cells through compromised membranes. After photo cross-linking, the intercalation between EMA and cellular DNA becomes irreversible due to the formation of intermolecular covalent bonds. The EMA-bound DNA cannot be amplified by PCR, thus indicating membrane damage.

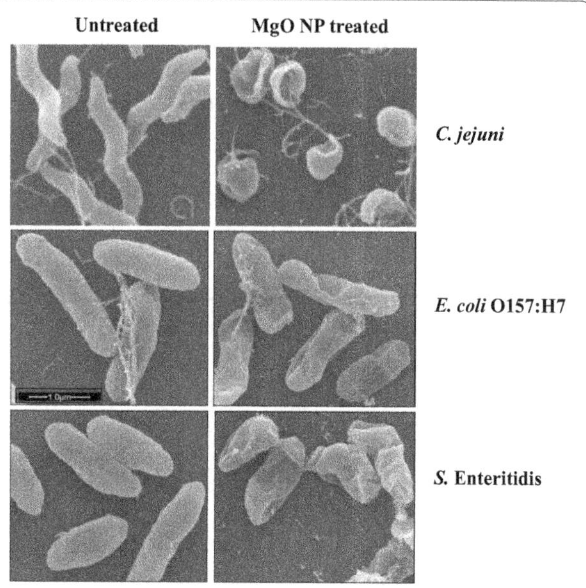

Fig. 3 Scanning electron micrographs of *C. jejuni*, *E. coli* O157:H7, and *S.* Enteritidis. SEM images were taken from the bacterial cells of *C. jejuni*, *E. coli* O157:H7, and *S.* Enteritidis treated (*right panel*) with 2 mg/ml MgO nanoparticles for 8 h. The control cells (*left panel*) were incubated under the same conditions without adding nanoparticles

The membrane permeability of *C. jejuni* after exposure to 1 and 2 mg/ml MgO nanoparticles for 4 h was assessed by EMA-qPCR assay. The results in Fig. 4 show that cells exposed to MgO nanoparticles had a nearly 1-log reduction in DNA amplification, indicating EMA penetration via damaged membranes. Similar experiments were performed on *E. coli* O157:H7 and *S.* Enteritidis cells after exposure to 2 and 4 mg/ml MgO nanoparticles. The effects of membrane leakage by MgO nanoparticles were less noticeable compared to *C. jejuni* (data not shown). Together, these results indicate that MgO nanoparticles increase cell membrane permeability and that *C. jejuni* is more susceptible to the membrane damage than *E. coli* O157:H7 and *S. Enteritidis.*

MgO nanoparticles induce the expression of oxidative stress response genes in *C. jejuni*

To study the molecular mechanism of MgO nanoparticle action on bacteria, expression of the genes involved in oxidative and general stress defenses was examined in *C. jejuni.* Late-log phase cells exposed to 1 mg/ml MgO nanoparticles for 30 min were collected for transcription analysis. Transcripts/mRNAs of these genes were prepared and quantified by reverse transcription-qPCR. In response to the treatment, the expression of oxidative stress response genes *katA* (encoding catalase), *ahpC* (encoding alkyl hydroperoxide reductase), and *dps* (encoding bacterioferritin) were upregulated 44-, 5- and 4-fold, respectively (Fig. 5). In addition, general stress response gene *spoT* [encoding a bifunctional (p) ppGpp synthetase/hydrolase] was also expressed 22-fold higher. As controls, transcriptions of housekeeping genes *gyrA*, *tsf*, and 16 s rRNA were not significantly affected. *katA* is the sole catalase gene present in *C. jejuni* and is essential for resistance to hydrogen peroxide [22]. *spoT* is a stringent response gene required for survival under high O_2 [23]. Dramatic increase of the catalase gene (*katA*) and stringent response gene (*spoT*) expression in response to treatment strongly suggests MgO nanoparticles induce oxidative stress in *C. jejuni* cells. To survive, bacteria regulate their detoxification system by increasing their level of oxidative stress defense proteins as well as some of the general stress response proteins.

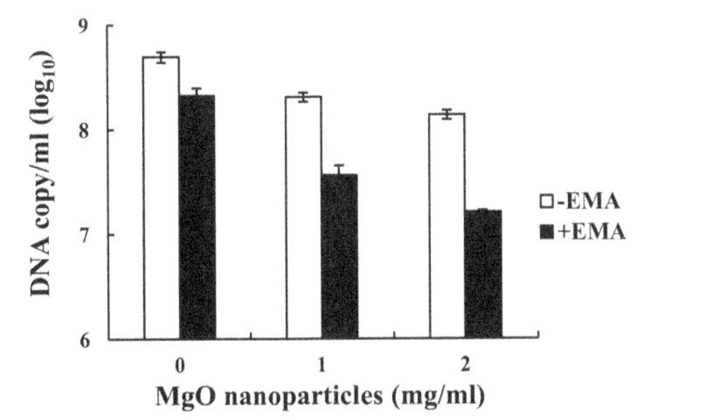

Fig. 4 EMA-qPCR analysis of *C. jejuni* membrane permeability. *C. jejuni* cells exposed to 0, 1 or 2 mg/ml of nanoparticles were measured for the inhibition of DNA amplification caused by the penetrated EMA. The inhibition of DNA amplification was monitored by qPCR of the *hipO* gene. In the presence of EMA, reduced DNA amplification indicates increased membrane permeability in the cells

Nanoparticle suspensions generate hydrogen peroxide

To further investigate the causes of oxidative stress in bacteria by MgO nanoparticles, we measured hydrogen peroxide (H_2O_2) produced in nanoparticle suspensions by using the highly sensitive Red Hydrogen Peroxide assay kit (with sensitivity as low as 10 pmol). H_2O_2 contains a highly reactive oxygen species and is able to penetrate into cells to cause oxidative stress. Since H_2O_2 is the most unstable form of reactive oxygen species, all the nanoparticle suspensions and H_2O_2 standard solutions were freshly prepared for the study. The levels of H_2O_2 produced in nanoparticle suspensions were determined by scanning the absorbance of resorufin fluorescence between 475 and 600 nm wavelengths and shown in Fig. 6. Referenced to the fluorescent absorbance of known concentrations of H_2O_2 standards (0.12, 0.37, 1.1, 3.3 and 10 μM), the H_2O_2 released in MgO nanoparticle suspension was estimated to be ca. 1.1 μM. In the same assay, a low level of H_2O_2 (ca. 0.12 μM) was also detected in ZnO nanoparticle suspensions, suggesting the H_2O_2 generated from MgO/ZnO nanoparticles contributed to antibacterial activity by inducing oxidative stress in cells.

Discussion

In this study, we have shown that MgO nanoparticles have a strong antimicrobial activity against three major foodborne pathogens from both the microplate-based resazurin assay and viable cell count method. The use of visible color change of resazurin from blue to red as an indicator of cell growth not only made the detection simple and fast, but allowed us to avoid the turbidity problem of insoluble nanoparticles interfering with cell optical density measurements. By using this assay, we determined that the minimal inhibitory concentrations of MgO nanoparticles against 10^4 CFU/ml of *C. jejuni* and *E. coli* O157:H7/*Salmonella* were 0.5 and 1 mg/ml, respectively. This is reasonably close to the result reported by Krishnamoorthy et al. [24] which showed the MIC of MgO nanoparticles (average size 25 μM) to 5×10^5 CFU/ml of *E. coli* was 0.5 mg/ml. The slight variances of the MICs found in these studies might be due to the size and/or shape variations of the nanoparticles or the different numbers of cells used.

By viable cell count, we found at least 2 mg/ml nanoparticles (equivalent to 1.3×10^{14} of 20 nm size and sphere-shaped MgO nanoparticles per ml) were required

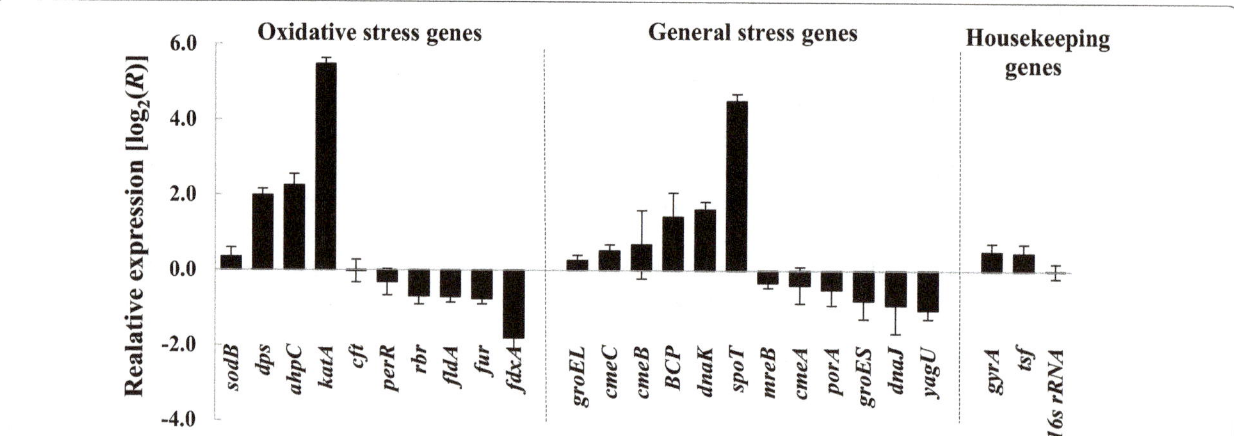

Fig. 5 Difference of stress gene expression between MgO nanoparticle treated and untreated *C. jejuni*. Late-log phase cells treated with 0 or 1 mg/ml MgO nanoparticles for 30 min were quantified for mRNA of stress genes by reverse transcription-qPCR. Relative expression ratio (ΔΔCt) of each gene is presented in a log_2 value

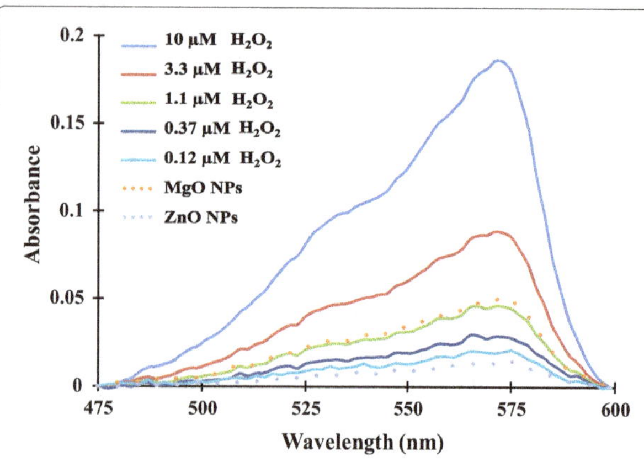

Fig. 6 Production of H_2O_2 in MgO/ZnO nanoparticle suspensions. The production of H_2O_2 in nanoparticle suspensions was measured by spectral absorbance scanning of resorufin, a fluorescent compound formed from a peroxidase substrate by H_2O_2 reduction. The solid curves represent the absorbance of various known concentrations of H_2O_2 standards, and the dotted curves show the levels of H_2O_2 released from MgO/ZnO nanoparticle suspensions after baseline subtraction. The maximum excitation and emission wavelengths of resorufin fluorescence are at 570 and 585 nm, respectively

to completely inactivate 10^8 CFU/ml of *C. jejuni*, and 4 mg/ml was the minimal amount of nanoparticles needed to kill 10^9 CFU/ml of *E. coli* O157:H7 or *Salmonella*. Apparently, higher concentrations of MgO nanoparticles were required to inactivate increased numbers of bacterial cells. Also, *C. jejuni* cells were found to be more susceptible to the nanoparticles than *E. coli* O157:H7 or *Salmonella*.

In this study, we chose *C. jejuni*, a microaerophilic bacterium, as a model organism to study the expression of stress defense genes. *C. jejuni* is extremely sensitive to oxidative stress due to the lack of SoxRS and OxyR, the most important regulatory proteins of oxidative stress defense in *E. coli* and *Salmonella*. It has been known that *C. jejuni* uses a number of enzymes including KatA, SodB, AhpC, Dps, Tpx, and Bcp to detoxify a low level of endogenous H_2O_2 produced during cell metabolism and as a defense to the oxidative stresses potentially encountered in a host and the environment [25]. To better understand bacterial cell responses to the nanoparticle treatment, we selected a number of genes associated with oxidative stress and general stress defenses as targets of transcriptional study. Of the important anti-oxidative stress proteins in *C. jejuni*, KatA is the only catalase and primary enzyme for decomposing H_2O_2 to H_2O and O_2; AhpC is the secondary enzyme for reducing H_2O_2 and is responsible for removal of low level of H_2O_2 from cells; and Dps protects cellular DNA from oxidative damage [26–28].

From transcriptional analysis of the stress response genes, we found the expression of *katA*, *ahpC*, and *dps* were significantly increased in response to a sub-lethal concentration of MgO nanoparticle treatment. Interestingly, the same set of stress response genes (*katA*, *ahpC*, and *dps*) were also found expressed at higher levels in the *C. jejuni* cells treated with 1 mM H_2O_2 [29]. H_2O_2 is highly reactive to cell biomolecules (e.g. DNA, proteins, and lipids) and causes oxidative stress in bacteria when present at high concentrations. Severe or continuous oxidative stress could result in failure of the cellular defense system and cell death. By using a highly sensitive cell-free assay, we detected approximately 0.12 and 1.1 µM H_2O_2 in ZnO and MgO nanoparticle suspensions, respectively. The concentrations of H_2O_2 produced in these nanoparticle suspensions seems to be relatively low, so it is possible that the production of H_2O_2 from nanoparticles is not the sole mechanism for the antibacterial activity of MgO/ZnO nanoparticles. SEM analysis showed considerable cell morphology change and membrane disruption in the cells treated with MgO nanoparticles. Furthermore, the EMA-qPCR results provided from this study confirmed that MgO nanoparticles increased membrane permeability in the bacteria which likely resulted in the leakage of cell content.

Our previous study showed ZnO nanoparticles had remarkable anti-*Campylobacter* activity with the MIC 8–16 fold lower than *E. coli* and *Salmonella* [13] In this study, we did not find significant differences of MgO nanoparticles in inhibiting growth or inactivating cells between *C. jejuni*, *E. coli* O157:H7, and *S.* Enteritidis. However, the effects of ZnO and MgO nanoparticles on triggering cell morphology change, membrane leakage, and oxidative stress, were found to be similar, suggesting the modes of actions of these nanoparticles on bacteria might be similar but not exactly the same.

On the basis of these findings, we propose multiple mechanisms for the action of MgO nanoparticles on bacteria: (1) MgO nanoparticles continuously generate a certain level of H_2O_2 while in suspension, which induces oxidative stress in cells; (2) physical interaction between nanoparticles and cell surface disrupts bacterial membrane integrity and causes membrane leakage; (3) higher concentrations of nanoparticles lead to severe membrane damage, cell content release, irreversible oxidization of biomolecules (e.g. DNA. proteins, and lipids), and ultimately cell death.

Conclusions

This study demonstrated that MgO nanoparticles have strong antibacterial activity against three important foodborne pathogens and addressed the underlying mechanisms of MgO nanoparticles' deleterious action on bacteria. The distinct evidence of H_2O_2 production in suspension of MgO nanoparticles, nanoparticle-induced expression of oxidative stress defense genes, alteration of cell morphology, and membrane leakage strongly suggests that the antibacterial mechanism of MgO nanoparticles is due to the induction of oxidative stress and disruption of membrane integrity in bacterial cells.

Abbreviations

CFU: colony forming unit; *C. jejuni*: *Campylobacter jejuni*; *E. coli*: *Escherichia coli*; *S.* Enteritidis: *Salmonella* Enteritidis; KatA: catalase; ROS: reactive oxygen species; G+: Gram positive; G−: Gram negative; SEM: scanning electron microscopy; EMA: ethidium monoazide; qPCR: quantitative real-time PCR; MH: Mueller–Hinton; MIC: minimum inhibitory concentration; rpm: revolution per minute; ATCC: American Type Culture Collection; Ct: threshold cycle; HRP: horseradish peroxidase; UV: ultraviolet.

Authors' contributions

YH designed all of the experiments, performed all calculations and statistical analyses, participated in running most of the experiments and prepared the manuscript. PI and SI performed the experiments and statistical data analyses. SR and TS performed the experiments and helped to draft the manuscript. AG assisted in certain aspects of the experiments as well as drafting the manuscript. All authors read and approved the final manuscript.

Author details

[1] Molecular Characterization of Foodborne Pathogens Research Unit, Eastern Regional Research Center, Agricultural Research Service, United States Department of Agriculture, 600 East Mermaid Lane, Wyndmoor, PA 19038, USA. [2] ICAR Research Complex for NEH Region, Umiam 793103, Meghalaya, India.

Acknowledgements

This research was supported by the US Department of Agriculture, Agricultural Research Service (USDA-ARS), and the overseas associateship award from the Department of Biotechnology, the Government of India Ministry of Science and Technology.

United States Department of Agriculture is an equal opportunity provider and employer.

Competing interests

The authors declare that they have no competing interests.

References

1. Scallan E, Hoekstra RM, Angulo FJ, Tauxe RV, Widdowson M-A, Roy SL, et al. Foodborne illness acquired in the United States—major pathogens. Emerg Infect Dis. 2011;17:7–15.
2. Silva J, Leite D, Fernandes M, Mena C, Gibbs PA, Teixeira P. *Campylobacter* spp. as a foodborne pathogen: a review. Front Microbiol. 2011;2:200.
3. Young KT, Davis LM, Dirita VJ. *Campylobacter jejuni*: molecular biology and pathogenesis. Nat Rev Microbiol. 2007;5:665–79.
4. Ferens WA, Hovde CJ. *Escherichia coli* O157:H7: animal reservoir and sources of human infection. Foodborne Pathog Dis. 2011;8:465–87.
5. Kingsley RA, Baumler AJ. Host adaptation and the emergence of infectious disease: the *Salmonella* paradigm. Mol Microbiol. 2000;36:1006–14.
6. O'Connell KM, Hodgkinson JT, Sore HF, Welch M, Salmond GP, Spring DR. Combating multidrug-resistant bacteria: current strategies for the discovery of novel antibacterials. Angew Chem Int Ed Engl. 2013;52:10706–33.
7. Neu HC. The crisis in antibiotic resistance. Science. 1992;257:1064–73.
8. Dizaj SM, Lotfipour F, Barzegar-Jalali M, Zarrintan MH, Adibkia K. Antimicrobial activity of the metals and metal oxide nanoparticles. Mater Sci Eng C Mater Biol Appl. 2014;44:278–84.
9. Hajipour MJ, Fromm KM, Ashkarran AA, Jimenez de Aberasturi D, de Larramendi IR, Rojo T, et al. Antibacterial properties of nanoparticles. Trends Biotechnol. 2012;30:499–511.
10. Tang ZX, Lv BF. MgO nanoparticles as antibacterial agent: preparation and activity. Braz J Chem Eng. 2014;31:591–601.
11. Leung YH, Ng AM, Xu X, Shen Z, Gethings LA, Wong MT, et al. Mechanisms of antibacterial activity of MgO: non-ROS mediated toxicity of MgO nanoparticles towards *Escherichia coli*. Small. 2014;10:1171–83.
12. Li X, Robinson SM, Gupta A, Saha K, Jiang Z, Moyano DF, et al. Functional gold nanoparticles as potent antimicrobial agents against multi-drug-resistant bacteria. ACS Nano. 2014;8:10682–6.
13. Xie Y, He Y, Irwin PL, Jin T, Shi X. Antibacterial activity and mechanism of action of zinc oxide nanoparticles against *Campylobacter jejuni*. Appl Environ Microbiol. 2011;77:2325–31.
14. Ge S, Wang G, Shen Y, Zhang Q, Jia D, Wang H, et al. Cytotoxic effects of MgO nanoparticles on human umbilical vein endothelial cells in vitro. IET Nanobiotechnol. 2011;5:36.
15. Love SA, Maurer-Jones MA, Thompson JW, Lin YS, Haynes CL. Assessing nanoparticle toxicity. Annu Rev Anal Chem. 2012;5:181–205.

16. Reddy KM, Feris K, Bell J, Wingett DG, Hanley C, Punnoose A. Selective toxicity of zinc oxide nanoparticles to prokaryotic and eukaryotic systems. Appl Phys Lett. 2007;90:2139021–3.
17. Chen C-Y, Nace GW, Irwin PL. A 6 × 6 drop plate method for simultaneous colony counting and MPN enumeration of *Campylobacter jejuni*, *Listeria monocytogenes*, and *Escherichia coli*. J Microbiol Methods. 2003;55:475–9.
18. He Y, Chen CY. Quantitative analysis of viable, stressed and dead cells of *Campylobacter jejuni* strain 81–176. Food Microbiol. 2010;27:439–46.
19. Suo B, He Y, Tu SI, Shi X. A multiplex real-time polymerase chain reaction for simultaneous detection of *Salmonella* spp., *Escherichia coli* O157, and *Listeria monocytogenes* in meat products. Foodborne Pathog Dis. 2010;7:619–28.
20. He Y, Frye JG, Strobaugh TP, Chen C-Y. Analysis of AI-2/LuxS-dependent transcription in *Campylobacter jejuni* strain 81–176. Foodborne Pathog Dis. 2008;5:399–415.
21. Livak KJ, Schmittgen TD. Analysis of relative gene expression data using real-time quantitative PCR and the 2[−Delta Delta C(T)] method. Methods. 2001;25:402–8.
22. Day WA Jr, Sajecki JL, Pitts TM, Joens LA. Role of catalase in *Campylobacter jejuni* intracellular survival. Infect Immun. 2000;68:6337–45.
23. Gaynor EC, Wells DH, MacKichan JK, Falkow S. The *Campylobacter jejuni* stringent response controls specific stress survival and virulence-associated phenotypes. Mol Microbiol. 2005;56:8–27.
24. Krishnamoorthy K, Manivannan G, Kim SJ, Jeyasubramanian K, Premanathan M. Antibacterial activity of MgO nanoparticles based on lipid peroxidation by oxygen vacancy. J Nanopart Res. 2012;14:1063.
25. Kim JC, Oh E, Kim J, Jeon B. Regulation of oxidative stress resistance in *Campylobacter jejuni*, a microaerophilic foodborne pathogen. Front Microbiol. 2015;6:751.
26. Baillon ML, van Vliet AH, Ketley JM, Constantinidou C, Penn CW. An iron-regulated alkyl hydroperoxide reductase (AhpC) confers aerotolerance and oxidative stress resistance to the microaerophilic pathogen *Campylobacter jejuni*. J Bacteriol. 1999;181:4798–804.
27. Huergo LF, Rahman H, Ibrahimovic A, Day CJ, Korolik V. *Campylobacter jejuni* Dps protein binds DNA in the presence of iron or hydrogen peroxide. J Bacteriol. 2013;195:1970–8.
28. Ishikawa T, Mizunoe Y, Kawabata S, Takade A, Harada M, Wai SN, et al. The iron-binding protein Dps confers hydrogen peroxide stress resistance to *Campylobacter jejuni*. J Bacteriol. 2003;185:1010–7.
29. Palyada K, Sun YQ, Flint A, Butcher J, Naikare H, Stintzi A. Characterization of the oxidative stress stimulon and PerR regulon of *Campylobacter jejuni*. BMC Genom. 2009;10:481.

17

Interaction of γ-Fe_2O_3 nanoparticles with *Citrus maxima* leaves and the corresponding physiological effects via foliar application

Jing Hu[1†], Huiyuan Guo[3†], Junli Li[1,3*], Yunqiang Wang[2], Lian Xiao[1] and Baoshan Xing[3]

Abstract

Background: Nutrient-containing nanomaterials have been developed as fertilizers to foster plant growth and agricultural yield through root applications. However, if applied through leaves, how these nanomaterials, e.g. γ-Fe_2O_3 nanoparticles (NPs), influence the plant growth and health are largely unknown. This study is aimed to assess the effects of foliar-applied γ-Fe_2O_3 NPs and their ionic counterparts on plant physiology of *Citrus maxima* and the associated mechanisms.

Results: No significant changes of chlorophyll content and root activity were observed upon the exposure of 20–100 mg/L γ-Fe_2O_3 NPs and Fe^{3+}. In *C. maxima* roots, no oxidative stress occurred under all Fe treatments. In the shoots, 20 and 50 mg/L γ-Fe_2O_3 NPs did not induce oxidative stress while 100 mg/L γ-Fe_2O_3 NPs did. Furthermore, there was a positive correlation between the dosages of γ-Fe_2O_3 NPs and Fe^{3+} and iron accumulation in shoots. However, the accumulated iron in shoots was not translocated down to roots. We observed down-regulation of ferric-chelate reductase (FRO2) gene expression exposed to γ-Fe_2O_3 NPs and Fe^{3+} treatments. The gene expression of a Fe^{2+} transporter, Nramp3, was down regulated as well under γ-Fe_2O_3 NPs exposure. Although 100 mg/L γ-Fe_2O_3 NPs and 20–100 mg/L Fe^{3+} led to higher wax content, genes associated with wax formation (WIN1) and transport (ABCG12) were downregulated or unchanged compared to the control.

Conclusions: Our results showed that both γ-Fe_2O_3 NPs and Fe^{3+} exposure via foliar spray had an inconsequential effect on plant growth, but γ-Fe_2O_3 NPs can reduce nutrient loss due to their the strong adsorption ability. *C. maxima* plants exposed to γ-Fe_2O_3 NPs and Fe^{3+} were in iron-replete status. Moreover, the biosynthesis and transport of wax is a collaborative and multigene controlled process. This study compared the various effects of γ-Fe_2O_3 NPs, Fe^{3+} and Fe chelate and exhibited the advantages of NPs as a foliar fertilizer, laying the foundation for the future applications of nutrient-containing nanomaterials in agriculture and horticulture.

Keywords: γ-Fe_2O_3 nanoparticles, Nano-enabled fertilizer, Foliar spray, Wax, Gene expression

Background

Iron deficiency in plants is widespread and can lead to reduction in crop yields and even complete crop failure [1]. Due to rapid conversion of iron into plant-unavailable forms when applied to calcareous soils, soil application of inorganic iron fertilizers to Fe-deficient soils is usually ineffective [2]. In comparison, synthetic Fe-chelates for amelioration of iron deficiency in plants is more effective, but more uneconomical [3]. It was reported that most foliar-applied micronutrients are not efficiently transported toward roots, which may remain deficient [2]. Nowadays, nanomaterials become a hotspot of research interests and attract the attention of many researchers.

*Correspondence: lijunli0424@sina.com
†Jing Hu and Huiyuan Guo are co-first authors
[1] School of Chemistry, Chemical Engineering and Life Sciences, Wuhan University of Technology, Wuhan 430070, People's Republic of China
Full list of author information is available at the end of the article

A variety of nanoparticles (NPs) have been studied on human cells [4, 5], animal cells [6] and plants [7] about their toxicity or applications. As one of the most widely explored and applied nanomaterials, iron oxide nanoparticles (γ-Fe_2O_3 NPs) are widely used in medical diagnostics, controlled drug release, separation technologies and environmental engineering [8]. Iron dynamically released from γ-Fe_2O_3 NPs may be a potential nutritional source for plants. It is likely that γ-Fe_2O_3 NPs could be an effective fertilizer for alleviation of Fe-deficiency in plants. Several studies have reported that root applied γ-Fe_2O_3 NPs have positive effects on plant growth. For instance, γ-Fe_2O_3 NPs can physiologically enhance seed germination, root growth, chlorophyll content in watermelon (*Citrullus lanatus*) planted in quartz sand [9] and Chinese mung bean (*Vigna radiata* L.) grown in silica sediment [10]. Rui et al. [11] reported that γ-Fe_2O_3 NPs increased root length, plant height, biomass, and chlorophyll levels of peanut (*Arachis hypogaea*) plants, indicating that γ-Fe_2O_3 NPs can possibly replace traditional iron fertilizers in the cultivation of peanut plants. To our knowledge, few researchers reported the effects of γ-Fe_2O_3 NPs on plants via foliar application yet.

Root is the major pathway for plants to absorb water and inorganic ions [12], through which NPs can be taken up and translocated to upper tissues [13–15]. When NPs were exposed to plants' leave surface, several studies have observed that plants can absorb NPs through the leaves as well. Corredor et al. [16] reported that carbon coated iron NPs were capable of penetrating pumpkin (*Cucurbita pepo* L.) leaves and migrating to other plant tissues. Larue et al. [17] found that Ag NPs were effectively trapped on lettuce (*Lactuca sativa*) leaves and taken up by cells after foliar exposure. It is hypothesized that there are two pathways for leaves to take up NPs and their solutes: for hydrophilic compounds via aqueous pores of the cuticle and stomata, and for lipophilic ones by diffusion through the cuticle [17]. Since the wax lipids may quickly adsorb on the large surface of NPs [18], particles might be trapped by the cuticular wax and then diffuse in the leaf tissue (after dissolution or translocation through the cuticle) [19]. For example, Birbaum et al. [18] reported that large agglomerates were trapped on the surface wax, whereas smaller particles might be taken up by the leaf. At the molecular level, wax inducer1 (WIN1), an ethylene response factor-type transcription factor, can activate wax deposition in overexpressing plants and influence wax accumulation through the direct or indirect regulation of metabolic pathway genes [20]. Alabdallat et al. [21] reported that WIN1 gene could modulate wax accumulation and enhance drought tolerance in tomato (*Solanum lycopersicum*) plants. Several plant ATP-binding cassette sub-family G member (ABCG) proteins are known or suspected to be involved in synthesis of extracellular barriers, among which ABCG12 is required for lipid export from the epidermis to the protective cuticle [22]. The interactions between γ-Fe_2O_3 NPs and plant leaves were inevitably affected by cuticular wax due to the fact that the plant cuticles form the outermost barrier between plant leaves and their local environment. Therefore, it is of great significance to study the changes of cuticular wax in plant leaves induced by foliar sprayed γ-Fe_2O_3 NPs. However, to our knowledge, the effects of foliar application of γ-Fe_2O_3 NPs on cuticular wax loads and related gene expression have not been reported. In the present study, in order to show the in-depth interactions between γ-Fe_2O_3 NPs and cuticular waxes in *Citrus maxima* leaves, wax content and wax synthesis or transport related genes, including WIN1 and ABCG12 were analyzed at the molecular level.

Additionally, in order to figure out the effects of γ-Fe_2O_3 NPs on plant growth and physiology, the corresponding parameters, including biomass, chlorophyll, soluble protein content, root activity, lipid peroxidation and activity of antioxidant enzymes were measured. *C. maxima* plants were exposed to 20, 50 and 100 mg/L γ-Fe_2O_3 NPs or Fe^{3+} by foliar application at an early growth stage. The latter treatment was set to study the phytotoxicity of Fe^{3+} ions by dissolving $FeCl_3{\cdot}6H_2O$. This is the first report on the γ-Fe_2O_3 NPs uptake and translocation in plants via foliar application, and the transcriptional modulation of genes involved in iron uptake or transport viz. ferric-chelate reductase (FRO2) and natural resistance-associated macrophage protein (Nramp3).

Methods

Materials and experimental setups

The γ-Fe_2O_3 NPs of 99.5% purity were purchased from Macklin Inc. (Shanghai, China). The shape and size were determined by a Tecnai G2 20 TWIN transmission electron microscope (FEI, USA). The hydrodynamic diameter and zeta potential were determined by a Zetasizer Nano ZS90 dynamic light scattering spectrometer (Malvern Instruments Ltd., United Kingdom). The characteristics of γ-Fe_2O_3 NPs are shown in Additional file 1: Figure S1 of the supplementary materials. γ-Fe_2O_3 NPs are spherical with an average diameter size of 20.2 ± 2.7 nm (Additional file 1: Figure S1A). The average hydrodynamic diameter and the zeta potential of γ-Fe_2O_3 NPs were 164.5 ± 11.3 nm and −11.7 ± 0.1 mV, respectively (Additional file 1: Figure S1B, C). *Citrus maxima* seeds were immersed in distilled water and germinated in moist perlite at 28 °C. Then the uniform seedlings were transferred to a hydroponic system amended with 1/2

Hoagland's nutrient solution without iron. 18 of seedlings were planted in each hydroponic container. Plants were sprayed with 50 mL of deionized water (control), 20, 50 and 100 mg/L γ-Fe_2O_3 NPs suspended in deionized water, 20, 50 and 100 mg/L Fe^{3+} (dissolved from $FeCl_3{\cdot}6H_2O$) solutions, and 50 μM Fe(II)-EDTA in the morning. During all the treatments, An iron-deficient control and a Fe(II)-EDTA treatment were set up for comparison. The concentrations of Fe^{3+} are calculated according to the containing iron content of γ-Fe_2O_3 NPs at same concentration. Therefore, γ-Fe_2O_3 NPs and Fe^{3+} treatments marked with the same concentration denote they have same iron content. Suspensions were sprayed with a hand-held sprayer bottle every 5 days when the plants had two true leaves. To facilitate foliar infiltration, all plants were sprayed with deionized water once per hour for 10 h to avoid early evaporation of the solutions and consequent precipitation of solutes on the leaf surface [23]. The plants were grown in an environmentally controlled growth chamber at 28/18 °C with a 16 h/8 h light/dark cycle; the light intensity was 2000 lx. The air was pumped into the hydroponic system every 3 h with 30 min each time. The nutrient solution was replaced every 5 days. After 30 days of exposure, representative parameters including chlorophyll, fresh biomass, soluble protein content, root activity, lipid peroxidation, antioxidant enzyme activities, iron content, iron-related gene expression, wax content, and wax-related gene expression were measured.

Fresh biomass measurement

Citrus maxima plants were carefully removed from the hydroponic system after 30 days. The fresh biomass of *C. maxima* including roots and shoots was weighed by using a FA1004C electronic analytical balance (Shanghai Yueping Scientific Instrument Co., Ltd, China).

Measurement of physiological and biochemical parameters

Chlorophyll content was determined by a modified procedure according to Lichtenthaler [24]. Soluble protein content was estimated according to a dying method using Coomasie Brilliant Fluka G-250 [10]. Measurement of root activity was according to the triphenyltetrazolium chloride method [25]. Malonaldehyde (MDA) was determined by the thiobarbituric acid method according to Heath and Packer [26]. The activity of superoxide dismutase (SOD) was evaluated by the ability to inhibit photochemical reduction of nitroblue tetrazolium according to Wang et al. [27]. The activity of catalase (CAT) was analyzed as described by Gallego et al. [28]. The activity of peroxidase (POD) was estimated by guaiacol colorimetric method as described by Zhang et al. [29].

Metal uptake analysis

Harvested leaf tissue was rinsed with deionized H_2O thrice to remove the surface retained γ-Fe_2O_3 NPs. All shoot and root samples were dried at 60 °C for 48 h in a drying oven. 100 mg of oven-dried shoot and root tissues were separately digested in 3 mL of concentrated HNO_3 at 115 °C on a hot block for 1 h. After cooling to room temperature, 0.5 mL of 30% H_2O_2 was added to the digestions at 100 °C for 0.5 h. The iron content was analyzed by an Avanta M atomic absorption spectrophotometer (GBC, Australia).

Measurement of wax loads

The content of cuticular waxes was determined using chloroform extraction as described by Premachandra et al. [30]. Leaf samples were immersed in 20 mL of chloroform in a Petri dish of 90 mm diameter for 5 s. The solvent was evaporated in a fume hood under a dry air stream, and the residue was allowed to dry for 24 h at room temperature. After drying, the content of cuticular waxes was weighed by using a FA1004C electronic analytical balance and expressed on the basis of FW (fresh weight).

Regulation of gene expression by RT-PCR

The isolation of total RNA, the synthesis of cDNA and RT-PCR analysis were conducted according to our previous study [31]. Primers for FRO2, Nramp3, ABCG12 and WIN1 genes were designed based on the sequences available in NCBI genbank using the PrimerQuest (Integrated DNA Technologies, Coralville, IA) as described in Table 1.

Statistical analysis

Each treatment was conducted with three replicates, and the results were presented as mean ± SD (standard deviation). The statistical analysis of experimental data was verified with the one-way ANOVA followed by Duncan's multiple comparison ($p < 0.05$) in the statistical package IBM SPSS Version 22.

Results

Effect of γ-Fe_2O_3 NPs and Fe^{3+} treatments on plant growth

The influence of γ-Fe_2O_3 NPs and their counterpart Fe^{3+} solutions (20–100 mg/L) on the growth of *C. maxima* leaves is shown in Fig. 1A. No visible signs of phytotoxicity are evident in *C. maxima* leaves under all treatments. As show in Fig. 1B, chlorophyll contents of all treatments showed no significant differences. The fresh biomass of Fe-exposed *C. maxima* plants had no significant differences from that of the control, except for 50 mg/L Fe^{3+} treatment, which had 15.4% higher fresh biomass (Fig. 1C). On the other hand, no positive effect of fresh

Table 1 Primers of genes used in this study

Gene	Primer sequence (Forward-5′–3′)	Primer sequence (Reverse-5′–3′)
Actin	CAGCTGTGGAGAAGAGCTATG	CGATCATGGATGGTTGGAAGA
Nramp3	GCGTGTTGATTGCTACTGTTATT	GATGAGCACGCCAACTAGAA
FRO2	GTGTCTGTTGAAGGACCCTATG	GCTCGCGGACTATGGAAATAA
ABCG12	GGAAGGGCTGGAAATTGAAATC	GCCCAGTAATATCCCACATCTC
WIN1	GCTCCTCATCATCATCACCTAC	GCCTCAGACAAGTCATAGAAGG

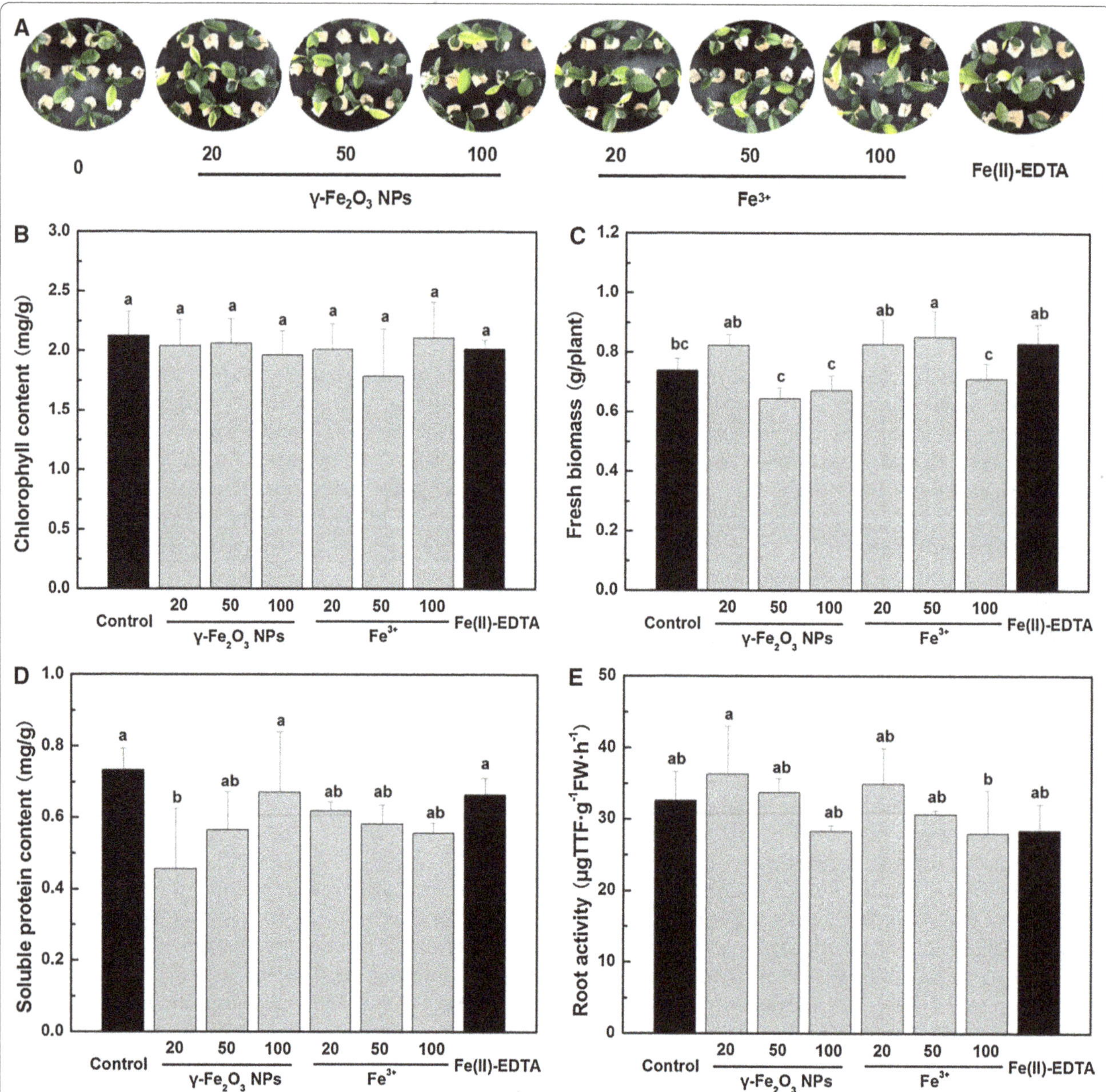

Fig. 1 **A** Images of *C. maxima* leaves exposed to different concentrations of γ-Fe_2O_3 NPs and Fe^{3+}. **B–E** Chlorophyll content, fresh biomass, soluble protein content in leaves, and root activity of *C. maxima* plants treated with different concentrations of γ-Fe_2O_3 NPs and Fe^{3+}, respectively. Data are shown as mean ± SD of three replicates. Values followed by *different lowercase letters* are significantly different at $p \leq 0.05$

biomass under the exposure of γ-Fe_2O_3 NPs and Fe^{3+} was induced compared with Fe(II)-EDTA treatment. Instead, fresh biomass of *C. maxima* seedlings was decreased by 22.1, 18.7 and 14.3% under the exposure of 50 and 100 mg/L γ-Fe_2O_3 NPs, and 100 mg/L Fe^{3+}, respectively.

Soluble protein amounts at various concentrations of γ-Fe_2O_3 NPs and Fe^{3+} exposure were unaffected compared to the control and Fe(II)-EDTA treatment, except for 20 mg/L γ-Fe_2O_3 NPs, which had lower soluble protein content (Fig. 1D). Root activity is a comprehensive assessment index that reflects the metabolic activity level and the ability of roots to absorb nutrients and water [32]. As Fig. 1E depicted, all foliar applied γ-Fe_2O_3 NPs and Fe^{3+} treatments had no impact on root activity as compared to the control and Fe(II)-EDTA treatment.

Lipid peroxidation and antioxidant enzyme activities of *C. maxima* plants

The oxidative stress induced by γ-Fe_2O_3 NPs and subsequent reactive oxygen species (ROS) scavenging by SOD, CAT and POD are presented schematically in Fig. 2A. In *C. maxima* shoots, no elevated lipid peroxidation by γ-Fe_2O_3 NPs was observed compared to both the control and Fe(II)-EDTA treatment (Fig. 2B). 20 and 100 mg/L Fe^{3+} treatment had a higher MDA formation by 26.0 and 49.1%, respectively as compared with the control. Also, MDA level of 100 mg/L Fe^{3+} treatment was 33.2% higher than Fe(II)-EDTA treatment. In *C. maxima* roots, MDA production remained unchanged regardless of the treatments.

Compared with the control and Fe(II)-EDTA treatment, the activities of SOD did not increase in both *C. maxima* shoots and roots after the γ-Fe_2O_3 NPs and Fe^{3+} treatments (Fig. 2C). As depicted in Fig. 2D, CAT activity of γ-Fe_2O_3 NPs treated *C. maxima* shoots numerically increased in a dose-dependent manner. Statistically, 100 mg/L γ-Fe_2O_3 NPs had 35.4% higher CAT activity than the control, and 21.1% higher than Fe(II)-EDTA treatment. On the other hand, CAT activity in Fe^{3+}-treated shoots was not significantly different from the control, but 100 mg/L Fe^{3+} treatment resulted in 31.0% lower CAT activity than Fe(II)-EDTA treatment. POD activities in shoots treated with 20 and 50 mg/L γ-Fe_2O_3 NPs, and 20 mg/L Fe^{3+} treatment were unaffected, while those of 100 mg/L γ-Fe_2O_3 NPs, 50 and 100 mg/L Fe^{3+} treatment were increased significantly, as compared to the control (Fig. 2E). In addition, no increase of POD activity in shoots under γ-Fe_2O_3 NPs or Fe^{3+} treatments was observed compared to that of Fe(II)-EDTA treatment. In *C. maxima* roots, both the CAT and POD activities remained unchanged, no matter what concentrations of γ-Fe_2O_3 NPs or Fe^{3+}were used (Fig. 2D, E).

Iron distribution and iron-related gene expression in *C. maxima* plants

The possible pattern of transformation and uptake of iron in *C. maxima* leaves is shown in Fig. 3A. Unfortunately, iron regulated transporter (IRT1) gene in citrus has not been sequenced yet. As expected, Fe concentration of *C. maxima* shoots exposed to both γ-Fe_2O_3 NPs and Fe^{3+} treatment increased rapidly with the increase of applied dosages (Fig. 3B). After exposure to 20, 50 and 100 mg/L γ-Fe_2O_3 NPs, Fe content in shoots was increased by 1.34, 3.78 and 6.77 times, respectively, relative to the control plants. Fe level of Fe^{3+} treatments was elevated by 2.33, 4.38, 8.62 times, respectively. In addition, the total Fe content in *C. maxima* shoots was not significantly different between Fe(II)-EDTA and control plants. In *C. maxima* roots, no obvious difference of Fe levels was noted between all Fe treatments and control plants.

FRO2 gene encodes a ferric chelate reductase, which can be activated when plants lack available Fe. As seen in Fig. 3C, the relative FRO2 gene expression of control was at a high level. γ-Fe_2O_3 NPs and Fe^{3+} treatments led to 29.4–91.4% lower levels of FRO2 gene expression than that of untreated control plants. Especially, 50 mg/L Fe^{3+} treatment significantly decreased FRO2 expression to a much lower level than other treatments. Meanwhile, FRO2 expression level of Fe(II)-EDTA treatment was also lower than untreated control, but not less than that of γ-Fe_2O_3 NPs and Fe^{3+} treatments. Nramp3 protein, which localizes in the vacuolar membrane (Fig. 3A), can transport Fe^{2+} and is upregulated by iron starvation. As depicted in Fig. 3D, 20–100 mg/L γ-Fe_2O_3 NPs had relatively lower expression levels of Nramp3 gene than control by 62.5–81.7%, but that of 100 mg/L of Fe^{3+} treatment was much higher by 1.58 times. Interestingly, Fe(II)-EDTA treatment had a quite higher level of Nramp3 gene expression.

Wax content and wax-related gene expression of *C. maxima* leaves

The potential interactions between γ-Fe_2O_3 NPs and cuticular wax as well as genes involved in the intracellular wax synthesis and transport to the outside of cell walls are presented schematically in Fig. 4A. Wax, which is composed of long-chain, aliphatic hydrocarbons derived from very-long-chain fatty acids (VLCFAs) [33], is the protective material on leaf epidermis [34] and plays an important role in particle incorporation. As seen from Fig. 4B, 20 and 50 mg/L γ-Fe_2O_3 NPs had no impact on wax content compared with the control, while 100 mg/L γ-Fe_2O_3 NPs exhibited significantly higher wax content by 2.1-fold. 20, 50 and 100 mg/L Fe^{3+} treatment had higher wax contents than the control by 1.17, 1.04 and 1.57 times, respectively. Wax content of Fe(II)-EDTA

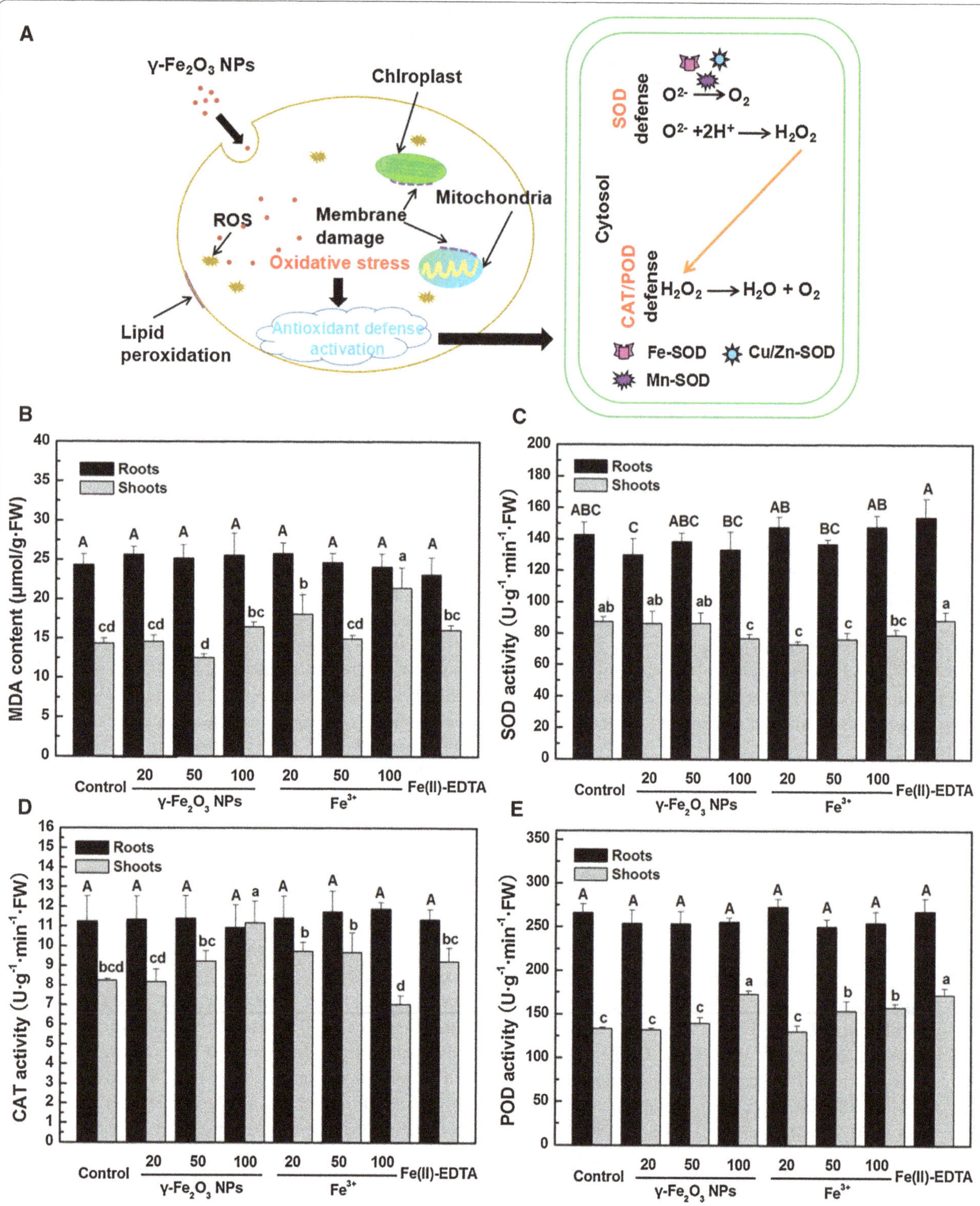

Fig. 2 A Schematic illustration of the activation of antioxidant enzymes in plants to scavenge excessive ROS production induced by γ-Fe_2O_3 NPs. **B–E** MDA content, activity of SOD, CAT and POD in roots and shoots of *C. maxima* plants treated with different concentrations of γ-Fe_2O_3 NPs and Fe^{3+}, respectively. Data are shown as mean ± SD of three replicates. Values of MDA content and antioxidant enzyme activities followed by *different lowercase* and *uppercase letters*, respectively, are significantly different at $p \leq 0.05$

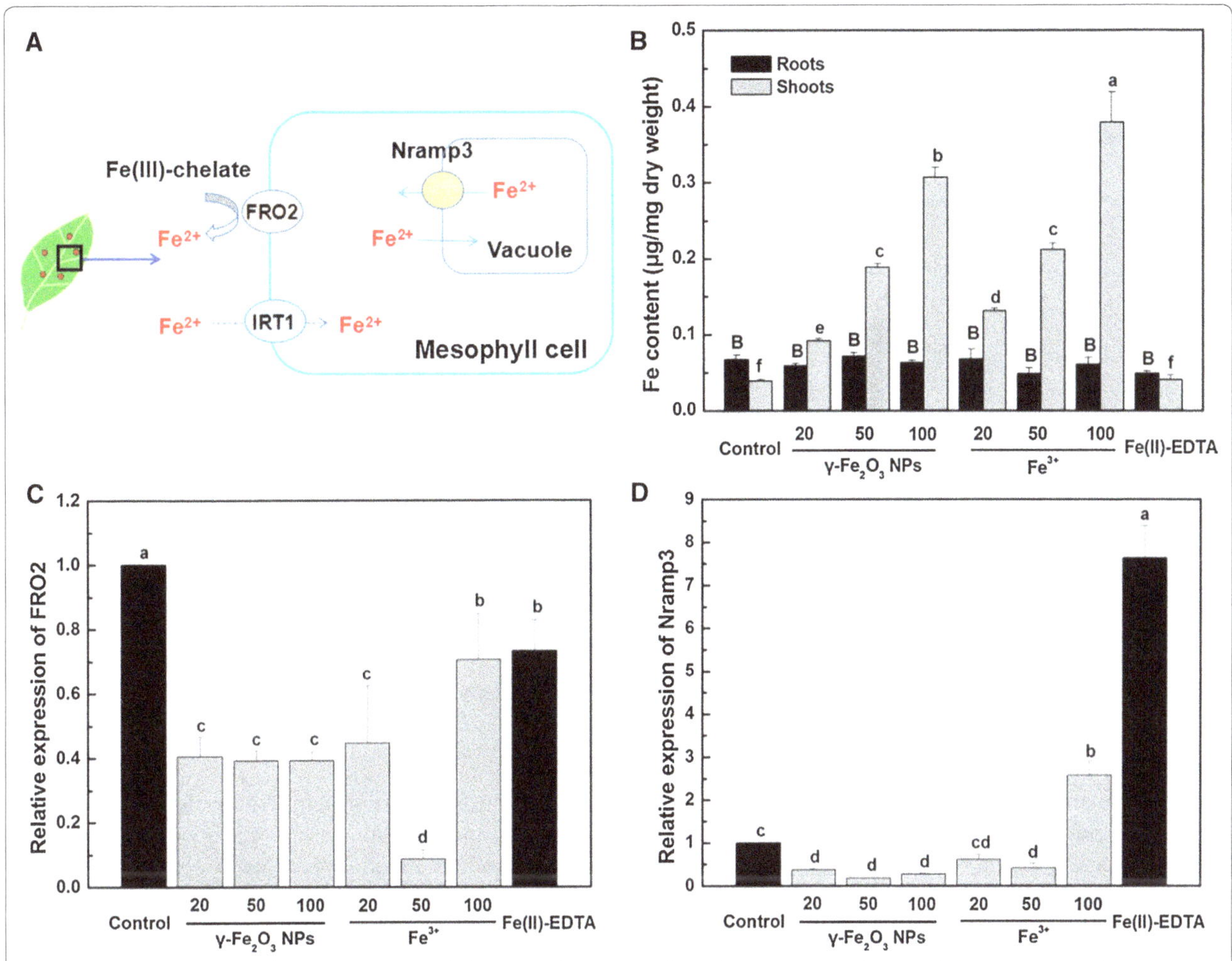

Fig. 3 **A** Schematic diagram of genes associated with the absorption and transformation of iron in plant leaves. **B–D** Iron content of *C. maxima* including roots and shoots, relative expression of FRO2 and Nramp3 of *C. maxima* leaves treated with different concentrations of γ-Fe_2O_3 NPs and Fe^{3+}, respectively. Data are shown as mean ± SD of three replicates. Values of Fe content and relative expression of each gene labelled by *different lowercase* and *uppercase letters*, respectively, are significantly different at $p \leq 0.05$

treatment was in a notably higher level compared with other treatments.

The relative expression levels of WIN1 gene under the exposure of all γ-Fe_2O_3 NPs and Fe^{3+} treatments were significantly lower than the control but not less than that of Fe(II)-EDTA treatment (Fig. 4C). In Fig. 4D, the relative expression levels of ABCG12 gene treated by γ-Fe_2O_3 NPs and Fe^{3+} were lower or unaffected as compared to untreated control. However, Fe(II)-EDTA treatment had a much higher ABCG12 gene expression level by contrast with other treatments.

Discussion

Growth and physiological effects of γ-Fe_2O_3 NPs and Fe^{3+}

Fe(II)-EDTA, as one of the most widely used supplements for improving Fe availability to plants [35], showed no evident promotion to plant growth via foliar application in our study. Meantime, γ-Fe_2O_3 NPs did not exhibit any superiority in overcoming Fe deficiency-induced chlorosis. We did not observe any evident difference of chlorophyll levels among treatments of γ-Fe_2O_3 NPs, Fe^{3+}, control and Fe(II)-EDTA, although iron content in *C. maxima* shoots of γ-Fe_2O_3 NPs and Fe^{3+} treatments was higher than control and Fe(II)-EDTA treatment. It is possible that iron was mainly used in other physiological reactions and thus no significant changes in chlorophyll content were observed. It is noteworthy that previously we found that through root exposure in a hydroponic system, 0–100 mg/L of γ-Fe_2O_3 NPs and Fe^{3+} had a dose-dependent effect on chlorophyll synthesis of *C. maxima* [31]. 50 mg/L γ-Fe_2O_3 NPs and all Fe^{3+} treatments notably increased chlorophyll levels. Fe(II)-EDTA treatment also had higher chlorophyll content as compared to the untreated control. However, foliar applications of

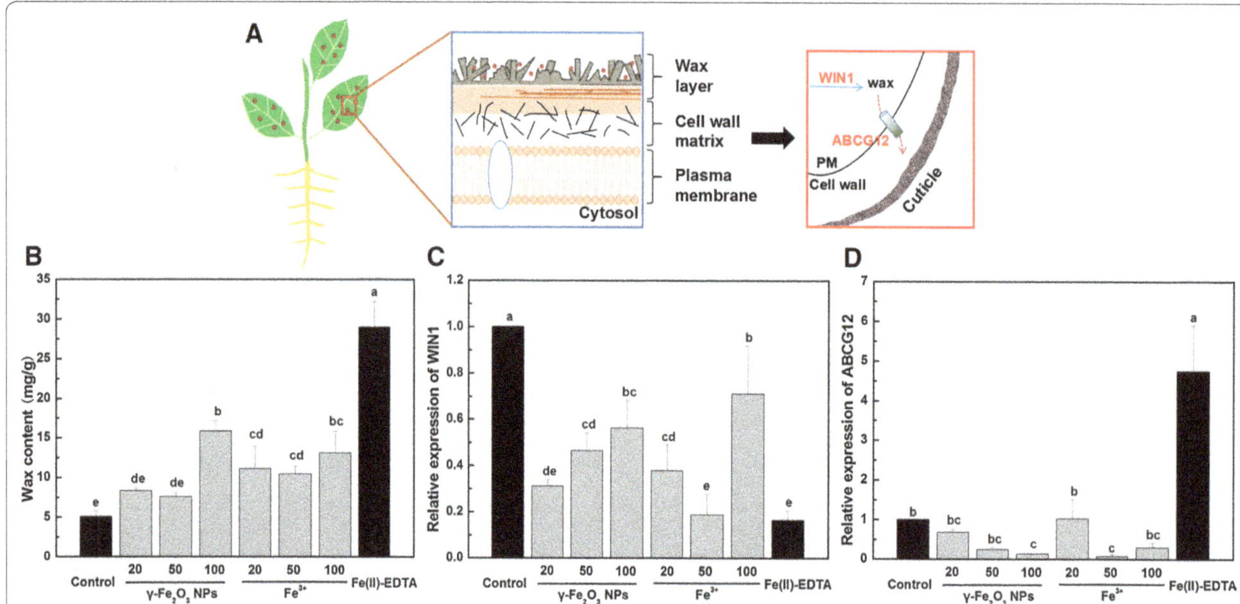

Fig. 4 **A** Schematic diagram of the interactions between NPs and cuticular waxes in leaves, and genes involved in wax synthesis and secretion in this study (PM: plasma membrane). **B–D** represent wax content, relative expression of WIN1 and ABCG12 genes of *C. maxima* leaves treated with different concentrations of γ-Fe_2O_3 NPs and Fe^{3+}, respectively. Data are shown as mean ± SD of three replicates. Values of wax content and relative expression of each gene followed by *different lowercase letters* are significantly different at $p \leq 0.05$

γ-Fe_2O_3 NPs and Fe^{3+}, as well as Fe(II)-EDTA, appeared to have no positive effect on chlorophyll synthesis and no obvious amelioration of chlorosis was observed, indicating that foliar application was less efficient than root application. However, Alidoust and Isoda [23] observed more pronounced positive effects of γ-Fe_2O_3 NPs on physiological performance of soybean (*Glycine max* (L.) Merr.) via foliar application than by soil treatment. They used different parameters, including γ-Fe_2O_3 NP size and concentrations, growth condition, treatment time as well as plant species, which may explain the contradictory results from ours.

To demonstrate if γ-Fe_2O_3 NPs altered the plant health at physiological level, we analyzed the change of soluble proteins, which is an important indicator of plants' defense. Plants could adapt themselves to various stresses by producing soluble proteins as osmolytes [36], antioxidants, or scavengers for eliminating free radicals in plants [37]. For example, Afaq et al. [38] observed an increase in the antioxidant enzymes after TiO_2 NPs treatment as indicated at the transcriptional or protein level. Meanwhile, it is known that various abiotic stresses lead to the overproduction of ROS in plants which are highly reactive and toxic, ultimately resulting in oxidative stress and protein damage [39]. Nevertheless, in this study, no oxidative stress was induced in plants exposed to 20 mg/L γ-Fe_2O_3 NPs, based on the results of MDA content and the antioxidant enzyme activities (Fig. 2B–E), indicating that the lower soluble protein level could be caused by an alternative mechanism, instead of protein damage caused by overproduction of ROS. Meantime, the unchanged soluble protein contents under other treatments might be a result of self-regulation by plants.

Oxidative stress caused by γ-Fe_2O_3 NPs and Fe^{3+} on plants

In this study, no elevated MDA level in shoots under γ-Fe_2O_3 NPs exposure was induced, suggesting that either foliar applied γ-Fe_2O_3 NPs did not induce lipid peroxidation even at high exposure concentrations or the plant's detoxification pathways were sufficient to address and remedy the induced stress [40]. Activities of three antioxidant enzymes in plants treated with 20 and 50 mg/L γ-Fe_2O_3 NPs were unaffected, while 100 mg/L γ-Fe_2O_3 NPs significantly increased the activity of CAT and POD. Higher activity of CAT and POD can contribute to the detoxification of excessive amounts of H_2O_2 [41]. Given the results of MDA levels and antioxidant enzymes, it is clear that 20 and 50 mg/L γ-Fe_2O_3 NPs did not induce oxidative stress in plant shoots, while 100 mg/L γ-Fe_2O_3 NPs might initially cause ROS generation but then the plant's defense systems remedied the induced stress. As for Fe^{3+} treatments, 20 and 100 mg/L treated shoots showed a much higher MDA content, while that of 50 mg/L Fe^{3+} treatment was unaffected compared with the control. However, no elevated activities of three antioxidant enzymes under 20 mg/L Fe^{3+}

treatment were observed, indicating that the increase of MDA level under 20 mg/L Fe^{3+} was an abnormal result. Combined MDA content with the higher POD activity of 50 and 100 mg/L Fe^{3+} treatments in shoots, it could be deduced that *C. maxima* treated with 50 mg/L Fe^{3+} could address and remedy the induced oxidative stress, while plants treated with 100 mg/L Fe^{3+} were not sufficient to deal with stress induced by Fe^{3+} at a high concentration. The unchanged MDA production and antioxidant enzyme activities in *C. maxima* roots among all the treatments indicated that no oxidative stress occurred in plant roots.

Uptake and translocation of γ-Fe_2O_3 NPs

Iron content of *C. maxima* shoots exposed to different concentrations of γ-Fe_2O_3 NPs showed a dose-dependent trend. The higher Fe level of γ-Fe_2O_3 NPs in shoots indicated that significant uptake had occurred. Several studies demonstrated that Fe_2O_3 NPs in a hydroponic system could enter plants through roots [42, 43], or silica sediment [10]. However, to our knowledge, few studies investigated whether foliar applied γ-Fe_2O_3 NPs could enter plant leaves and further translocate to roots or not. We observed the uptake of iron into shoots but no difference of iron content in *C. maxima* roots between all treatments, suggesting that no downward transport of iron occurred in *C. maxima* plants. In our previous study, we observed that root-applied γ-Fe_2O_3 NPs had no translocation from roots to shoots [31]. Therefore, either foliar spray or root supply of γ-Fe_2O_3 NPs alone cannot meet the requirement of the whole plants. A combination of both application methods may improve the effectiveness of iron fertilization in agricultural and horticultural production.

Generally, when plants are deprived of iron, the new leaves become chlorotic and young lateral roots show the characteristic Fe-deficient stress-response mechanisms: enhanced Fe(III) reducing capacity, subapical swelling and acidification of the medium [44]. However, previous studies showed that leaf mesophyll cells also display plasma membrane ferric reductase activity [44, 45]. When a plant suffers from iron shortage, the reductive system is strongly activated [45], with FRO2 gene encoding a ferric chelate reductase. The down-regulation of FRO2 gene expression under γ-Fe_2O_3 NPs and Fe^{3+} treatments compared to the control and Fe(II)-EDTA treatment indicated that *C. maxima* could utilize the supplied iron in γ-Fe_2O_3 NPs and Fe^{3+} via foliar application. The relative level of FRO2 expression exposed to 50 mg/L Fe^{3+} was the lowest. The supply of iron is not only dependent on applied dosage, but also plants' utilization ability. Based on the MDA data, 20 and 100 mg/L Fe^{3+} treatments had higher MDA formations than 50 mg/L Fe^{3+}, which indicates that 50 mg/L leads to less oxidative stress than the other two dosages. Therefore, plants could better utilize Fe^{3+} at 50 mg/L, which explains why the activation of FRO2 gene of 50 mg/L was lower than 20 and 100 mg/L Fe^{3+} treatments. Taken together, 50 mg/L Fe^{3+} can supply higher amount of iron than 20 mg/L Fe^{3+}. Meanwhile, the toxicities of 20 mg/L and 100 mg/L Fe^{3+} are higher than that of 50 mg/L and likely inhibits the leaf ability to absorb and utilize iron. Also, γ-Fe_2O_3 NPs and Fe^{3+} had a higher ability to supply iron to plants than Fe(II)-EDTA, except for 100 mg/L Fe^{3+}. In addition, the lower level of Nramp3 gene expression at all γ-Fe_2O_3 NPs concentrations indicated that plant was in iron-sufficient status. The much higher level of Nramp3 gene expression of 100 mg/L Fe^{3+} and Fe(II)-EDTA treatment than the control suggested that Fe(II)-EDTA and Fe^{3+} at high concentrations cannot alleviate iron deficiency via foliar spray. It was reported that Fe-chelates are more effective in soil than in foliar applications, and foliar Fe chelate-fertilization cannot yet be considered as a reliable strategy to control plant Fe-deficiency [46]. Previous study showed that γ-Fe_2O_3 NPs is a suitable adsorbent for effectively extracting pollutants from the environment due to their high specific surface area and accessible surface adsorption sites, which make them well applicable for the adsorption of pollutants [47, 48]. Given this, the strong adsorption ability of γ-Fe_2O_3 NPs contributed to their stable attachment on the leaf surface and further absorption by plants. In agricultural production, most of the applied fertilizers are frequently lost due to the degradation by photolysis, leaching, hydrolysis, and decomposition [49]. It is essential to reduce nutrient losses in fertilization and increase the crop yield through the development of nanomaterials-based fertilizers [49]. In this regard, our results revealed that γ-Fe_2O_3 NPs have the potential to be an effective nanofertilizer and reduce nutrient loss during and after application.

Interaction between γ-Fe_2O_3 NPs and cuticular wax

In this study, iron distribution indicated that γ-Fe_2O_3 NPs may be tightly attached to the leaf surface and/or taken up by the *C. maxima* leaves. Cuticular wax is a protective barrier on leaf epidermis, which could adsorb and trap intrusive NPs. Once NPs translocate through the cuticle, NPs could diffuse in the leaf tissue. We observed a significantly lower expression levels of WIN1 gene under all Fe exposures. No upregulation of ABCG12 gene expression treated with γ-Fe_2O_3 NPs and Fe^{3+} treatments was observed as well. However, wax contents of 100 mg/L γ-Fe_2O_3 NPs and 20–100 mg/L Fe^{3+} treatment were significantly enhanced. Such an increase of wax content could hinder the uptake of high levels of γ-Fe_2O_3 NPs and ionized iron (Fe^{3+}). Wax content is closely correlated with stress resistance of plants [50]. According to Fig. 2B–E, Fe^{3+} treatments induced stress in plant shoots. The higher wax levels of plant leaves under Fe^{3+} treatments might be a result of anti-stress. The high Fe content in shoots of 100 mg/L γ-Fe_2O_3 NPs suggested that most NPs were trapped on the surface wax as a result of the formation of clusters and large agglomerates [18]. Strangely, Fe(II)-EDTA treatment had a lower WIN1 gene expression level but a much higher ABCG12 gene expression level, while wax content of Fe(II)-EDTA treatment was at a notably high level. Fernández et al. [46] reported that sprayed Fe-chelates could be taken up via the cuticle due to the comparable sizes of Fe-compounds and the pores. The significantly higher wax content of Fe(II)-EDTA might be a mechanism of defending against alien substances. In addition to WIN1, there are many genes involved in the synthesis and secretion of surface wax [51]. For instance, CUT1, an Arabidopsis gene required for cuticular wax production, encodes a VLCFA condensing enzyme [33]. Therefore, the biosynthesis of wax is a collaborative and complicated process, which explain why 100 mg/L γ-Fe_2O_3 NPs and 20–100 mg/L Fe^{3+} led to higher wax content without inducing higher expression levels of WIN1 and ABCG12, as well as Fe(II)-EDTA treatment had the lower expression of WIN1 but higher level of wax. Moreover, Jetter et al. [52] reported that cuticular wax is typically a complex mixture of dozens of compounds with diverse hydrocarbon chain or ring structures. How much each of the wax compounds contributes to the overall biological functions of the cuticular wax is largely unknown [53]. Therefore, further explorations should be made to figure out the processes and mechanisms underlying the interactions between NPs and cuticular waxes.

Conclusions

Based on the growth and physiological parameters, it is clear that foliar sprayed γ-Fe_2O_3 NPs and Fe^{3+} at the concentrations used in this study had an inconsequential effect on plant growth as shown in chlorophyll content, fresh weight, and root activity. However, the expression of genes associated with the absorption and transformation of iron in leaves showed that plants were in iron-sufficient status. Further analysis of iron content shows no downward transport of iron from shoots to roots in all treated forms via foliar application. It is well known that iron is hard to transport from leaves. As for lipid peroxidation, all γ-Fe_2O_3 NPs exposures showed insignificant changes as compared with the control. Antioxidant analysis indicated that 20 and 50 mg/L γ-Fe_2O_3 NPs induced no oxidative stress while 100 mg/L γ-Fe_2O_3 NPs may induced stress initially but plants were sufficient to deal with it. Moreover, the higher wax content of 100 mg/L γ-Fe_2O_3 NPs as compared with the control would hinder the uptake of high levels of γ-Fe_2O_3 NPs. Results of WIN1 and ABCG12 gene expression revealed that the biosynthesis of wax is a collaborative and complicated process and more than one gene are involved in this process. Commendably, foliar applied γ-Fe_2O_3 NPs have the ability to reduce nutrient loss probably due to the strong adsorption ability and gradual Fe release. Given that no phytotoxicity of γ-Fe_2O_3 NPs at lower concentrations (20 and 50 mg/L) was observed, it is possible that using γ-Fe_2O_3 NPs at lower doses is feasible to enhance the utilization and efficiency of inorganic iron fertilizer in agricultural production. Moreover, in real applications, foliar sprayed γ-Fe_2O_3 NPs may be utilized together with soil supplied γ-Fe_2O_3 NPs to alleviate chlorosis and improve the iron use efficiency. Our findings provide a novel perspective to the interactions between foliar-applied NPs and plants, and will inspire further critical efforts to systemically explore the potential applications of γ-Fe_2O_3 NPs in agronomic production.

There is still much unknown about the speciation change of γ-Fe_2O_3 NPs during plant foliar interactions. Further efforts should be made to determine (1) if the γ-Fe_2O_3 NPs are absorbed as NPs directly, or dissolution occurs inside or outside plant leaves with free iron ions available for uses by plant leaves; (2) if γ-Fe_2O_3 NPs pass through leaf epidermis as NPs, what is their final speciation after interacting with leaf organelles? In addition, 20 mg/L may not be the lowest concentration to supply sufficient iron for plants. Concentrations of γ-Fe_2O_3 NPs lower than 20 mg/L should be tested in the future.

Abbreviations

ABCG: ATP-binding cassette sub-family G member; CAT: catalase; FRO: ferric-chelate reductase; FW: fresh weight; IRT: iron regulated transporter; MDA: malonaldehyde; NPs: nanoparticles; Nramp: natural resistance-associated macrophage protein; PM: plasma membrane; POD: peroxidase; ROS: reactive oxygen species; SOD: superoxide dismutase; WIN: wax inducer.

Authors' contributions

The study was planned by HJ and LJ. Plants were cultured by HJ, WY and XL. Data analysis was done by HJ. The manuscript was written by HJ, LJ and GH. XB helped revised the manuscript. All authors read and approved the final manuscript.

Author details

[1] School of Chemistry, Chemical Engineering and Life Sciences, Wuhan University of Technology, Wuhan 430070, People's Republic of China. [2] Institute of Economic Crops, Hubei Academy of Agricultural Sciences, Wuhan 430064, People's Republic of China. [3] Stockbridge School of Agriculture, University of Massachusetts, Amherst, MA 01003, USA.

Acknowledgements
LJ gratefully acknowledges the support from the China Scholarship Council (201406955053) to study at UMass Amherst.

Competing interests
The authors declare that they have no competing interests.

Funding
This work was supported by the National Natural Science Foundation of China (Grant No. 31301735); the Fundamental Research Funds for the Central Universities (WUT: 2017IB006); the International Science &Technology Cooperation Program, Science and Technology Department of Hubei Province (Grant No. 2016AHB028); and USDA-NIFA Hatch Program (MAS 00475).

References

1. Guerinot ML, Yi Y. Iron: nutritious, noxious, and not readily available. Plant Physiol. 1994;104(3):815–20.
2. Rengel Z, Batten GD, Crowley DE. Agronomic approaches for improving the micronutrient density in edible portions of field crops. Field Crop Res. 1999;60:27–40.
3. Wallace GA, Wallace A. Micronutrient uptake by leaves from foliar sprays of EDTA chelated metals. In: Nelson SD, editor. Iron nutrition and interactions in plants. Basel: Marcel Dekker; 1982. p. 975–8.
4. Erik G, Michael S, Christian MG, Piet H, Karen H, Doreen W, et al. Quantification of silver nanoparticle uptake and distribution within individual human macrophages by fib/sem slice and view. J Nanobiotechnol. 2017;15(1):21–31.
5. Peñaloza JP, Márquez-Miranda V, Cabaña-Brunod M, Reyes-Ramírez R, Llancalahuen FM, Vilos C, et al. Intracellular trafficking and cellular uptake mechanism of PHBV nanoparticles for targeted delivery in epithelial cell lines. J Nanobiotechnol. 2017;15(1):1–15.
6. Zhai X, Zhang C, Zhao G, Stoll S, Ren F, Leng X. Antioxidant capacities of the selenium nanoparticles stabilized by chitosan. J Nanobiotechnol. 2017;15(1):4–15.
7. Li J, Hu J, Xiao L, Gan Q, Wang Y. Physiological effects and fluorescence labeling of magnetic iron oxide nanoparticles on citrus (*citrus reticulata*) seedlings. Water Air Soil Pollut. 2017;228(1):52–60.
8. He S, Feng Y, Ren H, Zhang Y, Gu N, Lin X. The impact of iron oxide magnetic nanoparticles on the soil bacterial community. J Soils Sediments. 2011;11(8):1408–17.
9. Li J, Chang P, Huang J, Wang Y, Yuan H, Ren H. Physiological effects of magnetic iron oxide nanoparticles towards watermelon. J Nanosci Nanotechnol. 2013;13(8):5561–7.
10. Ren H, Liu L, Liu C, He S, Huang J, Li J, et al. Physiological investigation of magnetic iron oxide nanoparticles towards chinese mung bean. J Biomed Nanotechnol. 2011;7(5):677–84.
11. Rui M, Ma C, Hao Y, Guo J, Rui Y, Tang X, et al. Iron oxide nanoparticles as a potential iron fertilizer for peanut (*Arachis hypogaea*). Front Plant Sci. 2016;7:815–24.
12. Hong J, Wang L, Sun Y, Zhao L, Niu G, Tan W, et al. Foliar applied nanoscale and microscale CeO_2 and CuO alter cucumber (*Cucumis sativus*) fruit quality. Sci Total Environ. 2016;563–564:904–11.
13. Cifuentes Z, Custardoy L, de la Fuente J, Marquina C, Ibarra M, Rubiales D, et al. Absorption and translocation to the aerial part of magnetic carbon-coated nanoparticles through the root of different crop plants. J Nanobiotechnology. 2010;8(1):26–33.
14. Ghafariyan MH, Malakouti MJ, Dadpour MR, Stroeve P, Mahmoudi M. Effects of magnetite nanoparticles on soybean chlorophyll. Environ Sci Technol. 2013;47(18):10645–52.
15. Zhu H, Han J, Xiao J, Jin Y. Uptake, translocation, and accumulation of manufactured iron oxide nanoparticles by pumpkin plants. J Environ Monit. 2008;10(6):713–7.
16. Corredor E, Testillano PS, Coronado MJ, Gonzálezmelendi P, Fernándezpacheco R, Marquina C, et al. Nanoparticle penetration and transport in living pumpkin plants: in situ subcellular identification. BMC Plant Biol. 2009;9:45–55.
17. Larue C, Castillo-Michel H, Sobanska S, Cécillon L, Bureau S, Barthès V, et al. Foliar exposure of the crop *Lactuca sativa* to silver nanoparticles: evidence for internalization and changes in Ag speciation. J Hazard Mater. 2014;264:98–106.
18. Birbaum K, Brogiolo R, Schellenberg M, Martinoia E, Stark WJ, Günther D, et al. No evidence for cerium dioxide nanoparticle translocation in maize plants. Environ Sci Technol. 2010;44:8718–23.
19. Schreck E, Foucault Y, Sarret G, Sobanska S, Cécillon L, Castrec-Rouelle M, et al. Metal and metalloid foliar uptake by various plant species exposed to atmospheric industrial fallout: mechanisms involved for lead. Sci Total Environ. 2012;427–428:253–62.
20. Broun P, Poindexter P, Osborne E, Jiang CZ, Riechmann JL. WIN1, a transcriptional activator of epidermal wax accumulation in Arabidopsis. Proc Nat Acad Sci USA. 2004;101:4706–11.
21. Alabdallat AM, Aldebei H, Ayad JY, Hasan S. Over-expression of SlSHN1 gene improves drought tolerance by increasing cuticular wax accumulation in tomato. Int J Mol Sci. 2014;15(11):19499–515.
22. Mcfarlane HE, Shin J, Bird DA, Samuels AL. Arabidopsis ABCG transporters, which are required for export of diverse cuticular lipids, dimerize in different combinations. Plant Cell. 2010;22(9):3066–75.
23. Alidoust D, Isoda A. Effect of γFe_2O_3 nanoparticles on photosynthetic characteristic of soybean (*Glycine max* (L.) Merr.): foliar spray versus soil amendment. Acta Physiol Plant. 2013;35(12):3365–75.
24. Lichtenthaler HK. Chlorophylls and carotenoids: pigments of photosynthetic biomembranes. Methods Enzymol. 1987;148:350–82.
25. Li HS. Principles and techniques of plant physiological experiment. Beijing: Higher Education Press; 2000.
26. Heath RL, Packer L. Photoperoxidation in isolated chloroplasts. I. Kinetics and stoichiometry of fatty acid peroxidation. Arch Biochem Biophys. 1968;125:189–98.
27. Wang YH, Ying Y, Chen J, Wang XC. Transgenic arabidopsis overexpressing Mn-SOD enhanced salt-tolerance. Plant Sci. 2004;167:671–7.
28. Gallego SM, Benavídes MP, Tomaro ML. Effect of heavy metal ion excesson sunflower leaves: evidence for involvement of oxidative stress. Plant Sci. 1996;121(2):151–9.
29. Zhang J, Cui S, Li J, Kirkham MB. Protoplasmic factors, antoxidant responses, and chilling resistance in maize. Plant Physiol Biochem. 1995;33:567–75.
30. Premachandra GS, Hahn DT, Joly RJ. A simple method for determination of abaxial and adaxial epicuticular wax loads in intact leaves of *Sorghum bicolor* L. Can J Plant Sci. 1993;73:521–4.
31. Hu J, Guo H, Li J, Gan Q, Wang Y, Xing B. Comparative impacts of iron oxide nanoparticles and ferric ions on the growth of *Citrus maxima*. Environ Pollut. 2017;221:199–208.
32. Taniguchi T, Kataoka R, Futai K. Plant growth and nutrition in pine (*Pinus thunbergii*) seedlings and dehydrogenase and phosphatase activity of ectomycorrhizal root tips inoculated with seven individual ectomycorrhizal fungal species at high and low nitrogen conditions. Soil Biol Biochem. 2008;40:1235–43.
33. Millar AA, Clemens S, Zachgo S, Giblin EM, Taylor DC, Kunst L. CUT1, an Arabidopsis gene required for cuticular wax biosynthesis and pollen fertility, encodes a very-long-chain fatty acid condensing enzyme. Plant Cell. 1999;11(5):825–38.
34. Goyal S, Lambert C, Cluzet S, Merillon JM, Ramawat KG. Secondary metabolites and plant defence. In: Merillon JM, Ramawat KG, editors. Plant defence: biological control. Berlin: Springer; 2012. p. 109–38.
35. Abadía J, Vázquez S, Rellánálvarez R, Eljendoubi H, Abadía A, Alvarezfernández A, et al. Towards a knowledge-based correction of iron chlorosis. Plant Physiol Biochem. 2011;49(5):471–82.
36. Singh NK, Bracken PM, Hasegawa PM, Handa AK, Buckel S, Hermodson MA, et al. Characterization of osmotin: a thaumatin-like protein associated with osmotic adjustment in plant cells. Plant Physiol. 1987;85:529–36.

37. Shang F, Zhao X, Wu C, Wu L, Qiou H, Wang Q. Effects of chlorpyrifos stress on soluble protein and some related metabolic enzyme activities in different crops. J China Agric Univ. 2013;18:105–10.
38. Afaq F, Abidi P, Matin R, Rahman Q. Cytotoxicity, pro-oxidant effects and antioxidant depletion in rat lung alveolar macrophages exposed to ultrafne titanium dioxide. J Appl Toxicol. 1998;18:307–12.
39. Gill SS, Tuteja N. Reactive oxygen species and antioxidant machinery in abiotic stress tolerance in crop plants. Plant Physiol Biochem. 2010;48(12):909–30.
40. Ma C, Chhikara S, Xing B, Musante C, White J, Dhankher O. Physiological and molecular response of *Arabidopsis thaliana* (L.) to nanoparticle cerium and indium oxide exposure. ACS Sustain Chem Eng. 2013;1:768–78.
41. Bowler CH, Van Montagu M, Inzé D. Superoxide dismutase and stress tolerance. Annu Rev Plant Biol. 1992;43:83–116.
42. Li J, Hu J, Ma C, Wang Y, Wu C, Huang J, et al. Uptake, translocation and physiological effects of magnetic iron oxide (γ-Fe_2O_3) nanoparticles in corn (*Zea mays* L.). Chemosphere. 2016;159:326–34.
43. Van Nhan L, Ma C, Rui Y, Cao W, Deng Y, Liu L, et al. The effects of Fe_2O_3 nanoparticles on physiology and insecticide activity in non-transgenic and bt-transgenic cotton. Front Plant Sci. 2016;6:1263–74.
44. de la Guardia MD, Alcántara E. Ferric chelate reduction by sunflower (*Helianthus annuus* L.) leaves: influence of light, oxygen, iron-deficiency and leaf age. J Exp Bot. 1996;47(5):669–75.
45. Brüggemann W, Maaskantel K, Moog PR. Iron uptake by leaf mesophyll cells: the role of the plasma membrane-bound ferric-chelate reductase. Planta. 1993;190:151–5.
46. Fernández V, Orera I, Abadía J, Abadía A. Foliar iron-fertilisation of fruit trees: present knowledge and future perspectives—a review. J Hortic Sci Biotechnol. 2009;84(1):1–6.
47. Asfaram A, Ghaedi M, Hajati S, Goudarzi A. Synthesis of magnetic γ-Fe_2O_3-based nanomaterial for ultrasonic assisted dyes adsorption: modeling and optimization. Ultrason Sonochem. 2016;32:418–31.
48. Ozin GA, Arsenault AC, Cademartiri L. Nanochemistry: a chemical approach to nanomaterials. London: Royal Society of Chemistry; 2009.
49. Shankramma K, Yallappa S, Shivanna MB, Manjanna J. Fe_2O_3 magnetic nanoparticles to enhance *S. lycopersicum* (tomato) plant growth and their biomineralization. Appl Nanosci. 2016;6:983–90.
50. Zhou L, Ni E, Yang J, Zhou H, Liang H, Li J, et al. Rice OsGL1-6 Is involved in leaf cuticular wax accumulation and drought resistance. PLoS ONE. 2013. doi:10.1371/journal.pone.0065139.
51. Suh MC, Samuels AL, Jetter R, Kunst L, Pollard M, Ohlrogge JB, et al. Cuticular lipid composition, surface structure, and gene expression in arabidopsis stem epidermis. Plant Physiol. 2005;139(4):1649–65.
52. Jetter R, Kunst L, Samuels AL. Composition of plant cuticular waxes. In: Riederer M, Müller C, editors. Biology of the plant cuticle. Oxford: Blackwell; 2007. p. 145–81.
53. Buschhaus C, Jetter R. Composition differences between epicuticular and intracuticular wax substructures: how do plants seal their epidermal surfaces? J Exp Bot. 2011;62(3):841–53.

18

Block copolymer conjugated Au-coated Fe_3O_4 nanoparticles as vectors for enhancing colloidal stability and cellular uptake

Junbo Li[1*], Sheng Zou[1], Jiayu Gao[1], Ju Liang[1], Huiyun Zhou[1], Lijuan Liang[1] and Wenlan Wu[2]

Abstract

Background: Polymer surface-modified inorganic nanoparticles (NPs) provide a multifunctional platform for assisting gene delivery. Rational structure design for enhancing colloidal stability and cellular uptake is an important strategy in the development of safe and highly efficient gene vectors.

Results: Heterogeneous Au-coated Fe_3O_4 (Fe_3O_4@Au) NPs capped by polyethylene glycol-*b*-poly1-(3-aminopropyl)-3-(2-methacryloyloxy propylimidazolium bromine) (PEG-*b*-PAMPImB-Fe_3O_4@Au) were prepared for DNA loading and magnetofection assays. The Au outer shell of the NPs is an effective platform for maintaining the superparamagnetism of Fe_3O_4 and for PEG-*b*-PAMPImB binding via Au–S covalent bonds. By forming an electrostatic complex with DNA at the inner PAMPImB shell, the magnetic nanoplexes offer steric protection from the outer corona PEG, thereby promoting high colloidal stability. Transfection efficiency assays in human esophageal cancer cells (EC109) show that the nanoplexes have high transfection efficiency at a short incubation time in the presence of an external magnetic field, due to increased cellular internalization via magnetic acceleration. Finally, after transfection with the magnetic nanoplexes EC109 cells acquire magnetic properties, thus allowing for selective separation of transfected cells.

Conclusion: Precisely engineered architectures based on neutral-cationic block copolymer-conjugated heterogeneous NPs provide a valuable strategy for improving the applicability and efficacy of synthesized vectors.

Keywords: Block copolymer, Heterogeneous nanoparticles, Magnetofection, Colloidal stability, Gene vector

Background

Gene therapy has emerged as a promising approach for delivering foreign nucleic acid into target cells for the treatment of cancer [1, 2]. Successful gene therapy requires safe and effective gene delivery systems to introduce genetic material into tissues or cells without causing harmful side effects [3, 4]. Although viral vectors remain the primary gene delivery system utilized in gene therapy, they present important limitations, including considerable immunogenicity, toxin production and limited size transgenic capacity [5]. Synthetic gene vectors offer unique advantages that resolve some of these technical hurdles, for instance, they have a modifiable structure, low immunogenic response and the capability to carry large inserts [6–8]. However, as most synthetic gene vectors have low colloidal stability and poor transfection efficiency, they have limited applications in vivo [9–11]. More recently, inorganic nanoparticles have provided attractive scaffolds to assist gene delivery due to their unique physical and chemical properties. Moreover, the controlled synthesis, assembly and modification of these particles have seen remarkable technical advances [12, 13]. Gold nanoparticles (Au NPs) show great potential for applications in gene delivery because they have controlled size, excellent biocompatibility, tailored surface chemistry and can be easily synthesized [14, 15]. It has been demonstrated that a variety of Au NPs with polycationic modifications can increase siRNA and plasmid DNA payload and thus enhance transfection efficiency, when compared to the surface materials alone [16, 17].

*Correspondence: Lijunbo@haust.edu.cn
[1] School of Chemical Engineering & Pharmaceutics, Henan University of Science & Technology, Luo Yang 471023, China
Full list of author information is available at the end of the article

However, the exposed positive surface of these vectors causes undesired aggregation and extensive vector accumulation in the physiological media [18, 19], resulting in impaired intracellular uptake and hence unsuitability for in vivo applications [20]. To overcome these limitations, block copolymers have been introduced which enhance the colloidal stability of Au NP-based vectors by means of a novel three layer micelle-like structure [21]. For instance, poly(*N*-2-hydroxypropyl methacrylamide-*block*-*N*-[3-(dimethylamino)propyl] methacrylamide) [P(HPMA-*b*-DMAPMA)] stabilized Au NPs were developed by McCormick's group for siRNA delivery [22]. The cationic PDMAPMA was used for binding siRNA, and the neutral PHPMA shell for protecting it against enzymatic degradation. Kataoka's group fabricated a similar NP small vehicle (~50 nm) that showed significant colloidal stability in vivo and high accumulation of siRNA in a cancer model [23]. Moreover, our group studied a process of DNA loading and the influence of PEG-*b*-PAMPImB-capped Au NPs on colloid stability during delivery [24]. These vectors exhibited a mono-disperse state to translocate across the cell membrane and then partly entered the nucleus, thus inducing high and efficient gene expression. The neutral outer corona significantly promoted high colloidal stability, however, it also had a negative effect on cellular uptake due to the reduced interaction between vector and cell membrane, called "PEG dilemma" [25, 26].

Iron oxide magnetic nanoparticles (Fe_3O_4 NPs) have also become attractive nanomaterials with promising application prospects for gene delivery due to their recently discovered superparamagnetic behavior [27, 28]. With their polycationic surface and magnetic core, these vectors can carry nuclear acids to target cells within minutes when an external magnetic field is applied [29, 30]. Notably, cellular internalization is activated independently of the magnetic NPs' surface properties, including neutral or negative charges [31, 32]. The efficiency of magnetofection is several hundredfolds higher than that of conventional transfections [33–35], however, Fe_3O_4 NPs present some disadvantages. For instance, while surface covalent bond modifications are difficult to achieve, surface oxidation and corrosion occur easily in physiological environments [36, 37]. Au-coated Fe_3O_4 heterogeneous NPs (Fe_3O_4@Au) have been widely explored for increasing Fe_3O_4 core stability and biocompatibility. These NPs provide a platform for easily achieving covalent modifications via Au–S chemistry, while maintaining their magnetic targeting function [38]. Moreover, Fe_3O_4@Au NPs may have applications in bioseparation [39], bioimaging [40], and photodynamic therapy of cancer [41].

Here, heterogeneous Fe_3O_4@Au NPs conjugated to block copolymer PEG-*b*-PAMPImB were prepared for DNA delivery and magnetofection assays. The vector was constructed with four layers: (i) a Fe_3O_4 NP core, to increase cellular internalization through magnetic acceleration; (ii) an outer Au shell, to facilitate binding with block copolymer via Au–S bonds; (iii) an inner PAMPImB block, for condensing DNA; and (iv) a PEG corona, to increase colloidal stability. The synthesis of the magnetic vector and magnetofection process are illustrated in Scheme 1.

Results and discussion

Preparation and characterization of PEG-*b*-PAMPImB-Fe_3O_4@Au

We prepared heterogeneous Fe_3O_4@Au NPs by using a widely reported and efficient approach [42]. APTES functional Fe_3O_4 NPs were used as heterogeneous nucleation sites and $AuCl_4^-$ was then directly reduced on the interface of those Fe_3O_4 NPs. Next, we converted the dithioester of PEG-*b*-PAMPImB [previously synthesized by reversible addition fragmentation chain transfer (RAFT) polymerization] into a thiol end-group with hydrazine, and obtained PEG-*b*-PAMPImB-Fe_3O_4@Au NPs via a strong binding affinity between gold and thiol ending.

The hydrodynamic diameter and distribution of Fe_3O_4, Fe_3O_4@Au and PEG-*b*-PAMPImB-Fe_3O_4@Au NPs were first measured by dynamic light scattering (DLS). The average hydrodynamic diameter (D_h) of the Fe_3O_4 NPs is approximately 21 nm and the size distribution range is 12–34 nm (Fig. 1a). The Fe_3O_4@Au NPs has an approximate D_h at 29 nm, which is larger than that of Fe_3O_4 NPs, thus confirming the formation of a thin Au shell. The mean thickness of the Au shell was estimated to be 4 nm. The larger D_h of PEG-*b*-PAMPImB-Fe_3O_4@Au NPs (83 nm) and their size distribution range (61–107 nm) are indicative of a successful polymer conjunction. PEG-*b*-PAMPImB-Fe_3O_4@Au NPs were further characterized by TEM (Fig. 1b). The clear covered polymer layers observed on the surface of Fe_3O_4@Au NPs (magnified image insert in Fig. 1b) further confirmed that PEG-b-PAMPImB attached onto Fe_3O_4@Au NPs, as these images are clearly different from TEM images of Fe_3O_4 (Additional file 1: Figure S1a) and Fe_3O_4@Au (Additional file 1: Figure S1b). UV–vis absorption of Fe_3O_4, Au, Fe_3O_4@Au and PEG-*b*-PAMPImB-Fe_3O_4@Au NPs revealed different spectrum properties (Fig. 1c). Indeed, no significant absorption peaks were detected in the visible light curve of Fe_3O_4 NPs. Pure Au NPs have clear surface plasmon resonance (SPR) absorption (525 nm). A similar SPR absorption (548 nm) was observed on the spectrum of Fe_3O_4@Au NPs, confirming that the Au shell formed successfully on the surface of Fe_3O_4 NPs [42]. Upon coating with PEG-*b*-PAMPImB, Fe_3O_4@Au NPs showed a minor blue shift in the SPR, from 548 to

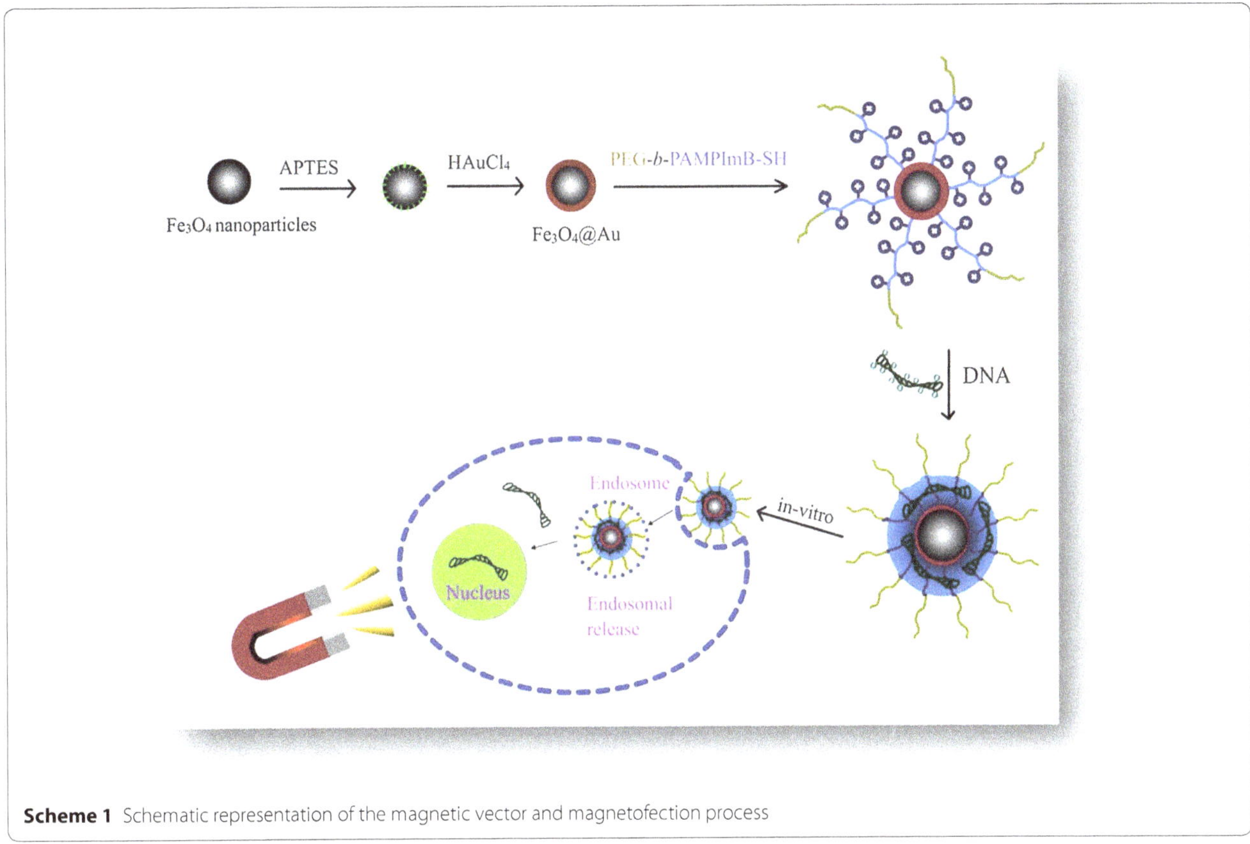

Scheme 1 Schematic representation of the magnetic vector and magnetofection process

542 nm, likely due to their enhanced dispersion in water. The superparamagnetic properties of Fe_3O_4, Fe_3O_4@Au and PEG-*b*-PAMPImB-Fe_3O_4@Au NPs were further assessed by Magnetization curves (M–H loop) measured at RT (Fig. 1d). The saturation magnetization value (Ms) for Fe_3O_4 NPs is 48.05 emu/g. After coating with the Au shell and PEG-*b*-PAMPImB, the Ms for Fe_3O_4@Au and PEG-*b*-PAMPImB-Fe_3O_4@Au NPs was lower, at 26.45 and 12.33 emu/g, respectively. This decrease in Ms has been attributed to the packaging of non-magnetic Au shell and PEG-*b*-PAMPImB on the periphery of Fe_3O_4 NPs [43]. The inserted photograph showed that PEG-*b*-PAMPImB-Fe_3O_4@Au NPs form a purple aqueous solution with homogeneous dispersion. This solution showed a typical macroscopic appearance of aggregation after positioning with a magnet for 30 min, demonstrating that PEG-*b*-PAMPImB-Fe_3O_4@Au NPs possess magnetic responsiveness.

DNA loading capability of PEG-*b*-PAMPImB-Fe_3O_4@Au

PEG-*b*-PAMPImB-Fe_3O_4@Au NPs are designed to load DNA via an electrostatic attraction between the positively charged PAMPImB and the negatively charged DNA phosphate groups. Agarose gel retardation assays were performed to assess the gene condensing capacity of the magnetic NPs and of the surface polymer. The migration of naked DNA, PEG-b-PAMPImB/DNA and PEG-b-PAMPImB-Fe_3O_4@Au NPs/DNA complexes at weight ratios ranged from 0/1 to 11/1 were shown in Fig. 2a. Both polymer and magnetic particles could condense pDNA efficiently at low weight ratios. The migration of DNA in agarose gels was significantly retarded and remained above the weight ratio (±) of two for PEG-*b*-PAMPImB, and above four for PEG-*b*-PAMPImB-Fe_3O_4@Au NPs. In agreement with these results, the zeta potentials (Fig. 2b) appear to increase with the weight ratio. At the weight ratio of five, the zeta potential of PEG-*b*-PAMPImB-Fe_3O_4@Au NPs with DNA shifts to positive. At higher ratios, the zeta potentials reach a maximum plateau, because the DNA negative charges are rapidly neutralized by the NPs' excess positive charges.

Size and colloidal stability of magnetic nanoplexes

Vector stability under physiological conditions has significant effects on gene expression in vitro and on further applications in vivo [44]. Our previous reports demonstrated that polyplexes [45] and nanoplexes [24] with a PEG shell, and which condense DNA via PAMPImB, are highly stable in physiological media. We assessed magnetic nanoplex stability in different media

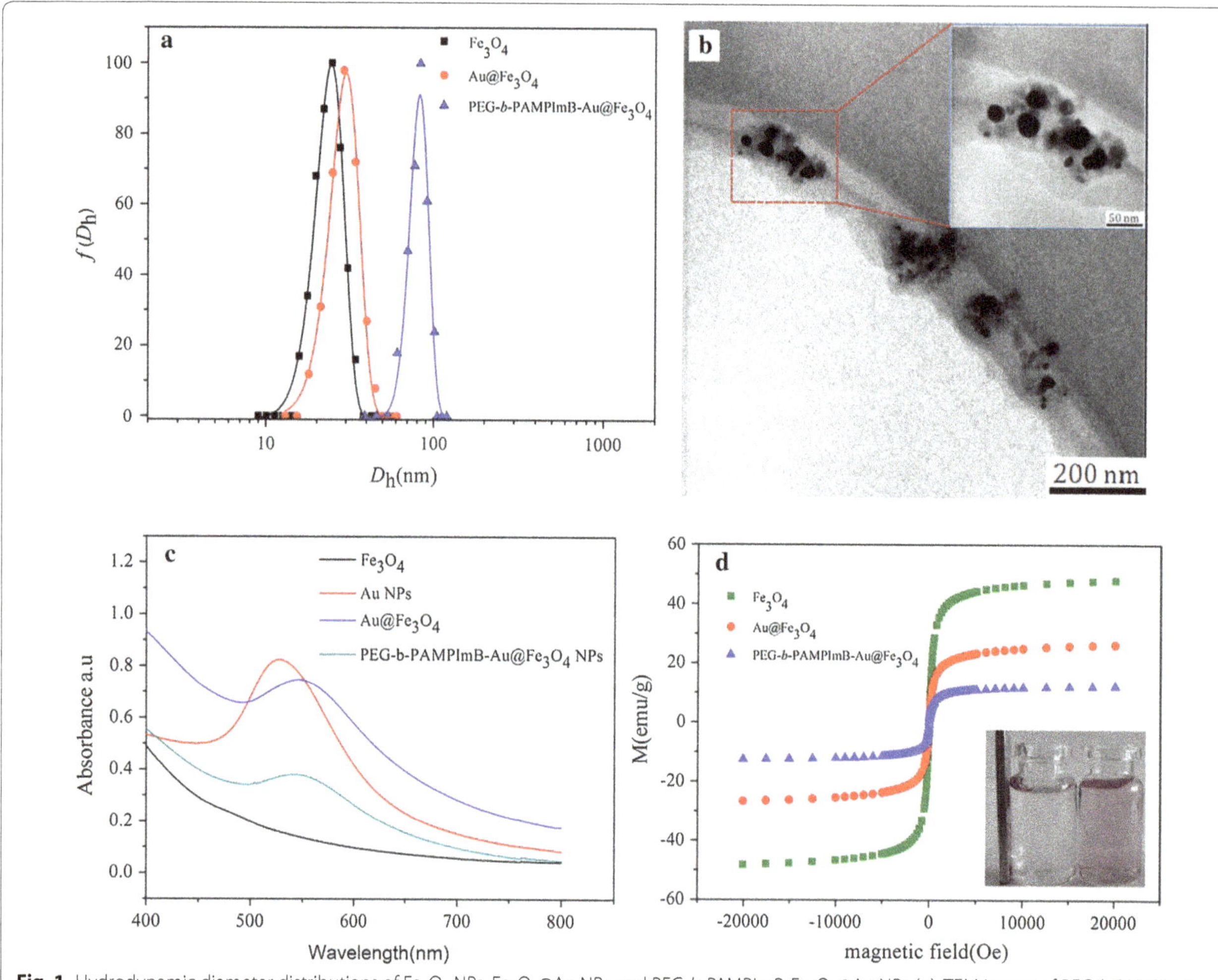

Fig. 1 Hydrodynamic diameter distributions of Fe_3O_4 NPs, Fe_3O_4@Au NPs and PEG-*b*-PAMPImB-Fe_3O_4@Au NPs (**a**); TEM image of PEG-*b*-PAMPImB-Fe_3O_4@Au NPs (**b**); the inserted image is a magnification of the NPs in **b**; UV–vis spectra of Fe_3O_4 NPs, Au NPs, Fe_3O_4@Au NPs and PEG-*b*-PAMPImB-Fe_3O_4@Au NPs (**c**); magnetization measurements as a function of applied field for Fe_3O_4 NPs, Fe_3O_4@Au NPs and PEG-*b*-PAMPImB-Fe3O4@Au NPs (**d**). The inserted image in **d** is a photograph of a solution of PEG-b-PAMPImB-Fe_3O_4@Au NPs before (*right bottle*) and after (*left bottle*) applying a NdFeB magnet during 30 min

by DLS (Table 1). PEG-b-PAMPImB-Fe_3O_4@Au NPs were mixed with pGFP-C1 at weight ratios of 20/1, 10/1 or 5/1 in pure water, and then each mix was transferred into PBS buffer (20 mM, pH 7.4), 150 mM NaCl or 10% FBS solution. The DNA weight was fixed at 200 ng in every sample. Three magnetic nanoplexes show a small hydrodynamic size at 75, 87 and 101 nm, and narrow particle size distribution in pure water. The magnetic nanoplexes have the most similar size (±18 nm) to magnetic NPs without DNA (Fig. 1a), indicating that PEG-*b*-PAMPImB-Fe_3O_4@Au NPs can be used as template, based on core–shell structure, to monitor nanoplex size. The D_h and PDI of the nanoplexes, regardless of their weight-ratio, were nearly unchanged in PBS buffer (pH 7.4) and 150 mM NaCl solution, when compared to pure water. This result shows that the nanoplexes are stable in these physiological media. However, in 10% FBS the D_h of the magnetic nanoplexes increased slightly and the PDI broadened at every weight-ratio, which may be due to protein adsorption onto the nanoplexes' surface. These data demonstrate that the periphery of magnetic nanoplexes covered by electrostatically neutral PEG provides high colloid stability for vectors in physiological conditions (i.e. presence of salt and serum).

Cytotoxicity of magnetic nanoparticles and nanoplexes

Low cytotoxicity is a highly desired property for carriers in drug and gene delivery. We assessed cytotoxicity of PEI25k, PEG-*b*-PAMPImB and PEG-*b*-PAMPImB-Fe_3O_4@Au NPs in human esophageal cancer cells (EC109) with MTT assays, at various concentrations of vector (Fig. 3a). The blank test, which was considered as a

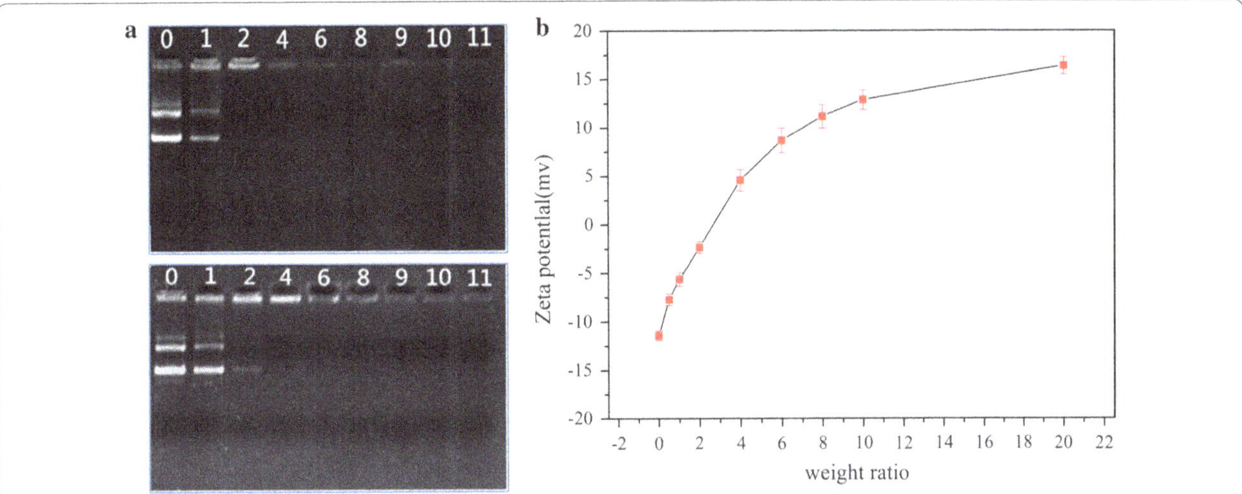

Fig. 2 Gel migration assay for PEG-*b*-PAMPImB (*up*) and PEG-*b*-PAMPImB-Fe_3O_4@Au (*down*) complex with DNA at various weight ratios (**a**), where the number on *each lane* represents the ratio of PEG-*b*-PAMPImB and PEG-*b*-PAMPImB-Fe_3O_4@Au to DNA in complexes; zeta potentials of PEG-*b*-PAMPImB-Fe_3O_4@Au/DNA complexes in pure water at various weight ratios (**b**). Values represent mean (±SD [n = 3])

Table 1 The hydrodynamic diameter and polydispersity index of PEG-b-PAMPImB/DNA-Fe_3O_4@Au NPs at different ratios and media

Weight ratio	Pure water		PBS (pH 7.4)		150 mM NaCl		10% FBS	
	D_h (nm)	PDI	D_h (nm)	PDI	D_h (nm)	PDI	D_h (nm)	PDI
20/1	75	0.15	73	0.20	72	0.16	89	0.25
10/1	87	0.17	84	0.19	81	0.19	116	0.24
5/1	101	0.21	98	0.18	96	0.19	122	0.27

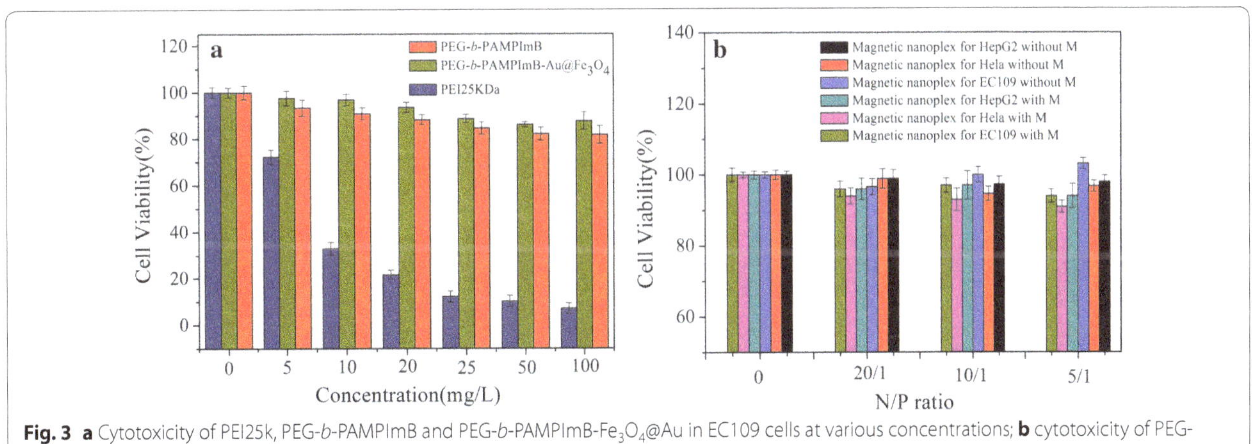

Fig. 3 **a** Cytotoxicity of PEI25k, PEG-*b*-PAMPImB and PEG-*b*-PAMPImB-Fe_3O_4@Au in EC109 cells at various concentrations; **b** cytotoxicity of PEG-*b*-PAMPImB-Fe3O4@Au/DNA in HepG2, HeLa and EC109 cells at different weight ratios. Values represent mean (SD [n = 3])

positive control, showed 100% cell viability. PEG-*b*-PAMPImB and PEG-*b*-PAMPImB-Fe_3O_4@Au NPs exhibited significantly lower cytotoxicity than PEI25k, and PEG-*b*-PAMPImB-Fe_3O_4@Au NPs showed slightly higher biocompatibility than PEG-*b*-PAMPImB (Fig. 3a). The cytotoxicity of composite nanoparticles often depends on their size and surface properties [46]. PEG-*b*-PAMPImB is a low cytotoxic polymer because PEG and the cationic histamine-like segment have high biocompatibility, as shown in our previous report [45]. The formation of micelle-like core–shell structures upon attachment to Fe_3O_4@Au further decreases cytotoxicity, as the PEG outer corona shields the inner shell of the cationic histamine-like segment.

Finally, we assessed the cytotoxicity of the PEG-*b*-PAMPImB-Fe_3O_4@Au/DNA nanoplexes in HepG2, HeLa and EC109 cells in the presence or absence of a magnetic field (Fig. 3b). The magnetic nanoplexes showed significantly low cytotoxicity whether or not a magnetic field was applied to the cells, suggesting that exposure to a magnetic field does not affect normal cell proliferation.

Magnetofection efficiency

pEGFP-C1 was employed as a reporter gene for evaluating the transfection efficiency of the magnetic nanoplexes (at weight ratio of 5/1) in EC109 cells with or without application of a static magnetic field. Transfection of pEGFP-C1 with Lipofectamine2000 or PEI25K (N/P of 5/1) was used as positive control. Transfection efficiency was assessed by quantifying the GFP-expressing cells with flow cytometry. The cells were first incubated with the vectors in serum-free culture medium at 37 °C. This medium was then replaced with fresh medium containing 10% serum and the cells were incubated for 24 h at 37 °C. Figure 4a shows the percentage of cells transfected, at different incubation times. Notably, the magnetic nanoplex under a magnetic field shows significantly higher transfection efficiency at a shorter incubation time (0.5 and 1 h) than Lipofectamine2000, PEI25K and the magnetic nanoplex without application of a magnetic field. The magnetic transfection efficiency reached a maximum of about 43% after 1 h of incubation, which is fivefold, threefold, and fourfold higher than the transfection efficiency of the nanoplexes without a magnetic field, Lipofectamine2000, and PEI25K, respectively. Moreover, the transfection efficiency of pEGFP-C1 with Lipofectamine2000, PEI25K and the magnetic nanoplex without a magnetic field increased with the incubation time, consistent with standard transfections [47]. Direct observation of the transfected cells with an inverted fluorescence microscope revealed that magnetofection at 1 h of incubation has the highest transfection activity when compared to transfections in the other conditions (Fig. 4b). These results suggest that the rapid accumulation of DNA-carrying magnetic vectors around cells increases transfection efficiency.

Cell uptake and magnetic separation

The mean mass of Fe and Au in total EC109 cells was measured by inductive coupled plasma-mass spectrometry (ICP-MS) at different incubation times (Fig. 5a). Before transfection, the magnetic nanoplexes contained 22.8 pg Fe and 13.8 pg Au. At 0.5–2 h of incubation time, the cells transfected using magnet-assisted transfection had a higher mass of Fe and Au than those transfected using a standard transfection method. The content of Fe and Au in the magnetotransfected cells peaked at 1 h and then remained unchanged, suggesting that internalization in the presence of a magnetic field was completed at this time point. Consistent with the continuous increase in uptake rate over time that is typical of standard transfections, the mass of Fe and Au in cells transfected with Lipofectamine2000, PEI25K or magnetic nanoplexes without a magnetic field at 4 h was comparable to the internalization rates of magnetic transfections at 1 h. Confocal laser scanning microscopy (CLSM) revealed that there is significantly more internalization of magnetic nanoplexes in transfections with the application of an external magnetic field (Fig. 5b) than without (Fig. 5c), in agreement with the ICP-MS results above. The accumulation of magnetic NPs in transfected cells confers them magnetic properties, thereby allowing for selective cell separation by application of a magnetic field [48]. After harvesting with an external magnetic field, the transfected cells were incubated overnight and then observed on an inverted fluorescence microscope.

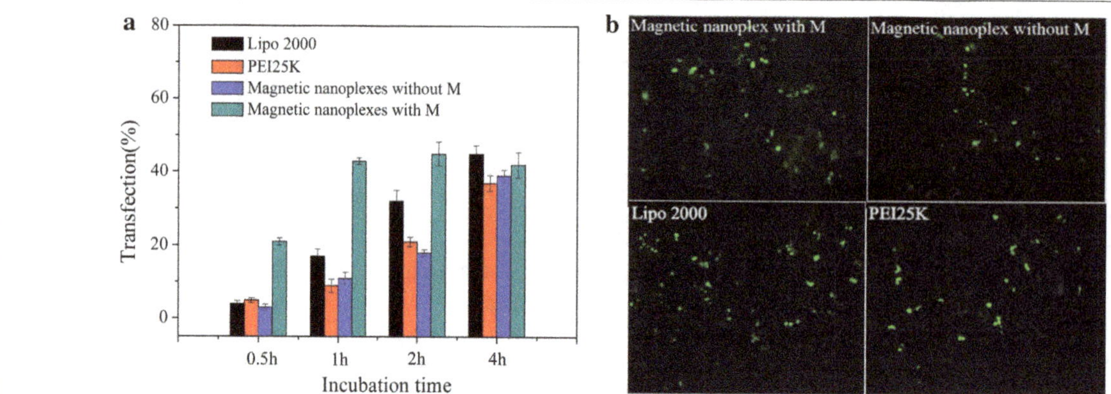

Fig. 4 Transfection efficiency (% of cells transfected) of EC109 cells treated with pEGFP-C1 and magnetic nanoplexes with or without application of a magnetic field, PEI (25 kD) and Lipo2000 at various incubation times (**a**); fluorescence microscope images of the EC109 cells in **a** at 1 h of incubation (**b**). Values represent mean (SD [n = 3])

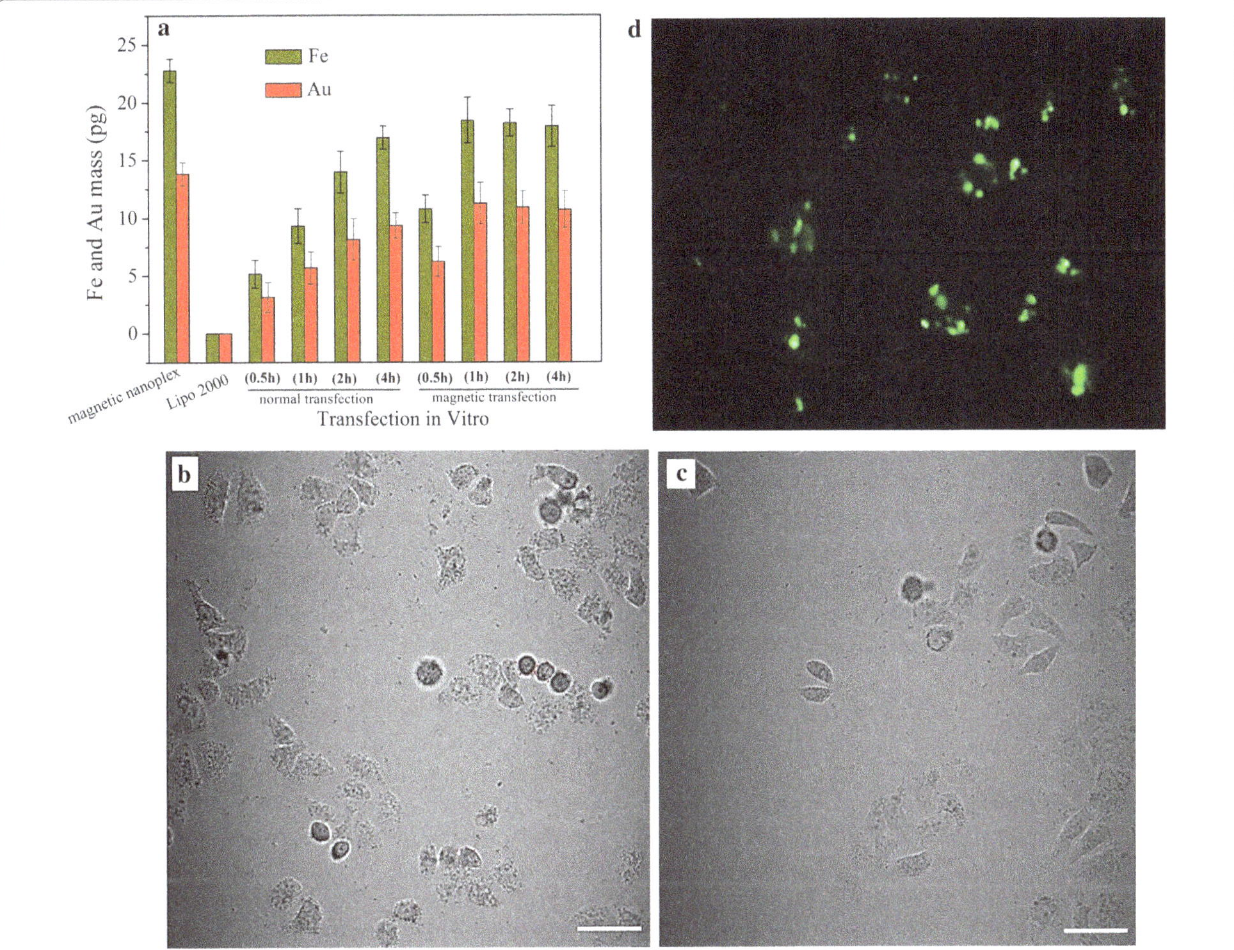

Fig. 5 The Fe and Au content in transfected cells at various incubation times (**a**); confocal microscopic images of EC109 cells transfected with magnetic nanoplex with (**b**) and without the application of a magnetic filed (**c**) at 1 h incubation time (*scale bar* is 50 μm); fluorescence microscope image of EC109 transfected cells after magnetic collection and growth overnight (**d**). Values represent mean (SD [n = 3])

Compared with the result in Fig. 4b, a clear increase in fluorescence intensity could be detected due to the magnetic agglomeration of transfected cells (Fig. 5d).

Magnetic gene delivery systems have attracted wide interesting because of their potential to achieve selective and efficient delivery of therapeutic genes to the target site/cells. Commonly, most formulations were fabricated with a magnetic core for magnetic target function and external coating cationic polymer for loading therapeutic gene. For example, Shi et al. conjugated plasmids on the surface of PEI modified Au–Fe_3O_4 dumbbell nanoparticles and obtained high efficiency in transfecting adherent mammalian cells under the magnetic attraction [49]. Zamanian' group prepared chitosan coated Fe_3O_4 nanoparticles and demonstrated the particles can enhance magnetofection efficiency due to the advantages posed by its magnetic properties and DNA-binding ability [33]. Those formulations have been widely demonstrated a high transfection efficiency in vitro by application of external magnetic field. However, the undesired aggregation was also brought due to random interactions between the vector's positive surface and the negative charges of biomacromolecules or components in physiological media [19], as these aggregated vectors would be unacceptable for medical applications in vivo.

In order to improve the applicability of magnetic delivery system, we developed Fe_3O_4@Au heterogeneous NPs capped neutral-cationic block copolymer as DNA vectors for magnetofection assays. In contrast to reported magnetic formulations, our vector shows clear benefits. First, we used Au-coated Fe_3O_4 heterogeneous NPs (Fe_3O_4@Au) as magnetic core instead of single Fe_3O_4 nanoparticles. The Au shell of Fe_3O_4@Au would increase Fe_3O_4 core stability and biocompatibility, while provide a platform for easily achieving covalent modifications via Au–S chemistry [50]. Second, we designed a neutral-cationic block copolymer as external coating polymer rather than only cationic polymer monolayer. The outer neutral PEG can provide high colloidal stability of vector in physiological conditions [45]. Moreover, the magnetic agglomeration of transfected cells has been proven to be feasible.

Conclusion

We have developed Fe_3O_4@Au heterogeneous NPs capped with neutral-cationic block copolymer via Au–S covalent bonds, and assessed their feasibility as DNA vectors in magnetofection assays. This vector has a four-layer microstructure comprising a Fe_3O_4 core, an Au shell, an inner cationic polymer and an outer neutral PEG. These different layers provide well-defined functions for magnetic response, polymer conjunction, DNA loading and colloidal stability. We show that the magnetic nanoplexes have high stability in physiological conditions and are rapidly internalized in magnet-assisted transfections, thereby allowing for efficient separation of transfected cells. Thus, precisely engineered architectures based on neutral-cationic block copolymer-conjugated heterogeneous NPs provide a valuable strategy for improving the efficacy and applicability of synthesized vectors.

Methods

Materials

Borane-tert-butylamine complex (BTBA), hydrogen tetrachloroaurate ($HAuCl_4 \cdot 3H_2O$) and branched poly(ethylenimine) (PEI25k) were purchased from Sigma-Aldrich and used as received. Ferrous chloride tetrahydrate ($FeCl_2 \cdot 4H_2O$, >99% purity) and ferric chloride hexahydrate ($FeCl_3 \cdot 6H_2O$, >99% purity) were bought from Shanghai Chemical Reagent Co., Ltd. (Shanghai, China). PEG_{112}-*b*-$PAMPImB_{96}$ was synthesized by RAFT polymerization by using PEG-CTA (Mw 5000) as macromolecular chain transfer agent and AMPImB as monomer. The detailed synthesis and characterization of this polymer was previously described [45]. Dulbecco's modified Eagle's medium (DMEM), penicillin–streptomycin, trypsin, fetal bovine serum (FBS), 3-[4,5-dimethylthiazol-2-yl]-2,5-diphenyltetrazolium bromide (MTT), and Dubelcco's phosphate buffered saline (DPBS) were purchased from Thermo Fisher Scientific. The reporter plasmid, enhanced green fluorescent protein gene (pEGFP-C1), was amplified in *E. coli* and purified by E.Z.N.A.® Endo-free plasmid DNA maxi kit (Omega). Purified pEGFP-C1 was stored at −20 °C and thawed at RT for the transfection assays.

Preparation of Fe_3O_4@Au nanoparticles

Magnetic nanoparticles of Fe_3O_4 were prepared with a chemical coprecipitation method according to a previously reported procedure [38]. Briefly, $FeCl_2 \cdot 4H_2O$ (0.398 g, 2.5 mmol) and $FeCl_3 \cdot 6H_2O$ (1.352 g, 5 mmol) were dissolved in 100 mL of Milli-Q water containing 20 μL of concentrated HCl. The solution was heated to 80 °C under a nitrogen atmosphere and then 150 mL of sodium hydroxide (1 M) was added dropwise with vigorous stirring. After stirring for an hour, the magnetic nanoparticles were harvested by using an NdFeB magnet, and washed with Milli-Q water 3–4 times until the supernatant liquor reached neutrality. Finally, the resulting Fe_3O_4 NPs were dried under vacuum at 60 °C for further use.

Au-coated Fe_3O_4 NPs (Fe_3O_4@Au) were prepared by directly reducing $HAuCl_4$ on the surface of APTES-functionalized Fe_3O_4 NPs. 0.2 g of Fe_3O_4 NPs were ultrasonically dispersed in 50 mL anhydrous ethanol, and then 0.1 mL of APTES as added at room temperature (RT). The mixture was vigorously stirred for 24 h and then acidized by adding 0.05 mL of a concentrated HNO_3 solution. APTES functionalization of Fe_3O_4 NPs was carried out in three cycles of separation and wash by using an NdFeB magnet and ethanol, respectively. The product was mixed with a 1% $HAuCl_4$ ethanol solution followed by dropwise addition of BTBA (0.05%, w/v). The color of the solution changed from brown to reddish-brown and then purple as the Au content was increased. The Fe_3O_4@Au NPs were purified by magnet separation and washed with 0.1 M HCl to remove the free Au and Fe_3O_4 NPs. Finally, the particles were re-dispersed in ethanol for further use.

Preparation of PEG-*b*-PAMPImB-Fe_3O_4@Au nanoparticles

PEG-*b*-PAMPImB-modified Fe_3O_4@Au (PEG-*b*-PAMPImB-Fe_3O_4@Au) NPs were prepared via formation of Au–S covalent bonds between the terminated group of polymer and an Au shell layer of Fe_3O_4@Au. We used a classic procedure as follows: PEG_{112}-*b*-$PAMPImB_{96}$ (0.1 g) was added to a 100 mL round-bottom flask containing 10 mL of an Fe_3O_4@Au ethanol suspension; 1 mL of 0.1 M aqueous hydrazine solution was then added with vigorous stirring to reduce its dithioester-terminated group with thiol. After 3 days of equilibrium at RT, the formed PEG-*b*-PAMPImB-Fe_3O_4@Au NPs were collected with a magnet and washed with water to remove unbound polymer. After vacuum drying, 0.1 g of PEG-*b*-PAMPImB-Fe_3O_4@Au was dissolved in 100 mL of Milli-Q water for further use.

Preparation of PEG-*b*-PAMPImB-Fe_3O_4@Au/DNA nanoplexes

PEG-*b*-PAMPImB-Fe_3O_4@Au NPs bound to DNA at various weight ratios were prepared by adding different volumes of PEG-*b*-PAMPImB-Fe_3O_4@Au (1 mg/mL) and 36 μL of DNA plasmid (200 ng/μL) into an aqueous solution. The nanoplexes (PEG-*b*-PAMPImB-Fe_3O_4@

Au/DNA) were gently vortexed and then incubated for 30 min at RT to ensure stable formation of nanoplexes. The nanoplexes were then subjected to a centrifuging–redispersing process to remove free unbound DNA.

Cell culture

EC109 cells were cultured in DMEM medium supplemented with 10% (v/v) heat-inactivated FBS and 1% penicillin–streptomycin at 37 °C in a humidified atmosphere containing 5% CO_2.

Agarose gel retardation assay

The nanoplexes' condensation capability was assessed by agarose gel electrophoresis. NP/DNA nanoplexes with serial weight ratios ranging from 0/1 to 10/1 were prepared according to the conditions described above. After 30 min of incubation at RT, the nanoplex solutions were analyzed by 1% agarose gel electrophoresis (100 V, 30 min) in TAE buffer. The DNA bands were visualized with UV light and analyzed with Cam2com software.

Cytotoxicity assay

EC109 cells were seeded into 96-well plates at 5000 cells/well and cultured 24 h in 200 μL of DMEM containing 10% FBS. A range of concentrations of PEG-*b*-PAMPImB, PEG-*b*-PAMPImB-Fe_3O_4@Au and PEI25k were prepared in PBS solution (pH 7.4). To estimate the influence of an external magnetic field on cell viability, PEG-*b*-PAMPImB-Fe_3O_4@Au/DNA nanoplexes with DNA weight ratios of 20/1, 10/1 and 5/1 were prepared and added into the wells. A magnetic sheet was placed under well plate to apply the magnetic field. 20 μL of each solution was added to the corresponding well, followed by 24 h of incubation. Then, the medium was replaced with 200 μL of fresh medium. MTT (20 μL, 5 mg/mL in PBS) stock solution was then added to each well. After 4 h, unreacted dye was carefully removed, and the formazan crystals were dissolved in DMSO (200 μL/well). The plates were incubated for another 10 min before measuring the absorbance at 570 nm with an ELISA microplate reader (Bio-Rad). Cell viability (%) was calculated as previously described.

In vitro transfection

To assess the transfection activity of the nanoplexes, EC109 cells were seeded in 24-well plates with an initial density of 5×10^4 cells/well in 1 mL DMEM containing 10% FBS and then incubated at 37 °C for 24 h in 5% CO_2 (to reach 70% confluence at the time of transfection). The magnetic nanoplexes, PEI25k/DNA and Lipofectamine2000/DNA were added to wells containing serum-free culture medium at 37 °C. After incubation, the medium was replaced with fresh medium containing 10% serum and transfected for 48 h at 37 °C. Magnetofection was performed by placing a magnetic sheet under the plates. The cells were monitored for expression of green fluorescence protein (GFP) with a fluorescence microscope. For observation with a confocal laser scanning microscope (LSM 780, Zeiss), the cells were washed with PBS three times, fixed with 4% paraformaldehyde for 30 min. Transfection efficiency was determined by flow cytometry to quantify the percentage of GFP-expressing cells. After transfection for 48 h at 37 °C, the harvested cells were washed with PBS, detached with 0.25% trypsin and then resuspended in 500 μL PBS (pH 7.4) for flow cytometry (FC500, Beckman Coulter).

To quantify the intracellular uptake of magnetic NPs, ICP-MS was performed to measure the concentration of Fe and Au in total cells. After incubation, the medium was removed and the cells were washed with PBS three times, and then treated with trypsin solution (containing 0.25% EDTA). The cell pellets were sorted into a 20 mL silicon glass vial and completely digested with 500 μL of Aqua regia. The digested solution was diluted to 5 mL with 1% Aqua regia and filtered with 0.22 μm filters (Millipore, USA). For estimating percentage of uptake, the same dosage of magnetic nanoplexes was directly digested by Aqua regia and diluted to the same volume for measuring Fe and Au concentration with ICP-MS (7500A, Agilent, USA).

Magnetic collection of transfected cells

EC109 cells were transfected at weight ratio 5/1, as described above. After transfection for 48 h at 37 °C, the harvested cells were washed with PBS, detached with 0.25% trypsin and then resuspended in 500 μL PBS (pH 7.4). Magnetic collection of transfected cell was performed by placing a magnetic sheet under the plates during the free sedimentation of cells. After 15 min, the medium was carefully removed and the cells were washed with PBS three times. Finally, the transfected cells were further incubated overnight and observed directly with a fluorescence microscope.

Characterization

Dynamic laser scattering (DLS) measurements were performed using a laser light scattering spectrometer (BI-200SM) equipped with a digital correlator (BI-9000AT) at 532 nm at RT. Transmission electron microscopy (TEM) measurements were conducted using a JEM-2100 electron microscope at an accelerating voltage of 200 kV; a small drop of solution was deposited onto a carbon-coated copper EM grid and dried at RT under atmospheric pressure. The UV–vis spectra were recorded on a Cary 50 Bio UV–Visible Spectrophotometer (Varian, USA) equipped with two silicon diode detectors and a xenon flash lamp. Zeta-potentials were measured using a temperature-controlled Zetasizer 2000 (Malvern Instruments. Ltd. UK).

Abbreviations

Fe_3O_4@Au: Au coating Fe_3O_4 NPs; PEG-*b*-PAMPImB-Fe_3O_4@Au: polyglycol-*b*-poly1-(3-aminopropyl)-3-(2-methacryloyloxy propylimidazolium bromine); EC109: human esophageal cancer cell line; Au NPs: gold nanoparticles; P(HPMA-*b*-DMAPMA): poly(*N*-2-hydroxypropyl methacrylamide-*block*-*N*-[3-(dimethylamino)propyl] methacrylamide); Fe_3O_4 NPs: iron oxide magnetic nanoparticles; RAFT: reversible addition–fragmentation chain transfer; DLS: dynamic light scattering; D_h: average hydrodynamic diameter; TEM: transmission electron microscopy; SPR: surface plasmon resonance; Ms: saturation magnetization value; ICP-MS: investigated by inductively coupled plasma mass spectrometry; CLSM: confocal laser scanning microscopy; BTBA: borane-tert-butylamine complex; PEI25k: branched poly(ethylenimine); DMEM: Dulbecco's modified Eagle's medium; FBS: fetal bovine serum; MTT: 3-[4,5-dimethylthiazol-2-yl]-2,5-diphenyltetrazolium bromide; DPBS: Dubelcco's phosphate buffered saline; pEGFP-C1: plasmid enhanced green fluorescent protein; RT: room temperature.

Authors' contributions

JL initiated the study, supervised data interpretations and drafted the manuscript. SZ performed the cell culture and gene expression experiments. JG carried out the cytotoxicity assays and revised the manuscript. JLiang performed the synthesis and characterization of the nanoparticles. HZ and LL performed the transfection efficiency and cell internalization assays. WW revised the manuscript. All authors read and approved the final manuscript.

Author details

[1] School of Chemical Engineering & Pharmaceutics, Henan University of Science & Technology, Luo Yang 471023, China. [2] School of Medicine, Henan University of Science & Technology, Luo Yang 471023, China.

Acknowledgements

Not applicable.

Competing interests

The authors declare that they have no competing interests.

Funding

This work was supported by the National Natural Science Foundation of China (Nos. 51103035 and 51403055).

References

1. Ji W, Sun B, Su C. Targeting MicroRNAs in cancer gene therapy. Genes. 2017;8:21.
2. Hendricks WPD, Yang J, Sur S, Zhou S. Formulating the magic bullet: barriers to clinical translation of nanoparticle cancer gene therapy. Nanomedicine. 2014;9:1121–4.
3. Grigsby CL, Ho Y-P, Leong KW. Understanding nonviral nucleic acid delivery with quantum dot-FRET nanosensors. Nanomedicine. 2012;7:565–77.
4. Cavallaro G, Licciardi M, Scirè S, Giammona G. Microwave-assisted synthesis of PHEA-oligoamine copolymers as potential gene delivery systems. Nanomedicine. 2009;4:291–303.
5. Nayerossadat N, Maedeh T, Ali PA. Viral and nonviral delivery systems for gene delivery. Adv Biomed Res. 2012;1:27.
6. Ullah I, Muhammad K, Akpanyung M, Nejjari A, Neve AL, Guo J, Feng Y, Shi C. Bioreducible, hydrolytically degradable and targeting polymers for gene delivery. J Mater Chem B. 2017;5:3253–76.
7. Fisicaro E, Compari C, Bacciottini F, Contardi L, Pongiluppi E, Barbero N, Viscardi G, Quagliotto P, Donofrio G, Krafft MP. Nonviral gene-delivery by highly fluorinated gemini bispyridinium surfactant-based DNA nanoparticles. J Colloid Interface Sci. 2017;487:182–91.
8. Long X, Zhang Z, Han S, Tang M, Zhou J, Zhang J, Xue Z, Li Y, Zhang R, Deng L, Dong A. Structural mediation on polycation nanoparticles by sulfadiazine to enhance DNA transfection efficiency and reduce toxicity. ACS Appl Mater Interfaces. 2015;7:7542–51.
9. Banga RJ, Krovi SA, Narayan SP, Sprangers AJ, Liu G, Mirkin CA, Nguyen ST. Drug-loaded polymeric spherical nucleic acids: enhancing colloidal stability and cellular uptake of polymeric nanoparticles through DNA surface-functionalization. Biomacromolecules. 2017;18:483–9.
10. Li Q, Hao X, Lv J, Ren X, Zhang K, Ullah I, Feng Y, Shi C, Zhang W. Mixed micelles obtained by co-assembling comb-like and grafting copolymers as gene carriers for efficient gene delivery and expression in endothelial cells. J Mater Chem B. 2017;5:1673–87.
11. Xing H-B, Pan H-M, Fang Y, Zhou X-Y, Pan QIN, Li DA. Construction of a tumor cell-targeting non-viral gene delivery vector with polyethylenimine modified with RGD sequence-containing peptide. Oncol Lett. 2014;7:487–92.
12. Rancan F, Gao Q, Graf C, Troppens S, Hadam S, Hackbarth S, Kembuan C, Blume-Peytavi U, Rühl E, Lademann J, Vogt A. Skin penetration and cellular uptake of amorphous silica nanoparticles with variable size, surface functionalization, and colloidal stability. ACS Nano. 2012;6:6829–42.
13. Chithrani BD, Ghazani AA, Chan WCW. Determining the size and shape dependence of gold nanoparticle uptake into mammalian cells. Nano Lett. 2006;6:662–8.
14. Niu J, Chu Y, Huang Y-F, Chong Y-S, Jiang Z-H, Mao Z-W, Peng L-H, Gao J-Q. Transdermal gene delivery by functional peptide-conjugated cationic gold nanoparticle reverses the progression and metastasis of cutaneous melanoma. ACS Appl Mater Interfaces. 2017;9:9388–401.
15. Stobiecka M, Hepel M. Double-shell gold nanoparticle-based DNA-carriers with poly-L-lysine binding surface. Biomaterials. 2011;32:3312–21.
16. Cebrián V, Martín-Saavedra F, Yagüe C, Arruebo M, Santamaría J, Vilaboa N. Size-dependent transfection efficiency of PEI-coated gold nanoparticles. Acta Biomater. 2011;7:3645–55.
17. Xiao T, Cao X, Shi X. Dendrimer-entrapped gold nanoparticles modified with folic acid for targeted gene delivery applications. J Control Release. 2013;172:114–5.
18. Chuang CC, Chang CW. Complexation of bioreducible cationic polymers with gold nanoparticles for improving stability in serum and application on nonviral gene delivery. ACS Appl Mater Interfaces. 2015;7:7724.
19. Figueroa ER, Lin AY, Yan J, Luo L, Foster AE, Drezek RA. Optimization of PAMAM-gold nanoparticle conjugation for gene therapy. Biomaterials. 2013;35:1725–34.
20. Yan X, Blacklock J, Li J, Möhwald H. One-pot synthesis of polypeptide-gold nanoconjugates for in vitro gene transfection. ACS Nano. 2012;6:111.
21. Muddineti OS, Ghosh B, Biswas S. Current trends in using polymer coated gold nanoparticles for cancer therapy. Int J Pharm. 2015;484:252–67.
22. Kirklandyork S, Zhang Y, Smith AE, York AW, Huang F, Mccormick CL. Tailored design of Au nanoparticle-siRNA carriers utilizing reversible addition–fragmentation chain transfer polymer. Biomacromolecules. 2010;11:1052–9.
23. Kim HJ, Takemoto H, Yi Y, Zheng M, Maeda Y, Chaya H, Hayashi K, Mi P, Pittella F, Christie RJ. Precise engineering of siRNA delivery vehicles to

tumors using polyion complexes and gold nanoparticles. ACS Nano. 2014;8:8979–91.

24. Li J, Wu W, Gao J, Liang J, Zhou H, Liang L. Constructing of DNA vectors with controlled nanosize and single dispersion by block copolymer coating gold nanoparticles as template assembly. J Nanopart Res. 2017;19:86.
25. Hatakeyama H, Akita H, Harashima H. A multifunctional envelope type nano device (MEND) for gene delivery to tumours based on the EPR effect: a strategy for overcoming the PEG dilemma. Adv Drug Deliv Rev. 2011;63:152–60.
26. Hao S, Yan Y, Ren X, Xu Y, Chen L, Zhang H. Candesartan-graft-polyethyleneimine cationic micelles for effective co-delivery of drug and gene in anti-angiogenic lung cancer therapy. Biotechnol Bioprocess E. 2015;20:550–60.
27. Xiao S, Castro R, Rodrigues J, Shi X, Tomás H. PAMAM dendrimer/pDNA functionalized-magnetic iron oxide nanoparticles for gene delivery. J Biomed Nanotechnol. 2015;11:1370.
28. Wang Y, Cui H, Yang Y, Zhao X, Sun C, Chen W, Du W, Cui J. Mechanism study of gene delivery and expression in PK-15 cells using magnetic iron oxide nanoparticles as gene carriers. Nano Life. 2014;04:1441018.
29. Alvizo-Baez CA, Luna-Cruz IE, Vilches-Cisneros N, Rodríguez-Padilla C, Alcocer-González JM. Systemic delivery and activation of the TRAIL gene in lungs, with magnetic nanoparticles of chitosan controlled by an external magnetic field. Int J Nanomed. 2016;11:6449–58.
30. Yiu HH, Mcbain SC, Lethbridge ZA, Lees MR, Palona I, Olariu CI, Dobson J. Novel magnetite-silica nanocomposite (Fe3O4-SBA-15) particles for DNA binding and gene delivery aided by a magnet array. J Nanosci Nanotechnol. 2011;11:3586.
31. Bajaj A, Samanta B, Yan H, Jerry DJ, Rotello V. Stability, toxicity and differential cellular uptake of protein passivated-Fe3O4 nanoparticles. J Mater Chem. 2009;19:6328–31.
32. Calatayud MP, Sanz B, Raffa V, Riggio C, Ibarra MR, Goya GF. The effect of surface charge of functionalized Fe3O4 nanoparticles on protein adsorption and cell uptake. Biomaterials. 2014;35:6389–99.
33. Sohrabijam Z, Saeidifar M, Zamanian A. Enhancement of magnetofection efficiency using chitosan coated superparamagnetic iron oxide nanoparticles and calf thymus DNA. Colloid Surf B. 2017;152:169.
34. Lo YL, Chou HL, Liao ZX, Huang SJ, Ke JH, Liu YS, Chiu CC, Wang LF. Chondroitin sulfate–polyethylenimine copolymer-coated superparamagnetic iron oxide nanoparticles as an efficient magneto-gene carrier for microRNA-encoding plasmid DNA delivery. Nanoscale. 2015;7:8554.
35. Ma Y, Zhang Z, Wang X, Xia W, Gu H. Insights into the mechanism of magnetofection using MNPs-PEI/pDNA/free PEI magnetofectins. Int J Pharm. 2011;419:247.
36. Salado J, Insausti M, Lezama L, Muro IGD, Moros M, Pelaz B, Grazu V, Fuente JMDL, Rojo T. Functionalized Fe3O4@Au superparamagnetic nanoparticles: in vitro bioactivity. Nanotechnology. 2012;23:315102.
37. Czugala M, Mykhaylyk O, Böhler P, Onderka J, Stork B, Wesselborg S, Kruse FE, Plank C, Singer BB, Fuchsluger TA. Efficient and safe gene delivery to human corneal endothelium using magnetic nanoparticles. Nanomedicine. 2016;11:1787.
38. Lyon J, Fleming D, Stone M, Schiffer P, Williams M. Synthesis of Fe oxide core/Au shell nanoparticles by iterative hydroxylamine seeding. Nano Lett. 2004;4:719–23.
39. Xie HY, Zhen R, Wang B, Feng YJ, Chen P, Hao J. Fe3O4/Au core/shell nanoparticles modified with Ni^{2+}—nitrilotriacetic acid specific to histidine-tagged proteins. J Phys Chem C. 2010;114:4825–30.
40. Ahmadi A, Shirazi H, Pourbagher N, Akbarzadeh A, Omidfar K. An electrochemical immunosensor for digoxin using core–shell gold coated magnetic nanoparticles as labels. Mol Biol Rep. 2014;41:1659–68.
41. Li L, Nurunnabi M, Nafiujjaman M, Yong YJ, Lee Y, Kang MH. A photosensitizer-conjugated magnetic iron oxide/gold hybrid nanoparticle as an activatable platform for photodynamic cancer therapy. J Mater Chem B. 2014;2:2929–37.
42. Liang RP, Wang XN, Wang L, Qiu JD. Enantiomeric separation by microchip electrophoresis using bovine serum albumin conjugated magnetic core–shell Fe_3O_4@Au nanocomposites as stationary phase. Electrophoresis. 2014;35:2824.
43. Cui YR, Hong C, Zhou YL, Li Y, Gao XM, Zhang XX. Synthesis of orientedly bioconjugated core/shell Fe3O4@Au magnetic nanoparticles for cell separation. Talanta. 2011;85:1246–52.
44. Smith AE, Sizovs A, Grandinetti G, Xue L, Reineke TM. Diblock glycopolymers promote colloidal stability of polyplexes and effective pDNA and siRNA delivery under physiological salt and serum conditions. Biomacromolecules. 2011;12:3015–22.
45. Li J, Zhao J, Gao J, Liang J, Wu W, Liang L. A block copolymer containing PEG and histamine-like segments: well-defined functions for gene delivery. New J Chem. 2016;40:7222–8.
46. Altunbek M, Çulha M, Baysal A. Influence of surface properties of zinc oxide nanoparticles on their cytotoxicity. Colloid Surf B. 2014;121:106–13.
47. Peng LH, Huang YF, Zhang CZ, Niu J, Chen Y, Chu Y, Jiang ZH, Gao JQ, Mao ZW. Integration of antimicrobial peptides with gold nanoparticles as unique non-viral vectors for gene delivery to mesenchymal stem cells with antibacterial activity. Biomaterials. 2016;103:137.
48. Majewski AP, Schallon A, Jérôme V, Freitag R, Müller AH, Schmalz H. Dual-responsive magnetic core–shell nanoparticles for nonviral gene delivery and cell separation. Biomacromolecules. 2012;13:857–66.
49. Shi W, Liu X, Wei C, Xu ZJ, Sim SS, Liu L, Xu C. Micro-optical coherence tomography tracking of magnetic gene transfection via Au–Fe_3O_4 dumbbell nanoparticles. Nanoscale. 2015;7:17249.
50. Singh D, Mcmillan JM, Liu XM, Vishwasrao HM, Kabanov AV, Sokolsky-Papkov M, Gendelman HE. Formulation design facilitates magnetic nanoparticle delivery to diseased cells and tissues. Nanomedicine. 2014;9:469.

Permissions

All chapters in this book were first published in JON, by BioMed Central; hereby published with permission under the Creative Commons Attribution License or equivalent. Every chapter published in this book has been scrutinized by our experts. Their significance has been extensively debated. The topics covered herein carry significant findings which will fuel the growth of the discipline. They may even be implemented as practical applications or may be referred to as a beginning point for another development.

The contributors of this book come from diverse backgrounds, making this book a truly international effort. This book will bring forth new frontiers with its revolutionizing research information and detailed analysis of the nascent developments around the world.

We would like to thank all the contributing authors for lending their expertise to make the book truly unique. They have played a crucial role in the development of this book. Without their invaluable contributions this book wouldn't have been possible. They have made vital efforts to compile up to date information on the varied aspects of this subject to make this book a valuable addition to the collection of many professionals and students.

This book was conceptualized with the vision of imparting up-to-date information and advanced data in this field. To ensure the same, a matchless editorial board was set up. Every individual on the board went through rigorous rounds of assessment to prove their worth. After which they invested a large part of their time researching and compiling the most relevant data for our readers.

The editorial board has been involved in producing this book since its inception. They have spent rigorous hours researching and exploring the diverse topics which have resulted in the successful publishing of this book. They have passed on their knowledge of decades through this book. To expedite this challenging task, the publisher supported the team at every step. A small team of assistant editors was also appointed to further simplify the editing procedure and attain best results for the readers.

Apart from the editorial board, the designing team has also invested a significant amount of their time in understanding the subject and creating the most relevant covers. They scrutinized every image to scout for the most suitable representation of the subject and create an appropriate cover for the book.

The publishing team has been an ardent support to the editorial, designing and production team. Their endless efforts to recruit the best for this project, has resulted in the accomplishment of this book. They are a veteran in the field of academics and their pool of knowledge is as vast as their experience in printing. Their expertise and guidance has proved useful at every step. Their uncompromising quality standards have made this book an exceptional effort. Their encouragement from time to time has been an inspiration for everyone.

The publisher and the editorial board hope that this book will prove to be a valuable piece of knowledge for researchers, students, practitioners and scholars across the globe.

List of Contributors

Daniel N. Freitas, Zoe N. Amaris and Korin E. Wheeler
Department of Chemistry and Biochemistry, Santa Clara University, Santa Clara, CA 95053, USA

Andrew J. Martinolich
Department of Chemistry and Biochemistry, Santa Clara University, Santa Clara, CA 95053, USA
Department of Chemistry, Colorado State University, Fort Collins, CO 80523-1872, USA

Xiaomei Li
Department of Environmental Toxicology, Eawag, Swiss Federal Institute of Aquatic Science and Technology, 8600 Dübendorf, Switzerland
School of Architecture, Civil and Environmental Engineering, École Polytechnique Fédérale de Lausanne, 1015 Lausanne, Switzerland

Laura Sigg
Department of Environmental Toxicology, Eawag, Swiss Federal Institute of Aquatic Science and Technology, 8600 Dübendorf, Switzerland
Department of Environmental Systems Science (D-USYS), ETH-Zürich, 8092 Zürich, Switzerland
Wattstrasse 13a, 8307 Effretikon, Switzerland

Marc J-F Suter, Smitha Pillai, Renata Behra and Kristin Schirmer
Department of Environmental Toxicology, Eawag, Swiss Federal Institute of Aquatic Science and Technology, 8600 Dübendorf, Switzerland
Department of Environmental Systems Science (D-USYS), ETH-Zürich, 8092 Zürich, Switzerland

Yang Yue
Department of Environmental Toxicology, Eawag, Swiss Federal Institute of Aquatic Science and Technology, 8600 Dübendorf, Switzerland
School of Architecture, Civil and Environmental Engineering, École Polytechnique Fédérale de Lausanne, 1015 Lausanne, Switzerland
Department of Basic Sciences and Aquatic Medicine, Norwegian University of Life Sciences (NMBU), Oslo 0454, Norway

Philipp Reichardt, Jutta Tentschert Andrea Haase and Andreas Luch
Department of Chemical and Product Safety, German Federal Institute for Risk Assessment (BfR), Max-Dohrn-Strasse 8-10, 10589 Berlin, Germany

I-Lun Hsiao
Department of Chemical and Product Safety, German Federal Institute for Risk Assessment (BfR), Max-Dohrn-Strasse 8-10, 10589 Berlin, Germany
Department of Biomedical Engineering and Environmental Sciences, National Tsing Hua University, Hsinchu, Taiwan

Yuh-Jeen Huang
Department of Biomedical Engineering and Environmental Sciences, National Tsing Hua University, Hsinchu, Taiwan

Frank S. Bierkandt and Norbert Jakubowski
Division of Inorganic Trace Analysis, German Federal Institute for Materials Research and Testing (BAM), Berlin, Germany

Irina B. Alieva, Olga S. Strelkova, Oxana A. Zhironkina and Varvara D. Cherepaninets
A.N. Belozersky Institute of Physico-Chemical Biology, Moscow State University, Moscow, Russia 119992

Igor Kireev
A.N. Belozersky Institute of Physico-Chemical Biology, Moscow State University, Moscow, Russia 119992
Biology Faculty, Moscow State University, Moscow, Russia 119992

Anastasia S. Garanina
Biology Faculty, Moscow State University, Moscow, Russia 119992

Natalia Alyabyeva and Antoine Ruyter
GREMAN, UMR CNRS 7347, Université François Rabelais, 37200 Tours, France

Viatcheslav Agafonov
GREMAN, UMR CNRS 7347, Université François Rabelais, 37200 Tours, France
MISiS, Leninskiy prospekt 2, Moscow, Russia 119049

Alexander G. Majouga
Chemistry Faculty, Moscow State University, Moscow, Russia 119992
MISiS, Leninskiy prospekt 2, Moscow, Russia 119049

Valery A. Davydov
Institute of High Pressure Physics RAS, Troitsk, Moscow region, Russia 142190

Valery N. Khabashesku
Center for Technology Innovation, Baker Hughes Inc., Houston, TX 77040, USA

Rustem E. Uzbekov
Laboratoire Biologie Cellulaire et Microscopie Electronique, Faculté de Médecine, Université François Rabelais, 37032 Tours, France
Faculty of Bioengineering and Bioinformatics, Moscow State University, Moscow, Russia 119992

Luis A. Velásquez and Carolina Otero
Center for Integrative Medicine and Innovative Science, Facultad de Medicina, Universidad Andrés Bello, Santiago, Chile

Juan P. Peñaloza, Mauricio Cabaña-Brunod, Rodrigo Reyes-Ramírez and Felipe M. Llancalahuen
Center for Integrative Medicine and Innovative Science, Facultad de Medicina, Universidad Andrés Bello, Santiago, Chile
Escuela de Bioquímica, Facultad de Ciencias Biológicas, Universidad Andrés Bello, Santiago, Chile

Valeria Márquez-Miranda and Fernando D. González-Nilo
Center for Bioinformatics and Integrative Biology, Facultad de Ciencias Biológicas, Universidad Andrés Bello, Echaurren #183, 8370071 Santiago, Chile

Cristian Vilos
Center for Integrative Medicine and Innovative Science, Facultad de Medicina, Universidad Andrés Bello, Santiago, Chile
Center for Bioinformatics and Integrative Biology, Facultad de Ciencias Biológicas, Universidad Andrés Bello, Echaurren #183, 8370071 Santiago, Chile

Fernanda Maldonado-Biermann
Departamento de Ciencias Físicas, Facultad de Ciencias Exactas, Universidad Andrés Bello, Santiago, Chile

Juan A. Fuentes
Laboratorio de Genética y Patógenesis Bacteriana, Facultad de Ciencias Biológicas, Universidad Andrés Bello, Santiago, Chile

Maité Rodríguez-Díaz
Escuela de Química y Farmacia, Facultad de Medicina, Universidad Andrés Bello, Santiago, Chile

Alexander Gogos, Thomas D. Bucheli and Franco Widmer
Agroscope, Institute for Sustainability Sciences ISS, 8046 Zurich, Switzerland.

Janine Moll
Agroscope, Institute for Sustainability Sciences ISS, 8046 Zurich, Switzerland.
Plant-Microbe-Interactions, Department of Biology, Utrecht University, 3508 TB Utrecht, The Netherlands

Marcel G. A. van der Heijden
Agroscope, Institute for Sustainability Sciences ISS, 8046 Zurich, Switzerland
Plant-Microbe-Interactions, Department of Biology, Utrecht University, 3508 TB Utrecht, The Netherlands
Institute of Evolutionary Biology and Environmental Studies, University of Zurich, Winterthurerstrasse 190, 8057 Zurich, Switzerland

Mona I. Shaaban
Pharmaceutics and Pharmaceutical Technology Department, College of Pharmacy, Taibah University, PO Box 30040, Al Madina, Al Munawara, Saudi Arabia.
Microbiology and Immunology Department, Faculty of Pharmacy, Mansoura University, PO Box 35516, Mansoura, Egypt

Mohamed A. Shaker
Pharmaceutics and Pharmaceutical Technology Department, College of Pharmacy, Taibah University, PO Box 30040, Al Madina, Al Munawara, Saudi Arabia

Pharmaceutics Department, Faculty of Pharmacy, Helwan University, PO Box 11795, Cairo, Egypt

Fatma M. Mady
Pharmaceutics and Pharmaceutical Technology Department, College of Pharmacy, Taibah University, PO Box 30040, Al Madina, Al Munawara, Saudi Arabia
Pharmaceutics Department, Faculty of Pharmacy, El-Minia University, El-Minia, Egypt

Freya Joris, Stefaan C. De Smedt and Koen Raemdonck
Lab of General Biochemistry and Physical Pharmacy, Department of Pharmaceutics, Faculty of Pharmaceutical Sciences, Ghent University, Ottergemsesteenweg 460, 9000 Ghent, Belgium

Daniel Valdepérez, Wolfgang J. Parak and Beatriz Pelaz
Department of Physics, Philipps University of Marburg, Renthof 7, 35037 Marburg, Germany.

Stefaan J. Soenen and Bella B. Manshian
Biomedical MRI Unit/MoSAIC, Department of Medicine, KULeuven, Herestraat 49, 3000 Louvain, Belgium

Branislav Ruttkay-Nedecky, Olga Krystofova, Lukas Nejdl and Vojtech Adam
Department of Chemistry and Biochemistry, Mendel University in Brno, Zemedelska 1, 613 00 Brno, Czech Republic
Central European Institute of Technology, Brno University of Technology, Technicka 3058/10, 616 00 Brno, Czech Republic

Kata Kenesei and Kumarasamy Murali
School of PhD Studies, Semmelweis University, Üllői Street 26, Budapest 1085, Hungary
Institute of Experimental Medicine, Hungarian Academy of Sciences, Szigony Street 43, Budapest 1083, Hungary

Emília Madarász
Institute of Experimental Medicine, Hungarian Academy of Sciences, Szigony Street 43, Budapest 1083, Hungary

Árpád Czéh
Soft Flow Hungary Kft., Kedves u. 20, Pecs 7628, Hungary

Jordi Piella and Victor Puntes
Catalan Institute of Nanoscience and Nanotechnology, Campus UAB, Bellaterra, 08193 Barcelona, Spain

Yunyun Chen and Yuan Yue
Materials Science and Engineering, Texas A&M University, College Station,TX 77843-3123, USA

Hong Liang
Materials Science and Engineering, Texas A&M University, College Station, TX 77843-3123, USA
Mechanical Engineering, Texas A&M University, College Station, TX 77843-3123, USA.

Carlos Sanchez
Mechanical Engineering, Texas A&M University, College Station, TX 77843-3123, USA

Mauricio de Almeida and Jorge M. González
Department of Plant Science, California State University, Fresno, CA 93740, USA

Dilworth Y. Parkinson
Advanced Light Source, Lawrence Berkeley National Laboratory, Berkeley, CA 94720, USA

Blanka Halamoda-Kenzaoui, Mara Ceridono, Patricia Urbán, Alessia Bogni, Jessica Ponti, Sabrina Gioria and Agnieszka Kinsner-Ovaskainen
European Commission Joint Research Centre, Directorate for Health, Consumers and Reference Materials, Via E. Fermi 2749, TP 127, 21027 Ispra, VA, Italy

Luisa H. A. Silva, Jaqueline R. da Silva, Ricardo B. Azevedo and Daniela M. Oliveira
IB-Departamento de Genética e Morfologia, Universidade de Brasília-UNB, Campus Universitário Darcy Ribeiro-Asa Norte, Brasília, DF CEP 70910-970, Brazil

Guilherme A. Ferreira and Emilia C. D. Lima
Instituto de Química, Universidade Federal de Goias, Goiânia, GO, Brazil

Renata C. Silva
Instituto Nacional de Metrologia, Rio de Janeiro, RJ, Brazil

Janine Moll, Thomas D. Bucheli, Florian Klingenfuss and Marcel van der Heijden
Agroscope, Institute for Sustainability Sciences ISS, 8046 Zurich, Switzerland.

Renato Zenobi
Department of Chemistry and Applied Biosciences, ETH Zurich, 8093 Zurich, Switzerland

Alexander Gogos
Agroscope, Institute for Sustainability Sciences ISS, 8046 Zurich, Switzerland.
Department of Chemistry and Applied Biosciences, ETH Zurich, 8093 Zurich, Switzerland

Fahmida Irin
Department of Chemical Engineering, Texas Tech University, Lubbock, TX, USA

Micah J. Green
Artie McFerrin Department of Chemical Engineering, Texas A&M University, College Station, TX, USA

Xiaona Zhai, Chunyue Zhang, Guanghua Zhao, Fazheng Ren and Xiaojing Leng
Beijing Advanced Innovation Center for Food Nutrition and Human Health, Beijing Laboratory for Food Quality and Safety, Beijing Dairy Industry Innovation Team, College of Food Science & Nutritional Engineering, China Agricultural University, Beijing 100083, China

Serge Stoll
Group of Environmental Physical Chemistry, F.-A. Forel Institute, University of Geneva, Geneva, Switzerland

Yiping He, Sue Reed, Andrew Gehring, Terence P. Strobaugh Jr. and Peter Irwin
Molecular Characterization of Foodborne Pathogens Research Unit, Eastern Regional Research Center, Agricultural Research Service, United States Department of Agriculture, 600 East Mermaid Lane, Wyndmoor, PA 19038, USA

Shakuntala Ingudam
ICAR Research Complex for NEH Region, Umiam 793103, Meghalaya, India

Jing Hu and Lian Xiao
School of Chemistry, Chemical Engineering and Life Sciences, Wuhan University of Technology, Wuhan 430070, People's Republic of China

Yunqiang Wang
Institute of Economic Crops, Hubei Academy of Agricultural Sciences, Wuhan 430064, People's Republic of China

Huiyuan Guo and Baoshan Xing
Stockbridge School of Agriculture, University of Massachusetts, Amherst, MA 01003, USA

Junli Li
School of Chemistry, Chemical Engineering and Life Sciences, Wuhan University of Technology, Wuhan 430070, People's Republic of China
Stockbridge School of Agriculture, University of Massachusetts, Amherst, MA 01003, USA

Junbo Li, Sheng Zou, Jiayu Gao, Ju Liang, Huiyun Zhou and Lijuan Liang
School of Chemical Engineering & Pharmaceutics, Henan University of Science & Technology, Luo Yang 471023, China

Wenlan Wu
School of Medicine, Henan University of Science & Technology, Luo Yang 471023, China

Index

www.ingramcontent.com/pod-product-compliance
Lightning Source LLC
Chambersburg PA
CBHW082056190326
41458CB00010B/3512

* 9 7 8 1 6 4 1 1 6 0 8 9 6 *